PRENTICE-HALL BIOLOGICAL SCIENCE SERIES

William D. McElroy and Carl P. Swanson, Editors

DYNAMIC

ECOLOGY

Boyd D. Collier
George W. Cox
Albert W. Johnson
Philip C. Miller

California State University
San Diego

PRENTICE-HALL, INC., ENGLEWOOD CLIFFS, N.J.

Library of Congress Cataloging in Publication Data

Main entry under title:

Dynamic ecology.

     Includes bibliographies.
      1. Ecology. I. Collier, Boyd.
QH541.D9   574.5'2   72-10189
ISBN 0-13-221283-8

DYNAMIC ECOLOGY
by B. D. Collier, G. W. Cox,
A. W. Johnson, P. C. Miller

© 1973 by Prentice-Hall, Inc.,
Englewood Cliffs, New Jersey

10 9 8 7 6 5 4 3 2 1

Printed in the United States of America

Prentice-Hall International, Inc., *London*
Prentice-Hall of Australia, Pty. Ltd., *Sydney*
Prentice-Hall of Canada, Ltd., *Toronto*
Prentice-Hall of India Private Limited, *New Delhi*
Prentice-Hall of Japan, Inc., *Tokyo*

CONTENTS

Preface

PREFACE

Ecology has followed a pattern of historical development similar to that of many other branches of science. Ecology began – in the late 1800s and early 1900s – as a primarily descriptive field. Ecologists sought patterns in the appearance and structure of organisms from different environments; these patterns – often named after their discoverers – became the biogeographic "rules" of ecology. These ecologists developed techniques for quantitatively describing the structure and taxonomic composition of communities of organisms. They erected comprehensive and increasingly detailed systems for the naming and classification of these communities. Armed with these tools they resolved patterns of geographical variation in community structure and composition into discrete, named units – life zones, associations, and biomes. They recognized patterns of temporal change in community characteristics and resolved these patterns into seres, seral stages, and various types of climaxes. Throughout the field they built an elaborate and imposing terminology. At the height of this phase they earned for ecology the reputation of being a field that elaborated on the obvious, that engaged in "superdescription," and that was so preoccupied with terminology that a spade was termed a "geotome."

Not all ecologists were so occupied, nor were many individuals concerned only with descriptive work, however. The roots of most current areas of research in modern ecology may be traced well back into this descriptive period. This is especially true for areas such as population ecology, physiological ecology, and aquatic ecology, where efforts to go beyond descriptive study began early.

Nevertheless, ecology as a field has undergone a major transition. Modern ecology has become primarily concerned with understanding how ecological systems *function,* and with understanding the reciprocal interrelationships between the structure and composition of a system and its pattern of function. For ecology this transition from a descriptive to a dynamic approach has perhaps occurred more rapidly than for any of the other sciences. In part, the transition has been forced by the need for solutions of major problems of environmental quality and productivity. In part, it also simply reflects the fact that the achievements of descriptive ecology have made it possible for ecologists to move on to studies of function.

Today, however, ecologists are asking questions, not of function or dynamics alone, but of structure *and* function, composition *and* dynamics. The answers to some of these questions have already demonstrated that there exists a body of theory uniting all of the major branches of the field. This body of theory includes principles that apply to all taxonomic groups of organisms—bacteria, green plants, invertebrates, and vertebrates. Other aspects of ecological theory apply to ecological systems regardless of whether they occupy freshwater, marine, or terrestrial environments.

Our objective in this text is to examine this body of theory. We shall be concerned not with structure alone, nor with function alone. Rather, we shall be concerned with the *dynamics* of ecological systems—the interrelations between structure and function. In following this approach we have deliberately omitted surveys, or systematic descriptions, of factors of the physical environment, major habitat types, and regional community or ecosystem types. Instead, we have concentrated on examination of the principles that underly these patterns. When considering relationships between organisms and the physical environment, for example, we have discussed major patterns of adaptation and response by organisms, and then have selected certain of these for detailed functional examination. At the population, community, and ecosystem levels we have emphasized aspects of structure and function that are of wide applicability. At all levels we have attempted to illustrate our discussions with the best examples available, regardless of the taxonomic or environmental relationships of the organisms involved. We have sought these examples in ecological literature covering most groups of organisms and both terrestrial and aquatic environments.

In attempting this treatment we have been fortunate in being able to draw upon the ideas and suggestions of many of our colleagues in ecology at California State University, San Diego, as well as at other institutions. We especially wish to thank Richard F. Ford and William E. Hazen for their contributions to our understanding of ecology. We must also acknowledge the contributions of the many workers from whose thought and research we have taken our ideas and examples, and the anonymous reviewers of our original manuscript for their many suggestions. Finally, we extend our special appreciation to a group of individuals who, in a variety of ways, and often with short deadlines and poor copy, helped with preparation of the manuscript: Rita Collier, Marjorie Graupman, Marguerite Jackson, Susan Johnson, and Patsy Miller.

B.D.C., G.W.C., A.W.J., P.C.M.

DYNAMIC ECOLOGY

Part I

INTRODUCTION

Chapter 1

ECOLOGY AND THE ENVIRONMENTAL CRISIS

INTRODUCTION

Man is faced with an ecological crisis. This crisis has developed as a consequence of increasing mismanagement of the world environment and unrestrained growth of human populations. It not only threatens his chances for achieving an adequate standard of living for the present human population, but also threatens his chances for continued existence as a species.

The warning signs of this crisis appear in specific problems such as the imbalance of food production and human population growth, the reduction in productivity of major areas of land and water due to pollution and mismanagement, the gradual change of regional and global climates resulting from urban activities and agricultural practices, the destruction of important wildlife species and disturbance of natural biotic communities, and the increase in number of pest and disease organisms. These problems are symptomatic of disturbances of processes operating at the level of the biosphere as a whole—disturbances capable of reducing the quality and productivity of the world environment below existing levels. Furthermore, many of these problems involve relationships of major concern in the world political sphere. At the very least, their existence contributes to much of the current political discontent in the world.

Many specific factors–cultural, economic, and historical–have contributed to the emergence of these problems (White, 1967; Cole, 1970; Marx, 1970).* At the roots of these problems, however, lie man's increasing exploitation of environmental resources, his ignorance of the natural processes affected by this exploitation, and his abiding faith in the ability

*See references at end of chapter.

3

of human technology to solve those problems that develop to major proportions.

The human population has now reached a level at which needs for environmental resources have required heavy exploitation of all major world environments—terrestrial, freshwater, and marine. Exploitation of food resources of certain major types, such as marine fisheries (Borgstrom, 1970), is approaching the maximum level possible. At the same time, man is largely ignorant of the basic factors responsible for the production of these resources, and of the long-range consequences of his patterns of exploitation.

Human technology has resulted in major problems beyond those of over-exploitation. Agricultural, industrial, and urban activities have now become the agents of global patterns of pollution, some of which threaten to disrupt basic processes of the biosphere. Human technology has now reached a point at which new developments may lead to worldwide effects of a detrimental nature before the possibility of occurrence of such effects is even recognized (Commoner, 1966).

Ecology has traditionally been defined as the branch of biology concerned with the study of interrelationships between organisms and their environment. In the context of the developing environmental crisis, however, ecology has indeed become something more—a "science of survival." To illustrate this point, and to show the diversity and complexity of the ecological effects of current human activities, we shall examine one example in detail. This example, only one of the many that could have been chosen, threatens to become a classic case of ecological ignorance and shortsightedness. It concerns the Aswan Dam, which has recently been completed on the Nile River in the United Arab Republic (UAR).

THE ASWAN DAM

The entire economy of the UAR is tied to the ecology of the Nile. The Nile, the longest river in the world, enters the UAR from the south, carrying water, silt, and nutrients from the Ethiopian highlands over 1,500 miles to the south (Fig. 1–1). It flows northward through the Sahara Desert, where the annual rainfall ranges from 10 inches down to nearly zero, and then empties into the Mediterranean Sea via a rich delta about 100 miles in width.

The Nile Valley, comprising perhaps 1/30th of the total area of the UAR, is consequently the center of agriculture and population. Almost all of the UAR's 33 million people live here. Until construction of the Aswan Dam, this population was dependent on agricultural production on about 6 million acres of river flood plain, augmented by an expanding Mediterranean fishery and by importation of food.

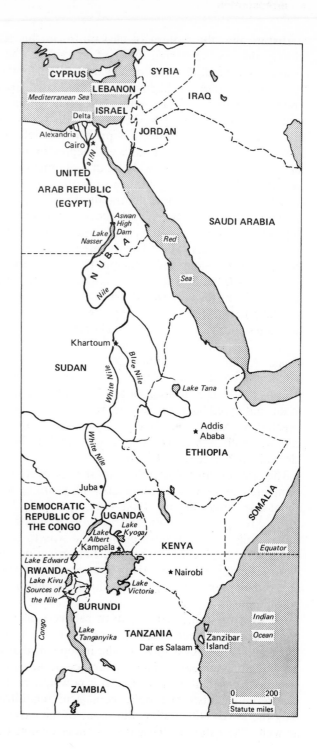

FIGURE 1–1. The Nile River system.

Historically, the pattern of agriculture in the Nile Valley has been closely tied to the annual cycle of the river. The flow of the Nile is strongly seasonal. August through November is the period of seasonal rains in the Ethiopian highlands and thus also the period of maximum volume of river flow. In the lower Nile Valley, this is a period of flooding of much of the low-lying land bordering the river. Although this flooding can be destructive at times, it is primarily beneficial. It restores the soil water supply, flushes away salts, and deposits a new layer of rich organic silt which acts as fertilizer. It is, in fact, this annual flooding that has kept the Nile Valley one of the most fertile land areas on earth in spite of continuous cultivation over thousands of years.

The inhabitants of the Nile Valley have never been ignorant of their dependence on the Nile River, and even on the annual floods that watered much of their farmland. The pressures of an exploding population, though, have encouraged attempts to expand and intensify agriculture along the Nile, mainly through the control of the river by dams and diversion of waters for year-round irrigation. The Nile is, in fact, dammed at several points. At Aswan, in the south, the Aswan High Dam is the fourth in a series of dams that were begun in 1902.

The Aswan High Dam (Fig. 1–2), promoted by the late president Gamal Abdul Nasser and financed almost completely by Russian foreign aid, was intended, in a single stroke, to modernize the agriculture and permit the industrialization of the UAR. It was to add 1.3 million acres to the cultivated land area and permit year-round cropping of much of the existing cultivated land. It was to provide cheap electrical power to encourage industrialization. Nasser intended it to be a living monument to his leadership of the UAR.

The Aswan High Dam, the largest dam of its kind in the world, was completed in 1970 after more than 11 years of work and at a cost of about one billion dollars. Lake Nasser, behind the dam, was projected to be 310 miles in length and to have a surface area of about 2,000 square miles. The purpose of the lake was to store 163 million cubic meters of water—enough to provide an irrigation reserve for a period of up to several years of low river flow (Sterling, 1971). The 12 generators at the dam were designed to produce 10 billion kilowatts of electrical power annually.

All this was projected to permit recovery of the cost of the dam in two years and effect a doubling of the national economy in ten years (Sterling, 1971). As it now stands, it appears that the Aswan Dam will become a monument to man's ignorance of the ecological effects of massive environmental intervention. It is now apparent that the Aswan Dam has seriously disturbed basic ecological relationships throughout not only the lower Nile Valley, but much of the eastern Mediterranean Sea as well. The nature of these disturbances demonstrates clearly the need

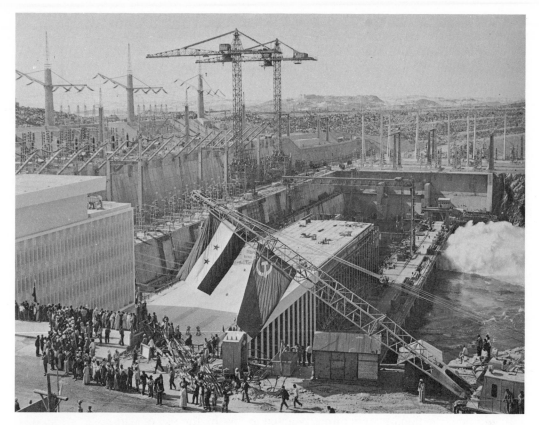

FIGURE 1–2. The Aswan Dam and Lake Nasser at the time of dedication ceremonies in January 1968. (Tass from Sovfoto.)

for an understanding of the basic principles of ecology by all responsible persons.

The specific problems occurring in this area result primarily from the effects of the Aswan Dam on patterns of water, nutrient, and silt supply to the area below the dam. One of the dam's primary objectives was to assure an abundant supply of water. Storage of water behind the dam was begun in 1964; the target date for filling of the lake was 1970. By early 1971, however, the lake was still less than one-half full, and revised estimates suggest that it may require anywhere from 12 to 200 years to complete the filling (Sterling, 1971).

The slow rate of filling is attributed to two factors. First, it is now estimated that annual losses by underground seepage are about 15 billion cubic meters—principally into highly porous sandstones forming the entire western lake shore. Originally, it was expected that silt would fill the pores in these sandstones and quickly cut off such losses. However, it now appears that almost all of the sediment deposition occurs in the central portion of the lake, along the old river bed, and that seepage losses are remaining high along the western shores. Second, evaporation losses have been higher than projected. Major losses were, of course, expected since the Aswan Dam is located in one of the hottest and driest places on earth. The presently measured annual loss of 15 billion cubic meters is about 50% above the estimates that had been made during planning stages. Apparently these preliminary estimates failed to adequately take into account the influence of wind action on evaporation from the lake surface. These combined seepage and evaporation losses approximately equal the total annual discharge of the Nile into the Mediterranean prior to construction of the dam. Thus, the dam appears to be losing the very water it was designed to preserve.

However, problems in the areas below the dam go far beyond considerations of quantity of water. After the dam was completed about 700,000 acres of land were converted from flood irrigation to canal irrigation and about 300,000 acres of new land were reclaimed. These areas can now be cultivated year-round. Although this has significantly increased agricultural production, several serious complications have developed. First, in canal irrigated areas, there is no longer any silt deposition during the flood period. Almost all of the Nile silt is now deposited on the bottom of Lake Nasser. A major requirement for use of artificial fertilizers on irrigated lands is consequently developing. This need is not, however, being paralleled by increased fertilizer use, so that both the quantity and quality of the harvested crops are declining in many areas.

In addition, the flood waters formerly acted to "flush" the soil of salts, especially in the delta area, where soils tend to be waterlogged. In such soils, capillary movement of water to the soil surface tends to transport and concentrate salts from deeper layers in the soil, This "flushing" action no longer occurs, and large areas of irrigated land, both in the delta region and in upper parts of the valley, are experiencing increases in salinity. It has been estimated that unless preventive measures at a minimum cost of one billion dollars are undertaken, much of this land will quickly become useless (Sterling, 1971).

Still a third problem, related to the reduced load of silt in river waters below the dam, is erosion of the banks, channel, and delta coast. The silt-free water below the dam flows faster and tends to reacquire a normal silt load. The resulting erosion threatens to undermine the foundations of many of the three dams and 550 bridges between the Aswan Dam and the sea. A project, known as the Nile Cascade, has been proposed to reduce this erosion. This project would involve the construction of 10 new barrier dams to slow the river flow and would cost about 250 million dollars. The delta coastline, formerly protected by the annual deposition of millions of tons of silt, is now receding — in some areas at a rate of several yards per year (Sterling, 1971).

On top of all this, the expansion of canal irrigation has increased the incidence of several diseases, including bilharzia, malaria, and trachoma, which are carried by aquatic invertebrates (Wagner, 1971). Bilharzia is a disease caused by the flatworm parasites *Schistosoma hematobium* and *S. mansoni*. These parasites have life cycles alternating between man and various species of freshwater pulmonate snails (Fig. 1–3). The parasite infects man in the *cercaria* stage. Cercaria are minute, free-swimming stages that are released from the snails and can attach to and penetrate the skin of humans wading in the water. In the human body, these cercaria penetrate vessels of the bladder, rectum, and intestine. They show special preference for the portal veins carrying blood from the intestine to the liver. Here, after several months, they grow into adult worms up to 2 cm in size. Eggs produced by these adults pass out of the human body in urine or feces, and if these reach fresh water, they hatch into tiny, ciliated *miricidia* capable of infecting aquatic snails.

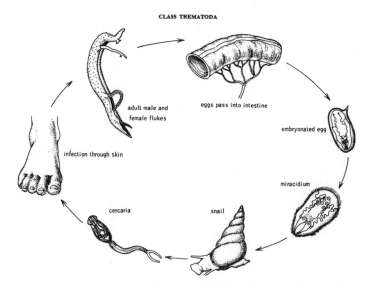

CLASS TREMATODA

adult male and female flukes

eggs pass into intestine

embryonated egg

infection through skin

miracidium

cercaria

snail

FIGURE 1–3. The life cycle of *Schistosoma mansoni*. (From R. D. Barnes, 1968.)

Bilharzia is not usually fatal to man but is a debilitating disease. In chronic cases the human becomes weak and emaciated and develops a fluid-filled, swollen abdomen reflecting impaired circulation of the intestinal venous system. Bilharzia has always been a common disease of the Nile Valley. Prior to construction of the Aswan Dam, its incidence was estimated to be about 47%. With construction of permanent irrigation systems, however, it has increased in frequency. This has occurred because freshwater snails have spread into the permanent irrigation ditches in areas formerly flood-irrigated. Previously, these snails were excluded by seasonal drying up of ditch and pond habitats. Where this spread has occurred, the incidence of bilharzia has increased from a level close to zero to as much as 80% (Sterling, 1971). The increased incidence of bilharzia has added several million new cases to the burden of a disease already causing an economic loss of about $560 million annually (Benarde, 1970).

These same irrigation ditches are also providing breeding areas for malaria mosquitos and for a fly which carries trachoma, a disease of the eye (Wagner, 1971).

The influence of the Aswan Dam is not limited to the boundaries of the UAR. Before the construction of the dam, an average of 30 billion cubic meters of silt-laden flood water was discharged annually into the Mediterranean Sea. This discharge occurred only during the flood season, since, at other times of the year, temporary earthen dams closed the two major mouths of the Nile. These dams permitted irrigation of major portions of the delta lands. When flood runoff began to reach the delta, first one, and later the other, of these dams was opened (Oren, 1969).

The Nile discharge influenced the entire ecology of the eastern Mediterranean. Salinity was markedly affected in a 600-mile stretch of continental shelf waters from the delta eastward as far as Lebanon. Decreases in salinity of up to 25% have been recorded off the coast of Israel (Oren, 1969). Blooms of both phytoplankton and zooplankton occurred at the time of the Nile discharge (ibid.). These blooms were almost certainly the consequence of nutrient input through the river discharge.

That this effect was also an important feature in the ecology of fish species is evidenced by the immediate decline of the fish harvest following closure of the Aswan Dam. In 1964, the last year that Nile flood water reached the sea, the total fish harvest by boats of the UAR was 135,000 tons. By 1967, this had dropped to 85,000 tons (Anon., 1970). Prior to 1965, catches of the sardine alone averaged 15,000 tons annually (ibid.). By 1968, this had declined to 500 tons, and by 1971, the sardine fishery is reported to have disappeared completely (Mayhew). Although development of a freshwater fishery in Lake Nasser is anticipated, this is projected to supply only about 12,000 tons when in full operation (Wagner, 1971). Loss of these marine fisheries is a serious matter for peoples of the UAR – a country in which per capita consumption of protein is only 10 kg annually (Anon, 1970).

The prediction and solution of such problems obviously requires an understanding of ecological relationships at several levels of complexity. It requires knowledge of how environmental factors operate to control the distribution of individual species, such as the specific invertebrate vectors of bilharzia and malaria. It requires an understanding of the dynamic interactions within and between populations of different species, e.g., those of man, schistosome flatworms, and aquatic snails. It requires an understanding of the dynamics of complex assemblages of species, such as those of phytoplankton, zooplankton, and fish in the eastern

Mediterranean. Even beyond this, it requires an appreciation of the patterns of input and cycling of nutrients, salts, and soil materials, and it requires an appreciation of how these control the basic fertility of the lands and offshore waters of the Nile Valley.

The Aswan Dam is only one example of the extent to which human technology is now modifying the world environment. In the following chapters it is our objective to examine the basic aspects of ecological theory which must be brought to bear on these problems. We will consider ecological relationships at each of the levels of complexity noted above.

LEVELS OF COMPLEXITY OF ECOLOGICAL SYSTEMS

In recent years, ecologists have focused their attention on the structure and dynamics of *ecological systems*. A system, by definition, is a group of objects possessing some form of regular interaction or interdependence. Ecological systems thus consist of one or more organisms, together with the various components of their physical and chemical environment with which they are functionally interrelated. In studying these systems, ecologists place primary emphasis on quantitatively measuring specific processes and in relating these processes to the structure of the system and to its pattern of change in structure through time.

Ecological systems distinguished in this manner vary in their degree of complexity (Evans, 1956). These systems may be regarded as existing on different levels of organization, with each level possessing unique structural and functional characteristics and presenting unique problems for study. The levels of organization usually recognized in ecological systems are those of the individual organism, the population, the biotic community, and the ecosystem.

At the *organismal level,* the system under examination is the relatively simple one consisting of the individual and its immediate biotic and abiotic environment. The objectives of study at this level relate primarily to the manner in which the various morphological, physiological, and behavioral characteristics of organisms function in growth, maintenance, and reproduction under various environmental conditions. The specialized fields of functional morphology, physiological ecology, and behavioral ecology are concerned primarily with relationships at this level.

In nature, individuals of a particular species exist as members of populations. A *population* consists of the interacting members of a particular species that occur together in a particular habitat. These assemblages show characteristics, such as density, pattern of distribution of individuals in space, age and sex composition, and rates of natality and mortality, that are measurable only at the population level. These characteristics, as well as many others, are influenced by conditions of the physical environment and by interactions with other species. The specialized field of population ecology is concerned with the study of ecological systems at this level of organization.

Populations of various species, however, are distributed not as distinct and separate units but as components of complex multispecies assemblages known as *biotic communities*. Again, such assemblages possess unique characteristics such as degree of "richness" in species, degree of organization into vertical strata, and relative frequency of particular patterns of morphology and behavior. Indirectly at least, all of the species occurring together in a given area of habitat are related by biotic interactions. Through time, changes in the characteristics of these assemblages may occur; one community may, for example, be ultimately replaced by another. The branch of ecology known as *community ecology* is concerned with relationships at this level of organization.

The *ecosystem level* of organization presents the greatest degree of complexity. An ecosystem may be defined as an environmental unit consisting of various biotic and abiotic components which are related through interchanges of chemical nutrients and energy. Or, stated more simply, it consists of the community of organisms together with its physical habitat.

For most ecosystems, the major biotic components include the green plants or *producers,* the various animal groups or *consumers,* and the fungi and bacteria or *decomposers*. The abiotic components consist of the various functional constituents or portions of the physical environment. In a lake ecosystem, for example, the water mass and the bottom sediments constitute two of the major abiotic components. The objectives of study at the ecosystem level of organization relate to the flow of energy and the exchange of chemical nutrients among ecosystem components, and they relate to the relationship of these processes to the structure of the system and pattern of change in structure through time.

Some authors (Rowe, 1951; Schultz, 1967) have argued that only the organismal and ecosystem levels can be recognized as true *levels of integration*. The four levels of ecological organization, however, represent steps of increasing complexity and may be treated as systems from the standpoint of techniques of study. Furthermore, these have been the levels around which specialization has tended to occur in modern ecology. For this reason, we shall use these levels of organization as the basis for our approach to the examination of ecology in this text.

A still more important reason for recognizing these different levels of organization is to place the diverse kinds of studies carried out by ecologists into a meaningful perspective (Bartholomew, 1964; Schultz, 1967). The tendency toward specialization and lack of communication among workers in a field such as ecology often results in the fact that individual workers come to feel that their area of specialization holds the key to solution of all major problems in the field. For example, certain ecologists may tend to feel that the primary goal of ecology is to explain the operation of ecological systems in terms of physiological processes at the suborganismal level. Others may feel, however, that the main objective is to evaluate all relationships in terms of their quantitative effects on ecosystem nutrient and energy exchange. The levels of integration approach not only recognizes the existence of unique characteristics at each level, but it also suggests a meaningful relationship between the different levels. This relationship is based on the fact that each level finds its mechanistic explanations in processes at the levels below and its significance in processes at higher levels (Bartholomew, 1964; Schultz, 1967). At the organismal level, for example, the mechanistic explanation of patterns of growth, maintenance, and reproduction is sought through examination of the effects of environmental factors on suborganismal features of morphology, physiology, or behavior. The significance of these organismal patterns is evident, however, at higher levels, where they determine the structure and behavior of populations, communities, and ecosystems. Likewise, mechanistic explanations of ecosystem function are sought through the study of processes involving individuals, populations, communities, or other ecosystem components, while the significance of ecosystem processes is evident at the level of major regional ecosystem types, or, perhaps, the biosphere as a whole.

FUNCTIONAL CHARACTERISTICS OF ECOLOGICAL SYSTEMS

Although ecological systems and system components of different degrees of complexity show unique structural and functional characteristics, they also possess certain common features. To a greater or lesser degree, all exhibit *homeostasis,* the ability to reestablish a normal state following disturbance. At the organismal level, homeostasis is seen in regulatory processes that maintain the internal environment of the organism in an optimal functioning state in the face of fluctuating external conditions. At higher levels, it is seen in compensatory responses in reproduction or other activities in populations subjected to disruptive processes such as predation. Homeostasis is also seen in the process of biotic succession that restores a stable, self-reproducing system following some major environmental disturbance and in the operation of various feedback mechanisms that tend to maintain a balanced flow of energy and nutrients into and out of major ecosystem components. One of the common objectives of ecologists studying systems of differing complexity is to determine the nature and capabilities of homeostatic mechanisms of these different systems.

Likewise, systems of all levels of complexity show change resulting from evolutionary processes. At the organismal level, studies may involve analysis of the intensity of selection and the rate of change of genetically based characteristics of species under different conditions.

At the community level, the evolutionary origin of major patterns of adaptation to environmental conditions may be examined. Studies at the ecosystem level may be concerned with evolutionary changes that influence overall system characteristics, such as the nature and degree of homeostasis. To the worker concerned with the study of ecological systems, the areas of ecology and evolution are inseparable. The dynamic processes occurring within ecological systems inevitably lead to evolutionary changes in the member species of these systems, as well as in the overall patterns of system function. This relationship has been aptly expressed as one of the "ecological theater and the evolutionary play" (Hutchinson, 1965).

THE STUDY OF ECOLOGICAL SYSTEMS

The environmental crisis has forced ecology to become a "problem-oriented" science (Cox, 1970). At the same time, it has drawn together aspects of ecological study that were formerly distinguished as "theoretical" and "applied." Protection of our world environment will require us to gain an understanding of the principles governing function of ecological systems at all levels of organization. In this sense, the "theoretical" and "applied" aspects of ecology are identical. Environmental disturbances, such as those resulting from the Aswan High Dam, have made them so.

Study of the dynamics of complex ecological systems necessitates improvement in the training of ecologists in the physical sciences and mathematics. Analysis of processes in ecological systems requires both evaluation of biotic characteristics of these systems and the accurate quantitative measurement of many physical and chemical factors. The evaluation of quantitative data on system relationships, however simple these relationships may appear to be, requires a basic knowledge of statistical procedures. For the study of complex systems, multivariate analysis techniques for sorting out the effects of several or many simultaneously operating variables are essential. Familiarity with techniques for formulating and testing predictive mathematical models of ecological systems has also become an important skill for functional ecologists. These models may incorporate rigorous assumptions of functional relationships or empirically determined relationships stated in mathematical form. The purpose of these models may be to test particular assumptions about system dynamics, give insight into relationships difficult to measure or test under actual conditions, or indicate the specific kinds of data needed for a more complete understanding of system function. The involved nature of techniques such as multivariate analysis and modelling frequently necessitates the use of computers and makes knowledge of computer programming procedures essential to ecologists interested in the study of complex systems.

The growth of this approach to ecology represents one of the most important recent developments in science. This approach holds the promise for solution of many of the most important problems facing man today. Realization of this potential can only occur, however, through the rapid growth of a body of ecologists trained in the study of complex ecological systems. This growth, in turn, will occur only as a result of increased awareness by the members of society at large that the well-being of man is dependent on that of the world ecosystem.

REFERENCES

Anonymous. 1970. Fisheries affected by Aswan Dam. *Comm. Fish. Res.,* **32:** 64.

Barnes, R. D. 1968. *Invertebrate Zoology,* 2nd ed. Philadelphia: W. B. Saunders.

Bartholomew, G. A. 1964. The roles of physiology and behaviour in the maintenance of homeostasis in the desert environment. In *Homeostasis and Feedback Mechanisms,* pp. 7-29, Eighteenth Symposium of the Society of Experimental Biology.

Benarde, M. A. 1970. *Our Precarious Habitat.* New York: Norton. 362 pp.

Borgstrom, G. 1970. The harvest of the seas: How fruitful and for whom? In H. W. Hefrich, Jr. (ed.), *The Environment Crisis,* pp. 65–84. New Haven: Yale University Press.

Coale, A. J. 1970. Man and his environment. *Science,* **170:** 132-36.

Commoner, B. 1966. *Science and Survival.* New York: Viking. 155 pp.

Cox, G. W. 1970. Lecture and laboratory approaches to the teaching of ecology. *BioScience,* **20:** 755–60.

Evans, F. C. 1956. Ecosystem as the basic unit in ecology. *Science,* **123:** 1127–28.

Hutchinson, G. E. 1965. The ecological theater and the evolutionary play. New Haven: Yale University Press. 139 pp.

Marx, L. 1970. American institutions and ecological ideals. *Science,* **170:** 945-52.

Mayhew, W. W. 1971. Personal communication.

Oren, O. H. 1969. Oceanographic and biological influence of the Suez Canal, the Nile and the Aswan Dam on the Levant Basin. *Progr. Oceanogr.,* **5:** 161–67.

Rowe, J. S. 1951. The level of integration concept and ecology. *Ecology,* **42:** 420–27.

Schultz, A. M. 1967. The ecosystem as a conceptual tool in the management of natural resources. In S. V. Ciriacy-Wantrup and J. J. Parsons (eds.), *Natural Resources: Quality and Quantity,* pp. 141–61. Berkeley: University of California Press.

Sterling, C. 1971. Aswan Dam looses a flood of problems. *Life,* **70**(5): 46–47.

Wagner, R. H. 1971. *Environment and Man.* New York: Norton. 491 pp.

White, L., Jr. 1967. The historical roots of our ecological crisis. *Science,* **155:** 1203-7.

Part II
THE ORGANISMAL LEVEL OF ORGANIZATION

Chapter 2
PATTERNS OF ADAPTATION

INTRODUCTION

Whenever a plant or an animal species demonstrates that it is able to live and reproduce in its surroundings, we conclude that it is *adapted* to the conditions of its environment. The organism itself is an environmental indicator that tells us better than any series of measurements of temperature, precipitation, soils and the like can that all of the needs of its life history are being met in this place. Those organism characteristics that are concerned with its perpetuation and continuity in its habitat are called *adaptations*. In one way or another the ecologist addresses himself to the relationship between the organism (which in this context is the embodiment of the adaptations) and its *environment,* which simply means anything that affects it directly.

Environmental conditions change from moment to moment and, of course, over longer time periods, but ordinarily not so greatly as to introduce conditions which the adapted organism has not experienced (and adjusted to) at some earlier time. In Chapter 7 we consider the possible outcomes of environmental change to which the organism is not adapted, but in this chapter we assume that the adaptations shown by organisms and selected from the naturally occurring variability that nearly all plants and animals possess are those that are advantageous to the organism in nature.

Judging what is and what is not an adaptation is difficult, and the literature of ecology and evolution is replete with misinterpretations and subjective opinions. A part of the problem is that we have tended to assume that all characteristics are adaptive in one way or another, when, in fact, this may not be true. For example, some characteristics may have become established by chance and are neither harmful nor beneficial to the organism. Others may have been adaptive at some earlier time in the evolution of the organism but are, at present, of no advantage. Still other characteristics may be indirectly correlated with characteristics that are adaptive but are themselves neutral.

The other part of the problem of interpreting adaptations lies with the observer, who because of human or technical deficiencies may not be able to unravel what is the fundamental importance of a particular characteristic to the organism.

The purpose of this chapter is to show that the variability that one sees in nature, both within and among species, is related to differences in the environments to which organisms are exposed. Furthermore, because the total range of expression of characteristics is limited, organisms living in similar habitats often exhibit similar habitat-related characteristics. For example, fishes that live on the ocean bottom are often flattened, and we assume that this characteristic is important to fish survival in this particular habitat.

We shall not attempt to explain here either specifically or in general how organism characteristics evolve. Genetic mechanisms that are well understood are responsible for naturally occurring variability; environmental pressures have been shown to be sufficiently great as to account for the survival of one variety with respect to others in a relatively few generations; together they account for adaptation and evolution.

THE INDIVIDUAL ORGANISM: THE BASIC ECOLOGICAL UNIT

The emphasis in this chapter is on the individual organism and its relationship with the environment. Although it is recognized that different individuals in a population of organisms contribute unequally to the next generation and are, therefore, differentially successful biologically, the ecologist faces the problem of interpreting the ecological success of organisms from what he can observe in the current generation. The organism's existence is evidence of its ecological success.

In our discussion of adaptations we shall concentrate on individual organisms rather than populations, communities, or ecosystems because in our opinion, natural selection works at that level. The student should be aware, however, that some ecologists believe that selection also works on higher units such as those mentioned above. Although it is premature to pursue this argument at length so early in this book, its importance ought to be emphasized because the difference lies at the base of some important controversies in ecology.

For example, one school of thought holds that the individual organism is the evolutionary unit; i.e., whatever one observes in nature can be explained by considering the influence of natural selection on individual organisms. The population, community, and the ecosystem have no properties of their own except for abstract ones. A birthrate is an example of an abstract property. Thus, in order to understand how these higher units function one needs only to know the individual functions of the organisms that comprise them. That individual organisms are present in the same area is coincidental and is due to their similar environmental tolerances. Beyond that they have few if any direct relationships among themselves.

The opposing view is that populations, communities, and ecosystems have evolved as such; i.e., the different species of a community or ecosystem and the way they are arranged have value to the community and any other arrangement or composition would have less value from an evolutionary point of view. The species are tied together very tightly in a variety of symbiotic relationships such that the removal of one species has repercussions that are felt elsewhere in the community. Thus, in a sense, the community is the unit of selection and has other properties that are analogous to those of the individual organism.

PATTERNS OF ADAPTATION

Ecologists and biogeographers have known for a long time that many morphological and physiological characteristics of organisms are related to the environment. Although finding this correlation does not necessarily indicate cause and effect, ecologists have often assumed that it does and have proposed ecological generalizations to describe it. We shall refer to these kinds of generalizations as *ecogeographical rules*.

Another pattern of organism-environment relations is shown by plants and animals living in similar climatic zones. In this case, it is assumed that the environment acts on genetically variable populations so that similar kinds of adaptations are selected, resulting in at least a superficial similarity between groups of organisms that may actually be rather distantly related. Thus some of the desert plants of the Old and New World look very much alike, e.g., the Cactaceae in North American deserts and the Euphorbiaceae in African deserts. This phenomenon, referred to by evolutionary biologists as *convergence,* is seen over and over again when areas of similar climates are compared as to their vegetations. In fact, the classification systems of the world's vegetation are a recognition of this similarity.

Ecogeographical Rules

The best known and most discussed of the ecogeographical rules are the related Allen's and Bergmann's rules. Bergmann's rule states that the size of *homeothermic* (temperature regulating) animals in a single or closely related evolutionary lines increases along a temperature gradient from warm to cold temperatures. Among the mammals, deer, rabbits, gophers, foxes, and man presumably demonstrate the rule, as do penguins, horned larks, and hummingbirds among the birds. The explanation of this size change is based on theoretical considerations of surface to volume ratios. As a homeothermic animal increases in size, its tendency to lose heat through its surface decreases in proportion to its increase in volume. Specifically, the surface of a globular object increases as the square, while the mass increases as the cube of its diameter.

Allen's rule has to do with the length of appendages in homeothermic animals. Structures such as ears, tails, and legs are shorter in the north and longer in the south. The need to conserve heat in the north and dissipate it in the south are presumed to be responsible for the observed trends. Examples that support this generalization include wing lengths in gulls and ear length in hares and foxes. Schreider (1964) reports that Eskimos have shorter arms and legs than do people living in warmer climates.

Neither of these rules is universally accepted by ecologists and physiologists. There are important exceptions to both. Scholander (1955, 1956), who is among the most outspoken of the critics of Allen's and Bergmann's rules, has objected to the interpretation of biometrical changes along the gradient on the grounds that there has been no careful analysis of the selective pressures that might be associated with changes from north to south. In the case of man, for example, the thermal environment of the Eskimo, i.e., inside his clothing, has not been shown to be measurably lower than that of other peoples. Regarding other animals, Scholander says, "There is no physiological evidence . . . that the minor and erratic subspecific trends expressed in Bergmann's and Allen's rules reflect phylogenetic pathways of heat conserving adaptation" and "cold climates do not produce a fauna tending towards large-sized globular forms with small protruding parts. The phylogenetic adaptation to heat conservation took place through other means, which have been measured, which are understood, and which apply to every individual of every species."* Among these "other means," Scholander includes behavioral adaptations by which animals avoid cold temperatures, e.g., dense layers of insulation such as fat, feathers, and fur, and vascular adaptations which allow organisms to function even if their appendages are at near-ambient temperatures. These adaptations will be considered in more detail on page 61.

Other ecogeographical rules include the following:

1. Gloger's rule states that races of birds or mammals living in cool, dry regions are lighter in color (have less melanin pigment) than races of the same species living in warm, humid ones.
2. Jordan's rule states that fishes inhabiting cold water tend to have more vertebrae than those living in warm water.
3. An observation that has not been associated with any single investigator states that the frequency of species of flowering plants that are polyploid (having more than two sets of chromosomes in

*Scholander, 1955, p. 24.

their somatic cells) increases toward the north (e.g., Johnson, Packer, and Reese, 1965).

Clearly, none of the rules and generalizations described above explains what is happening in nature. That the size of a structure decreases along a gradient of decreasing heat is not evidence that the most important selective agent is temperature. The rules belong in the same category as other scientific observations. Hypotheses must be proposed to account for them, and they must be tested in both laboratory and field conditions. In the case of the ecogeographical rules, most of the testing remains to be done. Today's observations are only the current expressions of long-term evolutionary processes; it is difficult to be certain of the most important selective forces that operated in the past.

Convergence

Like the ecogeographical rules, almost all of the early studies of convergence were descriptive rather than experimental. One of these, Raunkiaer's biological spectrum, will be discussed to illustrate this point. Raunkiaer's thesis was that the earliest flowering plants evolved under tropical conditions where the plant was not required to survive unfavorable climatic periods. As they migrated out of the tropics, plants encountered seasonal periods in which one or more environmental factors were limiting. In response to these conditions, structural and functional mechanisms that allowed for more flexibility in terms of habitats occupied were selected. Of particular interest to Raunkiaer was the position of buds in relation to the ground surface. He reasoned, for example, that a plant whose buds are buried in the soil is better able to endure seasonal cold than one whose buds are fully exposed to the atmosphere.

Raunkiaer randomly selected 1,000 higher plant species and classified them by the position of the buds. The frequency of plants in each category constituted the Raunkiaer "normal" spectrum. Deviations from the normal agreed with each other if they were drawn from similar areas. Thus, all plants native to tundra areas show similar biological spectra. Mostly their buds are buried in the ground. Plants of tropical areas, on the other hand, have exposed buds, presumably because they have no period of inactivity nor need of special protection. Raunkiaer proposed several categories of bud position (Table 2–1) and also recognized plants that reproduce only by seeds (annuals) as a separate category characteristic of deserts.

In Table 2–1 the following categories of "overwintering" structures are recognized:

Phanerophytes (Ph)—trees and shrubs, buds least protected; common in the tropics.

Chamaephytes (Ch)—buds at ground level during unfavorable seasons; become more common farther away from the equator.

TABLE 2-1. LIFE-FORM COMPOSITION OF THE VEGETATION OF
SEVERAL DIFFERENT CLIMATIC REGIONS.

Locality	No. of species	Species distribution (%)				
		Ph	Ch	H	Cr	Th
Phanerophytic*						
St. Thomas and St. Jan	904	61	12	9	4	14
Therophytic*						
Death Valley, Cal.	294	26	7	18	7	42
Hemicryptophytic*						
Georgia	717	23	4	55	10	8
Chamaephytic*						
St. Lawrence Isl., Alaska	126	0	23	61	15	1
Therophytic[†]						
Santa Catalina, Cal.	391	18	5	27	5	41
Normal spectrum	1,000	46	9	26	6	13

*Data from Oosting, 1956.
[†]Data from Thorne, 1967.

Hemicryptophytes (H)—dormant buds just beneath or in the soil surface; temperate climates primarily.

Cryptophytes (Cr)—buds deeply buried and food storing, e.g., bulbs;
especially in areas of extreme climates.

Therophytes (Th)—depend on seeds for survival.

Convergence also operates at a physiological level and occurs in organisms that do not necessarily otherwise resemble each other very closely.
Desert mammals, for example, must solve problems of temperature regulation and water balance. In many cases it has been shown that the
physiological mechanism that allows for adjustment to temperature is
almost the same in rather different groups. This will be discussed at
greater length later in this chapter.

Mooney and Dunn (1970) have studied convergence in the evergreen
shrubs that grow in the mediterranean climatic type. This kind of climate
has relatively warm, dry summers and cool, moist winters (Fig. 2–1).
They propose a model that summarizes selective agents and the evolutionary "strategies" that apply generally to vegetation in all mediterranean
climatic regions (Fig. 2–2). Much of what is known about the model
comes from the many studies of the chaparral vegetation of coastal California. The intriguing aspect of this model is that given certain environmental parameters (selective forces), the vegetation will respond in fairly
predictable ways in evolving solutions to the imposed selective forces.

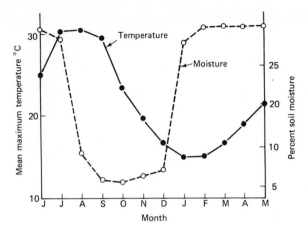

FIGURE 2–1. The seasonal course of air temperatures and soil moisture at a southern California chaparral site. (From Mooney and Dunn, 1970.)

This in turn creates the framework for evolution of the other organisms. Since the number of solutions to the environmental stresses appears to be limited, there is a tendency toward convergence both in form and function.

The Ecotype Concept

The first experimental approach to the problem of habitat-related variation in organisms was initiated by Turesson (1922). Turesson was impressed that populations of certain species differed among themselves and that those populations that grew in the same kinds of habitats tended to resemble each other more than they looked like populations living in other habitats. By cultivating these different populations side by side in a uniform garden, he demonstrated that the differences among them were at least partly genetic; i.e., the differences were maintained. Turesson called these genetically different populations *ecotypes*. In practice, the term has come to refer to those populations within a species that show morphological and physiological habitat-related and genetically produced differences. Ecotypes of the same species are not isolated from each other reproductively, but geographic or ecological isolation may substantially limit the exchange of genes.

The most ambitious program of investigations at the intraspecific level relating plant characteristics to environment was carried out at Stanford University. Clausen, a cytogeneticist, Keck, a plant taxonomist, and Hiesey, a physiologist, initiated a series of investigations lasting thirty years and which, with somewhat different emphases, continue today.

The Stanford Group established uniform gardens at three altitudes: the coast at Stanford (30 m), Mather (1,400 m), and Timberline (3,050 m) in the Sierra Nevada. By careful attention to techniques involving reciprocal transplants, uniform garden methods, controlled breeding experiments, and the like they demonstrated the nature of the genetic and environ-

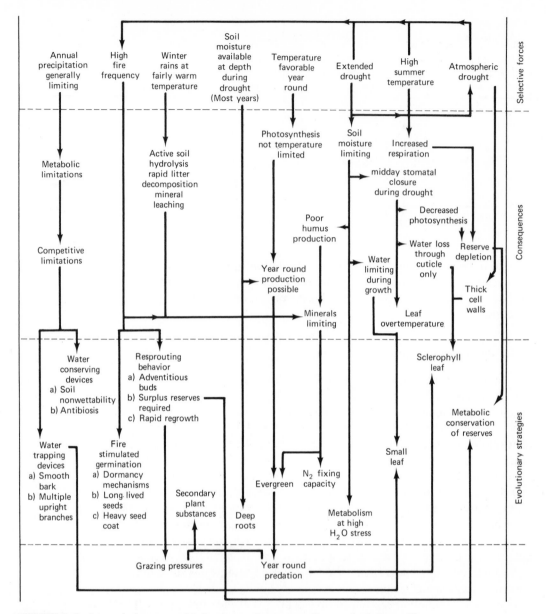

FIGURE 2–2. An evolutionary model for the mediterranean-climate shrub form. (From Mooney and Dunn, 1970.)

mental effects on plant populations. Although they investigated many species, they are best known for their work with *Potentilla glandulosa* and *Achillea millefolium*. Only the former will be discussed here.

At this point we should consider the kinds of observations that led to this work, the kinds of questions that were asked, and the nature of the

TABLE 2–2. CHARACTERISTICS OF THE ECOTYPIC SUBSPECIES OF *POTENTILLA GLANDULOSA* ALONG THE CENTRAL CALIFORNIAN TRANSECT.*

	typica	*reflexa*	*hansenii*	*nevadensis*
Distribution	Coast Ranges and lower Sierra Nevada	Low and middle altitudes of Sierra Nevada	Meadows, mid-altitudes of Sierra Nevada	High altitudes of Sierra Nevada
Habitat	Soft chaparral and open woods	Dryish, open-timbered slopes	Moist meadows	Moist, sunny slopes
Climatic tolerance as experimentally determined	Coastal to middle altitudes	Coastal to middle altitudes	Middle and high altitudes (poor survival near coast)	Middle and high altitudes (poor survival near coast)
Seasonal periodicity at Stanford (alt. 30 m)	Winter- and summer-active	Winter-active or dormant; summer-active	Winter-dormant; summer-active	Winter-dormant; summer-active
Internal variation	Wide, probably several "ecotypes"	Wide, probably several "ecotypes"	Wide, at least two "ecotypes"	Moderate, at least two "ecotypes"
Self-compatibility	Self-fertile	Self-fertile	Undetermined	Self-sterile

*Data from Clausen and Hiesey, 1958.

answers to the questions. Ecologists and other field biologists are faced with the problem of dealing with variability among the populations with which they work. *P. glandulosa* exhibits a great deal of variability in nature, is very widespread, and occupies many different habitats. Some populations are so different in appearance that they have been classified as different species. How much of this variability can be explained by the direct influence of the environment on the plant? How much of it is genetically determined? When can one state that populations are sufficiently different that they should be called separate species? How precisely are natural populations adapted to their local environments? Why are some species so limited in their distribution and others so widespread?

Potentilla glandulosa occupies a great part of a transect starting at the coast and extending across the Coast Ranges and up to an altitude of about 3,350 m in the Sierra Nevada. It does not occupy the coastal strip, the drier east side of the inner Coast Range, or the bottomlands in the Great Valley. Four subspecies* are recognized in *P. glandulosa* along this transect; their distributions, habitats, and ecotypic differentiation are summarized in Table 2–2 (Heslop–Harrison, 1964). All of the subspecies

*A subspecies is a geographic subdivision of a species distinguished by morphological characters.

are fully interfertile yet maintain their distinctness because well-defined and differentiated climatic zones occur across the transect from the Coast to the Sierra Nevada. Flowering time of adjacent subspecies may vary by nearly a month.

Clones* of the four subspecies were planted at Stanford, Mather, and Timberline, and the response of each was evaluated. Except for the observation that the high altitude subspecies *nevadensis* is most vigorous at Mather, the transplant experiments showed that each subspecies grew and survived best at or near the altitude from which it originated.

Relatively few studies have analyzed the genetic basis of ecotypic differentiation. The studies by Clausen and Hiesey (1958) are the most complete of these. Using the subspecies and ecotypes of *P. glandulosa* referred to on page 27 the following crosses were made:

1. *P. glandulosa typica* × *P. g. nevadensis* (alpine ecotype)
2. *P. g. reflexa* × *P. g. nevadensis* (subalpine ecotype)

From cross (1) first (F_1) and second (F_2) generation plants were grown in a uniform garden; in cross (2) F_1, F_2, and F_3 generation plants were grown in a uniform garden and F_2 plants were compared at the three transplant stations. The 19 characters listed in Table 2–3 were studied and estimates were made of the numbers of genetic loci that are probably involved in their inheritance. Without carrying on an extensive breeding program involving many progeny from several repetitions of the crosses and backcrosses of the offspring with parents, it is not possible to do more than estimate how many genes are involved Many characters, such as height, are often controlled by several independent genetic loci; others, such as number of leaflets in the bracts, seem to be under the control of a single pair of genes. Clausen and Hiesey thought that they could account for the variability in the 19 characters listed in Table 2–3. Of the 91 possible combinations that exist between them, 67 showed some degree of correlation; in cross (2) 12 characters were analyzed and 38 of the 66 possible combinations showed significant correlation [Figs. 2–3(a) and (b)].

The significance of these character correlations, which Clausen and Hiesey refer to as *coherence,* is that the characters apparently do not recombine at random; some probably are linked together by virtue of being on the same pair of chromosomes ($n = 7$ in *P. glandulosa*). This degree of coherence together with rather sharp habitat differences are probably sufficient to preserve the identities of the races. One would expect in this case to find ecotypes rather than ecoclines because there would be less likelihood of overlap and intergradation between populations.

Clausen, Keck, and Hiesey make it clear that the differentiation of

*A clone is a group of genetically identical individuals.

TABLE 2-3. ESTIMATE OF MINIMUM NUMBER OF GENES GOVERNING THE INHERITANCE OF NINETEEN CHARACTERS IN TWO INTERECOTYPIC HYBRIDS OF *POTENTILLA GLANDULOSA*.*

Character, and action of genes	Estimated no. of gene pairs
1. *Orientation of petals:* 2 erecting, 1 reflexing	3
2. *Petal notch:* 1 producing notch, 2 inhibiting	3
3. *Petal color:* 2 whitening, 2 producing yellow, 1 bleaching	5
4. *Petal width:* 4 widening, 1 complementary, 1 narrowing	6
5. *Petal length:* 4 multiples	ca. 4
	(plus possible inhibitors)
6. *Sepal length:* 3 or 4 multiples for lengthening, 1 for shortening, 1 complementary	ca. 5
7. *Akene weights:* 5 multiples for increasing, 1 for decreasing	ca. 6
8. *Akene color:* 4 multiples of equal effect	4
9. *Branching, angle of*	ca. 2
	(also genes for strict to flexuous branching)
10. *Inflorescence, density of*	ca. 1
	(plus modifiers)
11. *Crown height*	ca. 3
	(also genes for presence or absence of rhizomes and for thickness of rhizomes, to which crown height is related)
12. *Anthocyanin:* 4 multiples (1 expressed only at Timberline), 1 complementary	5
13. *Glandular pubescence:* 5 multiples, in series of decreasing strength	5
14. *Leaf length:* transgressive segregation; many patterns of expression in contrasting environments; possibly different sets of multiples activated	ca. 10–20
15. *Leaflet number in bracts*	ca. 1
	(plus modifiers)
16. *Stem length:* transgressive segregation, 5 to 6 multiples plus inhibitory and complementary genes; many patterns of expression in contrasting environments	ca. 10–20
17. *Winter dormancy:* 3 multiples of equal effect	3
18. *Frost susceptibility:* slight transgression toward resistance	ca. 4
19. *Earliness of flowering:* strongly transgressive; many patterns of altitudinal expression; possibly different sets of genes activated	many

*From Clausen and Hiesey, 1958.

populations into climatic races depends fundamentally on the occurrence of well-marked environmental differences over relatively short distances. An extremely interesting and informative example of the results of this kind of change is seen in the work of McNeilly (1968).

McNeilly studied populations of a grass, *Agrostis tenuis* that were growing on and around small copper and other heavy metal mines in Wales. Previous work had shown that *A. tenuis* is able to tolerate the normally toxic amounts of copper and other heavy metals in the worked soils, and, further, that the tolerance is heritable. Nontolerant plants, however, are killed by the concentrations of heavy metals in the soils.

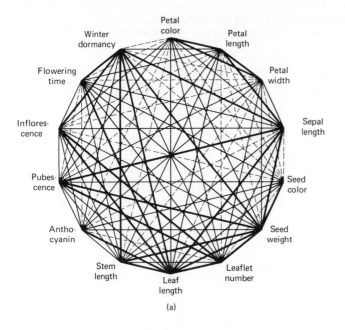

(a)

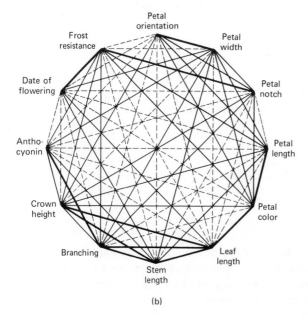

(b)

FIGURE 2-3. (a) Statistical correlations between pairs of characters in 992 F_2 progeny of *Potentilla glandulosa* subsp. *nevadensis* × *P. glandulosa* subsp. *typica*. (b) Statistical correlations between pairs of characters in 570 F_2 progeny of *potentilla glandulosa* subsp. *nevadensis* (subalpine form) × *P. glandulosa* subsp. *reflexa* (foothill form). In each figure a heavy line indicates $r = 0.25$–0.80; a light line, $r = 0.09$–0.25; and a broken line, $r = 0.00$–0.09 (insignificant.) (From Clausen and Hiesey, 1958.)

The mine boundaries are sharp and McNeilly demonstrated that the change from copper tolerance to nontolerance in the plants of the area occurred across an area about one meter wide (Transect 1 in Fig. 2–4). On the downwind side of the mine, however, tolerant individuals are

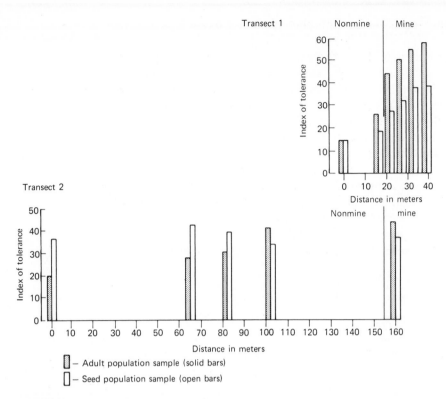

FIGURE 2–4. Mean index of copper tolerance in populations of *Agrostis tenuis*. Transect 1 sampled populations from the mine and the upwind adjacent nonmine area. Transect 2 sampled populations from the mine and the downwind adjacent nonmine area. (From McNeilly, 1968.)

found up to 150 m from the mine boundary (Transect 2 in Fig. 2–4). McNeilly explained this difference in distributions as follows:

> Selection pressures on toxic soils are strong and favour tolerant genotypes; selection pressure on normal soils are weak but favour normal individuals.

Although tolerant populations maintain their identity in spite of high gene flow because of strong selection pressures favoring tolerance, the selection pressures in nontolerant populations are not sufficient to maintain population divergence in the face of high gene flow (due to pollen from tolerant plants fertilizing downwind female flowers).

Ecotypes in Animals

The ecotype concept was developed in higher plants, and animal biologists have tended to pay less attention to it than botanists, perhaps because of the differences in mobility between plants and animals. Nevertheless, ecological differentiation occurs in animals as well as in plants, and some populations show discontinuous morphological and physiological variation.

Cade and Bartholomew (1959) discuss an interesting case of ecological differentiation in subspecies of the savannah sparrow, *Passercula sandwichensis*. *P. s. beldingi* and *P. s. rostratus* are restricted to salt marshes, and *P. s. brooksi, P. s. anthinus,* and *P. s. nevadensis* are migratory races* that do not ordinarily frequent areas of salt water. The two salt marsh races are distinct from the other three morphologically, although the latter are sometimes difficult to separate.

In order to determine the response of birds from each population to salt water, birds were captured from the field and fed water ranging in concentration from distilled water to 100% sea water. In general, the two salt marsh inhabitants were able to maintain their body weights when drinking water up to pure sea water, while the nonsalt marsh individuals lost weight on the two highest concentrations (Figs. 2–5 and 2–6). An-

FIGURE 2–5. The effects of drinking distilled water and various concentrations of sea water and of water deprivation on body weight of *P.s. anthinus, P.s. nevadensis,* and *P.s. brooksi.* The points represent weights measured at the end of a 5-day test period. The horizontal lines indicate means. (From Cade and Bartholomew, 1959.)

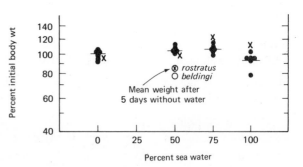

FIGURE 2–6. The effects of drinking distilled water and various concentrations of sea water and of water deprivation on body weight of *P.s. beldingi* (circles) and *P.s. rostratus* (x's). The points represent weights measured at the end of a 5-day test period. The horizontal lines indicate the means. (From Cade and Bartholomew, 1959.)

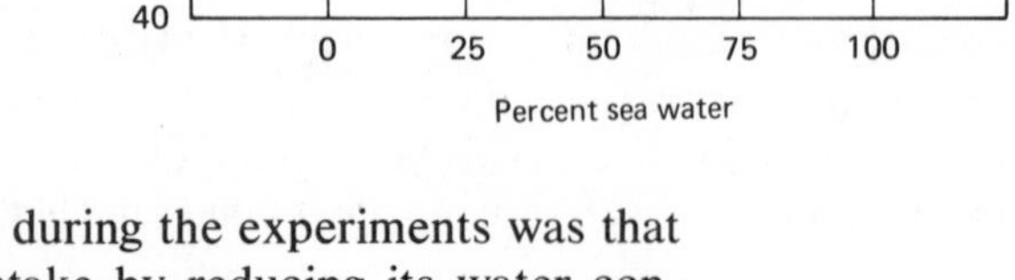

other surprising fact that came to light during the experiments was that *P. s. rostratus* could control its salt intake by reducing its water consumption to extremely low levels. Most birds that have been tested, including *P. s. beldingi* and the three nonsalt marsh races studied here, either increase their water intake when fed salt water or maintain high levels of water consumption (Figs. 2–7 and 2–8). During dehydration experiments, it developed that *P. s. rostratus* is able to maintain body weight for extended periods on a diet consisting solely of dry seeds.

*The term *race* is used in the sense of a geographically delineated population or subspecies.

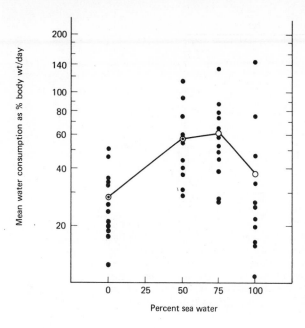

FIGURE 2–7. Ingestion of various concentrations of sea water by *P.s. anthinus, P.s. nevadensis,* and *P.s. brooksi.* Each point represents the mean daily consumption of one bird for a 5-day period. The lines connect the means of each sample. (From Cade and Bartholomew, 1959.)

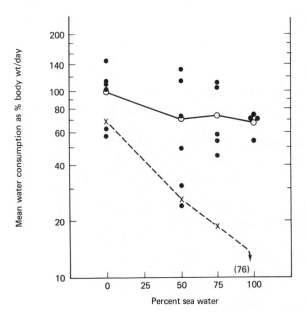

FIGURE 2–8. Ingestion of various concentrations of sea water by *P.s. beldingi* (circles) and *P.s. rostratus* (x's). Each point represents the mean daily consumption of one bird for a 5-day period. The hollow circles are the means for *beldingi.* (From Cade and Bartholomew, 1959.)

Cade and Bartholomew remark that "from the standpoint of salt intake the *beldingi* population can be considered an aggregation of individuals selected from one extreme of the range of salt tolerance shown by nonsalt marsh races." *P. s. rostratus,* however, does not overlap the others in terms of its salt tolerance and, in addition, is "the most clearly defined subspecies of *P. sandwichensis.*" Thus, in animals, as well as in plants, localized environmental conditions may allow for the development of ecological races.

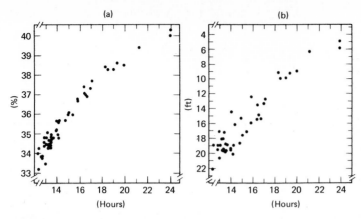

FIGURE 2–9. (a) Relationship between dry matter content and the length of daylight of the first day in the year with an average normal temperature of ± 6°C at the native habitats of 52 provenances of *Pinus sylvestris* grown at Stockholm. (From Langlet, 1959.) (b) Relationship between tree height at 17 years and the length of daylight of the first day in the year with an average normal temperature of +6°C at the native habitats of 46 provenances of *P. sylvestris* grown in New Hampshire, U.S.A. (Data of Wright and Baldwin, 1957; replotted by Langlet, 1959.)

Ecoclinal Variation

When environments change gradually, one should expect to see correspondingly gradual or clinal changes in adjacent populations. The work of Langlet (1959) and Wright and Baldwin (1957) with Scotch pine is a good example of ecoclinal variation and is of interest for other reasons as well.

Langlet and Wright and Baldwin analyzed variability in populations (provenances) of Scotch pine that had been collected over some 25° of latitude in Europe; seeds from each population were planted in uniform gardens in Sweden and in New Hampshire. Seventeen years later, in 1955, the populations were analyzed for several characters. Wright and Baldwin worked with the New Hampshire-grown populations and grouped the samples into a number of ecotypes that were based primarily on geographic criteria. Based on these groupings they calculated "regional means" of morphological and physiological characteristics (tree height, growth rate, etc.) and concluded that the variation in Scotch pine is discontinuous; i.e., each geographic group is distinct from others.

Langlet (1959) criticized the work of Wright and Baldwin and suggested that their failure to find continuous variation among the populations was due to the method of treating the data. In his analysis of the Stockholm populations, Langlet showed that the dry matter content of the leaves was correlated with the length of the first day in the year with an average temperature of 46°C at the native habitats of the populations [Fig. 2–9(a)]. He then replotted the height data of Wright and Baldwin against the same independent variable and found the same high degree of correlation as was the case using leaf dry matter [Fig. 2–9(b)]. Inasmuch as Langlet did not make any preliminary subjective groupings of the data, it would appear that his is the more objective method of data analysis.

Thus, at least with respect to some of the characters analyzed, Scotch pine is a good example of ecoclinal variation.

THE PHYSIOLOGICAL BASIS FOR ADAPTATION

We take it for granted that organisms are physiologically adapted to the habitats in which we find them. Depending on the ability of the organism to migrate and to vary genetically, it may establish itself in other habitats. Physiological ecology (or ecological physiology) attempts to explain how species can live in particular environments. Although this question has not been answered completely for any species, considerable information is available on physiological processes that are thought to be important in this respect. Some of these are germination, growth, differentiation, respiration, absorption, translocation, transpiration, reproduction, temperature regulation, osmotic concentration and the like.

Much of the emphasis in this field has been on comparative studies of populations of geographically widespread species. These studies have sought to determine whether such species have evolved ecological races (*evolutionary adaptation*) or whether they are able to adjust nongenetically to a broad spectrum of environmental conditions (*environmentally induced adaptation*). If an organism placed under environmental stress adjusts its physiology to the stressful condition, it is said to have *acclimated* to the change.

In order to determine whether the response of the organism is genetic or nongenetic, an approach similar to the one described on page 25 has been used:

1. Organisms representative of populations of one species from different habitats are grown in uniform environments.
2. Organisms from different habitats are transferred to environments representative of the gradient of environmental conditions to which the species as a whole is exposed.

Studies in physiological ecology have advanced very rapidly in recent years because of the development of controlled environment facilities in which the investigator may carefully regulate or control several environmental factors simultaneously, e.g., temperature, light, and humidity. These programmable environmental chambers and the development of complex and miniaturized electronic instruments used in the analysis of rate processes have substantially increased the data available in this area. If individuals of the same species from different habitats retain their differences when grown in uniform conditions, the differences between individuals are genetic; if they are minimized by exposure to uniform conditions, they are probably nongenetic.

The work of Mooney and Shropshire (1967) illustrates the difference between evolutionary adaptation and environmentally induced adaptation. *Encelia california* Nutt. is a shrub that grows along the immediate

FIGURE 2–10. *Encelia californica:* growth performance in uniform garden; plants at left were grown from seed of dwarf coastal dune population; plants at right were grown from seed of coastal hill population. All plants in figure about 1.5 m in height.

southern California coastline. It is found in a variety of habitats, e.g., coastal dunes, where it is prostrate and attains heights of no more than 15 cm, and coastal hills and valleys, where it may reach a height of one meter. In uniform garden plots plants from both habitats become shrubs of the same height (Fig. 2–10). Whether plants from the valley population would grow as a prostrate shrub if transplanted to coastal sand dunes is not known, but presumably it would from the evidence available.

To test the ability of the coastal and valley plants to adjust their rates of photosynthesis in response to the climate in which they have grown, Mooney and Shropshire placed individuals from the two populations in a cold-acclimation chamber which was programmed for a 15°C / 16-hour day and a 2°C / 8-hour night. Light intensity was 8,000 foot candles at plant height. After 12 days, the plants were removed from the cold-acclimation chamber; no new leaves were produced during this period. Photosynthesis measurements were taken immediately. Maximum net photosynthesis (gross photosynthesis minus respiration) occurred at less than 20°C on the cold-acclimated coastal plant and at less than 15°C for the valley plants (Fig. 2–11, top). At 30°C net photosynthesis was essentially zero for both populations.

Next the plants were left at 30°C in full light for 23 hours, and photosynthesis was measured again. Plants from both populations responded by a shift in the temperature of their maximum net photosynthesis to above 25°C. There was no apparent photosynthesis at 10°C. Absolute rates increased at most temperatures between treatments. When net photosynthesis is plotted on the basis of percent of the maximum rate,

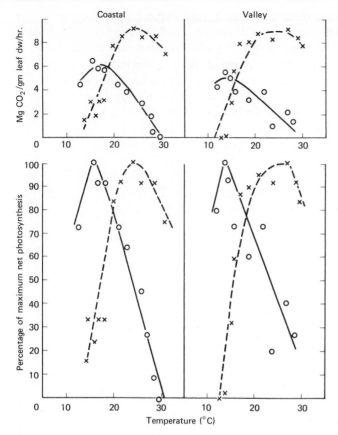

FIGURE 2–11. Temperature curves of photosynthesis of coastal and valley clones of *Encelia californica.* Top set of curves gives values of net photosynthesis on an absolute basis and bottom set on a relative basis. Circles indicate values measured immediately upon removal from 12 days in the cold-acclimation chamber (15°/2°C) and x's, values measured after the plants were maintained 23 hours in the light at 30°C. (From Mooney and Shropshire, 1967.)

the two curves are essentially mirror images of each other (Fig. 2–11, bottom).

In other experiments, cold-acclimated plants from both habitats showed lowered peak net photosynthesis rates after 24 hours exposure to cold temperatures, and warm-acclimated plants had elevated net photosynthesis rates at all temperatures after 23 hours exposure to warm temperatures. Respiration rates were "fairly uniform" after all of these treatments. Mooney and Shropshire conclude that "photosynthetic acclimation to warm temperatures thus appears to be the reverse of acclimation to cold."

In analyzing possible reasons for the essentially reversible changes in photosynthesis rate, Mooney and Shropshire state that changes in enzyme ability with temperature could be responsible. From what is known of this species then it appears that in both growth form and acclimation physiology the two populations are interchangeable.

When the *Encelia* work described above is compared to similar experiments using *Polygonum bistortoides,* a different picture emerges. In earlier work, Mooney (1963) had shown that coastal and subalpine populations of *P. bistortoides* maintained their dissimilar morphological and physiological characteristics when grown in a common environment.

Using acclimation conditions similar to the *Encelia* experiments *P. bistortoides* plants were acclimated to warm temperatures and showed similar responses to those found in *Encelia;* i.e., following exposure to warm temperatures, peak photosynthesis activity shifted to higher temperatures than for plants kept in the cold-acclimation chamber. The difference between *Encelia* and *Polygonum* in respect to their acclimation potential is that while the coastal and valley populations of *Encelia* show nearly identical responses to warm and cold acclimation, the coastal population of *Polygonum* shows large changes in temperature-related maximum net photosynthesis relative to those of the subalpine population (Fig. 2–12). Thus, the experiments on acclimation to warm temperatures confirm the differences found by Mooney in his earlier study.

Probably all plants and animals are capable of adjusting their environ-

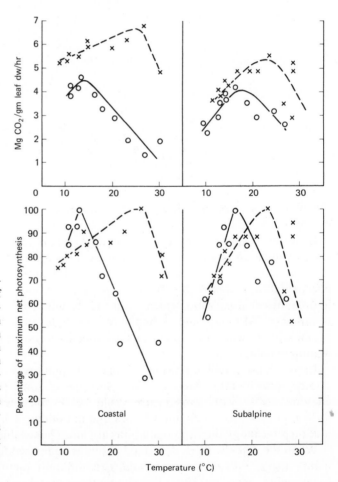

FIGURE 2–12. Photosynthetic acclimation to warm temperature of plants of two populations of *Polygonum bistortoides.* Top curves are expressed on an absolute basis, and those on the bottom on a relative basis. In the first column are the results from a coastal population and the second column, a subalpine population. The circles indicate values measured on plants immediately upon removal from 10 days in the cold acclimation chamber; the x's are values measured after plants were kept at 30°C and full light for 25 hours. (From Mooney and Shropshire, 1967.)

mentally related physiological optimum processes, but their ecological distribution depends more fundamentally on the adaptations they have evolved to diverse environmental conditions. In order to illustrate this more fully, we have selected two aspects of physiological ecology to discuss in detail: photosynthesis in higher plants and temperature regulation in vertebrate animals. Appreciation of the complex responses that have evolved for each of these processes will make it clear that organisms can evolve adaptations to the variety of environmental stresses in the biosphere.

Photosynthesis

Photosynthesis takes place in two steps, one requiring light and the other indifferent to it. The photochemical or light requiring process is in itself a two-stage process, but for our purposes it is sufficient to state that the light energy is absorbed by pigment systems involving chlorophyll with the result that the water molecule is split, the oxygen passing off as a gas. The energy received by the photochemical system is used through several steps to reduce carbon dioxide to carbohydrates. Thus, while we may write the equation for photosynthesis as:

$$6CO_2 + 12H_2O \xrightarrow[\text{Chlorophyll (or other pigment)}]{\text{673 kilocalories}} C_6H_{12}O_6 + 6O_2 + 6H_2O$$

we know that either the light or the "dark" systems can be modified in one or more ways. Clearly, the concentration of any of the ingredients, changes in the amount of energy entering the system, the efficiency of enzyme systems, and the external and internal environmental factors such as temperature and humidity can work singly or severally in ways to change the outcome of the process. This has been important in the evolutionary and ecological history of the green plants.

Environmental Limitation of Photosynthesis

Physiologists have demonstrated how the rate of photosynthesis is limited by environmental parameters. Temperature, light, carbon dioxide, and oxygen concentrations are among the most important extrinsic factors governing photosynthesis while chlorophyll content, water, anatomical features, and enzyme concentrations are among the intrinsic ones. Theoretically, any of the above can be limiting, but as will be seen below, two or more of these factors can operate interdependently to influence photosynthetic rate.

Carbon Dioxide. Carbon dioxide constitutes only about 0.03% by volume of the earth's atmosphere. The amount is rising slightly because of the burning of fossil fuels, but increases are buffered substantially because of the great capacity of CO_2 absorption in the oceans. It is probable that the CO_2 in the atmosphere and the oceans are in dynamic equilibrium.

According to Meyer et al. (1960), "The carbon dioxide concentration of the atmosphere is most frequently the limiting factor in photosynthesis for all photosynthetic tissues which are well exposed to light." The basis for this statement is the observation that the rate of photosynthesis is enhanced by concentrations of carbon dioxide up to at least several times its average value of 0.03% (Fig. 2–13). At concentrations 15–20

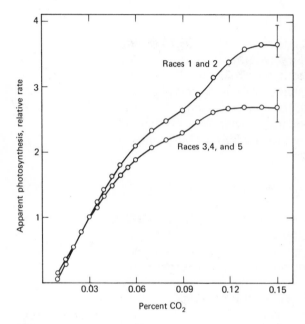

FIGURE 2–13. Effect of CO_2 concentration on photosynthetic rates of *Mimulus* races. Rate with 0.0300% CO_2 taken as unity in each case. (From Milner and Hiesey, 1964.)

times the usual atmospheric concentration, detrimental effects such as retarded photosynthetic rate often appear in time. Milner and Hiesey (1964), for example, showed that continuous exposure of *Mimulus cardinalis* to 0.15% CO_2 resulted in a high rate of photosynthesis that fell rapidly with time until values approximately the same as those measured at 0.0425% (Fig. 2–14) were reached.

Although carbon dioxide additions clearly increase photosynthetic rates, relatively little information is available that demonstrates adaptations of plants to differing amounts of CO_2 in nature.

Billings, et al. (1961) measured the effects of low concentrations of carbon dioxide on alpine (2,027 m) and arctic (sea-level) populations of *Oxyria digyna*. They found that the alpine plants were more effective in fixing CO_2 at all concentrations tested (Fig. 2–15). The significant discovery here is that the compensation concentration (where CO_2 uptake = CO_2 evolution) was not reached in the alpine population until near or

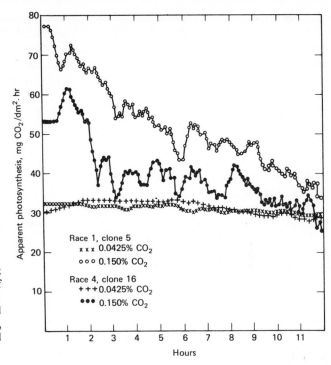

FIGURE 2–14. Change in rate during continuous photosynthesis by clones of races 1 and 4 with 0.150 and 0.0425% CO_2. Some points for race 4 are omitted where the curves for 0.0425% CO_2 nearly coincide. (From Milner and Hiesey, 1964.)

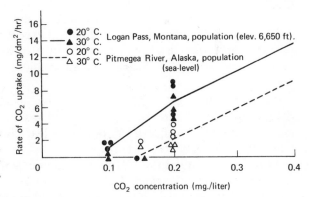

FIGURE 2–15. Apparent photosynthesis rates of *Oxyria* leaves at low CO_2 concentrations. Points are averages of two to four determinations on a single plant. (From Billings et al., 1961.)

below 0.1 mg of CO_2 / 1. None of the sea-level plants had any apparent photosynthesis at this level which approximates the naturally occurring CO_2 tension at 12,200 m. Billings et al. conclude that it is conceivable that low CO_2 pressures limit the upward distribution of plants of some species.

The Role of Light. Light varies in the environment in intensity, quality, and duration, but only intensity will be discussed here. The intensity of light that falls on the earth's surface varies considerably according to the time, the season, meteorological conditions, topographic conditions (slope and exposure), altitude, depth below the water's surface (for aquatic plants), and, in plants, the position of the leaf with respect to other light intercepting surfaces and the orientation of the leaf.

To some extent, the light absorbed by the leaf surface may be modified by the surface features of the leaves. Billings and Morris (1951), for example, compared the amount of light (from 400μ to 700μ) reflected by leaves from different environments ranging from shaded to open desert (Fig. 2-16). They concluded that "the greater the exposure to sunlight and the drier the habitat, the greater is the leaf reflectance in the visible spectrum." Moony and Johnson (1965) likewise showed that the leaves of the alpine race of *Thalictrum alpinum* reflect more light than the sea-level arctic race (Fig. 2-17). These observations suggest that there is a selective advantage for plants in high-radiation environments that evolve scales, hairs, or waxy layers that tend to reflect light. Whether light reflectance has to do with temperature control or the involvement of light in photosynthesis is not certain.

One of the most thorough and revealing studies that relates light to photosynthesis and plant adaptation is that of Björkman and Holmgren (1963, 1966) and Björkman (1968) on *Solidago virgaurea*. This species is widespread throughout Europe and northern Asia, from the subtropics to the arctic and from open heaths and meadows to dense forests. Many races have been described, among them one from shaded habitats and another from exposed habitats which Björkman and Holmgren worked.

Plants were cultivated from two shaded and two exposed habitats. All conditions of cultivation were identical except for light intensity; one pair of plants from each clone was cultivated at a low light intensity $(3 \times 10^4$ ergs cm^{-2}s$^{-1})$ and another pair at high light intensity $(15 \times 10^4$ ergs cm^{-2}s$^{-1})$, both at 400-700 μ.

After four to eight weeks of cultivation, plants were selected for photosynthesis measurements. An infrared gas analyzer was used to detect changes in CO_2 concentrations in a leaf chamber, in which light, humidity, and temperature were regulated.

Two different measures of photosynthetic activity were used. The steepness of the initial slope is considered to be a measure of the photochemical capacity (relation between light and photosynthesis at low light intensities). The point at which this linear relationship changes and becomes saturated, i.e., "Further increases of light do not increase photosynthetic rate" is an expression of processes other than photochemical. This can be explained by examining Fig. 2-18, which is based on the work with *S. virgaurea*.

Between plants of the same clone, the photosynthetic rate is modified by the intensity of the cultivation light, but the modifications differ with respect to the sun or shade origins of the clones. Initial slopes in clones from exposed habitats [(c) and (d) in Fig. 2-18] are nearly the same re-

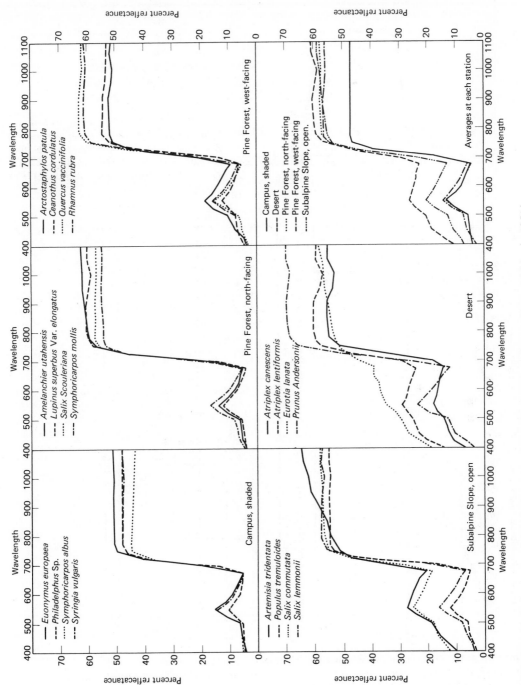

FIGURE 2–16. Percentage reflectance from leaves of various species. (From Billings and Morris, 1951.)

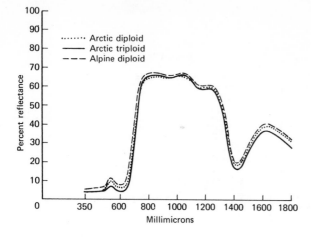

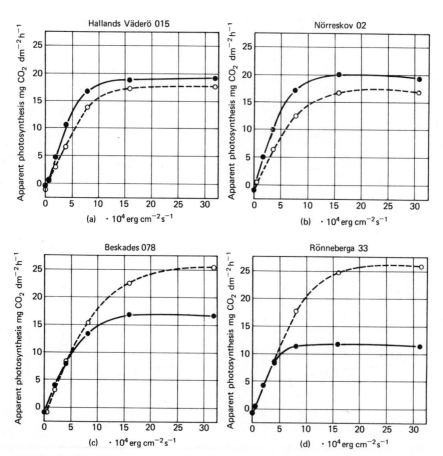

FIGURE 2–17. Percentage reflectance of radiant energy from leaflet upper surfaces in three populations of *Thalictrum alpinum*. Significantly different (1% level) at 600 and 1,600 mμ. (From Mooney and Johnson, 1965.)

FIGURE 2–18. The relation between apparent photosynthesis and light intensity in clone plants of the four populations grown at two different light intensities: (a) and (b) originated from shaded habitats; (c) and (d) originated from exposed habitats.

———————— cultivated at 3×10^4 ergs cm^{-2} s^{-1}.
– – – – – – – cultivated at 15×10^4 ergs cm^{-2} s^{-1}.
(From Björkman and Holmgren, 1963.)

44

gardless of the intensity of the cultivation light; in clones from shaded habitats [(a) and (b) in Fig. 2–18] the initial slope is steeper in plants that had been cultivated at a low light intensity than in plants cultivated at a high light intensity. The rates of photosynthesis are highest at light saturation in the plants that originated from exposed habitats and were cultivated at high light intensities. Plants cultivated at low light intensity, whether originating from shaded habitats or not, are saturated at lower light intensities [Fig. 2–18 (a)–(d)]. The data on initial slopes and saturation intensities are summarized in Table 2–4.

The biological interpretation of Björkman and Holmgren's data is that:

1. Plants from shaded habitats are better able to utilize weak light than are plants from exposed habitats.
2. High light intensities reduce the photochemical capacity of plants from shaded habitats (the implication here is that high light intensities tend to injure the photochemical apparatus).
3. In strong light, plants from exposed habitats have much higher rates, implying better utilization of light at high intensities.

Björkman and Holmgren also examined plant characters such as leaf thickness, chloroplasts, leaf size, amount of pigments, and leaf nitrogen. In general, plants grown at high light intensities tend to develop thicker leaves; plants originating from exposed habitats and grown in high light

TABLE 2–4. (A) "PHOTOCHEMICAL CAPACITY" (MEASURED AS INITIAL SLOPE OF THE RATE-INTENSITY CURVES) FOR PLANTS OF THREE CLONES FROM EACH OF FOUR LOCALITIES, TWO EXPOSED AND TWO SHADED, OF *SOLIDAGO VIRGAUREA,* GROWN AT LOW LIGHT INTENSITY. (B) PHOTOSYNTHETIC RATE AT LIGHT SATURATION IN PLANTS FROM THE SAME CLONES GROWN AT A HIGH INTENSITY.*

Shaded habitats			Exposed habitats		
	A	B		A	B
Locality 1			Locality 3		
clone a	3.10	18.9	clone a	2.40	31.8
clone b	2.98	18.6	clone b	2.17	17.3
clone c	2.58	15.2	clone c	2.27	27.6
Locality 2			Locality 4		
clone a	2.90	18.6	clone a	2.26	24.6
clone b	3.04	17.0	clone b	2.34	23.8
clone c	2.98	19.2	clone c	2.44	27.8

*From Heslp–Harrison, 1964.
Notes: Mean values for A: shaded 2.93 ± 0.08; exposed 2.31 ± 0.04.
Mean values for B: shaded, 17.9 ± 0.6; exposed, 25.5 ± 2.0.
(Data from Björkman and Holmgren, 1963.)
Values for B expressed in mg $CO_2\,dm^{-2}\cdot h^{-1}$.

intensities have leaves twice as large as those produced in low light; plants originating from shaded habitats and grown in low light, on the other hand, produce leaves twice as large as those produced in high light. Probably the most significant observation, relative to the experiments described above, is that pronounced destruction of chloroplasts occurred in plants from shaded habitats when cultivated at high light intensities.

The experiments described above show that differences in photosynthetic behavior are consistent with the light intensities prevailing in the natural habitats of the plants and, thus, are the result of genetic adaptation to habitat.

In later work, Björkman (1968) proposed that the lowered light saturated photosynthesis in shade plants could be caused by an enzyme deficiency that does not allow as much CO_2 fixation as in plants from exposed habitats. Björkman hypothesized that shade plants might be deficient in the enzyme carboxydismutase (ribulose diphosphate carboxylase), the enzyme that is thought to be responsible for the bulk of CO_2 fixation in green plants, and he, therefore, studied its activity in a number of sun and shade species grown in their natural habitats (Table 2–5).

Irradiance in fully lit habitats may be 100–200 times as great as in shaded habitats, and as seen in Table 2–5, light saturated rates of photosynthesis in the former are much higher than in the latter. Carboxydismutase activity is also higher in the leaves of the sun species. The shade leaves have more chlorophyll per unit of leaf fresh weight than the sun leaves but considerably less soluble protein which, Björkman suggests, is made up primarily of the enzyme. This observation led to the hypothesis that "the fraction of the chemical energy used for the synthesis of com-

TABLE 2-5. CARBOXYDISMUTASE ACTIVITY (CO_2/μ/mol/min) IN SUN AND SHADE PLANT SPECIES.*

| Species | Carboxydismutase activity, CO_2/μ/mol/min | | | | Mean irradiance of habitat 10^3 ergs/cm^2/s^1 $\lambda < 700$nm | Habitat |
	Per dm^2 leaf area	Per g fresh tissue	Per mg chlorophyll	Per mg soluble protein		
A. Shade species						
Adenocaulon bicolor	2.0	2.0	0.63	0.22	2.05	
Aralia californica	2.0	2.3	0.76	0.16		Redwood forests
Disporum smithii	1.0	1.1	0.40	0.20		
Trillium ovatum	2.0	1.7	0.50	0.15		
Viola glabella	1.0	1.2	0.38	—		
B. Sun species						
Atriplex patula	23.0	10.1	5.72	0.43	485.0	Edge of marsh
Echinodorus berteroi	15.0	7.6	3.26	0.24		Edge of lake
Mimulus cardinalis	17.0	5.4	3.28	0.35		Campus at Stanford
Plantago lanceolata	10.0	4.1	1.89	0.26		Campus at Stanford
Solidago spathulata	12.0	4.9	2.81	0.23		Campus at Stanford

*From Björkman, 1968.

ponents determining the efficiency of light absorption in relation to the fraction used for the synthesis of components determining the capacity of enzymic steps is larger among the shade than among the the sun plants." Ecologically, this can be interpreted as follows: the shade plants used in this study have evolved in such a way as to maximize their production of chlorophyll, but at the expense of producing those enzymes that convert light energy to chemical energy.

Supporting evidence comes from two sources. First, Björkman shows that among other plants the level of carboxydismutase activity and the rate of light saturated photosynthesis are approximately parallel; and second, a non-CO_2-fixing mutant of *Chlamydomonas reinhardii* lacks carboxydismutase.

We have discussed Björkman's work in detail here because it demonstrates the fundamental nature of genetic adaptations in plants. Based on our current knowledge of gene action, i.e., the production of biologically active molecules, it is clear that qualitative and quantitative changes in the production of specific enzymes determine plant functions and, thus, their distribution with respect to environmental stimuli.

Temperature Effects on Photosynthesis

If neither carbon dioxide nor light (nor any other factor) is limiting, the rate of photosynthesis increases with increasing temperature up to a point that varies somewhat from one population to another. Generally, a reasonably close relationship is found between the mean maximum ambient temperature and the optimum temperature for photosynthesis. As shown above (p. 42), the photosynthesis temperature optimum can be modified somewhat by acclimation.

Much attention has been devoted to studies of temperature and photosynthesis in local populations of wide-ranging species. Mooney and Johnson (1965) found that latitudinal races of *Thalictrum alpinum* show clear differences in their temperatures for maximum photosynthesis (Fig. 2–19). That this pattern is not always reflected along such climatic gradients is seen in the work of Milner and Hiesey (1964) who found a photosynthesis temperature optimum at 30°C for all six populations of *Mimulus cardinalis* whose native habitats range in altitude from 45 to 200 m.

When temperatures remain at optimum or above for long, the rate of photosynthesis drops off (Fig. 2–20). The same phenomenon has been noted under a variety of conditions (see Billings et al., 1966). The latter speculate that under intense solar radiation, leaf temperatures rise above ambient as has been shown by Salisbury and Spomer (1964), and this may cause a depression of enzyme activity and a substantial increase in respiration rate (Fig. 2–21).

It is important to emphasize that the relationship between temperature and photosynthesis holds only when no other environmental factor is

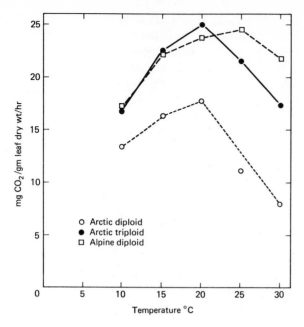

FIGURE 2–19. Temperature-related net photosynthetic curves for three populations of *Thalictrum alpinum*. (From Mooney and Johnson, 1965.)

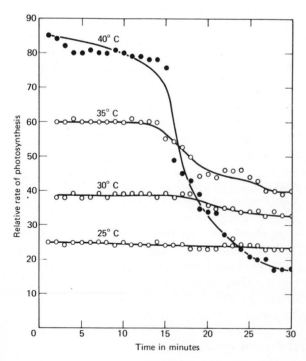

FIGURE 2–20. Relative rates of apparent photosynthesis in waterweed (*Anacharis canadensis*) at different temperatures over a period of 30 minutes. (From Meyer et al., 1960.)

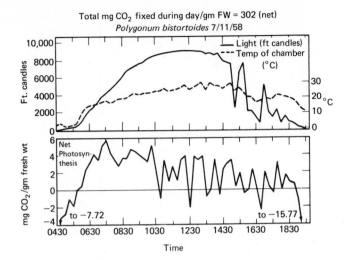

Total mg CO$_2$ fixed during day/gm FW = 302 (net)
Polygonum bistortoides 7/11/58

FIGURE 2-21. Midday depression in photosynthetic rate in *Polygonum bistortoides*. (From Mooney, 1963.)

limiting. As has already been described it is probable that either light or carbon dioxide is more often limiting than temperature.

The Effect of Oxygen on Photosynthesis

It has been known for many years that the rate of CO$_2$ fixation in some plants is inhibited by the atmospheric concentration of oxygen. The degree of inhibition decreases with decreasing oxygen concentration but is noted even below 2% O$_2$. It is now assumed that the reason for this apparent inhibition is due to the fact that most plants carry on respiration in the light.

In recent years, however, it has been found that some plants, especially tropical grasses do not have this "photorespiration." Hence, their apparent rate of photosynthesis may be nearly twice that of the plants that respire in the light. (Compare groups 1 and 3 with group 2 in Table 2–6.) This has been confirmed in experiments by Downton et al. (1969) in which blocking photorespiration by lowering the ambient oxygen concentration increases the rate of photosynthesis to approximately that of tropical leaves.

Irvine (1970) provided a sugar cane plant with ^{14}CO$_2$ and demonstrated that other grasses took it up later from the sugar cane respiration when the two were confined together. He concludes, therefore, that photorespiration does occur in these grasses. It decreases with light intensity, and the very low rates at high light intensities are due to a more efficient system of CO$_2$ capture (Fig. 2–22).

From an ecological point of view it would appear that the lack of photorespiration may be of advantage in competition because of the

TABLE 2-6. RATES OF APPARENT PHOTOSYNTHESIS, TRANSPIRATION, AND RESPIRATION FOR LEAVES OF DIFFERENT SPECIES.*

Plant species	(1) P_{310} CO_2 mg/dm²/h¹	(2) Transpiration in light H_2O g/dm²/h¹	(3) Transpiration in dark H_2O g/dm²/h¹	(4) Respiration in dark and normal air CO_2 mg/dm²/h¹	(5) $R^{dark}_{zero\ CO_2}$ CO_2 mg/dm²/h¹	(6) $R^{light}_{zero\ CO_2}$ CO_2 mg/dm²/h¹	(7) Column 5– Column 6 CO_2 mg/dm²/h¹	(8) $\frac{Column\ 7}{Column\ 5} \times 100$ %
Group 1								
Sorghum vulgare 'DD 38'	55 ± 6	3.26 ± 0.20	0.596 ± 0.10	1.6 ± 2	2.8 ± 0.1	0	2.8 ± 0.1	100
Zea mays 'De Kalb 805'	53 ± 6	3.71 ± 0.51	0.556 ± 0.10	2.9 ± 0.3	4.1 ± 0.7	0	4.1 ± 0.7	100
Group 2								
Gossypium hirsutum 'Acaia 4–42'	30 ± 3	3.00 ± 0.47	0.661 ± 0.08	3.2 ± 0.9	5.6 ± 0.8	2.2 ± 0.3	3.5 ± 0.5	61 ± 2
Gossypium barbaden-se 'Pima S–2'	29 ± 2	2.58 ± 0.35	0.623 ± 0.12	3.5 ± 0.6	6.1 ± 0.9	1.7 ± 1	4.1 ± 0.5	70 ± 6
Beta vulgaris 'MS NB₁ × NB₄'	24 ± 2	3.34 ± 0.58	0.636 ± 0.07	2.6 ± 0.4	4.9 ± 0.8	2.4 ± 0.3	2.5 ± 0.6	50 ± 3
Helianthus annus 'California No. 2'	36 ± 4	3.67 ± 0.36	0.769 ± 0.17	5.0 ± 1.3	6.1 ± 1.2	3.1 ± 0.9	3.0 ± 0.5	40 ± 4
Group 3								
Amaranthus edulis Giant pigweed	58 ± 5			5.2 ± 0.8	5.8 ± 1.0	0	5.8 ± 1.0	100

*From El Sharkawy et al., 1967.

Note: Means ± standard deviation.

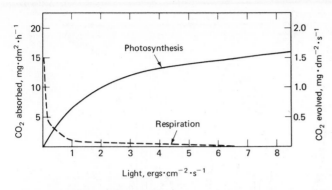

FIGURE 2–22. Photosynthesis and respiration in a tropical grass. (From J. E. Irvine, 1970.)

larger food resources laid down (Black et al., 1969). It is interesting to note that some of our more aggressive weeds, e. g., Bermuda grass, crab grass, and pigweed, lack photorespiration.

The Role of Water in Photosynthesis

Water is rarely a limiting factor in photosynthesis directly, i.e., as a raw material for the process. Its indirect effects, however, are noticeable, particularly during periods of water stress. Grieve and Hellmuth (1970) report that summer photosynthetic rates of western Australia plants dropped off by two-thirds or more from their spring maxima because water deficits activated stomatal closure. Strain (1970) showed that extreme water deficit (−60 bars) in the creosote bush, *Larrea divaricata,* and mesquite, *Prosopis,* occurred concomitantly with very low net photosynthesis rates (Fig. 2–23). Both of these plants have the ability to survive extended drought periods by avoidance (shedding leaves and branches). The importance of water in the maintenance of leaf temperature is discussed in Chapter 3.

Miscellaneous Factors

Chlorophyll content, leaf anatomy, soil nutrients, and other extrinsic and intrinsic factors also influence the rate of photosynthesis, but enough detail has been provided to demonstrate that plants have evolved a variety of adaptations which enable survival against environmental stresses and to explain how plants have managed to inhabit all but the most extreme climates on this planet.

TEMPERATURE REGULATION IN VERTEBRATE ANIMALS

Introduction

One of the ideas discussed thus far is that all adaptations, whatever their manifestations, have a genetic basis. Presumably this holds for those re-

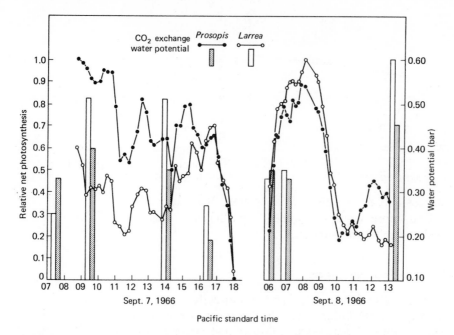

FIGURE 2–23. Relative net photosynthesis and tissue water potential on consecutive late-summer days. Values are expressed relative to the highest net photosynthesis rate (mg CO_2 g dry weight^{-1}h^{-1}) obtained in the two-day period. (From Strain, 1970.)

sponses that are learned as well as for those that we think are instinctive. The central dogma of molecular genetics is that the genes' ultimate function is to produce a biologically meaningful molecule or part of one, a polypeptide, that may individually or together with other such molecules play a catalytic role in the organism. Somehow the organism must adjust the circumstances of its life so that its ontogeny occurs in an environment sufficient to allow these temperature dependent biological processes to occur but to avoid, at least until reproduction has occurred, the upper and lower extremes that make this impossible.

In the first part of this chapter we discussed some of the mechanisms that have evolved in plants to enable their widespread occurrence in all but the most extreme climates. The completely barren (biologically) areas of the earth are probably limited to the (essentially) rainless deserts such as one finds in northern Chile and southern Peru, to the immediate ice fringes in Antarctica, or to areas where highly toxic concentrations of chemical substances have accumulated. The continued presence of plants, however, guarantees an energy supply that may be used during all or a part of the year by consumer organisms. This energy provides, in a sense, the raw material needed by the consumers to perfect their own adaptations to the environment. Higher forms of animal life occur in aquatic habitats whose temperatures fall mostly between $-2°C$ and $+40°C$ and in terrestrial situations from $-50°C$ or somewhat lower to upward of $60–70°C$. Excluding man, whose climate is portable, vertebrate animals have evolved in many different ways to survive these extremes. We shall now examine some of these survival strategies.

Organisms and Their Temperature Relations

Most animals have little internal control over their body temperatures in the sense of producing enough energy by their metabolic activity to maintain body temperatures higher than their surroundings. Such animals are called variously *conformers, ectothermic* (heat derived from outside the animal), or *poikilothermic* (variable body temperature). Birds and mammals most of the time, and a few other animals part of the time, produce enough energy to maintain body temperatures higher than the ambient temperature. These are called *regulators, endothermic* (heat derived from metabolic activity), or *homeothermic* (constant body temperature).

Animals of the first group are able to do relatively little internally to regulate their temperatures since their metabolic activities are themselves temperature controlled by environmental temperatures. Thus, net flow of heat is from the environment to the organism or vice versa unless the two are equal. In fact, the rates of all of the ectotherm's biochemical processes are environmentally controlled and approximately follow van't Hoff's or the Q_{10} rule* (double for each $10°C$ increase in temperature). It follows from this that the ectotherm must be able to sense when external temperatures reach close to critical limits and must attempt to avoid them somehow.

Endotherms, on the other hand, increase their metabolic rate as the ambient temperature declines in order to maintain body temperature. The steepness of the gradient between the core temperature of the animal and its environment determines the oxygen consumption of the animal (Figs. 2–24 and 2–25). Various measurements suggest that about 80 to 90% of the organism's oxidative energy is used in thermal homeostasis mechanisms. Thus, the price of thermal independence must be paid for in terms of a more or less regular consumption of energy. One must conclude, however, that the evolution of the homeotherms is the best kind of evidence that such energy sources have been constantly available.

*The formula for determining Q_{10} is:

$$\log Q_{10} = \frac{10(\log k_1 - \log k_2)}{t_1 - t_2}$$

where

k_1 = rate at one temperature (t_1)
k_2 = rate at second temperature (t_2)

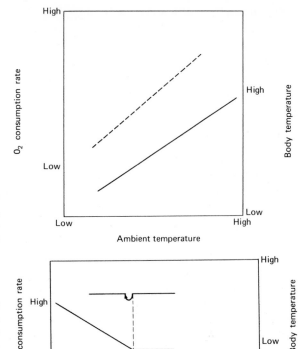

FIGURE 2-24. Generalized response of body temperature and oxygen consumption rate of a poikilotherm to a changing ambient temperature. (From Vernberg and Vernberg, 1970.)

FIGURE 2-25. Generalized response of body temperature and oxygen consumption rate of a homeotherm to a changing ambient temperature. At a critical temperature t^0 the body temperature drops and the oxygen consumption rate increases. As a result, the body temperature returns to the previous level. (From Vernberg and Vernberg, 1970.)

Acclimation and Adaptation

The terminology of physiological adjustments in animals is confused. Folk (1966) reviews past usages of the words adaptation, acclimation, and acclimatization. *Adaptation* as used here refers to evolutionary change; *acclimation* and *acclimatization* both refer to the organism's ability to compensate for external change, the former in the laboratory and the latter in natural situations. Bartholomew (1968) questions if the distinction is worth maintaining; in this text we use the term acclimation to refer to an environmentally induced change.

That acclimation to temperature stress occurs in ectotherms as well as in endotherms is evidence that these animals are not completely at the mercy of the external environment. A generalized response to a new thermal regime (Fig. 2-26) applies to ectotherms and endotherms alike. The time required for acclimation varies according to the organism and its previous thermal experience, but Hutchinson (1961) has shown that the *critical thermal maximum* (the temperature at which an organism is

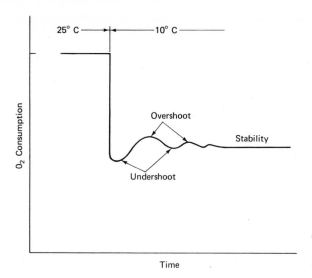

FIGURE 2–26. Generalized response of O_2 consumption rate of a homeotherm to a $15°$ temperature stress. (From Vernberg and Vernberg, 1970.)

so immobilized that it eventually dies) of salamanders begins to change immediately and by 48 hours has stabilized at a level nearly 3°C higher than initially. This compares favorably with the time required for changes in photosynthetic rate described by Mooney and Shropshire, p. 37.

There are distinct limits to what can be accomplished by acclimation. This process is obviously not sufficient to allow most animals, particularly those with narrow temperature tolerances (*stenothermal* animals), to survive sudden changes that are beyond those ordinarily experienced naturally, nor will it be adequate to permit substantial changes in geographic distribution. As a general rule (which is somewhat circular) animals that live in relatively stable temperature regimes have narrow tolerances, whereas those that live in fluctuating temperature regimes have wide tolerances—i.e., they are *eurythermal*.

Adaptations to Temperature

The range of strategies adopted by vertebrate animals in temperature regulation is impressive. The overriding evolutionary outcome of many different adaptations appears to be the selection by the organism of those temperatures that enable it to carry on its energy gathering and other activities without compromising the functional integrity of its biochemical systems. Ectothermic animals accomplish this especially by heat-seeking or avoiding behavior, endothermic ones by complex and often interrelated morphological and physiological solutions that are frequently combined with behavioral adaptations.

Temperature Regulation in Ectotherms

If fish are placed in a tank having a temperature gradient, they will select the temperatures they prefer (Fig. 2–27). Although this temperature is dependent on the thermal history of the fish, the fish will eventually migrate to a temperature in which the gradient temperature and the acclimation temperature are the same. This is called the *final preferendum* (Fry, 1947). Fish will leave the preferendum for feeding or presumably for other activities. Since fish have little opportunity for modifying their internal temperature, except perhaps transiently during periods of high activity, fish that inhabit areas where water temperatures change seasonally apparently must raise or lower their metabolic rates to compensate for the change. Those that have an opportunity to follow their temperature preferenda during migration will probably do so.

Those amphibians that are completely aquatic probably control their temperatures in the same way as fish. Apart from a few salamanders of the order Caudata, however, amphibians spend some or all of their existence on land. Brattstrom (1963) reported on temperature requirements of 99 species of amphibians from North and Central America. Salamanders had body temperatures from −2 to 26.7°C and anurans from 3.0 to 35.7°C. He found that salamanders' temperatures ordinarily match the temperature of the water in which they are found, or if terrestrial the temperature of the substrate where they are found. No ability to adjust temperatures was found in salamanders except for some ability to acclimate lethal temperatures.

Frogs and toads gain heat by basking in the sun and lose it by evaporating water from their skins. Thorson (1955) reported one case of a *Rana*

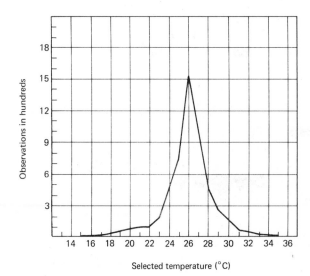

FIGURE 2–27. Temperature selection by 39 individuals of shore fish, the California opaleye (*Girella nigricans*). (From Norris, 1963.)

pipiens that maintained a body temperature of 36.8°C (by evaporative cooling) for over three hours while kept in a dessicator at 50°C. The behavioral selection of environmental temperatures is the method of primary significance in amphibian regulation of temperature.

Hock (1964) notes that "in a very real sense reptiles do not adapt to life in cool areas but infiltrate all areas where environmental factors allow their existence." Apparently, few of these areas are satisfactory because the evidence from modern distribution shows reptiles almost absent from the boreal and arctic portions of the world. In Europe where the Gulf stream ameliorates temperatures considerably, two snake and two lizard species reach above 66°N.

That life in rigorous cold environments is not beyond the capability of reptiles is shown by Pearson's (1954) studies of the lizard *Liolaemus multiformis* in the Peruvian Andes above 5,000 m. The lizard accomplishes this feat in the face of nightly subfreezing temperatures by basking in the bright sunlight until it reaches body temperatures of above 30°C.

The five reptiles that do reach high latitudes are all viviparous, an adaptation that is in itself probably in response to cold temperatures, the time of egg hatching in lizards being universally proportional to air temperature. A pregnant viviparous reptile, however, can maintain a higher body heat by behavioral thermoregulation. Hock (1964) also believes that the lack of suitable hibernation sites is responsible for the paucity of reptiles in the north.

Regulation of temperature in reptiles through behavior is more highly developed than in all other vertebrate ectotherms. By moving between sun and shade, changing its orientation to the sun, flattening or otherwise modifying its body shape, burrowing, or becoming nocturnal, reptiles regulate temperature to a remarkable degree. Norris (1953) for example shows that the desert iguana, *Dipsosaurus dorsalis* regulates its temperature to the extent that the standard error of its mean activity temperatures is ±0.005°C. Within a single species, behavioral control of temperature may involve several separate but interrelated responses. Heath (1965) shows that the horned lizard *(Phrynosoma coronatum)* engages in an ordered sequence of behavior depending on its body temperature (Fig. 2–28). It is interesting to note that at the highest temperature *Phrynosoma* engages in physiological mechanisms (panting and cloacal discharge) for temperature control.

Although reptiles have a dry cornified skin, they are able to regulate temperature to some extent by evaporative water loss through the skin and the respiratory tract (Fig. 2–29). The heat loss by these methods, however, is minor in comparison with the amount of heat the lizards would be expected to gain from being fully illuminated.

A few lizards have the ability to differentially control their rates of heating and cooling through regulation of the heart rate and presumably,

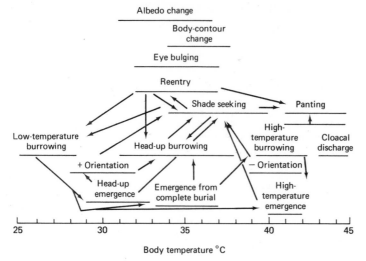

FIGURE 2–28. Diagram showing different patterns of behavioral thermoregulation and body temperature in the horned lizard (*Phrynosoma coronatum*). (From Heath, 1965.)

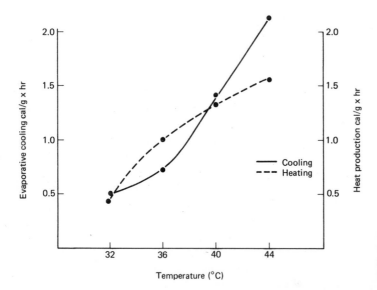

FIGURE 2–29. Evaporative water loss in relation to metabolism in *Dipsosaurus dorsalis*. (Modified from Templeton, 1960.)

the flow of blood through the tissues. Bartholomew and Lasiewski (1965) show that the Galapagos marine iguana *(Amblyrhynchus cristatus)* has this capability. The lizard spends most of its time basking on rocky shores where it maintains a temperature of about 37°C by behavioral means. It feeds on marine algae underwater at temperatures of 22 to 27°C. It was found to have heating rates about twice as fast as cooling rates (Fig. 2–30); and the heart rate of *A. cristatus* was found to be much more rapid when the lizard was heating than when it was cooling (Fig. 2–31.)

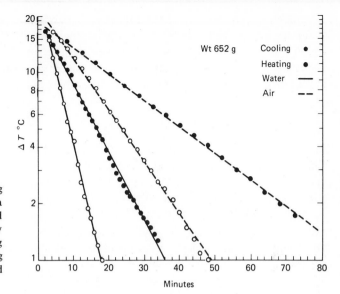

FIGURE 2–30. Heating and cooling rates of the Galapagos marine iguana (*Amblyrhyneus cristatus*) in water and air. ΔT is the difference between body and ambient temperatures (T_A). During heating $T_A = 40°C$; during cooling $T_A = 20°C$. (From Bartholomew and Lasiewski, 1965.)

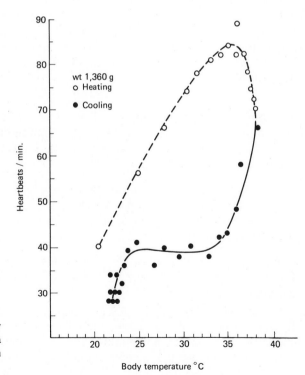

FIGURE 2–31. Relation of heart rate to body temperature in the Galapagos marine iguana during heating and cooling in water. (From Bartholomew and Lasiewski, 1965.)

Somewhat reminiscent of the reflective capability of desert plants are observations on reflectivity in desert lizards. Whether color changes, which tend to match background colors (Norris and Lowe, 1964), are primarily adaptations to temperature or function as protection against predation is not clear. Probably they serve both purposes.

Temperature Regulation in Endotherms

Birds and mammals are similar in that they tend to maintain a relatively constant temperature regardless of the ambient temperature. This is primarily accomplished by the metabolic activity of the animal. It is probably intuitive that the greater the difference between the body temperature T_B and the ambient temperature T_A, the more energy the animal must expend in maintenance. At low temperatures ($T_A < T_B$), the cost of maintenance to the animal is in terms of the calories lost per °C difference between T_B and T_A. At high temperatures ($T_A > T_B$), the animal can lose heat to the environment only by the evaporation of water from its skin and respiratory surfaces. The mobilization of water necessary to accomplish this also requires energy from the animal. Lying between these two energy requiring points lies an area, the *thermal neutral zone,* in which temperature regulation is accomplished almost incidental to the animal's total suite of energy requiring activities.

Bartholomew (1968) has discussed the complexity of temperature regulation in endotherms. He proposes one simple model (of which there are many interrelated ones) that involves control centers in the central nervous system; others are peripherally located, but work toward the same ends (Fig. 2–32).

Adaptations to Low Temperatures. The subject of adaptation to low temperature in birds and mammals has been reviewed in recent years by Irving (1964) and Folk (1966). Generally, the methods by which animals adapt to low temperatures are:

1. Decreasing the thermal conductance
2. Increasing internal heat production
3. Avoidance (behavioral adjustments)
4. Adaptive hypothermia

We shall return to the last of these following the discussion of adaptation to heat because the process applies to animals inhabiting both hot and cold climates.

Behavioral adjustments to cold are often of a general nature: most birds migrate away from intense cold, and most small mammals remain under the snow or soil where temperatures are warm, relative to the outside air. The brown lemming, for example, remains active under snow all winter as

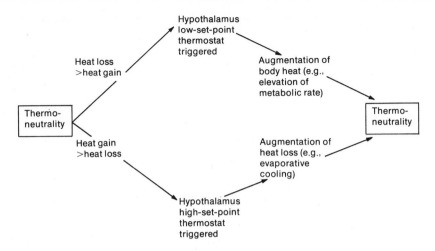

FIGURE 2–32. A simple cybernetic model illustrating temperature control in endotherms. (From Bartholomew, 1968.)

does its predator, the least weasel. The red squirrel *Tamiasciurus hudsonicus,* which lives in the coldest parts of interior Alaska, is active all winter but burrows into its snow-covered mounds of cones when the lowest temperatures occur.

The strategy in decreasing thermal conductance is to flatten the gradient from the body core to the environment. It is here that we return to Bergmann's rule (see p. 22). Since large animals have reduced surface areas relative to their volumes, they should be the most efficient in reducing their thermal conductance. Although the generalization seems to apply in certain cases, it is clear that there are other methods available for accomplishing the reduced heat loss. One is by producing thick layers of fur and feathers. Scholander et al. (1950b) measured the insulating properties of the hair of 18 arctic mammals ranging in size from a 30 g least weasel (*Mustela*) to a 500 kg moose (*Alces*) (Fig. 2–33) (see also p. 62). The layers are obviously effective for large mammals, which, according to Bergmann's rule, probably need it least, but are too thin to be of much value to lemmings and other small mammals, which as indicated above, have adopted different strategies. Irving (1964) expresses doubts that small birds can carry (and still fly) enough protective insulation to protect them from severe cold, but he does not explain how small birds such as chickadees (*Parus*) manage to endure temperatures of −50°C.

Appendages such as the feet, flippers, and bills have lower temperatures than does the body core (Fig. 2–34). Although it would appear that such structures should lose vast amounts of heat to the environment, it is

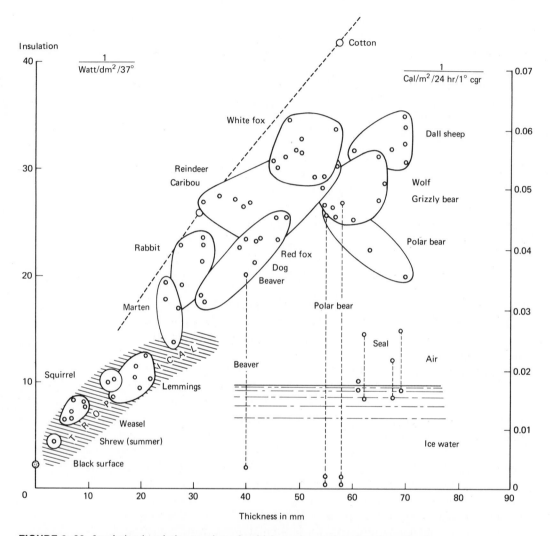

FIGURE 2–33. Insulation in relation to winter fur thickness in a series of arctic mammals. The insulation in tropical mammals is indicated by the shaded area. In the aquatic mammals (seals, beaver, polar bear) the measurements in 0°C air are connected by vertical broken lines with the same measurements taken in ice water. In all cases the hot plate guard ring unit was kept at 37°C and the outside air or water at 0°C. The two upper points of the lemmings are from *Dicrostonyx,* the others from *Lemmus.* (From Scholander et al., 1950b.)

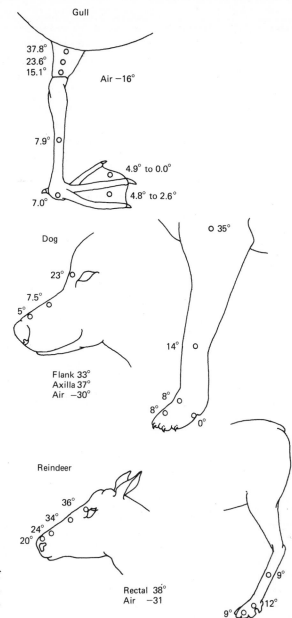

Gull

37.8°
23.6°
15.1°

Air −16°

7.9°

4.9° to 0.0°

4.8° to 2.6°

7.0°

Dog

○ 35°

23°

7.5°

5°

14°

Flank 33°
Axilla 37°
Air −30°

8°

8°

0°

Reindeer

36°

34°

24°

20°

Rectal 38°
Air −31

9°

12°

9°

FIGURE 2-34. Topographic distribution of superficial temperatures in a dog, a reindeer, and the leg of a gull. (From Irving and Krog, 1955.)

now known that countercurrent heat exchange occurs in order to reduce the heat flow to the appendages. The principle operating is seen in arterial vessels which are completely surrounded by veins. Heat transfer occurs between the vessels with the net effect of reducing peripheral heat loss. Solving one problem presents another; i.e., how do the extremities retain their functions at cold temperatures? The question has not been answered satisfactorily, although it is known that appendages retain sensitivity to stimuli at cold temperatures.

The most common response to cold temperatures is by an elevated metabolic rate. In a sense, all other answers to the heat problem have been accomplished as long-term evolutionary solutions that make it possible for the animal to avoid short-term costly energy expenditures. Heat is also generated by shivering and by a process called nonshivering thermogenesis. Although not fully understood, it apparently involves the hormonally induced enhanced metabolism of lipids and possibly carbohydrates.

Adaptations to High Temperatures. It has been noted by Scholander et al. (1950b) that the core temperature of homeotherms is nonadaptive to climate because all birds and mammals, whatever their origin, show only slight differences in this respect. As applied to animals of desert areas, this means that ambient temperatures during parts of many days during the year will be higher than body temperatures. As indicated above, the problem of moving heat against the gradient can be accomplished only by the evaporation of water by the animal, an energy requiring process, and one that is complicated by the paucity of water in desert regions. Of the two problems we have considered here, adaptation to heat and adaptation to cold, Bartholomew (1968) thinks that the former has been the more difficult for birds and mammals to solve.

All mammals, large and small, face the problem in hot, dry climates of how to obtain or retain water when free water is scarce or nonexistent and how to minimize heat flow to the animal from the environment. Behavior plays an important role in equipping animals for desert existence. Probably the most widespread behavioral adaptation of desert animals is burrowing. Nearly all desert mammals are small, and almost all of them remain in burrows during the day and emerge only at night when, in most deserts, temperatures are cooler. This small size poses problems, however, because of the relationship between size and the ability to cool evaporatively (Fig. 2–35). According to Schmidt-Nielsen (1964) most rodent burrows extend to depths of one-half to one meter or more, where soil temperatures are relatively constant and 10° to 15°C cooler than ambient air temperatures. Bartholomew (1968) believes that nocturnality and burrowing are of most value, not in terms of temperature regulation but in protection from predation.

Large diurnal mammals obviously do not dig burrows, but they can regulate temperatures to some extent by seeking shade. The advantage of large size in evaporative cooling has been noted above (Fig. 2–35). The point has also been made that large mammals are better able to travel long distances to surface water.

Diurnal mammals also decrease heat intake through pelage charac-

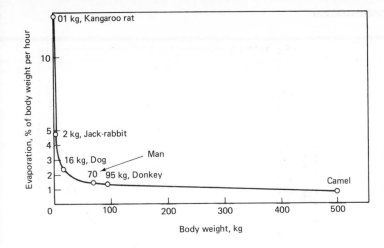

FIGURE 2-35. Relation between body size and the evaporation estimated to be necessary for the maintenance of a constant body temperature in a hot desert climate. (From Schmidt–Nielsen, 1954.)

teristics that reflect radiant energy and insulate the animal from the high external temperatures. The gradient from the skin surface to the surface of the outer hair has been measured by MacFarlane (1964) to be as much as 45°C (from 40°C to 85°C). Although desert mammals' skin pigments are often dark, presumably as protection against ultraviolet radiation, the hair tends to be light colored and glossy.

Hudson (1962) has demonstrated that the antelope ground squirrel, *Citellus leucurus,* solves its heat problem by accepting heat overloads briefly. The animal becomes hyperthermic for short periods, allowing its temperature to rise, then retreats to its burrow where it loses heat, and then it repeats the cycle.

The water relations of desert rodents that derive nearly all of their water from their (usually dry) food have been studied extensively. The kangaroo rat, *Dipodomys,* for example, gets most of its water from the oxidation of its food, although vegetative material, particularly if in a humid atmosphere, will contain much free water. Water is then conserved by kangaroo rats by the excretion of extremely concentrated urine and the elimination of feces that contain relatively little water. This example is discussed at greater length in a different context in Chapter 3.

Desert birds show typical avoidance type behavioral adaptations in hot, dry climates. Midday reduction in activity is common. A few species are active only at twilight, and at least one, the burrowing owl, uses underground burrows. Generally, however, desert birds have not used subterranean areas to reduce water loss.

The body temperatures of desert birds are about the same as non-desert birds, averaging from 42 to 44°C. Birds from both groups apparently die at about 47°C (Dawson and Bartholomew, 1968). The higher (by about 4 to 5°C) temperatures of desert birds than mammals means that at high T_A their thermal gradients are less steep than those of mammals. Consequently, their problems of heat regulation are somewhat less severe, at least on a purely physical basis.

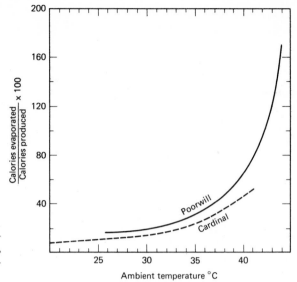

FIGURE 2–36. Comparison of the effectiveness of evaporative cooling by pant (cardinal), and gular flutter (poorwill). (From Bartholomew, 1968.)

When $T_A > T_B$ in birds, they can lose heat by panting or by fluttering the gular area (the floor) of the mouth. The latter is more effective than the former in reducing excess heat (Fig. 2–36). Bartholomew (1968) reports that by employing gular fluttering the poorwill can maintain a T_B of 45°C at a relative humidity of 20%.

Adaptive Hypothermia

Some homeothermic animals have the ability to allow their body temperatures to approximate those of their cooler surroundings. The process can be induced experimentally in animals that show it naturally. Hypothermia is a general term that includes torpor, hibernation, and estivation, all of which, according to Bartholomew (1968) have similar features:

1. $T_B \approx T_A$.
2. Oxygen consumption is reduced to as little as 5% of normal.
3. Breathing rate is greatly reduced.
4. A dormancy or torpor qualitatively different from sleep occurs.
5. Heart rate is reduced.
6. The animal retains the ability to arouse and reestablish its normal body temperature.

The major adaptive advantages of hypothermia are varied, but all serve the same general purpose: the conservation of energy during periods when it may be difficult to obtain.

Hypothermia occurs over periods of hours, days, or probably even months. Preparation for extended periods of dormancy are more elaborate than for the daily torpor experienced by some desert rodents. Food storage in the form of body fats occurs in many hibernators prior to the onset of hypothermia, and the lapse into the event itself may be gradual and prolonged.

Small birds and animals like hummingbirds, bats, and pocket mice experience daily torpor, the first two routinely, the mouse at times of short food supply. A few birds become torpid during cold parts of the year.

That these processes belong to the same set of events is suggested by the genus *Citellus*, the ground squirrels, different species of which hibernate, estivate, do both, or do neither.

These phenomena should be viewed as modifications of the basic process of endothermy and not as a separate class of events.

The great diversity of life on the earth is the best kind of evidence for belief in the ability of organisms to accommodate environmental change. This chapter has emphasized adaptations that higher organisms show in different environments. An understanding of adaptation by natural selection is fundamental to all other work in ecology.

Another side of evolution and adaptation is also becoming obvious to us. We now see that this universal quality of life that has contributed to its survival in the face of substantial changes in the earth's environment is a double-edged sword. In our efforts to eliminate organisms that we consider undesirable by using herbicides, pesticides, and the like, we have encountered as a major block to our efforts this same ability of organisms to adapt. Because of their newly acquired abilities to survive these stresses, we now find these resistant organisms growing in numbers and, ironically, posing a threat to animals higher in the trophic structure who consume them as food and, concomitantly, accumulate their chemical burdens. Man himself is now becoming an unintended victim of his own technology. The unfortunate conclusion is that the possibility in man and higher animals of adaptive change to these kinds of stresses is negligible because of the long generation time in these animals and the relatively high degree of sensitivity they exhibit to many environmental contaminants.

REFERENCES

Bartholomew, G. A. 1968. Body temperature and energy metabolism. In M. S. Gordon (ed.), *Animal Functions: Principles and Adaptations*. New York: Macmillan. 560 pp.

———, and R. C. Lasiewski. 1965. Heating and cooling rates, heart rate and simulated diving in the Galapagos marine iguana. *Comp. Biochem. Phys.,* **16**:573–82.

Billings, W. D., E. C. C. Clebsch, and H. A. Mooney. 1961. Effect of low concentrations of carbon dioxide on photosynthetic rates of *Oxyria. Science,* **133**:1834.

––––––. 1966. Photosynthesis and respiration rates of Rocky Mountain plants under field conditions. *Amer. Midl. Nat.,***75** (1):34–44.

Billings, W. D., and Robert J. Morris. 1951. Reflection of visible and infrared radiation from leaves of different ecological groups. *Am. J. Bot.,* **33**:327–31.

Björkman, O. 1968. Carboxydismutase activity in shade-adapted and sun-adapted species of higher plants. *Physiologia Plantarum,* **21**:1–10.

––––––, and D. Holmgren. 1963. Adaptability of the photosynthetic apparatus to light intensity in ecotypes from exposed and shaded habitats. *Physiol. Plant.* **16**:889–914.

––––––. 1966. Photosynthetic adaptation to light intensity in plants native to shaded and exposed habitats. *Physiol. Plantarum,* **19**:854–59.

Black, C. C., Jr., T. M. Chen, and R. H. Brown. 1969. Biochemical basis for plant competition. *Weed Sci.,* **17**:338–44.

Brattstrom, B. H. 1963. A preliminary review of the thermal requirements of amphibians.*Ecology,* **44**:238–55.

Cade, T., and G. A. Bartholomew. 1959. Sea water and salt utilization by savannah sparrows. *Physiol. Zoology,* **32**(4):230–38.

Clausen, J., and W. M. Hiesey. 1958. *Experimental studies on the nature of species. IV. Genetic structure of ecological races.* Carnegie Inst. of Wash. Publ. No. 615. 312 pp.

Dawson, W. R., and G. A. Bartholomew. 1968. Temperature regulation and water economy of desert birds. In G. W. Brown (ed.), *Desert Biology.* New York: Academic Press.

Dawson, W. R., and K. Schmidt-Nielsen. 1964. Terrestrial animals in dry heat: Desert birds. In D. B. Dill (ed.), *Handbook of Physiology,* pp. 181–92. Washington, D.C.: American Physiological Society.

Downton, J., J. Berry, and E. B. Tregunna. 1969. Photosynthesis: temperature and tropical characteristics within a single grass genus. *Science,* **163**:78–79.

El Sharkawy, M. A., R. S. Loomis, and W. A. Williams. 1967. Apparent reassimilation of respiratory carbon dioxide by different plant species. *Physiol. Plant,* **20**:171–86.

Folk, G. E., Jr. 1966. *Introduction to Environmental Physiology.* Philadelphia: Lea & Febiger. 308 pp.

Fry, F. E. J. 1947. Effects of the environment on animal activity. University of Toronto, *Biol. Ser.,* **55**:1–62.

Grieve, B. J., and D. O. Hellmuth. 1970. Eco-physiology of Western Australian Plants. *Oecol. Plant.,* **5**:33–68.

Heath, J. E. 1965. Temperature regulation and diurnal activity in horned lizards. University of California Publications in*Zoology,***64**:97–136.

Heslop-Harrison, J. 1964. Forty years of genecology. In J. B. Cragg (ed.), *Advances in Ecological Research.* London and New York: Academic Press.

Hock, R. J. 1964. Terrestrial animals in cold: Reptiles. In D. B. Dill (ed.), *Handbook of Physiology,* pp. 357–59. Washington, D.C.: American Physiological Society.

Hudson, G. W. 1962. The role of water in the biology of the antelope ground squirrel *Citellus leucurus*. University of California, (Berkeley), *Publ. Zool.*, **64**:1–56.

Hutchinson, V. H. 1961. Critical thermal maxima in salamanders. *Physiol. Zool.*, **34**:92–125.

Irvine, J. E. 1970. Evidence for photorespiration in tropical grasses. *Physiol. Plant.*, **23**(3):607–12.

Irving, L. 1964. Terrestrial animals in cold: Birds and mammals. In D. B. Dill (ed.), *Handbook of Physiology*, pp. 361–77. Washington, D.C.: American Physiological Society.

―――― , and J. Krog. 1955. Temperature of skin in arctic as a regulator of heat. *J. Appl. Physiol.*, **7**:355–64.

Johnson, A. W., J. G. Packer, and G. Reese. 1965. Polyploidy, distribution and environment. In H. E. Wright, Jr. and D. G. Frey (eds.), *The Quaternary of the United States*, pp. 497–508. Princeton, N.J.: Princeton Univ. Press.

Langlet, O. 1959. A cline or not a cline—a question of Scots Pine. *Sylvae Genetica*, **8**:13–22.

MacFarlane, W. V. 1964. Terrestrial animals in dry heat: Ungulates. In D. B. Dill (ed.), *Handbook of Physiology*, pp. 541–50. Washington, D.C.: American Physiological Society.

McNeilly, T. 1968. Evolution in closely adjacent plant populations. III. *Agrostis tenuis* on a small copper mine. *Heredity*, **23**:99–108.

Meyer, B. S., D. B. Anderson, and R. H. Bohning. 1960. *Introduction to Plant Physiology*. Princeton, N.J.: Van Nostrand. 541 pp.

Milner, H. W., and W. M. Hiesey. 1964. Photosynthesis in climatic races of *Mimulus*. II. Effect of time and CO_2 concentration on rate. *Plant Physiol.*, **39**(5):746–50.

Mooney, H. A. 1963. Physiological ecology of coastal, subalpine, and alpine populations of *Polygonum bistortoides*. *Ecology*, **44**:812–16.

―――― , and E. L. Dunn. 1970. Convergent evolution of mediterranean-climate evergreen sclerophyll shrubs. *Evolution*, **24**(?):292–303.

Mooney, H. A., and A. W. Johnson. 1965. Comparative physiological ecology of an arctic and an alpine population of *Thalictrum alpinum L. Ecology*, **46**(5): 721–27.

Mooney, H. A., and F. Shropshire. 1967. Population variability in temperature related to photosynthetic acclimation. *Oecol. Plant*, **2**:1–13.

Norris, K. S. 1953. The ecology of the desert iguana *Dipsosaurus dorsalis*. *Ecology*, **34**:265–87.

―――― . 1963. The functions of temperature in the ecology of the percoid fish *Girella nigricans* (Ayres). *Ecol. Monogr.*, **33**(1):23–62.

―――― , and C. H. Lowe. 1964. An analysis of background color-matching in amphibians and reptiles. *Ecology*, **45**:565–80.

Oosting, H. J. 1956. *The Study of Plant Communities*, 2nd ed. San Francisco: Freeman. 440 pp.

Pearson, O. P. 1954. Habits of the lizard, *Liolaemus multiformis multiformis* at high altitudes in southern Peru. *Copeia*, **1954**:111–16.

Salisbury, F. B., and G. G. Spomer. 1964. Leaf temperatures of alpine plants in the field. *Planta*, **60**:497–505.

Schmidt-Nielsen, K. 1954. Heat regulation in small and large desert mammals. In J. B. Cloudsley–Thompson (ed.), *Biology of Deserts,* pp. 182–87. London: Inst. of Biol.

———. 1964. *Desert Animals: Physiological problems of Heat and Water.* London: Oxford University Press. 277 pp.

Scholander, P. F. 1955. Evolution of climatic adaptation in homeotherms. *Evolution,* **9**:15–26.

———. 1956. Climatic rules. *Evolution,* **10**:339–40.

———, **R. Hock, V. Walters, and L. Irving.** 1950a. Climatic adaptations in arctic and tropical poikilotherms. *Physiol. Zool,* **26**(1):67–92.

———. 1950b. Body insulation of some arctic and tropical mammals and birds. *Biol. Bull.,* **99**:225.

Schreider, E. 1964. Ecological rules, body-heat regulation, and human evolution. *Ecology,* **18**:1–9.

Strain, B. R. 1970. Field measurements of tissue water potential and CO_2 exchange in the desert shrubs *Prosopis juliflora* and *Larrea divaricata. Photosynthetica,* **4**(2):118–22.

Templeton, J. R. 1960. Respiration and water loss in the desert iguana *Dipsosaurus dorsalis. Physiol. Zool.,* **33**:136–45.

Thorne, R. J. 1967. Flora of Santa Catalina Island, California. *Aliso,* **6**:1–77.

Thorson, T. B. 1955. The relationships of water company to terrestrialism in amphibians. *Ecology,* **36**:100–16.

Turreson, G. 1922. The genotypical response of the plant species to habitat. *Hereditas,* **3**:211–350.

Vernberg, F. J., and W. B. Vernberg. 1970. *The Animal and the Environment.* New York: Holt, Rinehart & Winston. 398 pp.

Wright, J. W., and H. J. Baldwin. 1957. The 1938 International Union Scotch Pine provenance test in New Hampshire. *Silvae Genetica,* **6**:2 – 14.

Chapter 3
ORGANISMS AND THE PHYSICAL ENVIRONMENT

THE ENVIRONMENT AND ENVIRONMENTAL FACTORS

Traditionally, the environment has been divided into three categories: physical, chemical, and biotic, with specific processes or properties being called environmental factors. For example, air temperature, light, soil moisture, and soil texture are regarded as factors of the physical environment; the concentrations of minerals in the soil, of phosphate in water, and of carbon dioxide in the air as factors of the chemical environment; and organisms of the same or different species as factors of the biotic environment. Organisms are influenced, of course, by many environmental factors. These factors interact in a complex manner to affect each other as well as the responses of the organism. On one hand, the seemingly infinite number of interactions has led some ecologists to seek correlations between single factors and single responses of organisms. On the other, it has led a few investigators to proclaim the environment a holistic system that cannot be meaningfully analyzed into factors. The former procedure has sometimes led to erroneous conclusions and conflicting results. The latter can lead to no scientific studies at all.

It is now recognized that when several environmental factors influence an organism process, the response to an individual factor, or a particular process, may vary greatly and even reverse, depending on the other factors. For example, transpiration will increase with wind if the atmospheric humidity is low, but it will decrease with wind if the humidity is high. When confronted with a multitude of factors influencing a particular relationship, the ecologist must carefully specify and control all but one, or else he must measure all factors and responses simultaneously. The latter approach has led to the development of large interdisciplinary ecological research projects.

Progress in developing a theory of the interaction of several environmental and organismal factors has been most rapid for physical factors of the terrestrial environment. For this reason the physical environment will be emphasized in this chapter. The traditional factors of the chemical environment will be discussed in Chapter 10, in conjunction with an examination of biogeochemical cycles. The biotic environment will be treated in other chapters (4–6) dealing with the dynamics of plant and animal populations. In this chapter environment will be discussed first on the geographic level. Then the interactions between organisms and their immediate physical environment will be discussed with reference to some of the morphological and physiological adaptations discussed in Chapter 2.

GEOGRAPHIC PERSPECTIVE ON CLIMATE

Differences in the orientation of the earth's surface to the sun produce patterns of differential heating and cooling that cause large-scale atmospheric movements. These air movements, together with different intensities of solar heating, produce climatic patterns over the earth's surface. In the tropics, where the surface of the earth is perpendicular to the rays from the sun for most of the year, heating of the earth by the sun produces a region of rising air. As the air rises, it cools; when it cools to the dew point, water condenses. Thus, due to these convectional processes, the tropics are characteristically cloudy and rainy. At the poles, because of the low intensity of incoming radiation, the air cools and the cooler air descends in altitude and moves near the surface of the earth back toward the equator (Fig. 3–1).

This large-scale pattern is broken up into smaller cells. At about latitude 30°, air descends and flows northward and southward, toward both poles and the equator. As the air descends it warms, and as it warms its capacity to hold water vapor increases. This descending air is dry, and these latitudes are typically deserts with clear skies, high solar radiation, high temperatures, and low humidity. Even over the ocean the air humidity is low. The North and South African, Australian, South American, and Mexican deserts are at these latitudes. The warm air flowing toward the polar regions meets the cold air flowing away from the poles at about latitude 60°. The meeting of these two air masses often results in violent weather and the frontal activities common in the midwestern United States. Colder, denser polar air flows under and raises the warmer, less dense air. If the warm air has picked up any moisture, it then condenses and rain occurs. The meeting of these two air masses, called a *front,* can often be recognized by a line of rain clouds extending southwest to northeast.

As the position of the earth's axis in relation to the sun changes during the year, the areas of heating and cooling change. During the summer in

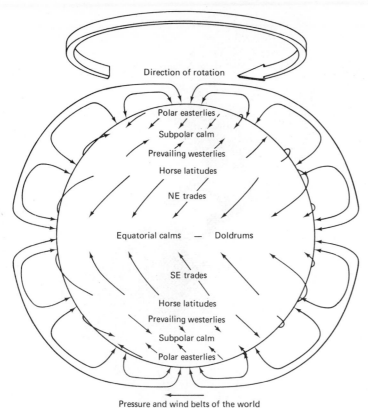

Direction of rotation

Polar easterlies

Subpolar calm

Prevailing westerlies

Horse latitudes

NE trades

Equatorial calms — Doldrums

SE trades

Horse latitudes

Prevailing westerlies

Subpolar calm

Polar easterlies

Pressure and wind belts of the world

FIGURE 3-1. General circulation of the atmosphere. (After Blair and Fite, 1965.)

the northern hemisphere when the sun is north of the equator, the regions of descending air and of frontal activity are shifted northward. During the winter in the northern hemisphere the sun has moved south of the equator, and the regions of descending air and frontal activity correspondingly shift southward.

The formation of convectional masses of rising air and rainfall from this rising air is greatest when the sun is directly overhead two times a year. As one progresses toward the poles, one enters the latitudes of descending air and year-long drought. On the poleward side of this area, one enters latitudes that may be reached by the frontal belt during the winter but not during the summer. Here there is a tendency for winter rain and summer drought. However, if the cold air masses that penetrate these regions were formed over continents, they will be dry and will contribute little rainfall, as in the case of the Great Plains region of the United States. In the Great Plains and midwestern United States warm moist air from the Gulf of Mexico is drawn in by the heating of the land surface. The rising of this air over land by the heating produces convectional storms that are the major source of moisture in the Midwest. However, if the polar air masses have been formed over water, they will be nearly

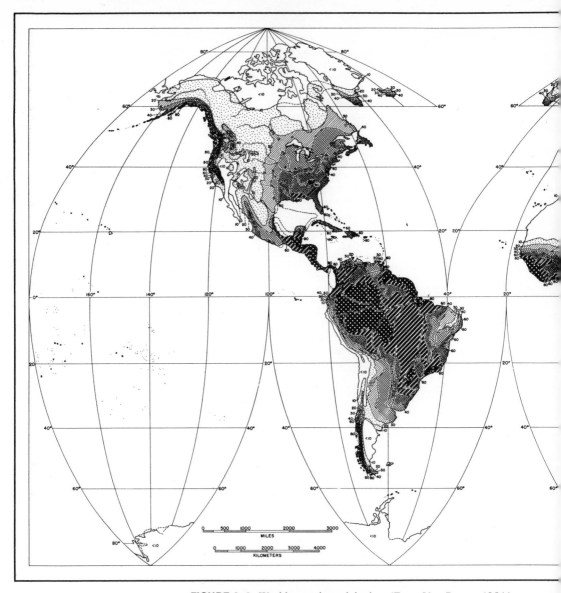

FIGURE 3–2. World annual precipitation. (From Van Royen, 1954.)

saturated and, when cooled by being forced to high altitudes, can yield water, as is shown along the northwestern coast of the United States. The western coasts of North America and Chile form precipitation gradients between areas of predominantly descending air, characterized by continual drought, and areas of penetration of predominantly polar marine air, characterized by continual precipitation (Fig. 3–2).

Dry areas are produced in ways other than by descending air near latitude 30°. As air rises to pass over a mountain range it expands and

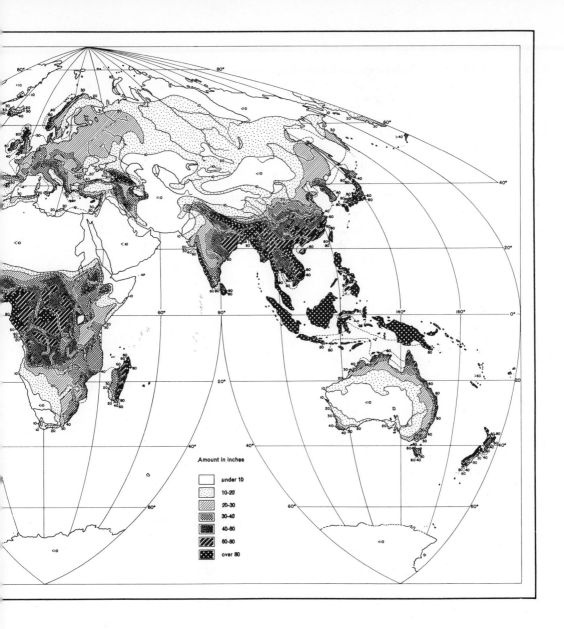

Amount in inches

	under 10
	10-20
	20-30
	30-40
	40-60
	60-80
	over 80

cools. Much of the moisture in the air condenses and falls as rain or snow. On the lee side of the mountains the dry air descends and warms. Thus, the lee sides of mountain ranges are often dry and are termed "rain shadows." This is the primary cause of the deserts of the western and southwestern United States and of the aridity of the Great Plains. The development of these arid areas can be traced in the fossil record made during geological periods of mountain uplifting.

The vegetation in the different climatic areas differs. In the area of continual rain at the equator, tropical rain forests occur. In areas of summer rain and winter drought, grasslands occur. Savannas are found in the transitions from forest to grasslands. When there is continual drought, deserts appear, and when the climatic pattern is one of winter rain and summer drought, shrubby forms, such as the chaparral of the south-western United States and the desert scrub of the Great Basin, predominate. These regions are best differentiated in the transect from central Africa northward into Europe. In North America the position of these regions is altered by the presence of major north-south oriented mountain ranges such as the Sierra Nevada and Rocky Mountains (Fig. 3–3). The ability to make generalizations about the kind of vegetation found in different climatic regions rests on the evolutionary convergence of the vegetation structure, as discussed in Chapter 2.

MICROCLIMATE

Organisms do not interact with the large-scale climatic patterns but with the climate close to the ground or within the vegetation, that is, with the *microclimate*. Early ecologists recognized that the physical environment was important in determining distribution of organisms, The approach taken by early ecologists to relate organisms to microclimate was largely descriptive and may be illustrated by several brief examples.

Shade tolerance has been considered important in determining the successional relations of plants. Clements (1916) ascribed a primary role to light as a cause of plant succession on an area that had been cleared. At first, species that required light occurred. Since these species could not reproduce in their own shade, they were eventually replaced by more shade tolerant species. This competition for light gave an advantage to the tallest plant in the community, since the tallest plant shaded the plants below. Thus, the sequence of plants was from short ones to the tallest that could live in the area. Warming (1909), on the basis of moisture, divided habitats into dry, wet, and in-between (xeric, hydric, and mesic, respectively) and recorded the different vegetation forms that occurred on these habitats. Geiger (1966) summarized much of the early literature on the interrelationships of microclimates and organisms.

This early work was largely directed toward the study of single environmental factors and single responses of organisms, although it was recognized that interactions occurred. Only recently, however, have studies been undertaken to clarify these complex interactions. The synthesis of basic theory about the interactions of organisms and physical factors was rapid because there existed a large body of theory in heat transfer engineering that could be borrowed and modified for ecological purposes. Rashke (1960) initiated studies of plant leaves by utilizing established physical relationships, and Gates (1962) further applied these principles to ecological problems.

INTERRELATIONS BETWEEN THE PHYSICAL ENVIRONMENT AND THE ORGANISM

Overview

Of the various environmental factors, those of the physical environment are the most readily quantified. Physical factors of the terrestrial environment that are usually measured are air temperature, humidity, visible radiation, and wind (Fig. 3–4). In ecological studies, interest is not with the environment per se, but with the responses of the organism to the environment. The ecologist may be interested in the physiological or behavioral responses of an organism or of a group of organisms to different environments, or he may be interested in explaining their spatial distribution. Or, perhaps, he may want to interpret the evolution of a type of organism from evidences in the fossil record. It becomes apparent that the various interactions between organisms and environment occur through time intervals of different duration.

The immediate response of the organism to factors of the physical environment is the production or alteration of the temperature of the body surface. Air temperature, radiation, and wind directly affect the temperature of the organism. Humidity affects the temperature of the organism by affecting the evaporation rate. Solar radiation also directly affects photosynthesis, and wind also directly affects evaporation. Other physical factors such as air temperature affect physiological processes by first affecting the temperature of the organism. The temperature and the physiological processes of water loss, respiration, and photosynthesis can change within seconds in response to changes in the physical environment. However, the temperature response of an organism will take longer if the organism has a greater mass or has temperature regulating ability. For example, Gates (1963) showed that temperatures of an oak leaf that was alternately sunlit and shaded fluctuated 10°C within a minute. But when Bartholomew and Tucker (1963) cooled large iguanid lizards in the laboratory, about 40 minutes were required to cool the lizard 10°C.

The characteristics of the organism which determine the specific interrelations between the physical factors and the immediate organism responses are termed *coupling factors* (Gates, 1968). An organism is tightly coupled to a factor of the physical environment if a change in the factor results in a relatively great change in the response of the organism. An organism is loosely coupled to a factor if a change in the factor results in a relatively small change in its response.

As the organism evaporates, respires, or photosynthesizes over a period of time, other responses may occur to preserve its water or heat balance. Under heat stress homeothermic animals will dilate the blood vessels of the skin in order to transport more heat from the center of the body to the surface where it can be lost to the air. Some homeothermic

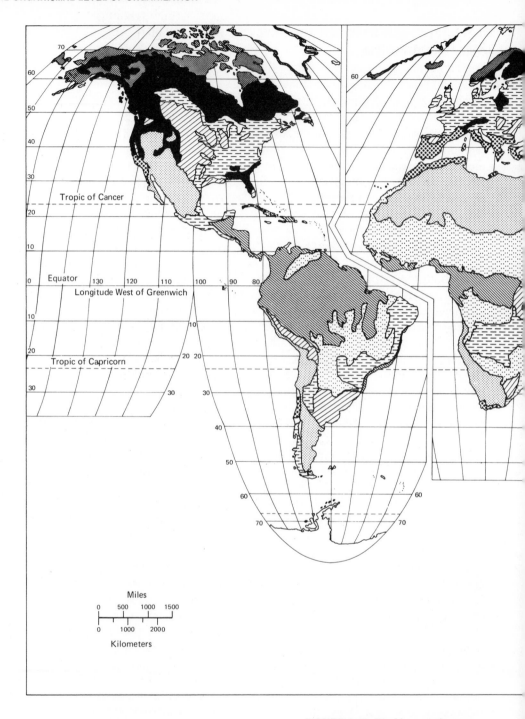

FIGURE 3–3. World vegetation map.

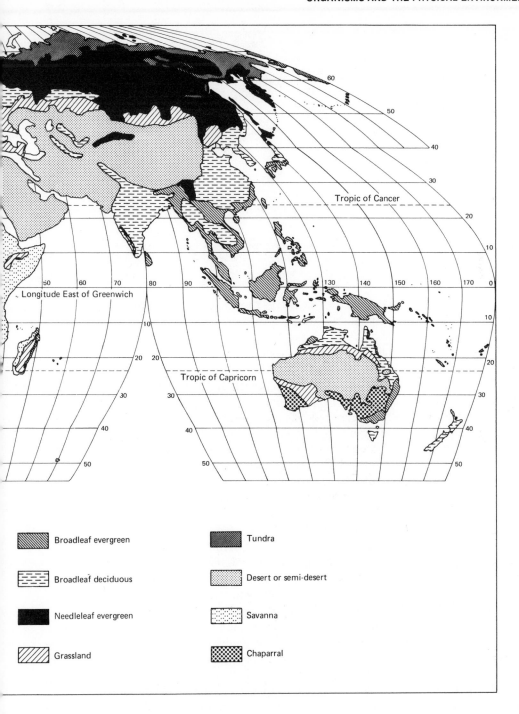

Tropic of Cancer

Longitude East of Greenwich

Tropic of Capricorn

Broadleaf evergreen

Broadleaf deciduous

Needleleaf evergreen

Grassland

Tundra

Desert or semi-desert

Savanna

Chaparral

(Modified from Espenshade, 1960.)

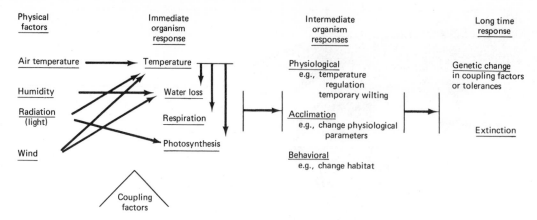

FIGURE 3-4. Factors of the physical environment and some of the organism responses that they induce.

animals possess sweat glands in the skin that enable them to dissipate heat by secreting water onto the skin surface where it then evaporates. However, prolonged water loss may be detrimental to the organism. Plants may close their stomates in order to conserve water when the amount of water loss becomes critical. The length of the transpiration period depends on the water stress to which the plant is subjected. Animals may also restrict sweating. When evaporative cooling is thus reduced, the organism's temperature will increase and the organism undergoes a thermal stress. Plants commonly undergo stomatal closure during midday if the rate of water loss exceeds the rate of water uptake or if high respiration rates create high carbon dioxide concentrations within the leaf. Since these stomatal closures occur during the hottest part of the day, it is evident that stomates are not adaptations to control leaf temperature.

In addition to the physiological regulation of body temperature and water content, an animal can behaviorally regulate its temperature and rate of water loss by moving to a new habitat or by avoiding undesirable habitats. During the heat of the day kangaroo rats conserve water in the desert environment by remaining in their burrows where temperatures are cooler and humidities higher than on the desert floor (Bartholomew, 1964). Antelope ground squirrels are active in the sun during the heat of the day but periodically cool themselves by seeking shade or prostrating themselves on cool rocks. These responses enable the organism to adjust to the physical environment within several minutes or hours.

Over a matter of years or decades, a species may change genetically in response to selection by factors of the physical environment. A slow secular change in the physical environment may cause all of the habitats available to the organism to become unfavorable—that is, the organism cannot exist in any habitat in the area with its existing physiological mechanisms. If under these changing conditions the organism does not change genetically, it will become extinct. Since the coupling relationships are partly due to genetically based characteristics of organisms, changes in organism properties which determine the coupling factors will involve genetic changes such as those shown in the fossil record for desert plants. For plant populations fossil evidence shows trends in properties or organisms that have been interpreted to be responses to the physical environment. Leaves of fossil floras in the North American deserts show a decrease in size during the geological period of uplift of the Sierra Nevada range. Changes in species characteristics along environmental gradients also suggest that evolutionary responses to the physical environment are continually occurring. Desert organisms are often protectively colored, matching the color of their substrates; but at the southern limit of their distribution the coloration often becomes lighter, presumably because of a more intense solar radiation and thermal stress (Bodenheimer, 1958).

Processes of Energy Exchange

The diverse physical factors and responses of organisms have a common basis in that they all involve an exchange of energy. By working with the processes of energy exchange we can integrate and quantify our picture of the interactions between the organism and the physical environment. The processes of energy exchange are convection, evaporation, radiation, and conduction, listed in the order in which ecologists have usually, implicitly, involved them (Fig. 3–5).

Convection refers to the process of replacing cells of warm air with cells of cool air. When the organism's surface is warmer than the air, the organism will continually warm the cooler air which is brought to its surface. During the process the organism is losing heat to the air. *Evaporation* is the change in state of water from a liquid to a gas. This change in state removes energy from the liquid and lowers its temperature. The rate of evaporation will depend partly on the diffusion of water vapor away from the surface, that is, evaporation cannot occur in an atmosphere saturated with water vapor at the temperature of the evaporating surface. If the concentration of water vapor over the surface is constant, the rate of evaporation will equal the rate of diffusion. In the literature the processes of evaporation and of water vapor diffusion are discussed as one process. *Radiation* is the transfer of energy through the electromagnetic spectrum. *Conduction* is the transfer of energy in molecular motion by one molecule hitting another.

The responses of organisms to each process of energy exchange are listed in the last column in Fig. 3–5. Each energy exchange process involves a temperature response in the organism. Evaporation also affects

Process of energy exchange	Environmental factors involved	Coupling factors	Organism properties involved in coupling factor	Organism response
Convection	Wind velocity and direction air temperature	Convection coefficient	Size Shape Texture Orientation	Temperature
Evaporation	Wind velocity and direction humidity	Boundary layer resistance	Size Shape Texture Orientation	Temperature Water loss
		Leaf resistance	Density and size of stomata Resistance of cuticle to water loss	
Radiation	Solar radiation		Absorbance Orientation	Temperature Photosynthesis
	Infrared radiation		Emittance	Temperature
Conduction	Matrix temperature	Conduction coefficient	Conductance	Temperature

FIGURE 3–5. Processes of energy exchange and factors of the physical environment.

the water loss from the organism, and radiation affects photosynthesis. The environmental factors involved in each process of energy exchange are listed in the second column in Fig. 3–5. Evaporation rate is influenced by the velocity and direction of the wind relative to the position of the organisms and by air temperature. Solar and infrared radiation from the sky and the surroundings are the specific processes of radiation exchange. The temperature of the soil, water, tree trunk, or anything that the organism is in direct contact with (called the *matrix*) is important in determining the exchange of energy to or from the organism by conduction.

The coupling factors are listed in the third column in Fig. 3–5, e.g., the convection coefficient, boundary layer resistance, leaf resistance, and conduction coefficient. The coupling factors can be considered constants of proportionality, relating the transfer of energy to a gradient of temperature or humidity. The convection coefficient relates the energy exchanged by convection to the temperature gradient between the organism's surface and the air. The conduction coefficient relates the energy exchanged by conduction to the temperature gradient between the organism's surface and the surrounding matrix. The boundary layer and leaf resistance relate the diffusion of water vapor and evaporation to the gradient in water vapor between the evaporating surface and the air. These processes will be discussed more fully later in this chapter. Table 3–1 summarizes the symbols used.

Those properties of the organism that are involved in the coupling factors are listed in the fourth column. The convection coefficient and boundary layer resistance are affected by the size and shape of the organism, its texture, and its orientation to the wind. Both the convection coefficient and boundary layer resistance involve the process of replacing cells of air at the organism's surface with cells of air from the free air, and, thereby, replacing the heat and water vapor content of the air at the surface of the organism. The density and size of stomates and the resistance of the cuticle to water loss are related to leaf resistance. The organism is coupled to its solar radiation environment by its absorptance and its orientation to the direct beam from the sun. The organism is coupled to the materials with which it is in direct contact by its conductance and the conductance of the matrix (Fig. 3–5).

The organism receives solar energy S by direct radiation from the sun, which is transmitted through the atmosphere, by diffuse radiation from

TABLE 3-1. SUMMARY OF COMMON SYMBOLS USED IN CHAPTER 3.

Symbol	Definition	Units	Common values
S	Solar radiation	cal/cm²/min	0 to 1.6
IR	Infrared radiation	cal/cm²/min	0.4 to 0.6
C	Convectional exchange of heat	cal/cm²/min	–0.1 to +0.2
E	Evaporation rate	g/cm²/min	0 to 5×10^{-4}
L	Latent heat of evaporation	cal/g	580
M	Metabolic rate	cal/cm²/min	0 to 0.2
G	Conductional exchange of heat	cal/cm²/min	–0.5 to +0.1
P	Rate of carbon dioxide exchange	g/cm²/min	
T	Temperature	°C	–1 to 50
rh	Relative humidity	decimal fraction	2 to 100
Ψ	Water potential	ergs/g	
e	Vapor pressure	mb	5 to 35
ρ_w	Water vapor density	g/cm³	5 to 30×10^{-6}
ρ_a	Air density	g/cm³	1.25×10^{-3}
c_p	Specific heat of air at constant pressure	cal/g/°C	0.24
μ	Wind velocity	cm/sec	0 to 1,000
D	Molecular diffusion coefficient	cm²/sec	0.24 for water
a	Absorptance to solar radiation	decimal fraction	0.4 to 0.6
E	Emittance to infrared radiation	decimal fraction	0.97
h_c	Convection coefficient	cal/cm²/min/°C	0.017
k	Thermal conductivity	cal/cm²/min/°C	

TABLE 3-1. (cont.)

Symbol	Definition	Units	Common values
VHC	Volumetric heat capacity	cal/cm^3/°C	
D_d	Diurnal damping depth	cm	10 to 15
K (subscript)	Eddy diffusivity or turbulent transfer coefficient	cm^2/sec	10^2 to 10^4
r	Resistance to water loss or carbon dioxide exchange	min/cm	0.01 to 1.0
z_0	Roughness length or roughness parameter	cm	0 to 10
d	Zero plane displacement	cm	0 to 300
K	Extinction coefficient to radiation	1/F	0.3 to 2.0
R	Universal gas constant	cal/g/°K	0.11
σ	Stefan-Boltzmann's constant	cal/cm^2/min/°K^4	8.13 × 10^{-11}
k	Von Karman's constant		0.4
z	Height above or depth below surface	cm	
t	Time	days or minutes	
zn	Zenith angle of the sun	degrees	0 to 90
β	Altitude of the sun	degrees	0 to 90
α	Inclination of a surface to the horizontal	degrees	0 to 90
ϕ	Azimuth of the sun	degrees	0 to ± 180
η	Azimuth of the surface	degrees	0 to ± 180

the sky resulting from the scattering of the direct beam by the molecules and dust particles present, and by reflected radiation from the surrounding surfaces. Infrared radiation IR is received by the organism from the sky and from the surroundings. The organism absorbs part of this available incident energy. This absorbed energy is usually lost through convection C, evaporation LE, and infrared radiation from the organism IR_o. However, in some situations the organism may gain energy by convection and if dew forms on the body surface, energy is added to the organism. Metabolism M contributes energy which is lost by these same processes. The processes of energy exchange can be summarized in the energy budget equation:

$$S + IR + M = IR_o \pm C + LE \qquad (3-1)$$

The equation is written with the energy absorbed or produced by the organism on the left side of the equality and the processes by which energy is usually lost from the organism on the right. All of the processes by which energy is lost are related to the temperature of the organism. If the energy absorbed by the organism does not equal the energy lost, the temperature of the organism will change, increasing or decreasing the energy lost, until the equation balances. The basic energy budget equation can be expanded in diverse ways to explore in more detail the inter-relations between an organism and its physical environment.

When specific equations for each process of energy exchange are inserted in Eq. (3–1), values for all but one of the variables can be chosen by the researcher to represent some environment situation and organism of interest, and the equation can be solved for the remaining factor. The relation between two variables in the energy budget equation can be studied by altering the value of one variable and calculating the value of the second, while the other variables are held constant. In this way the interrelationships between the organism and its environment can be studied rapidly and easily. The process by which these experiments are carried out is called *simulation;* an active real-world situation is being simulated by manipulating equations believed to describe real-world processes. Some insight into the interrelations between specific variables can be gained in this way although the results of such simulations can only be valid as the data and equations used. In the following discussion of interrelations between organisms and environment the influence of various environmental factors on some response will first be presented by using the results of a simulation experiment.

Temperature

Temperature is probably the most commonly measured parameter of the environment. Temperature may be defined as the condition of a body which determines the transfer of heat to or from other bodies. The fact that temperature alone is not an adequate measure of the environment for defining the interactions between an organism and its environment is easily visualized if one recalls stepping from air at 70°F into water at 75°F. The water, although warmer, feels colder because it removes heat from the skin surface at a faster rate than does the air. Across a given temperature gradient heat will be transferred five times faster from the skin through water than through air. This means that organisms in water will rarely have a temperature which differs much from that of the surrounding water. Two properties affecting the rates of transfer of heat are the thermal conductivity and the specific heat. The thermal conductivity relates the heat conducted through a substance to the temperature gradient. The heat capacity relates the energy added to or subtracted from a substance to its change in temperature. The difference in an organism's response to cold water or air is because water has a higher conductivity and a higher specific heat than air. Thus, the conductivity and the specific heat of air and water are as important as their temperatures in determining the interactions between organism and environment.

A brief excursion into some concepts of molecular activity may clarify the interrelations between the heat content and the temperature of a substance. The molecules in a substance are undergoing three kinds of motion:

1. Rotational, the motion of a molecule around its axis.
2. Vibrational, the relative motion between two atoms along a line connecting them.
3. Translational, the motion of molecules in a straight line.

The heat content of an object consists of the total energy associated with the randomly directed motion of the molecules that make up the object. The temperature of an object is a measure of the average transitional kinetic energy of the molecules within the object; thus the vibrational and rotational energies do not contribute to its temperature. Temperature is sometimes erroneously referred to as a measure of the heat content of an object. Only translational energy is conducted and thus temperature is defined as the condition of a body that determines the transfer of heat to or from other bodies. The commonly used unit of heat exchange is the calorie, which is the amount of energy required to raise a gram of water one degree Celsius at 14°C. Since heat results from the kinetic energies of translational, rotational, and vibrational motions of molecules, and since temperature results only from the energy of translational motion, the specific heat may be thought of as a measure of the relative amount of heat energy that is in the translational form. The specific heat of an object can be expressed as cal/g/°C or cal/cm^3/°C. The first is called the *specific heat* and the second, the *volumetric heat capacity*. The volumetric heat capacity is the expression that can most easily be used in ecology, since organisms are interacting with the environment across surfaces and through volumes.

The convectional exchange of energy tends to make the temperature of the organism equal to the temperature of the air. Small organisms or parts of organisms have a high potential for convectional energy exchange and are tightly coupled to the air temperature. Thus, their temperatures will be close to those of the air throughout the day. The convectional exchange of energy is directly proportional to the difference between surface and air temperatures and the velocity of the wind and is inversely proportional to the size of the organism. In concise terms the convectional exchange C can be expressed as follows:

$$C = h_c (T_o - T_a) \qquad (3\text{--}2)$$

where T_o is the temperature of the organism surface, T_a is the temperature of the air, and h_c is the convection coefficient that includes the wind velocity and sizes of the organism.

Values for the convection coefficient for organisms were first obtained from approximations made by heat transfer engineers for flat plates and cylinders (Rashke, 1960; Gates, 1962). For simulating the effects of air temperatures on an organism's temperature, the simplest unit considered will be one that approximates a flat plate, e.g., a leaf. Later, convection coefficients for more complex geometric forms such as pine and spruce branches and lizards were measured (Gates et al., 1965; Tibbals et al., 1964; Bartlett and Gates, 1967). Some of the values for the convection coefficients are summarized in Table 3–2. Recently, when convection coefficients for leaves were measured in the field under natural conditions, the engineering approximations were found to underestimate field values. It is thought that the air in natural situations is more turbulent than the air in the wind tunnels where the engineering values are obtained and that the transport in more turbulent air is higher (Pearman et al., unpublished).

The departure of the temperatures of flat leaves of various sizes from the temperatures of the air can be calculated by solving the energy budget equation for this departure while varying the environmental and organism parameters involved (Fig. 3–6). For simplicity, transpiration is assumed to be zero. The parameters involved are then absorbed radiation, wind, and leaf size. During the day and night, wider leaves depart more from air temperature than do narrower leaves. As transpiration increases the

TABLE 3-2. VALUES FOR THE CONVECTION COEFFICIENT FOR LEAVES OF DIFFERENT SHAPES UNDER DIFFERENT CONDITIONS.

Leaf	Air and leaf conditions	h_c	Author
Flat leaves	still air, leaf horizontal hotter than air, facing upward	$7.86(\Delta T/w)^{1/4}$	a*
	still air, leaf horizontal, hotter than air, facing downward	$3.87(\Delta T/w)^{1/4}$	a
	still air, leaf horizontal average of both surfaces	$5.86(\Delta T/w)^{1/4}$	a
	still air, leaf vertical	$6.04(\mu/w)^{1/2}$	a
	moving air	$5.73(\mu/w)^{1/2}$	a
Cylindrical leaves	still air	$6.00(\Delta T/w)^{1/4}$	a
	moving air	$6.17v^{1/3}w^{-2/3}$	a
Ponderosa pine branch	still air	$[11. + 3.6(\Delta T)^{0.3}]$	c†
	moving air	$(18.0 + 0.71\mu)$	c
Blue spruce branch	still air	$10.0(\Delta T)^{1/4}$	b‡
	moving air, any direction, $v < 30.0$ cm/sec	$(20.4 + 0.2\mu^{0.97})$	b
	moving air, any direction $v < 30.0$ cm/sec	$(0.95\mu^{0.97})$	b
White fir	still air	$10.4(\Delta T)^{1/4}$	b
	moving air, across branch	$(20.4 + 1.75\mu^{0.75})$	b
	moving air, along branch	$(20.4 + 2.79\mu^{0.75})$	b

*From Gates, 1962.
†From Gates et al., 1965.
‡From Tibbals et al., 1964.

Note: Wind velocity (μ) is given in cm/sec, the width or diameter of the organism (w) in cm, T in °C, and h_c in 10^{-3} cal/cm^2/min/°C.

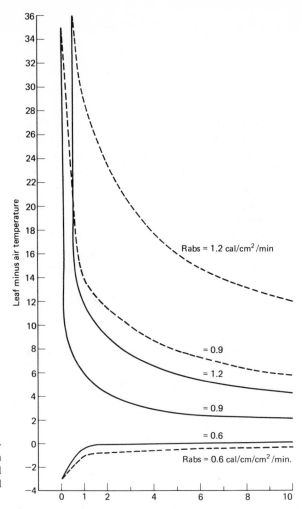

FIGURE 3–6. Departure of leaf from air temperature at an air temperature of 25°C with varying wind speeds, absorbed radiation, and leaf widths. Solid line is for a 1 cm leaf; dashed line is for a 10 cm leaf.

leaf temperature will approach and can be lower than air temperature. A graph of the daily courses of temperatures of several leaves shows how they tend to follow air temperature, although departing during the day when the leaves are heated by solar radiation and during the night when the leaves drop below air temperature (Fig. 3–7).

The application of these same principles to animals involves simplifying the shape of the animal to approximate it with a cylinder. Since the convection coefficient for cylinders is about the same as for flat plates, the trends shown in Fig. 3–6 for a leaf will be similar to those for cylinders. The assumption, however, is made that metabolism and evaporation are negligible energy exchange processes, which would not be the case for homeotherms.

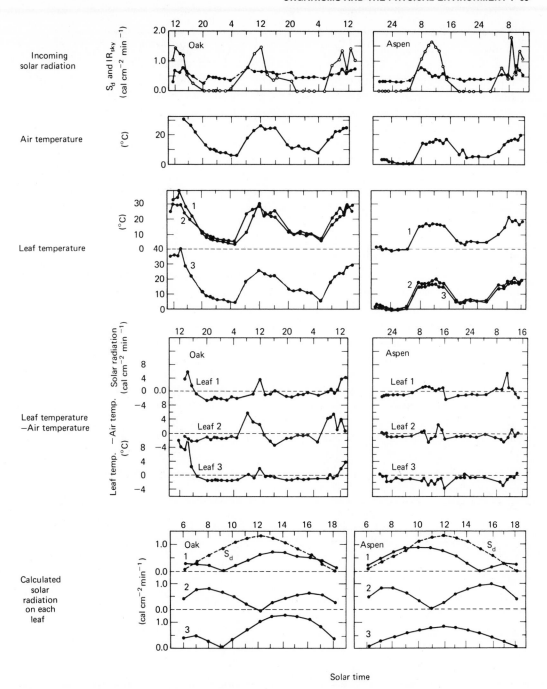

FIGURE 3–7. Daily courses of solar radiation, air and leaf temperatures, departures of leaf from air temperature, and potential direct solar radiation on three sunlit leaves of aspen, *Populus tremuloides,* and oak, *Quercus gambellii.* (From Miller, 1967.)

If an animal has no fur covering, the surface which exchanges energy with the air by convection is its skin; but if fur is present, the active surface is the outer edge of the fur. With fur-covered animals metabolic heat must be conducted from the skin through the fur to the outer surface, or heat from the air must be conducted through the fur to the skin. The thicker the fur or the more effective its insulating value, the steeper the temperature gradient can be between the animal's skin and the active surface where energy exchange with the environment occurs. The rate of heat conduction G between two points through a substance is proportional to the temperature difference between the two points and the thermal conductivity of the substance k, and inversely proportional to the distance between the points. Thus, the rate of heat conduction through the fur can be given by

$$G = k\frac{T_1 - T_2}{\text{Distance}} \qquad (3-3)$$

where T_1 is the temperature of the skin and T_2 is the temperature at the outer surface of the fur. Some thermal conductivities and thicknesses of fur of different animals are given in Table 3–3.

Scholander (1955) showed that animals from the tropics and arctic had the same body temperatures even though the air temperatures in their native environments differed by a factor of 10. Animals from the arctic have thicker fur than do tropical animals, which maintain a steeper temperature gradient (Fig. 3–8). The sloth is an exception, since it is tropical but has thick fur. However, the metabolic rate of sloths is about half that of other mammals, and thicker fur thus permits maintenance of a reasonable body temperature in spite of a low metabolic rate (Scholander, 1955).

TABLE 3-3. THERMAL CONDUCTIVITY, G, OF WINTER FUR FROM ARCTIC AND TROPICAL MAMMALS.*

	Thickness in cm^2	Thermal conductivity, k	k/z
Dall sheep	7	8.1	1.16
White fox	5	5.8	1.16
Wolf	6.5	9.0	1.39
Caribou	3.5	4.9	1.39
Red fox	4.5	7.8	1.74
Dog	4.0	7.0	1.74
Rabbit	3.0	5.2	1.74
Polar bear	6.5	11.3	1.74
Arctic ground squirrel	1.5	5.2	3.47
Weasel	0.8	5.6	6.95
Shrew (summer)	0.5	3.5	7.00

*Calculated from Scholander, 1955.

Note: $k = 10^{-3}$ cal/cm/min/°C.

$G = (k/z)(dT)$ where dT is the temperature difference in °C.

A somewhat different situation is shown by camels which are adapted to high desert temperatures by a thick fur covering that acts as insulation to reduce the effect of desert heat. The large mass of a camel results in a slow rise in body temperature through the day, while the low nighttime air temperatures in the desert allow the camel's body temperature to drop before it reaches a detrimental level (Schmidt–Nielsen, 1964).

A homeotherm, by definition, will be able to maintain its body temperature over a certain range of air temperatures by varying its metabolic rate. Some poikilotherms regulate their temperatures behaviorally. For example, hawk moths can raise the temperature of flight muscles from 32° to 36°C by vibrating their wings before takeoff. Gregarious caterpillars can raise their temperatures 1½° to 2°C by clustering together. Locust nymphs may increase their temperature 10°C by basking in the sun with their bodies perpendicular to incoming radiation. Ants move their larvae to warm or cool places within the nest. Bees control temperatures within their hives to a range between 13° and 25°C by fanning their wings to evaporate water droplets when it is too hot or releasing body heat through increased metabolic activity when too cold.

Arctic animals with thick fur must also dissipate metabolic heat produced when working or running. This heat dissipation problem is often solved by having little insulation on certain extremities, usually the legs. When excess metabolic heat is produced, it can easily be dissipated by passage of blood from the warm body through these appendages. These

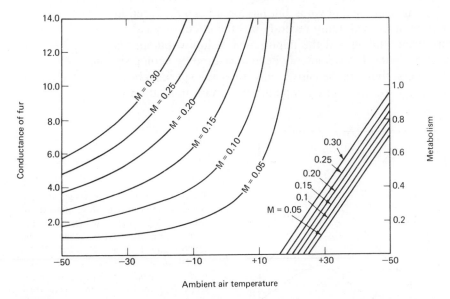

FIGURE 3–8. Conductance of fur (k/z) required to maintain a body temperature of 37°C with different ambient air temperatures and metabolic rates (M). Evaporation $= 0$; absorbed radiation $= 1.00$ cal/cm^2/min; wind velocity $= 2$ mph; body diameter $= 4$ cm. Conductance is in 10^{-3} cal/cm^2/min (°C); metabolism is in cal/cm^2/min.

extremities must be able to tolerate extreme cold temperatures, however, when the blood does not supply heat to them. Excessive transfer of body heat by the blood to the appendages, where it would be lost, is prevented by a specialized organization of blood vessels. The arteries are located close to the veins so that the warm arterial blood from the body gives up heat to the cold venous blood returning from the extremities. If the arteries and veins were separated, the warm arterial blood would lose heat to the environment in the extremities, and the cold venous blood would have to be warmed in the body.

As mentioned before animals can adjust to extreme regional temperatures by moving to different microhabitats where temperatures are more favorable for them. Figure 3–9 shows the range of temperatures permitting the development of several invertebrates, together with the yearly course of monthly mean temperatures for a subtropical area (New Delhi, India) and a temperate area (London, England). The temperatures required for development restrict some organisms to temperate areas and others to tropical areas. The housefly is restricted to warm places: temperatures are thus favorable for its development throughout the year in New Delhi, but only during the summer in London.

Wind

Wind affects the body temperatures of organisms as well as influencing their dispersion and causing mechanical damage. Some of these effects, such as mechanical damage, are obvious but have not been analyzed in detail and will not be discussed. Only recently has the effect of wind on body temperature been studied and the importance of low wind-speeds recognized (Fig. 3–6). Wind has commonly been measured by ecologists with cup anemometers that are sensitive only to wind speeds above 2 mph. However, wind speeds below 2 mph are important in relation to

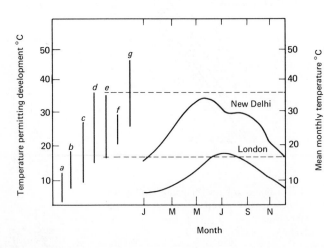

FIGURE 3–9. Graph of range of temperatures permitting development of different organisms and mean monthly temperature for New Delhi, India (29°N) and London, England (52°N) a: *Dendronotus frondosus* (mollusk); b: *Mytilus eduli;* c: *Smynthurus viridus* (eggs) (springtail); d: *Cirnex lectularius* (bedbugs); e: *Musca domestica* (housefly); f: *Lernea clegans* (crustacea); g: *Thermobia domestica* (firebrat). (After Andrewratha and Birch, 1954.)

the temperature of an organism. Furthermore, the wind speed measured two meters above the ground in a cleared area is usually not the same as that experienced by the organism at the ground surface or in the vegetation canopy. Although it seems obvious that the ecologists should measure wind that is actually experienced by the organism, many studies have utilized wind records from Weather Bureau stations. Usually, these measurements are taken well above the vegetated surface and at some distance from the study area.

Wind flow over a surface is either laminar or turbulent. In laminar flow the wind flows in layers over the surface and wind velocity increases in the layers farther from the surface. There is no air movement in the vertical direction in laminar flow. In order to leave the surface and diffuse into the free air, substances must pass by molecular diffusion from the surface into the free air. The equation for molecular diffusion is

$$W_t = D \frac{dc}{dz} \tag{3-4}$$

where W_t is the amount of material transported across the gradient, in $g/cm^2/sec$; c is the concentration of the substance, in g/gm^3; z is the distance from the surface, in cm; and dc/dz is the rate of change in concentration of the substance with respect to distance. The *diffusion coefficient* D depends on the properties of the substance and equals 0.239 cm^2/sec for water vapor and 0.11 cm^2/sec for CO_2 at 20°C.

Air flow in nature is usually turbulent; laminar flow occurs only within a few millimeters of the surface. The theory for the transport of material and energy in turbulent air was developed by analogy to the theory for molecular diffusion. In turbulent transfer cells or eddies of air leave the surface because of differences in the densities of the air at the surface and in the surrounding air. As the cells leave the surface they may be replaced by cells from a distance away with different heat, water vapor, or carbon dioxide content resulting in a net exchange of these substances with the surface. The rate of exchange of cells at the surface can be expressed as volumes per unit surface area per unit time ($cm^3/cm^2/sec$ or cm/sec). The equation for rate of exchange is written in a manner analogous to that for molecular diffusion. Instead of the diffusion coefficient, the coefficient is called the *eddy flux* or *eddy coefficient,* which is a measure of the rate of transfer of volumes of air, and usually symbolized K with some appropriate subscript to denote what is being transported.

The quantity of heat exchanged will depend on the difference in heat content of the air at the surface and the air above the surface. The difference in heat content H between two volumes of air depends on their temperatures and the volumetric heat content of air, and is given by

$$dH = c_p \, \rho_a \, (T_1 - T_2) \tag{3-5}$$

where c_p is the specific heat of air at constant pressure, ρ_a is the density of air, and T_1 and T_2 are air temperatures of two points. The product of the specific heat and the air density is the volumetric heat capacity of the air. The rate of exchange of heat between the two points is given by multiplying the difference in heat content dH by the rate of exchange of volumes of air K. Thus:

$$C = \frac{K_h\, dH}{dz} \qquad (3\text{–}6)$$

where C is the convectional exchange of heat and dz the distance between the points. Similarly, the fluxes of water or carbon dioxide from a surface are related to the differences in water or carbon dioxide content of the air next to the surface and the air above and the rate of transfer of volumes of air:

$$E = \frac{K_w d\rho_w}{dz} \qquad (3\text{–}7)$$

$$P = \frac{K_c\, d(CO_2)}{dz} \qquad (3\text{–}8)$$

The eddy coefficient, convection coefficient, and the boundary layer resistance are related. The eddy coefficient K in cm^2/sec refers to a flux of heat or materials across a distance dz. If the distance were undefined, the coefficient would be in cm/sec, which is obtained by dividing the eddy coefficient by the distance. The reciprocal is called the air resistance, in sec/cm. The air resistance is a measure of the resistance of the boundary layer of air next to the leaf to the diffusion of heat and gases. The air resistance r_a is related to the convection coefficient h_c by

$$r_a = \frac{c_p \rho_a}{h_c} \qquad (3\text{–}9)$$

where c_p is the specific heat of air, and ρ_a is the density of the air.

The influence of wind on leaf temperature can be approximated by defining reasonable values for the environmental variables in the energy budget equation and solving for the difference between leaf and air temperature for each of several wind speeds (Fig. 3–6). As wind speed increases from zero to 1 or 2 mph, the temperature of the leaf decreases abruptly, but with wind speeds greater than 2 mph, further increases in wind have little effect on leaf temperature.

Humidity

Several different ways of expressing the water content of the air are now in common use. *Absolute humidity* is defined either as vapor density, e.g., grams of water vapor per unit volume of air, or as the *partial pressure* of the water vapor in the air, in millibars (mb). At 20°C saturated air contains 17.3 g/m³ of water vapor, which exerts a partial pressure of 23.4 mb. Air is usually not saturated, however. The difference between the saturation vapor pressure and the actual vapor pressure is termed the *vapor pressure deficit*. The ratio of the actual vapor density to the saturation vapor density, or the actual vapor pressure to the saturation vapor pressure, is called the *relative humidity*. For example, air at 20°C, with a relative humidity of 50%, has a vapor density of 8.16 g/cm³ and a vapor pressure of 11.7 mb.

A fundamental characteristic of water is its free energy content, or ability to diffuse. This is termed *water potential* and is measured in ergs/gm. Water potential includes the effects of several forces that act to inhibit or promote the movement of water.

The expressions for humidity can be interrelated through a series of simple equations. Vapor density ρ_w and vapor pressure e_w are interrelated by

$$\rho_w = \frac{e_w}{RT} \qquad (3\text{-}10)$$

Water potential ψ and relative humidity rh are related by

$$\psi = RT \ \ln \ (rh) \qquad (3\text{-}11)$$

$$\psi = RT \ \ln \ \frac{\rho_w}{\rho_s} \qquad (3\text{-}12)$$

$$\psi = RT \ \ln \ \frac{e_w}{e_s} \qquad (3\text{-}13)$$

where the subscripts w and s refer to the actual water content and saturation water content of the air, T is the absolute temperature (degrees Kelvin) and R is the universal gas constant. The universal gas constant may be thought of as being proportional to the fraction of kinetic energy of one mole of gas which is due to translational motion, analogously to specific heat. The gas constant for water vapor is 4.62×10^6 ergs/g °K or 0.11 cal/g/°K.

The moisture content of the air has ecological significance because of its influence on the rate of evaporation from the bodies of organisms. Evaporation is a change in state of water from the liquid to the gaseous state. This change in state requires energy; approximately 580 gram-calories are required to evaporate one gram of water if the water is at 20°C. The evaporation rate from a surface is usually assumed to equal the rate of diffusion of water vapor from the surface. In the diffusion process, the movement of water vapor is impeded by resistances in the path of the gas. In plants and animals the evaporative surfaces are the surfaces of the living cells within the organism, and the loss of water is impeded by the nonliving layers of the outer integument together with the boundary layer of the air next to the organism. The greater the resistance

r of the integument and boundary layer the slower will be the rate of evaporation. Evaporation E can thus be expressed:

$$E = \frac{\rho_o - \rho_w}{r}$$

where ρ_o is the vapor density of the evaporating surface of the organism and ρ_w is the vapor density of the air. ρ_o is usually assumed to be the saturation vapor density at the temperature of the evaporating surface. The temperature of the evaporating surface can be higher than the temperature of the air, as discussed previously. When this occurs, it is possible for the surface to evaporate water into saturated air, a point that should be distinguished from the earlier statement that evaporation could not proceed if the air next to the surface, at the temperature of the surface, were saturated. Evaporation into saturated air is commonly observed, e.g., when the sun shines after a rain and clouds of water vapor form over forests or when cold air temperatures occur over a warm pond or lake and "steam" forms over the body of water. In each case water is evaporated from the surface that is warmer than the saturated air. The water evaporated from the surface supersaturates the air and the water vapor in the air condenses and becomes visible.

Over a pan of evaporating water, r equals the resistance of the air in the boundary layer over the surface. The boundary layer is a layer of relatively still air next to the surface which has a higher water vapor content than the free air and which acts to resist the diffusion of water vapor from the surface to the free air. In a plant leaf, the evaporating surfaces are the mesophyll cell walls. The path of the diffusing water molecules from the mesophyll cell walls to the free air is across the intercellular air spaces, either through the stomates or cuticle, and across the leaf boundary layer. Thus, the diffusion of water vapor is impeded by several resistances: the resistance of the intercellular spaces between the mesophyll cell walls and the epidermis of the leaf r_m; the resistance of the epidermis r_l; and the resistance of the leaf boundary layer r_a. The resistance of the epidermis consists of the resistances of the stomates r_s; and of the cuticle r_c. The resistances of the intercellular air spaces, epidermis, and boundary layer are in series since diffusing gas molecules must pass across each of these resistances; and the resistances of the stomates and cuticle are parallel since diffusing gas molecules can cross either of these resistances. Using an electrical analogue, the resistances are related as follows:

$$r = r_m + r_l + r_a \text{ and } r_l = \frac{r_s r_c}{r_s + r_c} \tag{3-15}$$

Milthorpe (1962) and Lee and Gates (1964) showed that the resistance of the intercellular spaces was negligible compared with the resistances of the leaf and boundary layer. Values for the leaf resistances of various plants are given in Table 3–4. Figure 3–10 shows the air resistance of various sized leaves with varying wind speeds.

As the leaf resistances increase or as leaf widths decrease [see discussion of Eq. (3–9)], air resistance becomes less important in controlling water loss relative to leaf resistance. Thus, experiments measuring the relation of stomate resistances to transpiration and experiments measuring the influence of wind on transpiration have yielded conflicting results because of other variables, e. g., leaf size.

TABLE 3–4. RESISTANCE OF LEAVES OF VARIOUS SPECIES TO TRANSPIRATION WITH A MINIMUM RESISTANCE OF THE BOUNDARY LAYER.

Species	r_{H_2O}	Reference (author)
Free water surface	0.0367	a*
Populus tremuloides (wet site)	0.05	b†
Populus tremuloides (dry site)	0.0667	b
Populus grandidentata	0.0667	b
Ammophila breviligulata	0.05	b
Betula papyrifera	0.1	b
Arctostaphylos uva-ursi	0.0834	b
Chamaedaphne calyculata	0.2	b
Vitis sp.	0.25	b
Acer rubrum	0.20	b
Pinus resinosa	0.50	b
Pinus strobus	0.67	b
Thuja occidentalis, shade	1.25	b
Thuja occidentalis, sun	1.25	b
Pteridium aquilinum	3.3	b
Conocarpus erectus	0.05	c‡
Avicennia tomentosa	0.04	c
Laguncularia racemosa	0.03	c
Rhizophora mangle	0.04	c
Coccoloba uvifera	0.04	c
Adenostoma fasciculatum	1.0	d§
Adenostoma fasciculatum	0.71	d
Quercus dumosa	0.196	d
Ceanothus crassifolius	0.35	d
Cotton, *Gosypium*		e‖
lower	0.020	
upper	0.148 (0.05–0.20)	e
mean	0.084	e

*From Gates, personal communication.
†From Miller and Gates, 1967.
‡From Miller, unpublished data.
§From Grieve and Went, 1965.
‖From Van Bavel et al., 1965.

Note: Resistances are in min/cm.

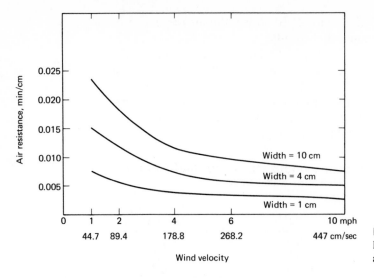

FIGURE 3–10. Air resistance of leaves relative to wind velocity and leaf size.

The effect of leaf resistance on transpiration can be approximated by assigning values to the environmental and plant variables in the energy budget equation and solving for transpiration and leaf temperature for each of several varying leaf resistances. Since evaporation is a cooling process, leaf temperatures will increase as the leaf and air resistance increases. The three curves in Fig. 3–11 represent the relationship of transpiration, leaf temperature, and leaf resistance at relative humidities of 75, 50, and 25% (and their corresponding absolute humidities) with an air temperature of 25°C and an absorbed radiation of 1.00 cal/cm²/min. We see that humidity of the air has little effect on the rate of evaporation and on the leaf temperature, especially with leaf resistances greater than about 0.2 min/cm. The response of leaf temperature to changes in resistance is greater when the leaf resistances are below about 0.2 min/cm.

Such calculations indicate that if transpirational cooling is to be important in reducing the temperatures of leaves, the resistances of these leaves should be below 0.2 min/cm.

An additional hypothesis is also suggested: if loss of heat is a problem for leaves there should be an inverse relation between leaf size and leaf resistance. Small leaves should be able to dissipate heat by convection with sufficient effectiveness that evaporative cooling need not be involved. Large leaves, on the other hand, should not be able to dissipate heat by convection at a sufficient rate not to require evaporative cooling. Thus, they should have low leaf resistances. This hypothesis is supported by an analysis of the relationship between leaf width and leaf resistance for a variety of species (Fig. 3–12). Large-leaved species have low leaf resistances; species with narrow leaves have high leaf resistances. There are narrow-leaved species with low leaf resistances, but no species have been found to have both a high leaf resistance and large leaves.

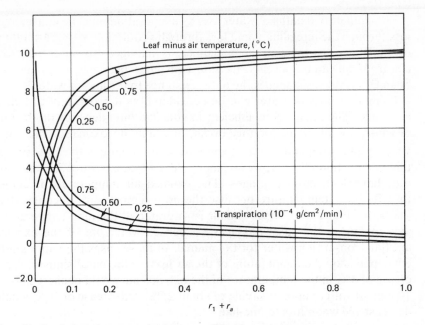

FIGURE 3–11. Relation of leaf and air resistance to the departure of leaf from air temperature and to transpiration at three relative humidities at an air temperature of 25°C. The absorbed radiation was set at 1.00 cal/cm²/min; wind speed at 2 mph; and leaf width at 4 cm.

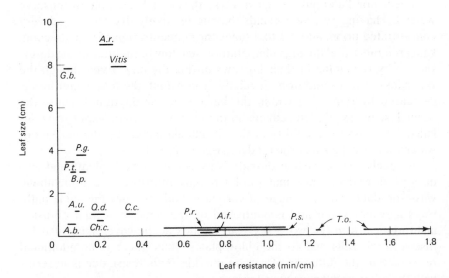

FIGURE 3–12. Range of minimum leaf resistances of various plant species plotted against leaf width to show the interrelationships between two plant properties. Initials indicate the species listed in Table 3–4. The line gives the leaf width and the range of measured resistances.

For animals two paths of water loss by evaporation must be considered: loss through the integument and loss through ventilation of the respiratory organ. Water loss through the integument is analogous to water loss from a leaf. If an animal sweats, its skin resistance becomes zero, and the boundary layer resistance is the only resistance to the loss of water.

Pulmonary loss of water can be considered as a process of turbulent exchange. The exchange coefficient in vols./cm²/min depends on the rate of ventilation of the respiratory organ. In a first approximation the water vapor gradient depends on the temperature of the lung or the air temperature and air humidity. However, many desert animals, such as kangaroo rats, have long nasal passages. The exhaled air temperature is cooler than its body temperature because the nose temperature is cooler than the body (Fig. 2–34). Since the air is saturated at the lung temperature, any cooling in the nasal passages condenses water out of the exhaled air, thus conserving water for the animal. In these cases the vapor gradient depends on the temperature of the air leaving the nasal sinuses. A decrease of 5°C in the temperature of the exhaled air would conserve about 10 g/m³ of water vapor, a saving of about 25%. A decrease of 10°C would decrease the water loss to one-half.

Because many animals are mobile they can choose their environment. Schmidt-Nielsen (1964) estimated the water budget of the kangaroo rat *Dipodomys merriami*. He found that 66% of the kangaroo rat's body weight is water and that this percentage remains constant. This water balance is maintained by several adaptations to the desert environment, both behavioral and physiological. First, the animals are able to conserve water by having no sweat glands, having relatively dry feces, and very concentrated urine, about 3 to 4 times the concentration of man. Second, kangaroo rats avoid the high temperatures and low humidities of the desert surface by remaining in their burrows during the day. Even though the water loss in feces and urine is relatively constant, the total water loss is decreased by remaining within the burrow during daytime because the water lost by exhalation remains in the burrow as water vapor to be inhaled again. As the humidity of the air inhaled increases, the net loss of water vapor by the kangaroo rat decreases (Fig. 3–13).

The exchange of carbon dioxide is essential for both plants and animals as an essential raw material for photosynthesis and as a metabolic waste product. The exchange of carbon dioxide follows the same pathways as water vapor and is controlled by the same molecular diffusion and turbulent transfer processes. Carbon dioxide, however, must also diffuse into the cells and to the chloroplasts; thus, there is an additional resistance to the diffusion of carbon dioxide. This resistance is in series with the other resistances to water loss.

All terrestrial organisms face the dilemma of inhibiting water loss while exchanging carbon dioxide. Waggoner (1967) discussed the problem of the evolution of plant and animal structures related to the exchange

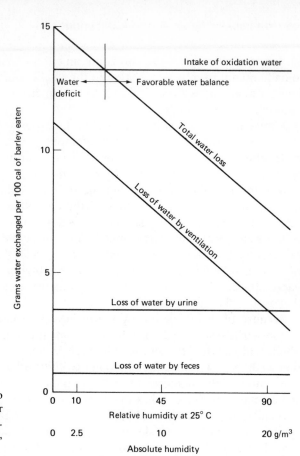

FIGURE 3–13. Water budget for the kangaroo rat, expressed in grams water exchanged per 100 cal of barley eaten, at different humidities. (From Schmidt–Nielsen and Schmidt–Nielsen, 1951.)

of water vapor and carbon dioxide. If water vapor and carbon dioxide are exchanged by molecular diffusion alone, water vapor will move 1.72 times faster than carbon dioxide, across the same concentration gradient, because of the difference in molecular diffusion coefficients of these two gases. However, in turbulent transfer water vapor and carbon dioxide are exchanged at equal rates, if the concentration gradients are equal, because the eddy diffusion coefficients are equal. Thus, it is advantageous, in terms of water conservation, for terrestrial organisms to exchange gases by mass transfer, such as through breathing or by wind. For animals this means avoiding the diffusion of these gases through the skin and maximizing their exchange through ventilation, and it is advantageous for plants to maximize mass transport and minimize boundary layer resistance when photosynthesis is taking place.

Solar Radiation

In this chapter a distinction will be maintained between solar radiation and light. Solar radiation is an energy flux from the sun which is almost wholly within the wavelengths of 0.3 to 3.0 microns. Light refers to the response of the human eye and is not an energy unit. The human eye responds nonuniformly to radiation in the visible portion of the spectrum, i.e., between the wavelengths 0.3 to 0.7 microns, with a maximum response at about 0.5 microns.

Above the earth's atmosphere a surface perpendicular to the sun's rays receives solar energy at a rate of about 2.00 cal/cm²/min. This flux, called the *solar constant,* varies only slightly during the year as the earth is alternately farther from or closer to the sun. As the solar radiation passes through the earth's atmosphere, it is depleted by water vapor, dust particles, and gas molecules of the air. When the sky is clear and relatively free of water vapor and dust, scattering is caused by molecules of the gases of the air. These molecules scatter the shorter wavelengths preferentially so that the sky appears blue. Water vapor scatters radiation of all wavelengths so that a sky with a high concentration of water vapor appears white. Dust scatters the long wavelengths preferentially so that dusty skies appear red.

Because of this depletion by scattering and reflection, solar radiation at the earth's surface is usually less than the solar constant. At solar noon, that is, when the sun is at its highest point in the sky, solar radiation might be 1.4 to 1.6 cal/cm² around latitude 40° in the summer.

Because of the scattering of solar radiation by dust and water vapor, some solar radiation reaches the earth's surface as diffuse light from the sky, called *skylight.* Although the intensity of skylight is greatest near the zenith (the point in the sky directly overhead), most studies have assumed that skylight comes from all parts of the sky equally. Skylight commonly amounts to about $\frac{1}{6}$ of the total solar radiation reaching the earth's surface, although it may vary from 5% in very clear skies to 100% in overcast skies.

Total solar radiation on the earth's surface can amount to more than the solar constant for brief periods when the direct solar beam is scarcely attenuated and diffuse radiation is reflected from clouds. This has been reported for high mountains such as in Colorado where the intensity of the direct beam is high and a site may temporarily receive reflected solar radiation. A common pattern is for solar radiation to increase to a high level as the cloud approaches, then drop abruptly as the cloud obscures the sun. Temperatures of leaves and surfaces may drop 10°C within a few minutes when this occurs. Invariably, humans respond by remarking how cold it has become, although air temperature has not changed. This sensory impression results from a drop in the temperature of the skin because of the reduction in solar radiation.

The reflectance of the earth's surface, expressed as the percentage of incoming solar radiation reflected back into space, is termed the *albedo.* Table 3–5 summarizes the albedo of various surfaces. The albedo of vegetated surfaces varies with the *solar altitude* (height of the sun above the horizon), increasing with low solar altitudes in the morning and eve-

TABLE 3-5. ALBEDO OF VARIOUS SURFACES FOR TOTAL
SOLAR RADIATION, WITH DIFFUSE REFLECTION.*

Surface	Albedo (%)
Fresh snow cover	75–95
Fresh snow	87
Old snow cover	40–70
Snow, several days old	70
Snow, old	46
Light sand dunes, surf	30–60
Bare ground	10–20
Sandy soil	15–40
Meadows and fields	12–30
Woods	5–20
Dark cultivated soil	7–10
Water surfaces, sea	3–10
Grass, high fresh	26
Grass, high dry	31–33
Grass, dry	15–25
Forest, green	3–10
Forest, snow-covered ground	10–25

*From List, 1963.

ning and decreasing with high solar altitudes at midday. For example a
3% variation in the albedo over conifers and 5% variation over an oak
woodland has been measured in Minnesota during the day (Bray et al.,
1966).

The reduction of solar radiation as it passes through the atmosphere
depends on the length of the path. The tilt of the earth causes this path
to be longer in winter than in summer, and thus the flux of solar radiation
at the surface is less intense in winter. As the sun rises in the sky in the
morning, the path length becomes shorter and the flux of solar radiation
increases. In the desert, where clear skies and low humidity occur in the
early morning, solar radiation on a surface perpendicular to the sun is
high soon after the sun rises.

Above the earth's atmosphere the intensity of solar radiation varies
symmetrically with season and latitude. In the tropics this seasonal variation is relatively small, varying between 790 and 896 cal/cm²/day. At
the poles the seasonal variation is large, varying from 0 to 1077 cal/cm²/
day (List, 1963). Solar radiation at the earth's surface is less intense in
the arctic tundra at Point Barrow, Alaska. Maximum solar radiation is
about 0.5 cal/cm²/min. The day length is 24 hours during the growing season, and daily totals of 300 to 500 cal/cm²/day are accumulated. In the
tropics solar radiation can be about 1.6 cal/cm²/min, but daily totals are
not as high as in polar regions during the summer because cloudiness reduces the intensity and days are only about 12 hours long. The greatest

daily amounts of solar radiation generally occur in the middle latitudes where skies are clearer than in the tropics and summer days are about 14 hours long.

At the earth's surface the annual symmetry of the intensity of solar radiation is altered because of seasonal patterns of dust and water vapor. In Colorado, for example, moist air from the Gulf of Mexico is drawn into the atmosphere in late summer, reducing the intensity of solar radiation; therefore, the most intense radiation is received early in the summer. Although air temperatures are higher in July and August, solar radiation is lower than in May and June (Fig. 3–14).

Spectral Distribution

The spectral distribution of solar radiation above the earth's atmosphere is shown in Fig. 3–15. About 50% of the sun's energy is in the visible spectrum and 50% is in the near infrared, between 0.7 and 3.0 microns. Skylight differs from the direct solar beam in having a greater proportion of energy in the infrared wavelengths. Light from an incandescent lamp has much of its energy in the infrared but light from a fluorescent lamp is almost totally in the visible spectrum.

Organisms have specific spectral distributions of absorptances to visible and near infrared energy (see below), and when individuals with different absorptances are placed in energy environments with the same total intensity, they will absorb quite different amounts of energy. Some plants, for example, may not grow under incandescent lights, even though the total radiation intensity is equal to the sun's, because the fraction in the visible spectrum may be so small that the stomates will not open.

The reflectance of a water surface varies with the altitude of the sun. When the sun is directly overhead, only 5% of the incident solar radiation is reflected, but when the sun is 5° above the horizon, 95% of the ra-

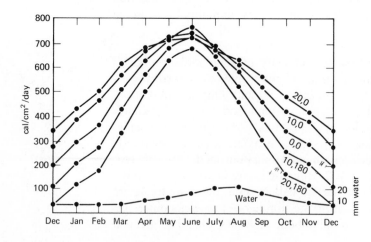

FIGURE 3–14. Seasonal course of potential solar radiation on surfaces of different orientation, together with the pattern of seasonal abundance of precipitable water vapor in the atmosphere for Colorado. The slope and azimuth of the surfaces are given on the graph. A south slope is 0 degrees and a north slope, 180 degrees.

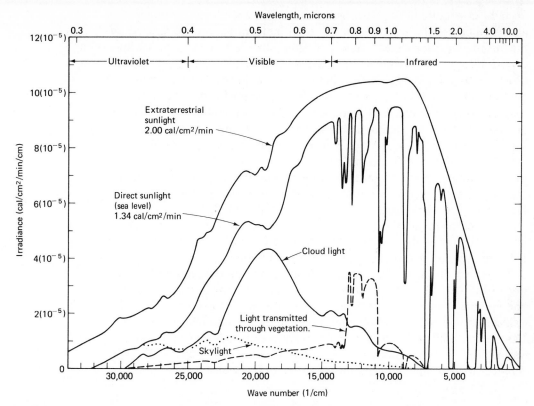

FIGURE 3–15. Spectral distribution of extraterrestrial solar radiation, of solar radiation at sea level for a clear day, of sunlight from a complete overcast, and of sunlight penetrating a stand of vegetation. Each curve represents the energy incident on a horizontal surface. (From Gates, 1965a.)

diation is reflected (List, 1963). This reflection decreases the intensity of solar radiation within the water during all parts of the day, but especially in the morning and evening. "Twilight" within the water thus lasts longer than above the water.

Solar radiation decreases with depth in the water and its spectral composition changes. If the decrease in intensity were due solely to reflection and absorption by water molecules, the intensity would decrease exponentially. Water attenuates the red wavelengths more than the blue so that there is a shift in the spectral composition of the light within the water. In clear water at 3.6 m 95% of the red light and less than 23% of the blue light is removed. Plankton tend to be distributed nonuniformly with depth so that departures of measured solar radiation profiles from those expected in pure water commonly occur because of their presence.

An organism is coupled to solar radiation by its absorptance and orientation to the sun. All the energy incident on an organism is either reflected, transmitted, or absorbed. The reflectance, transmittance, or absorptance are the fractions of the total incident energy exchanged by the respective processes. Table 3–6 gives the reflectances of various organisms to different wave bands. The spectral distribution of reflectance of leaves is given in Fig. 2–16. These organism properties vary with the environment and have a genetic basis. Billings and Morris (1951) showed that the reflectance of visible and near infrared wavelengths for subalpine and desert plants was about 10% higher than that for montane plants or for plants on the University of Nevada campus. This increase in reflectance is correlated with the high intensities of solar radiation occurring in desert and subalpine habitats. Birkebak and Birkebak (1964) showed that leaves may have different absorptances on their upper and lower surfaces. Dedykin and Bedenko (1961) showed that the reflectance and transmittance of leaves of plants grown on moist and on dry soil differ. Their data indicate that absorptance was from 10 to 20% higher for leaves of the same species grown on moist soil than for those grown on dry soil.

The effect of these differences in absorptance on the temperatures of organisms can be demonstrated by a simulation experiment with the energy budget equation and using leaves as a specific example. Realistic values for relevant environmental parameters are assumed in the equation, which is then solved for leaf temperature while the absorptance is varied (Fig. 3–16). Under the conditions illustrated in Fig. 3–16, a difference in absorptance of 0.2, in the range of 0.4 and 0.6, causes a tem-

TABLE 3-6. PERCENT REFLECTANCE OF VARIOUS ORGANISMS TO VARIOUS WAVELENGTHS.*

Substance	Wave band (microns)		
	0.3–0.7	0.7–3.0	8–12
Populus deltoides (leaf)	7	35	
Fungus	10–35	60 (35–80)	10
Liverwort	7	45	4
Pika fur (back)	10 (5–15)	40 (20–60)	5
Cat fur	40 (20–60)	65 (60–50)	5
Human skin	25 (5–45)	30 (3–50)	
Stellar Jay feathers (back)	3	40 (5–50)	

Substance	Wave band (microns)	
	0.4–0.7	0.7–1.1
Lizard		
ventral	42	30
dorsal	12	9

*From Gates et al., 1966.

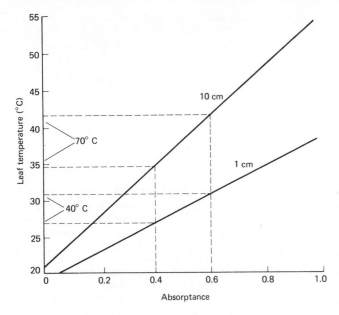

FIGURE 3–16. Effect of leaf absorptance on leaf temperatures with two widths of leaves. Environmental conditions are: $S_{dn} = 1.20$ cal/cm²/min; $S_r = 0.30$ cal/cm²/min; $IR_{\text{down}} = 0.60$ cal/cm²/min; $IR_{\text{up}} = 0.65$ cal/cm²/min; $V = 44.7$ cm/sec; $LE = 0.0$; $T_a = 20°C$.

perature difference of 7°C for a leaf 10 cm wide and a difference of 4°C in a leaf 1 cm wide. This reduction in temperature could be significant in preventing leaves from reaching lethal temperatures, in reducing transpiration rates, and in reducing respiration rates, thereby increasing net photosynthesis.

Orientation of Organism to Direct Solar Radiation

An organism's orientation to the direct solar radiation beam affects its temperature. Leaves are variously inclined toward the sun; butterflies and lizards warm themselves by variously orienting their bodies toward the sun. There is a tendency for steeply inclined leaves to occur in habitats with high solar intensities and more horizontally inclined leaves to occur under low solar intensities. This is observed not only in plants in different environments, but even for leaves in the canopy of a single tree.

The intensity of solar radiation incident on the organism will be least when the organism is oriented parallel to the sun's rays and greatest when the organism is oriented perpendicular to the sun's rays.

The influence of leaf orientation on leaf temperatures can be studied by simulating the effect of various leaf orientations, using the energy budget equation. This simulation is shown in Fig. 3–17. In this example the solar flux is divided into a direct beam of 1.0 cal/cm²/min with a zenith angle of 30° and a diffuse beam of 0.2 cal/cm²/min. The leaf absorptance was assumed to be 0.5. The horizontal axis is the cosine of the angle of

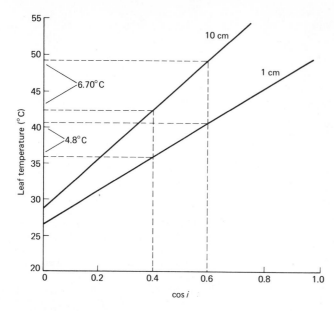

FIGURE 3–17. Effect of leaf orientation of leaf temperature with leaves of two widths. Environmental conditions are: $zn = 30°$; $a = 0.50$; $S_{dir} = 1.00$ cal/cm²/min; $S_{diff} = 0.20$ cal/cm²/min; $S_r = 0.30$ cal/cm²/min; $R_d = 0.60$ cal/cm²/min; $R_u = 0.65$ cal/cm²/min; $V = 44.7$ cm/sec; $LE = 0.0$ cal/cm²/min; $T_a = 20°C$.

incidence of the sun's rays. A cos i of 0.0 means that the leaf is parallel to the sun's rays; a cos i of 1.0 means that the leaf is perpendicular to the sun's rays. In nature, cos i varies between 0 and 1 and this variation may amount to a difference in leaf temperature of 25°C for a leaf 10 cm wide. The reduction of cos i could be significant in reducing the maximum leaf temperatures below lethal limits, reducing transpiration rates, or reducing respiration and hence increasing net photosynthesis.

The temperature of each leaf in a canopy varies according to its orientation to the sun, independently of the other leaves. Figure 3–7 showed the course of leaf temperatures of aspen, *Populus tremuloides,* and oak, *Quercus gambelli,* in northwestern Colorado during two-day periods. The pattern of incoming solar radiation is given in the solid line at the top of the figure. In the aspen measurements the first day was cloudless, but the second included some scattered cumulus clouds. In the oak measurements on the second day the sky was clear in the morning but was continuously overcast in the afternoon, and, therefore, the daily solar radiation curve is truncated. The second set of curves shows air temperature; the fourth curve shows the differences between leaf and air temperatures; and at the bottom, the calculated daily course of incident solar radiation is shown. Only the temperatures of leaves at the top of the canopy are given, and thus the effects of shading are minimal. Temperatures of leaves that are alternately sunlit and shaded vary widely during the day and have no regular pattern.

In general, the daily course of leaf temperatures follows the daily course of air temperatures. Characteristically, leaf temperatures are higher than air temperatures during the day and lower during the night because of the large amount of radiation absorbed during the day in contrast to the night (Fig. 3–6). Individual leaves deviate from air temperature differently during the day, however, depending on their orientation to the sun. For example, the difference between leaf temperature at the top of the canopy and air temperature was shown to be closely correlated with the incident solar radiation for both aspen and oak, indicating that the independence of leaf temperature was due to the differences of their orientation to the sun.

The same principles we have described for plants can also be applied to animals, especially to poikilotherms. For example, Bogert (1949) found that lizards in Florida and Arizona moved between sunny and shady places according to the temperature of their bodies and by this method were usually able to keep the temperature of their bodies within a range of about 3°C. Under particular weather conditions the lizards would remain abroad all day, but on a clear summer's day, with the sun near its zenith, nearly all the lizards would be found in shady places. Bartlett and Gates (1967) developed a detailed energy budget equation for lizards and calculated the temperature of a hypothetical lizard in various environmental situations. By assuming that the temperature range that the lizard could tolerate was limited and that the lizard could move from place to place in order to maintain a tolerable temperature range, they predicted the movements of the lizard and the positions it would take during the day. As Fig. 3–18 shows, their prediction agreed well with observations of lizards in the field.

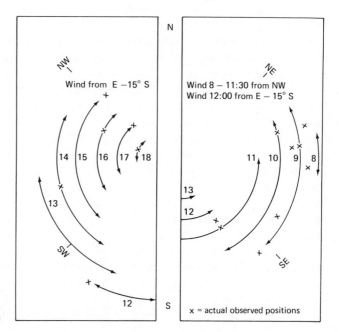

FIGURE 3–18. Predicted and observed hourly positions of a lizard, *Sceloporus occidentalis*, on a tree trunk for June 2, 1964. (From Bartlett and Gates, 1967.)

The surface of an animal or a pine needle, which can be approximated by a cylinder, is continuously inclined with respect to the sun and thus presents a more complex situation than does the flat-plate analogy we have used for leaves. If one wishes to calculate the radiation received by a whole animal, the animal can be divided into units approximating cylinders or spheres in shape and the intensity flux on each unit can be calculated. By doing so, one should begin to appreciate the role of extremities and circulation in the heat balance of animals. A detailed analysis of a whole animal, combining physiological and physical measurements has not yet been done.

Infrared Radiation

Almost all of the energy lost by an organism and, often, almost all of the energy absorbed by an organism are infrared radiation. A complete understanding of interactions between organisms and their environment requires an understanding of infrared radiation exchange. In the past, infrared radiation has been virtually ignored in ecological studies. This avoidance has probably been possible because the net radiation balance of organisms is usually small, about 0.0 to -0.1 cal/cm²min, and the infrared radiation load from the environment is relatively constant, varying within about 0.2 cal/cm²/min at any one place. In looking for correlations between an organism response and an environmental change, changes in the infrared environment, or net exchange, are marked by the larger changes in solar radiation, temperature, humidity, or wind. To understand the causal relationships involved, however, the infrared environment must be considered.

A distinction must be maintained between the radiation emitted by an organism and that reflected by the organism. Some of the early literature has confused these processes. The energy emitted by an organism occurs within the wavelengths 9 to 11 microns; that reflected is solar radiation, largely within the wavelengths from 0.3 to 3.0 microns. Common infrared cameras are responding to reflected infrared energy in the wavelengths 0.7 to 3.0 microns. The energy in these wavelengths originates with the sun and is largely reflected by organisms. Thus, the intensity recorded depends on the intensity of the incoming energy and the reflectance in these wavelengths. Because infrared energy in the wavelengths from 9 to 11 microns is absorbed by most substances, including glass, cameras needed to photograph objects in the far infrared require special lenses that are quite expensive. The color of organisms is caused by the spectral distributions of the incoming energy and of the organism's reflectance in the wavelengths from 0.3 to 0.7 microns. On the other hand, the energy emitted by an organism depends on the temperature of the organism's surface.

A fundamental physical principle is that a good absorber is a good emitter at the same wavelength, i.e., absorptance at a certain wavelength equals emittance at the same wavelength. However, the absorptance in different parts of the spectrum is essentially independent of the absorptance in other parts. The absorptance in the band from 0.3 to 0.7 microns is caused by the pigmentation of the organism. The absorptance in the band from 0.7 to 3.0 occurs in the cuticle and epidermis and is independent of pigmentation. The absorptance in the far infrared depends on properties of the surface (Gates, 1965b). The logical sequence of statements such as: "Polar bears are white because of their high reflectance in the visible" (true); "High reflectance in the visible means a low emittance in the far infrared" (absolutely false); "A low emittance in the far infrared means higher surface temperature" (true); "Low emittance in the far infrared would be advantageous to arctic animals whose problem is keeping warm" (probably true); and "Therefore, polar bears are white," occur in the literature and lead to false conclusions concerning the cause of white fur on arctic animals. Measurements on arctic animals have indicated emittances in the far infrared that are similar to those of other organisms (Hammel, 1956).

The spectral distribution of energy emitted also depends on the temperatures of the radiating surfaces. As a surface heats, the peak in the spectral distribution curve shifts toward the shorter wavelength. Thus, as an object heats it first radiates in the infrared and feels warm; but the object can be seen by the human eye only because of visible light reflected from it. As the surface gets hotter, the peak radiation shifts into the near infrared and visible. When an object is hot enough, it can be seen by the human eye because of visible energy emitted by the object. The sun emits radiation as a surface at 6,000°K so that almost all of the energy is in the visible and near infrared wavelengths. The earth's surface, plants, and animals, at about 293°K or 20°C, emit energy wholly in the far infrared, peaking at 9 to 11 microns, and thus are visible to the human eye only because of reflected light.

The infrared energy radiated by an object is related only to the surface temperature of the object. The relation between emitted radiation IR and the surface temperature T in degrees Kelvin is expressed in the Stefan-Boltzmann equation, which is

$$IR = \epsilon \, \sigma T^4 \qquad (3-16)$$

The Stefan-Boltzmann constant σ equals about 8.13×10^{-11}. The symbol ϵ designates the emittance of the surface. The value of ϵ is equal to the ratio of the infrared actually emitted by the object to that emitted by an ideal, perfect radiator at the same temperature. The emittance of almost all objects of ecological importance, including plant leaves, animals, and vegetated surfaces is between 0.95 and 0.98, but sand and rock have emittances as low as 0.60 (Gates, 1965b).

Water vapor, carbon dioxide, and ozone are the major absorbers and emitters of infrared radiation in the atmosphere. The infrared radiation from the sky depends mostly on the temperature and the water-vapor concentration of the atmosphere. Sky radiation is sometimes summarized as the sky radiant temperature, which is the temperature an ideal radiating body must have to radiate infrared at the same intensity as the sky. This temperature is often from $-40°$ to $+20°C$, corresponding to a radiation rate of from 0.2 cal/cm²/min in the arctic to 0.6 cal/cm²/min in the subtropics.

Various mathematical approximations using measurements of air temperatures and humidity at the earth's surface have been proposed in order to calculate infrared radiation from the sky, e.g., the Brunt formula (Gates, 1962):

$$IR_{sky} = (0.4 + 0.06 \sqrt{e_w}) \, \sigma T_a^4 \qquad (3-17)$$

where T_a is air temperature in degrees Celsius and e_w is humidity in millibars. This is an empirical relation using measurements at the earth's surface in a standard weather screen. The fact that the equation provides relatively good predictions, even though it does not include atmospheric carbon dioxide, indicates the relatively small effect of this gas on sky radiation.

The emittance of organisms varies little, from 0.95 to 0.98, and variations of this magnitude would cause only small variations in organisms' temperatures. However, when an organism is in a laboratory chamber with glass or metal walls, which have much lower emittances, its body temperature may be lower than the air temperature in the chamber because the organism emits radiation more effectively than does a glass wall at the same temperature. If the net radiation exchange between the organism and the wall is zero, and the other processes of energy exchange are negligible, the temperature of the organism will be lower than the temperature of the wall. Experiments conducted under these conditions have been reported in which the lower temperatures of organisms were interpreted, erroneously, as resulting from evaporation. Simple heat budget calculations using gravimetric water loss and radiation exchange would have shown that evaporation could not have been responsible for the observed temperature depression of the organism.

INTERRELATIONS BETWEEN THE PHYSICAL ENVIRONMENT, PHYSIOLOGY, AND BEHAVIOR

A fundamental concept of ecology is that the organism and its environment are inseparable. As we have seen, the processes of energy exchange are fundamental to the interactions between an organism and its environment. In the previous discussion the environment was defined so that measurements could be made for the calculation of flux or exchanges of energy and materials between an organism and its environment. In the following we shall discuss, in a rather descriptive way, microclimates of different habitats of ecological importance, but the concept of an organism, existing within the microclimate and exchanging energy and materials with it, must be constantly kept in mind.

Wind Profiles

Wind speed is reduced near the ground because of the drag on the flow of air produced by the ground. When a solid object slides over another solid object, the movement is resisted by frictional forces along the surface of contact between the objects; but when a fluid, such as air or water, moves over a solid surface the frictional forces resisting the movement are distributed through the fluid, producing a velocity gradient above the surface. The tangential force exerted at the top of the fluid causing the fluid to move against friction is called the *shearing stress*. This force is defined in laminar flow by the equation

$$\tau = \mu \frac{du}{dz} \tag{3-18}$$

Where τ is the shearing stress in dynes/cm^2, μ is the dynamic viscosity of the fluid, and du/dz is the velocity gradient. In nature, however, air flow is usually turbulent, not laminar, and under these conditions the shearing stress is defined by the equation

$$\tau = -\rho_a K_m \frac{du}{dz} \tag{3-19}$$

where K_m is the eddy diffusivity of momentum, or a transfer coefficient relating the transfer of momentum from the air to the surface, and ρ_a is the air density.

Under certain conditions discussed later the profile of wind velocity assumes a logarithmic form so that the wind speed u_i at any height z_i is given by

$$u_i = \frac{1}{k} \left(\frac{\tau}{\rho_a}\right)^{\frac{1}{2}} \ln\left(\frac{z_i}{z_0}\right) \tag{3-20}$$

where k is a constant equal to about 0.4 and z_0 is a constant of integration. A graph of wind speed over a flat surface plotted against height would show zero wind speed occurring at a height slightly above the surface. If the wind speed is plotted against the logarithm of height, the plot would form a straight line with a slope of

$$\frac{1}{k} \left(\frac{\tau}{\rho_a}\right)^{\frac{1}{2}}$$

and an intercept of

$$\frac{1}{k} \left(\frac{\tau}{\rho_a}\right)^{\frac{1}{2}} ln \left(z_o\right)$$

that is, zero wind speed occurs at the height z_o above the actual surface. This height is related to the roughness of the surface and is called the *roughness length* or *roughness parameter*. With tall vegetation another parameter (the zero plane displacement d) must be introduced, and the equation becomes

$$\mu_i = \frac{1}{k} \left(\frac{\tau}{\rho_a}\right)^{\frac{1}{2}} ln \left(\frac{z_i - d}{z_0}\right) \tag{3-21}$$

Representative values for z_o and d are given in Table 3–7.

These logarithmic equations are fundamental to many calculations and formulations of the transfer and redistribution of heat, water vapor, carbon dioxide, smog, and radioactive isotopes. These equations, however, are applicable, only under conditions of neutral stability of the air,

TABLE 3-7. REPRESENTATIVE VALUES OF z_0, d, AND τ_x FOR NATURAL SURFACES WITH NEUTRAL STABILITY AT DIFFERENT WIND SPEEDS.*

Surface	z_0(cm)	d (cm)	$\tau_x \left(\dfrac{dynes}{cm^2}\right)$
Very smooth (mud flats, ice)	0.001		0.32
Smooth mow on short grass	0.005		
Desert	0.03		
Snow surface, natural prairie	0.10		
Lawn, grass up to 1 cm high	0.1		0.84
Short grass, 1 to 3 cm	0.5	0	
Downland, thin grass up to 10 cm	0.7		1.6
Thick grass, up to 10 cm	2.3		2.5
Long grass, 60 to 70 cm	3.0	30	
Thin grass, up to 50 cm	5.0		3.8
Thick grass, up to 50 cm	9.0		4.9
Sea	0.03	4	
Sea, wind 600 to 700 cm/sec	0.6		
Mown grass 1.5 cm	0.2		
3.0 cm	0.7		
4.5 cm, wind at 2 in = 2 m/sec	2.4		
4.5 cm, wind at 2 in = 6–8 m/sec	1.7		
Long grass, wind = 1.5 m/sec	9.0		
3.5 m/sec	6.1		
6.2 m/sec	3.7		

*From Sutton, 1953, and Priestley, 1959.

that is, when there is almost no temperature gradient. At night, when the air near the ground is denser than the air above and the air is stable, and during the day when strong solar heating occurs and the air near the ground is less dense than above making the air unstable, the equations do not apply. The wind profile equations also assume level, uniform surfaces, which implies that all transport is vertical. Although agricultural crops have usually been found to fulfill these conditions, natural communities are more heterogeneous and often occur on rolling terrain, making the formalization of an analytical theory for these situations immensely more complex. Within natural vegetation canopies species differences in absorptance, leaf inclination, leaf area, and leaf resistance cause differences in the penetration of wind, the absorption of net radiation, and the conversion of the absorbed energy to heat or evaporation, which in turn causes horizontal heterogeneities in the air temperature and water vapor; and, consequently, horizontal transfers of energy and materials.

Although there are major difficulties in studying wind profiles in these complex real canopies, progress is being made through the extension and modification of theories for simpler situations. This development of our understanding will undoubtedly be important in extending our knowledge of why organisms are distributed as they are in nature. Many terrestrial invertebrates, for example, are sensitive to fluctuations in temperature and humidity, and they respond behaviorally to the profile of wind, air temperature, and humidity. The sheep tick in Australia is a good example. This parasite can reproduce only after a blood meal on sheep. To do this it must crawl up to the tip of a blade of grass to wait for a passing sheep on which it can feed. While it is on the tip of the blade, its evaporation rate under many environmental conditions is high, and it, therefore, can remain there for only a short time before it has to crawl down the blade into the vegetation in order to regain its moisture content. This it does periodically, but while within the vegetation its chance of catching a passing sheep is much reduced. Thus, the tick must migrate through profiles of wind, temperature, humidity, and radiation. The changes in temperature and evaporation rate of the tick can be calculated using the energy budget equation for a sphere. Values for the environmental variable can be calculated from various physical models of the environment, such as Eq. (3-21). One calculation using realistic values is illustrated in Fig. 3-19.

Air Temperature Profiles

The atmosphere does not absorb solar radiation effectively; rather, it is heated by absorbing infrared energy radiated from the earth and by convection of heat from the earth's surface to the air. Soil and vegetated surfaces, on the other hand, absorb solar and infrared radiation. Temperature fluctuations of the surface are usually greater than temperature

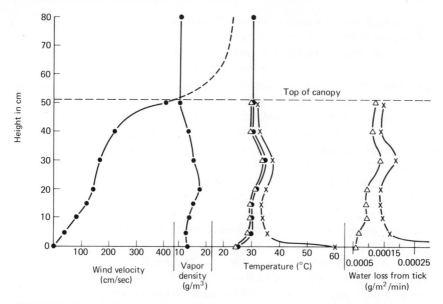

FIGURE 3–19. Calculated organism temperature and evaporation rate at different levels above the ground when wind, air temperature, and humidity vary and the organism can be in sun or shade. Δ = shaded tick, X = sun exposed tick, and ● = environmental values.

fluctuations of the air near the surface. Under conditions of strong incoming solar radiation the temperature of a bare ground surface may be 40°C greater than the temperature of the air two meters above it. For example, in Australia the temperature of the ground surface between acacia trees was found to reach 70°C (Turner, 1965). At night the situation is reversed and the ground surface is colder than the air above. The temperature measured in a standard face is not the same as that experienced by organisms at the ground surface or in a forest canopy. Within the first few centimeters above the ground surface the gradient in air temperature is very steep, perhaps up to 20°C. Many ground-dwelling insects in deserts have long legs which keep their bodies out of the extremely hot air next to the ground or they are able to bury themselves in the ground during the hottest parts of the day in order to escape the high ground surface temperatures.

When the ground is covered with vegetation, the temperature fluctuation at the surface becomes less extreme because the surfaces absorbing the incoming solar radiation are spread out through the canopy rather than all the energy being absorbed in the plane of the soil surface. The level of greatest temperature fluctuation in a vegetation canopy does not occur near its top, but within it. This layer of maximum temperature fluctuation is called an *active layer*. If the canopy is dense enough to impede the subsidence of colder, denser air at night, the hottest and coldest temperatures occur near the top, but still within the canopy. If the canopy is not very dense, the hottest temperatures may occur near the top; but the coldest temperatures will occur at the ground surface because the colder, dense air can settle down through the canopy.

The development of the migratory phase of migratory locusts illustrates the importance of studying an organism's microenvironment. These insects are well known for their destructive outbreaks and swarming behavior; and because of their economic importance, extensive research on them has been conducted. It is now generally accepted that a large population in an outbreak area is not sufficient for the gregarious phase and migratory behavior to develop. The change from individualistic to gregarious behavior of the locust *Chortoicetes terminifera,* for example, is a result of increasing responsiveness of individual locusts to the presence and movement of others (Clark, 1947, 1949). The crowding essential for this species to develop gregarious behavior and the capacity for mass migration comes about through the response of individuals to particular microclimatic conditions in a dense population. In an outbreak center in New South Wales, Australia, estimated densities of the first-instar nymphs of *C. terminifera* were up to 2,000 per square yard. These nymphs were distributed in relation to suitable food, and they were not gregarious, although their parents had been. Most nymphs spent the night in the vegetation or in other places where the temperatures were about 2°C at daybreak. At these low temperatures the nymphs remained motionless, but as the temperature increased they became more active, until, at 20°C, their movements were normal. In the laboratory the nymphs preferred a temperature of 42°C; in the field between 20°C and 45°C they sought out the warmest possible situations at ground level, orienting their bodies broadside to the sun and crowding closely together. On cool days these basking groups persisted until the late afternoon when solar radiation decreased and temperatures fell. On warmer days, as the temperatures increased to 45°C, the groups dispersed and individual nymphs sought shadier places; ordinarily the groups reformed later in the afternoon as temperatures dropped. Outbursts of jumping occurred in the basking groups at irregular intervals, and by the time the adult stage was attained, the gregarious form had developed. Thus, it appears that as the nymphs sought out locations of their preferred temperature they became sufficiently crowded to develop into the migratory phase.

Humidity Profile

Humidity gradients develop over evaporating surfaces so that the highest humidities occur near the surface. The water vapor must diffuse away from the surface in a direction of decreasing vapor concentration. Within a vegetation canopy the level of greatest humidity will vary depending on the relative rates of evaporation from the ground surface and rates of transpiration at different levels in the canopy. Commonly, the level of maximum vapor density will be lower than the level of maximum air temperature.

With increased humidity toward the ground, evaporation is reduced, and organism temperatures may increase. Calculations using the energy budget equation indicate that, if the surface is warm, surface-dwelling organisms will be exposed to both thermal and water stress. By moving into shaded areas within the canopy, both of these stresses are reduced and it may be possible for the organism to take up moisture.

Many species of terrestrial invertebrates show a humidity preferenda and these preferenda may differ among closely related species. Figure 3–20 shows the preferenda of six species of harvestmen (daddy-longlegs) as measured in the laboratory and related to the relative humidity of the microhabitats in which they were found in nature. All but one species occurred in locations having relative humidities near their preferenda; the one exceptional species was found to occupy a location having a higher relative humidity than its preferendum (Macfadyen, 1963).

In a study of the environmental effects on the behavior and distribution of mosquitoes, Haufe (1954) trapped mosquitoes over a prairie and in a forest. The forest had a luxuriant ground cover of grass and herbs 3–4 feet in height and a dense shrub layer that reached a height of 20 to 25 feet. The top of the tree canopy was at a level of about 50 feet. In both the forest and prairie, the traps were placed at heights of 25 and 50 feet.

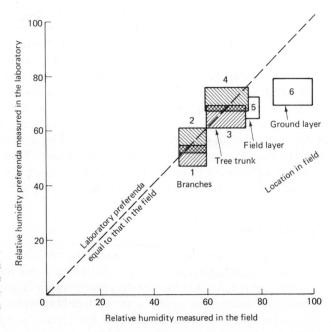

FIGURE 3–20. Relative humidity preferenda of six species of harvestmen measured in the laboratory and relative humidity measured in the field in microhabitats in which they occur. (From Macfadyen, 1963.)

The total number of females of all species caught over the prairie decreased with height, from 78% at 5 feet, 13% at 25 feet, to 9% at the top. In the forest 29% of the total catch of female mosquitoes was taken at 5 feet. In both locations catches were taken only in the period from late afternoon to midmorning. On the prairie large catches usually occurred on warm evenings with light winds. Similarly, in the forest large catches occurred on warm evenings, but the effect of wind was reduced.

Haufe (1952) found that variation in the activity of arctic mosquitoes was primarily dependent on temperature and humidity. In the case of two species, nearly saturated air induced a resting state, and winds above 18 mph and fog also affected activity, but light did not. On open tundra, when temperatures exceeded a critical level, mosquitoes ceased to fly and descended to the sparse moss and lichen layer where they actively crawled. At temperatures below a critical lower level flying ceased again, but they continued to beat their wings. Thus, sustained activity occurred only within a narrow range of temperature and humidity. In controlled laboratory experiments the amount of activity increased with increases in temperature and vapor pressure up to a maximum vapor pressure; and in the field this pattern was still evident, although other variables, such as degree of starvation and age distribution, were also influential. The relation of mosquito activity to temperature and humidity is summarized in Fig. 3–21 for the subtropic species *Aedes aegypti* (Haufe, 1966).

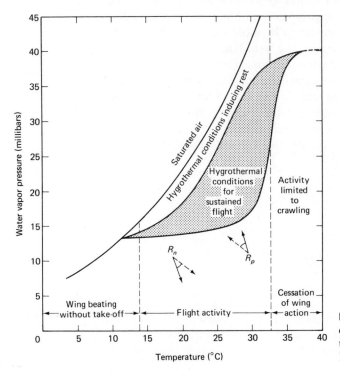

FIGURE 3–21. Reactions of the mosquito *Aedes aegypti* (*L.*) in relation to temperature and vapor pressure. (From Haufe, 1966.)

Processes Affecting Profiles

Vertical air temperature and humidity gradients in the air develop through processes of vertical exchange of energy and water vapor. These processes are related to the wind speed, the surface properties of roughness and height, and the instability of the air. Above a surface there must be a vertical flux of convectional and evaporational heat, and this flux is maintained by gradients of temperature and humidity that develop. The gradients depend on the amount of heat and water vapor to be transferred and the rate of exchange of volumes of air. The flux of heat C and water vapor E can be written

$$C = K_h c_p \rho_a \frac{dt}{dz} \tag{3-22}$$

$$E = K_w \frac{d\rho_w}{dz} \tag{3-23}$$

where the symbols are defined as before. The measurement of the exchange coefficients K_h and K_w for particular situations is difficult. This coefficient varies from 10^2 for red clover, to 10^3 for corn (Lemon, 1967), and 10^4 for pine (Denmead, 1964). Usually the fluxes of heat and water vapor are calculated from more easily measured quantities by combining the basic flux equations with either the energy budget equation or some variant of the wind profile equation. Both methods assume uniform homogeneous surfaces, so that the net exchange of quantities in horizontal directions is zero and all the exchange is vertical. The assumptions underlying the heat budget approach appear to be more easily satisfied in the field than the assumptions underlying the aerodynamic approach. The estimation of C and E depends on the measurement of K. In the aerodynamic method K is estimated from the equation for shearing stress and for wind profile. The wind profile equation is solved for shearing stress and this is substituted into the equation for shearing stress. The latter equation is then solved for K. These equations have been reorganized to express the flux of moisture as a gradient of humidity divided by an air resistance r_a (Monteith, 1963). If

$$E = \frac{\rho_{w_2} - \rho_{w_1}}{r_a} \tag{3-24}$$

by the aerodynamic equation, then

$$r_a = \frac{\left[\frac{ln(z_2 - d)}{(z_1 - d)} \right]^2}{k^2(u_2 - u_1)} \tag{3-25}$$

These equations show that the higher the wind speed and zero plane displacement, the lower the air resistance to evaporation over a surface. The formulation of air resistance permits a measurement of the evaporation power of different sites, although the air resistance is valid only when the logarithmic wind profile is valid.

To reach the free air above the canopy, water evaporated from the mesophyll surfaces of a leaf must diffuse against leaf resistance, leaf boundary layer resistance, and canopy air resistance. Minimum leaf resistance varies between 0.01 and 0.05 min/cm. Leaf boundary resistance varies with wind speeds, but is generally low (Fig. 3–10). For canopy resistance, a K of about 2,000 cm²/sec converts to a canopy resistance of about 0.0026 min/cm. The leaf boundary layer resistance is often negligible compared to the leaf resistances and can often be ignored. However, although air resistance in the canopy is smaller than the leaf or boundary layer resistances, air resistance in the canopy cannot be ignored because the quantities of sensible heat and water vapor transported through the canopy are usually many times greater than from a leaf even when temperature and humidity gradients are much smaller than those influencing water loss from leaves.

Solar Radiation within Vegetation Canopies

The solar radiation regime within canopies of terrestrial vegetation is complex because of the pattern of light penetration between leaves and reflection by the leaves. This produces discontinuous variations in solar radiation intensity between sunlit and shaded areas. Solar energy is depleted in the canopy by absorption and reflection by leaves and stems. Thus expressions for the diminution of light through the canopy usually require a measure of the effective leaf area of the canopy.

The leaf area in the canopy is measured by the *leaf area index, F,* which is the surface area of leaves per unit surface area of ground. A canopy with 1.0 cm² of horizontal leaf covering 1.0 cm² ground surface would have the leaf area index of 1.0. If the 1.0 cm² of leaf were tilted 45° to the horizontal, the leaf area index would still be 1.0, although an observer above the canopy could see about one-third of the ground below the canopy.

Monsi and Saeki (1953) measured and theoretically analyzed the diminution of solar radiation with depth in the canopies for various plant communities. They found that the decrease of light intensity with depth fit the equation

$$S_F = S_o e^{-KF} \tag{3–26}$$

where S_F is the solar radiation intensity, S_o is the intensity of the incoming solar radiation, and F is the leaf area index above the level at which S_F is being calculated. The extinction coefficient K is an empirically determined parameter with characteristic values for different vegetation types. This K is not related to the eddy coefficient (also symbolized K) used in the preceding section. With light intensity measured in foot candles, K ranges from 0.3 to 2.0 (1/F) under cloudy conditions. In grassland communities K is usually between 0.3 and 0.5 (1/F) and in broad-leaved communities, between 0.7 and 1.0. Values for the extinction coefficient in cal/cm²/min for a corn canopy have been estimated as 0.01, 0.006, and 0.008 for visible, near infrared, and far infrared radiation, respectively (Lemon, 1963). The canopy of a dense corn crop and a forest canopy show low transmission of visible energy and high transmission in the near infrared — a pattern similar to the spectral distribution of individual leaves (Lemon, 1963; Federer and Tanner, 1966; Gates, 1965b).

The extinction of light in a canopy depends on the altitude of the sun as well as on the leaf area index. If the leaves are steeply inclined, the penetration of light from directly overhead will be great, but the penetration of light from low in the sky will be small. One needs only to recall the lengthening of shadows as the sun sets in order to visualize the process of decreasing solar radiation in the canopy as the sun becomes lower in the sky. The extinction coefficient is the shade cast onto a horizontal surface by a leaf area index of one. If the leaves are horizontal, the shade cast by one leaf area index will be equal to the leaf area, regardless of the position of the sun, and the extinction coefficient will equal one. If the leaves are not horizontal, the shade cast onto a horizontal surface will vary with the position of the sun. A leaf area index of 1.0 with tilted leaves will cast a shadow greater than itself when the sun is low and less than itself when the sun is overhead. Thus, the extinction coefficient will be greater than 1.0 in the morning, and less than 1.0 when the sun is overhead (Fig. 3–22). When the sun's altitude β is greater than the leaf slope α, the extinction coefficient K is given by:

$$K = \cos \alpha \qquad (3\text{--}27)$$

Expressions for the extinction coefficient when the sun's altitude is less than the leaf slope vary, but they can be expressed by using trigonometric functions. The extinction of the direct beam varies with the altitude of the sun, the angle of the leaves, the density of the leaves, and the arrangement of the leaves. When the altitude of the sun is greater than the slope of the leaves, the extinction of radiation is constant. But when the altitude of the sun is less than the slope of the leaves, the extinction increases with increasing depth (Fig. 3–22). Commonly, the intensity of the direct beam is assumed to remain constant with increasing depth of the canopy in sunlit areas. However, total solar radiation in sunlit areas may increase because of reflection from leaves, especially at higher elevations or under very clear skies (Miller, 1969).

The spectral distribution of light within the canopy varies with depth because of the difference in the spectral reflection of individual leaves. Green light is reflected by leaves more than blue and red light, and near infrared radiation is reflected more than green light. This results in relatively more energy within the canopy in the photosynthetically inactive regions of the spectrum, although the total intensity may be the same.

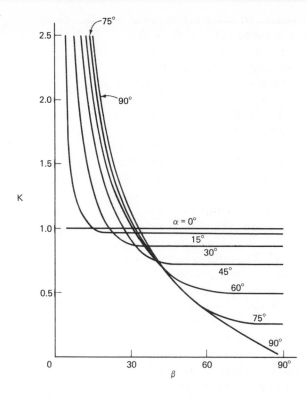

FIGURE 3-22. Extinction coefficients for different solar altitudes and leaf inclinations. (From Loomis et al., 1967.)

Topographic Effects on Solar Radiation

The orientation of a surface to the sun's rays affects the intensity of radiation received on the surface, with the maximum intensity being received by a surface oriented perpendicular to the sun's rays and a minimum intensity being received by a surface oriented parallel to the sun's rays. The ratio of solar radiation on any surface to solar radiation on a surface perpendicular to the sun's rays is equal to the cosine of the angle of incidence of the sun's rays to the surface. The angle of incidence is measured between the direction of the sun's rays and a line perpendicular to the plane of the surface, and it changes as the position of the sun and the orientation of the surface changes. The position of the sun can be given by the azimuth of the sun and either its altitude or its zenith angle, that angle between the direction of the sun and a point directly overhead. The surface orientation can be defined by the inclination of the surface from the horizonal and its azimuth. Azimuths, in this case, are measured from south, with east recorded as $-90°$, and west as $90°$ (north can be either $+180°$ or $-180°$). The cosine of the angle of incidence of the sun's rays to the surface can be calculated from the following equation:

$$\cos i = \cos \alpha \cos zn + \sin \alpha \sin zn \cos (\phi - \eta) \qquad (3\text{-}28)$$

where i is the angle of incidence of the incoming rays to the surface, α is the inclination of the surface to the horizontal, zn is the zenith angle of the sun, ϕ is the azimuth of the sun, and η is the azimuth of the surface.

Since surfaces of different orientation receive different amounts of solar radiation, different thermal and evaporation demands are placed on them. Calculations of the potential direct solar radiation on surfaces of different orientation give an indication of the magnitude of these demands. In summer at latitude 40° in the northern hemisphere the difference in total daily radiation between surfaces with less than 20° slope is not great. A south-facing slope receives more intense radiation at midday, but a north-facing slope, as long as its horizon is not obstructed by other hills, receives direct solar radiation longer during the day. In the winter at latitude 40° the solar radiation load on the two surfaces is much greater.

Vegetation is commonly different on north and south exposures. In the northern hemisphere, vegetation that is more typical of drier areas usually occurring on the south exposure and vegetation more typical of moister areas occurring on the north exposures. These differences may be caused by the different evaporation demands that occur in winter.

The microenvironment of valley bottoms and ridges differs markedly. First, the solar radiation load on a ridge is greater than that in a valley bottom. The solar radiation is intercepted by the valley sides so that the bottom receives solar radiation only when the sun is high in the sky. The full complement of sky radiation is not received in the valley bottom, but the bottom receives infrared radiation from the valley sides. Wind may be lower in the valley bottoms than on the ridges; however, the winds on the ridges may be of more constant direction. During the day, heating of the air creates an unstable condition, so that the air also follows the topography. In a valley running north-south and draining to the north, the predominate wind direction on the ridges during the day may be from the west, but in the valley bottom the winds will be to the south, i.e., up valley during the middle of the day, and to the north in the evening. In the morning when the sun shines directly on the valley floor, the air may be very calm and high heating of leaves and other surfaces may result. Air temperatures may vary more in the valley bottom than on the ridge tops because of the greater rising of air on the ridge tops and greater convectional cooling. The subsidence of cooler, more dense air at night causes the lower temperature in the valley bottom than on its sides and the ridge tops. The stability of the air at night causes the humidity to be greater in the bottom of the valley than at the top which may result in decreased fire danger in the valley bottom.

Extreme conditions may exist when the air drainage from the valley is blocked by a low ridge. Here frost may occur in the valley bottom when there is none on its sides. In a limestone area of Europe, for example, there is inversion of vegetation zones within large sink holes, with alpine flora occurring at the bottom of the depression and subalpine flora above the alpine. This inversion of vegetation is caused by a climatic inversion, with the coldest climate occurring at the bottom (Geiger, 1966).

TABLE 3-8. THERMAL PROPERTIES OF VARIOUS SUBSTANCES OF ECOLOGICAL INTEREST.*

Type of soil or material	Density (g/cm³)	Volumetric heat capacity (10⁻³ cal/cm³/°C)	Thermal conductivity (10⁻³ cal/cm/min/°C)	Diurnal damping depth (cm)
Soil materials				
Soil organic matter	1.3	600	40	5.5
Dry sand	1.4 – 1.7	100–580	24–42	5.8 – 10.5
Wet sand		200–600	120–360	16.6
Dry clay		100–400	12–90	7.4 – 10.2
Wet clay	1.7 – 2.2	300–400	120–300	13.5 – 18.5
Ice	1.7 – 2.3	460	300–420	14.1 – 20.5
Still water	1.0	1000	78–90	6.0 – 6.4
Still air	1.25 × 10⁻³	0.24–0.34	3.0–3.6	69.7 – 75.7
Rock		430–580	240–600	16.0 – 21.8
Sandstone			1,980–3,960	
Marble			3,468–4,956	
Granite			3,222–6,942	
Concrete		500	1,488–2,232	24.6
New snow			12–18	
Old snow			180–300	
Woods				
Dry wood		100–200	12–30	7.4 – 8.3
Balsa			81.7–144	
Cypress			166	
White pine			193	
Mahogany			224	
Virginia pine			243	
Oak			253	
Maple			273	
Other materials				
Glass			1,236–1,488	
Silver		590	60,000	215
Iron		820	126,000	265.5

*From List, 1963; Van Wijk, 1963; and Weast, 1967.

Soil Temperature

Part of the energy absorbed by a soil surface is conducted downward and stored, thus producing a rise in soil temperature. This conduction of heat takes place according to the conduction equation presented earlier (p. 90). The increase in temperature of the deeper layers of soil depends on the rate of heat conduction, the volumetric heat capacity of the soil, and length of time during which heat is being conducted. The thermal conductivity* of the soil, which depends on what the soil consists of, how porous it is, and how wet it is, determines the rate of heat conduction. Typical values for thermal conductivities are shown in Table 3–8.

The rate of change of soil temperature can be calculated from the rate

*Defined on p. 90.

at which energy is being supplied at the soil's surface G, the volumetric heat capacity of the soil VHC, and the surface to volume ratio according to the equation

$$\frac{dT}{dt} = \frac{G}{VHC} \cdot \frac{\text{Surface area receiving energy}}{\text{Volume of soil}} \qquad (3\text{-}29)$$

Values for the volumetric heat capacity of various materials are given in Table 3–8.

The time required for heat to be conducted through soil causes fluctuations in soil temperature at some depth in the soil to lag behind fluctuations occurring at the surface. The length of this lag and, in addition, the magnitude of the fluctuations decrease at progressively deeper levels in the soil. Because of this phenomenon, both the daily and seasonal oscillations in temperature that occur at the surface of the soil are delayed and damped out within the soil. An index relating various soil properties to the penetration of the temperature fluctuations is the *damping depth,* which may be defined here as the depth at which the temperature amplitude is 0.37 of the amplitude at the surface. At a depth of three times the damping depth, the temperature amplitude is 0.05 of the amplitude at the surface. Figure 3–23 demonstrates this phenomenon empirically; Fig. 3–24 shows idealized variations of soil temperatures at several depths predicted on the basis of the theory of heat transfer in soil. To facilitate comparison, the curves in Fig. 3–24 are shown as deviations from a mean temperature of 20°C.

The sand wasps' parasite *Dasymutilla* seeks out the wasps' burrows to lay its eggs on the wasps' larvae. This parasite is wingless and cannot evade the high surface temperature by flying. Instead, it evades lethal temperatures by crawling into the wasps' burrows or climbing up grass blades into cooler air. In addition, *Dasymutilla* can withstand higher temperatures than the sand wasp, showing the first signs of adverse high temperature effects at 52°C. Thus, these two organisms which share the same microhabitat have evolved physiological and behavioral adaptations that permit them to cope with temperature extremes near the surface of the soil.

Concluding Remarks

The state of knowledge of the effect of the physical environment on organisms has reached a fairly rigorous stage; more rigorous perhaps than many biologists wish. However, there is still much work to be completed in clarifying the environmental interactions in natural communities and in clarifying the physiological and population responses of the organisms. This clarification is essential if we are to predict the future human environment as we alter our physical and chemical surroundings. We must answer the questions, "How much?" "How many?" and "How long?" i.e., quantitative questions, when we consider the ecological consequences of man's additions and subtractions to the ecosystem. Some idea of the changes in the physical environment produced by man is given in Table 3–9. The student can utilize the concepts presented in this chapter and in the preceding chapter in order to predict some of the consequences if such changes in the physical environment became widespread.

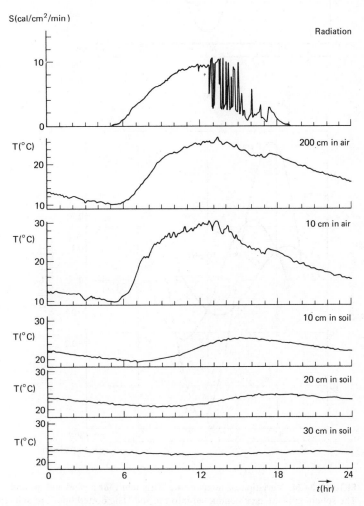

FIGURE 3–23. Short-wave solar radiation, air temperatures, and soil temperatures for a clay soil with grass cover at Wageningen, the Netherlands, on a day with broken sky in the afternoon. (From Van Wijk, 1963.)

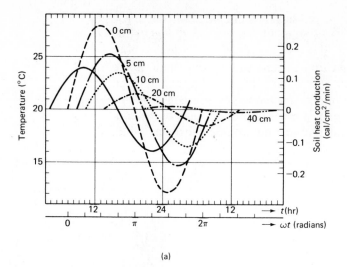

(a)

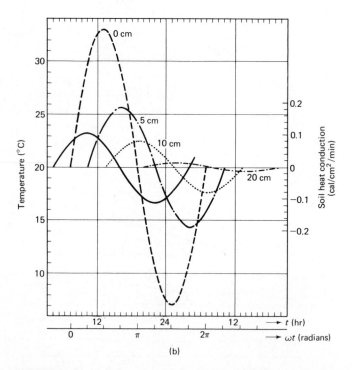

(b)

FIGURE 3–24. Variation of temperature with depth for a sand soil (a) and a peat soil (b). The solid line is the heat conducted into the soil. The dotted lines are soil temperatures at different depths. The values for thermal conductivity, volumetric heat capacity, and damping depth are 0.242, 0.5, and 15.2 for the sand and 0.042, 0.75, and 5.1 for the peat. Units are cal/cm/min °C, cal/cm³ °C, and cm. The equation used to calculate the temperatures is $T_{z,t}$ $T_z + {}^A T_{oe}^{-z/D\sin} (Wt - z/D)$. ${}^A T_0$ is the temperature amplitude at the soil surface; e is 2 $\pi/24$; and t is in hours. (Further discussion is given in Van Wijk, 1963.)

TABLE 3-9. CLIMATIC CHANGES PRODUCED BY CITIES.*

Element	Comparison with rural environment
Dust and pollution	10 to 25 times more
Radiation	15 to 20% less
Clouds	5 to 10% more
Precipitation	5 to 10% more
Temperature	1 to 2°F more
Relative humidity	3 to 10% less
Wind speed	20 to 30% less

*From Landsburg, 1958.

REFERENCES

Andrewartha, H. G., and L. C. Birch. 1954. *The Distribution and Abundance of Animals.* Chicago: University of Chicago Press.

Bartholomew, G. A. 1964. The roles of physiology and behavior in the maintenance of homeostasis in the desert environment. In *Homeostasis and Feedback Mechanisms,* pp. 7–29. Eighteenth Symposium of the Society of Experimental Biology.

———, and V. A. Tucker. 1963. Control of change in body temperature, metabolism, and circulation by the Agamid lizard, *Amphibolurus bachatus. Physiol. Zool.,* **36**:199–210.

Bartlett, P. N., and D. M. Gates. 1967. The energy budget of a lizard on a tree trunk. *Ecology,* **48**(1):315–22.

Billings, W. D., and R. J. Morris. 1951. Reflection of visible and infrared radiation from leaves of different ecological groups. *Amer. J. Bot.,* **38**:327–31.

Birkebak, R., and R. Birkebak. 1964. Solar radiation characteristics of tree leaves. *Ecology,* **45**(3):646–49.

Blair, T. A., and R. C. Fite. 1965. Weather Elements. Englewood Cliffs, N.J.: Prentice-Hall.

Bodenheimer, F. S. 1958. Climatic factors in arid zone animal ecology. In *Climatology: Reviews of Research.* UNESCO.

Bogert, C. M. 1949. Thermoregulation in reptiles, a factor in evolution. *Evolution,* **3**:195–211.

Bray, J. R., J. F. Sanger, and A. L. Archer. 1966. The visible albedo of surfaces in central Minnesota. *Ecology,* **47**(2):524–31.

Chapman, R. N., C. E. Mickel, J. R. Parker, G. E. Miller, and E. G. Kelly. 1926. Studies in the ecology of sand dune insects. *Ecology,* **7**:416–26.

Clark, L. R. 1947. An ecological study of the Australian plague locust *Chortoicetes terminifera* Walk. in the Bogan-Macquarie outbreak area in New South Wales. In *Bull. Counc. Scient. and Indust. Research, Australia,* No. 266.

———. 1949. Behavior of swarm hoppers of the Australian plague locust, *Chortoicetes terminifera* Walk. In *Bull. Counc. Scient. and Indust. Research, Australia,* No. 245.

Clements, F. E. 1916. *Plant Succession.* Carnegie Inst. Wash., Publ. No. 242.

Dedykin, V. P. and V. P. Bedenko. 1961. The connection of the optical properties of plant leaves with soil moisture. *Biol. Sci. Sec. Transl.,* **134**:212–14 (transl. from *Doklady Akd. Nauk, USSR,* **134**:965–68).

Denmead, O. T. 1964. Evaporation sources and apparent diffusivities in a forest canopy. *J. Applied Met.,* **3**:383–89.

Espenshade, E. B. 1960. *Goode's World Atlas.* Skokie, Ill.: Rand McNally.

Federer, C. A. and C. B. Tanner. 1966. Spectral distribution of light in the forest. *Ecology,* **47**:555–60.

Fry. F. E. J. 1947. Effects of the environment on animal activity. University of Toronto, *Stud. Biol.,* **55**:1–62.

Gates, D. M. 1962. *Energy Exchange in the Biosphere.* New York: Harper & Row. 151 pp.

———. 1963. Leaf temperature and energy exchange. *Archiv fur Meteorologie, Geophysik und Bioklimatologie,* **551**:321–36.

———. 1965a. Energy, plants, and ecology. *Ecology,* **46**:1–24.

———. 1965b. Characteristics of soil and vegetated surfaces to reflected and emitted radiation. *Proc. Third Symp. of Remote Sensing of Environment.* Ann Arbor: University of Michigan.

———. 1968. Energy exchange between organisms and environment. *The Australian Journal of Science,* **32**(2):67–74.

———, **H. J. Keegan, and V. R. Weidner.** 1966. Spectral reflectance and planetary reconnaissance. In *Scientific Experiments for Manned Orbital Flight,* Science and Technology Series, **4**:71–86.

———, **E. C. Tibbals, and F. Kreith.** 1965. Radiation and convection for ponderosa pine. *Amer. J. Bot.,* **52**(1):66–71.

Geiger, R. 1966. *The Climate Near the Ground.* Cambridge, Mass.: Harvard University Press.

Grieve, B. J. and F. W. Went. 1965. An electric hygrometer apparatus for measuring water-vapor loss from plants in the field. In F. E. Echardt (ed.), *Methodology of Plant Eco-physiology.* Paris, France: UNESCO.

Hammel, H. T. 1956. Infrared emissivities of some arctic fauna. *J. of Mammalogy,* **37**:375–81.

Haufe, W. O. 1952. Observations on the biology of mosquitoes *(Diptera: Culicidae)* at Goose Bay, Labrador. *Canad. Ent.,* **84**:254–63.

———. 1954. The effects of atmospheric pressure on the flight responses of *Aedes aegypti* (L.). *Bull. Ent. Res.* **45**:507–26.

———. 1966. *The significance of biometeorology in the ecology of insects. Int. J. Biometeor,* **10**:241–52.

Landsburg, H. E. 1958. Trends in climatology. *Science,* **128**:749–58.

Lee, R., and D. M. Gates. 1964. Diffusion resistance in leaves as related to their stomatal anatomy and micro-structure. *Amer. J. Bot.,* **51**:963–75.

Lemon, E. R. 1963. Energy and water balance of plant communities. In L. T. Evans (ed.), *Environmental Control of Plant Growth.* New York: Academic Press.

———. 1967. The impact of the atmospheric environment of the integument of plants. *Int. J. Biometeor.,* **3**:57–69.

List, R. J. 1963. *Smithsonian Meteorological Tables.* Washington, D.C.: Smithsonian Institution.

Loomis, R.S., W. A. Williams, and W. G. Duncan. 1967. Community architecture and the productivity of terrestrial plants. In A. San Pietro, F. A. Green, and

T. J. Army (eds.), *Harvesting The Sun*. New York: Academic Press. 342 pp.

Macfadyen, A. 1963. Animal ecology. London: Pitman. 344 pp.

Miller, P. C. 1967. Leaf temperatures, leaf orientation and energy exchange in quaking aspen (*Populus tremuloides*) and Gambell's oak (*Quercus gambellii*) in central Colorado. *Oecol. Plant.,* **2**:241–70.

————. 1969. Solar radiation profiles in openings in canopies of aspen and oak. *Science,* **164**:308–9.

————, **and D. Gates** 1967. Transpiration resistance of plants. *Amer. Midl. Natural,* **77**:77–85.

Milthorpe, F. L. 1962. Plant factors involved in transpiration. In *Plant-Water Relationships in Arid and Semi-Arid Conditions*. Proc. of the Madrid Symposium, UNESCO.

Monsi, M., and T. Saeki. 1953. Uber den Lichfactor in den Pflanzengesellschaften und seine Bedentung fur die Stoffproduktion. *Jap. J. Botan.,* **14**:22–52.

Monteith, J. L. 1963. Gas exchange in plant communities. In L. T. Evans (ed.), *Environmental Control of Plant Growth*. New York: Academic Press.

Pearman, G. I., H. L. Weaver, and C. B. Tanner. (n.d.) Boundary layer heat transport coefficients under field conditions. Mimeographed Report, Soils Dept., University of Wisconsin.

Priestley, C. H. B. 1959. *Turbulent Transfer in the Lower Atmosphere*. Chicago: University of Chicago Press. 130 pp.

Rashke, K. 1960. Heat transfer between the plant and the environment. *Ann. Review Plant Physiol.,* **11**:111–26.

Schmidt-Nielsen, B., and Schmidt-Nielsen, K. 1951. A complete account of water metabolism in the kangaroo rat and an experimental verification. *J. Cell Comp. Physiol.,* **38**:165–81.

Schmidt-Nielsen, K. 1964. *Desert Animals: Physiological Problems of Heat and Water*. London: Oxford.

Scholander, P. F. 1955. Evolution of climatic adaptation in homeotherms. *Evolution,* **9**:15–26.

Sutton, O. G. 1953. *Micro-meteorology*. New York: McGraw-Hill.

Tibbals, E. C., E. K. Carr, D. M. Gates, and F. Kreith. 1964. Radiation and convection in conifers. *Amer. J. Bot.,* **51**:529–38.

Turner, J. C. 1965. Some energy and microclimate measurements in a natural arid zone plant community. In F. F. Eckardt (ed.), *Methodology of Plant Ecophysiology*. Proc. of the Montpellier Symposium, UNESCO.

Van Bavel, C. H. M., F. S. Nakayama, and W. L. Ehrler. 1965. Measuring transpiration resistance of leaves. *Plant Physiol.,* **40**:535–40.

Van Royen, W. 1954. *The Agricultural Resources of the World,* Vol. 1 of *Atlas of the World's Resources*. Englewood Cliffs, N.J.: Prentice-Hall, Inc.

Van Wijk, W. R. 1963. *Physics of Plant Environment*. Amsterdam: North-Holland Publishing.

Waggoner, P. E. 1967. Moisture loss through the boundary layer. *Biometeorology,* **3**:41–52.

Warming, E. 1909. *Oecology of Plants*. London: Oxford University Press.

Weast, R. C. (ed.). 1967. *CRC Handbook of Chemistry and Physics,* 48th ed., Cleveland, Ohio: Chemical Rubber Company.

Part III
THE POPULATION LEVEL
OF ORGANIZATION

Chapter 4
THE STRUCTURE OF POPULATIONS

INTRODUCTION

One of the principal goals of ecologists is to understand how natural processes determine the size and composition of plant and animal populations, eventually with enough thoroughness to be able to predict accurately changes in these variables from information on changes in environmental factors. Although both theoretical and experimental research has been conducted in this area for half a century or more, and descriptive studies have been conducted for much longer, we are still far from this goal except for a small number of species. The slowness of progress in this area of ecology is largely due to the lack of extensive, accurate, pertinent data, but it also stems from the lack of adequate theory relating in a dynamic system all of the important factors that affect population size and composition. Obviously, developing sound theory and gathering pertinent data are mutually complementary activities.

In this chapter, however, we shall first concentrate on aspects of population ecology concerned with the description of population processes by using statistics such as birth and death rates; and second, we shall discuss how organisms in a population are distributed in their habitat. The concepts involved are especially important because they form the basis of more advanced theory that is even now being developed.

SURVIVORSHIP

The simplest (but least useful) index of mortality in a population is the *crude death rate,* which is the quotient of the total number of deaths during a unit of time divided by the population size. This is the index so commonly

135

reported for human populations in the popular press, e.g., "The fatality rate for the disease was 29.2 per 10,000 per year." It takes little insight to realize that mortality among old adults will affect a population very differently from mortality among juveniles. Consequently, an *age specific death rate* is a more useful statistic than the crude death rate. Age specific death rate is usually defined as the quotient of the number of organisms of age x dying in a short interval of time divided by the number of age x that were alive at the start of that interval of time. This quotient is usually symbolized as q_x.

To facilitate further discussion of age specific mortality, let us imagine a group of individuals starting out life together at one point in time. A group of individuals of the same age is called a *cohort*. We can show the diminishment of a cohort through time by means of a curve, known as a *survivorship* curve, as shown in Fig. 4–1. The number of the cohort remaining alive at any particular age x is symbolized as l_x, and thus survivorship curves are also commonly called l_x curves. If we let ΔN be the number of organisms dying in the small interval of time Δt, age specific mortality rate can be graphically interpreted as the ratio of $\Delta N / \Delta t$. This ratio is usually called d_x. As the interval Δt becomes very small, this ratio approaches the slope of the l_x curve at x. Dividing d_x by l_x will then give the mortality rate per individual of age x; this quotient is called q_x. Readers familiar with calculus will recognize that we can thus think of q_x as

$$q_x = \frac{1}{l_x}\frac{dl_x}{dt}$$

which is known as the *instantaneous age specific death rate*. It is often convenient to define q_x in this way when dealing with theoretical problems

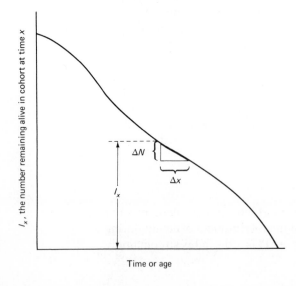

FIGURE 4–1. Survivorship curve showing the number of individuals remaining alive in a cohort as a function of age. ΔN is the number of animals dying in the small interval of time Δt.

in population dynamics. In reality, however, an l_x curve is not continuous (remember, we are talking about numbers of organisms), and for practical reasons data must be gathered at distinct points in time, rather than continuously.

Although there is an infinite variety of possible l_x curves, a few special cases can be distinguished for purposes of comparison and discussion. Figure 4-2 shows four curves plotted with an arithmetic scale on the l_x axis. (A logarithmic scale is also commonly used along the vertical axis.) The uppermost curve (a) shows the shape produced when the probability of death is small during early life, but it becomes large late in life. A curve of this kind may apply to laboratory animals that die of "old age," but it is rarely applicable to natural populations. The second curve (b)—actually a straight line—would result if a constant number of individuals died in each age interval. This is virtually unknown among real populations and is shown only for comparative purposes. The third curve (c) is produced when the mortality rate is the same irrespective of age. This curve would be a straight line sloping downward to the right if a logarithmic scale were used on the l_x axis. Even though constant mortality rate over the entire range of age is unrealistic for most species, since it implies that the probability of dying is independent of age, it is not unusual for q_x to remain constant for long intervals of age. If a log scale is used for the l_x axis, then the l_x curve will be straight during those age intervals. The bottom l_x curve (d) applies to cases in which the mortality rate, q_x, is heaviest for the young and decreases with age. This situation is somewhat closer to reality for many species than the preceding cases, but still it is not fully realistic. In many species showing such a pattern in early life, a time is reached when age becomes a liability, and the probability of dying increases.

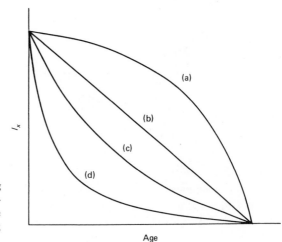

FIGURE 4-2. Survivorship curves resulting from several hypothetical patterns of age-specific mortality rates: (a) d_x increasing as age increases; (b) constant d_x; (c) constant q_x; (d) q_x decreasing as age increases.

If each of the preceding curves is unrealistic to some extent, then what does a realistic l_x curve look like? Survivorship curves for natural populations often have shapes something like those in Fig. 4–3: steep initially, implying a high mortality rate among the young; gradually sloping and upwardly concave in the center, implying a fairly constant mortality rate for the adults; steep toward the end, showing an increase in mortality among the old individuals.

In actual practice, the exact ages of individuals in a natural population can rarely be ascertained, and even if they could, age specific death rates and other population statistics could not be tabulated for all possible values of x. Instead, individuals are grouped into *age classes,* and all the individuals in an age class are treated as though they were of the same age. If the size of each age class is sufficiently small, inaccuracies resulting from this procedure are insignificant.

A table that summarizes population statistics is called a *life table*. Table 4–1 is a life table for adult honeybee workers (*Apis mellifera*). The first column shows the age x, of the cohort for each five-day interval. The second column, l_x, shows how many individuals from the cohort are still alive at the beginning of the corresponding age interval. These figures are adjusted to what they would be if the cohort had begun with 1,000 individuals. An adjustment of this kind has the advantage of simplifying comparisons of several life tables. (If the decimal point is shifted three places to the left, the resulting values can be thought of as being probabilities of an individual surviving to age x.) The column headed d_x shows how many organisms died during the corresponding age interval, and the fourth column shows q_x, the age specific death rate. The last column in

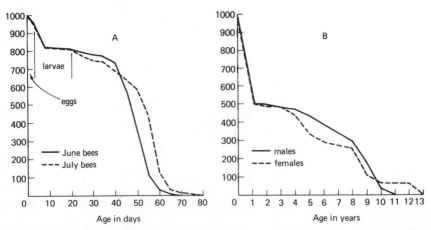

FIGURE 4–3. Survivorship curves for (A) worker honeybees and (B) Uganda defassa waterbuck. (From Sakagami and Fukuda, 1968, and Spinge, 1970.)

TABLE 4-1. LIFE TABLES FOR ADULT HONEYBEE WORKERS (*APIS MELLIFERA*).*

Life table for June adult bees (average longevity 28.345 days).				Life table for July adult bees (average longevity 32.424 days).			
x (days)	l_x	d_x	e_x	q_x	x (days)	l_x	d_x	e_x	q_x
0–5	1,000	17	28.3	0.0170	0–5	1,000	38	32.3	0.0380
5–10	983	23	23.8	0.0234	5–10	962	40	28.4	0.0416
10–15	960	17	19.3	0.0177	10–15	922	10	24.6	0.0108
15–20	943	33	14.6	0.0350	15–20	912	67	19.8	0.0735
20–25	910	183	10.0	0.2011	20–25	855	60	15.9	0.0702
25–30	727	272	6.9	0.3741	25–30	795	70	12.0	0.0880
30–35	455	316	4.6	0.6945	30–35	725	174	7.9	0.2400
35–40	139	93	4.4	0.6691	35–40	551	387	4.6	0.7024
40–45	46	40	3.6	0.8696	40–45	164	120	4.5	0.7317
45–50	6	6	2.5	1.0000	45–50	44	24	4.8	0.5454
					50–55	20	20	2.5	1.0000

*From Sakagami and Fukuda, 1968.

Table 4–1 shows the mean life expectancy for individuals attaining age x. This is not a particularly useful statistic to ecologists, and we shall not discuss its calculation, which can be found in Deevey (1947).

The literature of ecology contains many examples of life tables from a wide variety of taxonomic groups. In most instances these data have been obtained by sampling a population under one set of environmental conditions. If there is reason to believe that conditions are fairly constant through time, a life table made from such data is a reasonable way of summarizing a population's behavior under these conditions. Ecologists are usually more interested, however, in discovering how populations behave under different conditions, since this kind of information may reveal homeostatic mechanisms that allow the populations to persist when faced with environmental fluctuations. Ecologists are also interested in how two or more ecologically similar species respond to the same environment. When data of this kind are available, it may be possible to make inferences about the outcome of competition between species. Both of these interests suggest a comparative approach in which several life tables are used. Examples in which this has been done are few in number.

Lowe (1969) has provided two life tables for Red Deer (*Cervus elaphus*) (see Tables 4–2 and 4–3). These examples not only illustrate the value of a comparative approach, but also show two different ways of constructing life tables and indicate some of the difficulties involved. The data were gathered on the Isle of Rhum, Scotland, which is sufficiently

**TABLE 4-2. TIME SPECIFIC LIFE TABLES FOR
RED DEER ON RHUM IN 1957.***

		Stags		
x (Age, years)	l_x (Survivors at beginning of age class x)	d_x (Deaths)	e_x (Further expectation of life, years)	$100q_x$ (Mortality rate/1,000)
1	1000	282	5.81	282.0
2	718	7	6.89	9.8
3	711	7	5.95	9.8
4	704	7	5.01	9.9
5	697	7	4.05	10.0
6	690	7	3.09	10.1
7	684	182	2.11	266.0
8	502	253	1.70	504.0
9	249	157	1.91	630.6
10	92	14	3.31	152.1
11	78	14	2.81	179.4
12	64	14	2.31	218.7
13	50	14	1.82	279.9
14	36	14	1.33	388.9
15	22	14	0.86	636.3
16	8	8	0.5	1,000

		Hinds		
x (Age, years)	l_x (Survivors at beginning of age class x)	d_x (Deaths)	e_x (Further expectation of life, years)	$100q_x$ (Mortality rate/1,000)
1	1000	137	5.19	137.0
2	863	85	4.94	97.3
3	778	84	4.42	107.8
4	694	84	3.89	120.8
5	610	84	3.36	137.4
6	526	84	2.82	159.3
7	442	85	2.26	189.5
8	357	176	1.67	501.6
9	181	122	1.82	672.7
10	59	8	3.54	141.2
11	51	9	3.0	164.6
12	42	8	2.55	197.5
13	34	9	2.03	246.8
14	25	8	1.56	328.8
15	17	8	1.06	492.4
16	9	9	0.5	1,000

*From Lowe, 1969.

Note: Nomenclature follows that adopted by Southwood, 1966.

TABLE 4-3. DYNAMIC LIFE TABLES FOR RED DEER ON RHUM, 1 YEAR OF AGE IN 1957.*

		Stags		
x (Age, years)	l_x (Survivors at beginning of age class x)	d_x (Deaths)	e_x (Further expectation of life, years)	$100q_x$ (Mortality rate/1,000)
1	1000	84	4.76	84.0
2	916	19	4.15	20.7
3	897	0	3.25	0
4	897	150	2.23	167.2
5	747	321	1.58	430.0
6	426	218	1.39	512.0
7	208	58	1.31	278.8
8	150	130	0.63	866.5
9	20	20	0.5	1,000

		Hinds		
x (Age, years)	l_x (Survivors at beginning of age class x)	d_x (Deaths)	e_x (Further expectation of life, years)	$100q_x$ (Mortality rate/1,000)
1	1000	0	4.35	0
2	1000	61	3.35	61.0
3	939	185	2.53	197.0
4	754	249	2.03	330.2
5	505	200	1.79	396.0
6	305	119	1.63	390.1
7	186	54	1.35	290.3
8	132	107	0.70	810.5
9	25	25	0.5	1,000

*From Lowe, 1969.

Note: Nomenclature follows that adopted by Southwood, 1966.

small (10,684 ha) and lacking in dense vegetation to allow most of the deer living on it to be counted, and to allow the remains of a large fraction of dead deer to be found. Before 1957 the deer on Rhum had been hunted only lightly, and Lowe considered that most of the mortality was due to natural factors. In 1957 the population was censused, and in each year from 1957 to 1966 a count was made of the deer that had been shot or had died of other causes. These deer were aged and sexed, and since they accounted for nearly 92% of the deer alive in 1957, Lowe could determine the age structure of the population as it existed at that time.

If it is assumed that prior to a census, the size and age specific mortality rates of a population have not changed significantly for a period of time equal in length to the age of the oldest animals in the population, a life table can be constructed from age distribution data (Southwood, 1966). These assumptions imply that age classes are equal in size at birth, and the differences among them at the time of the census are due to differences in cumulative mortality. In other words, the differences in numbers in any pair of adjacent age classes at the time of the census are assumed to reflect the mortality in a group of individuals as they age. Thus, the data can be used to construct directly an l_x column. A life table obtained in this way is called *time specific* (Southwood, 1966). Table 4–2 shows a time specific life table constructed by Lowe from his estimation of the age structure of the Red Deer population in 1957. We shall return to this table after describing how Table 4–3 was formed.

Since the number of Red Deer in the first year age class on Rhum in 1957 was known from the census, and an excellent sample of deer dying from 1957 to 1966 was obtained, Lowe could directly assess mortality within the 1957 cohort. That is, data for a d_x column of a life table could be obtained, and, from these figures an l_x column could be constructed. The resulting life table, which is called a *dynamic life table* (Southwood, 1966), is shown in Table 4–3. This table shows data for only nine years, since this was as long as the cohort originating in 1957 had been followed.

Several interesting facets appear in these tables. The most obvious is shown in the time specific table; namely, the high death rate (q_x) among males in age classes 7–9 and among females in age classes 8–9. Lowe (1969) believes that this is a consequence of "highly selective age specific mortality factors," but there are other possible explanations. These must be discussed, not because we think Lowe is mistaken, but because they illustrate how various difficulties can arise in the construction of life tables. Recall that in a time specific life table, the l_x column is based on the relative frequencies of animals in succeeding age classes as observed at one point in time (1957 in this case). Anything that might reasonably cause a "hump" or "valley" in the age frequency distribution would give the appearance of a dip or peak in the q_x distribution. Taking a simple example, suppose reproduction were unusually high in one year, but age specific mortality rates remained unchanged for several years. If the age structure were then used to form an l_x column, the animals from the year of unusually successful reproduction would appear with a high relative frequency, and the corresponding value of q_x could appear to be quite low. Similarly, an unusually poor year for reproduction could be misinterpreted as an unusually high death rate for a later age class. These kinds of occurrences clearly are violations of the assumptions made in constructing a time specific life table—our point is, therefore, that if no other information is available, such violations may be made without our knowledge, and the resulting data may be misinterpreted. In the case of the Red Deer on Rhum, we do not know what the correct interpretation of the time specific life-table data is. If the dynamic life table extended for several more years, and if conditions had not changed on Rhum after 1957, we could compare the q_x data for the 7–9 year classes between the two tables. Unfortunately, this is not possible.

Other problems in constructing life tables mainly involve technical difficulties in obtaining data on numbers dying in each time interval and the ages of those animals dying. Lowe was fortunate in being able to find many of the deer that died; this is usually very difficult. Determining the age of the deer at death was done primarily by examining teeth, which is a reasonable but not error-free method.

A second noteworthy aspect of the life tables for Red Deer is the difference in survival rates between males and females, with the females apparently suffering higher mortality rates in the first few years of life than do the males. Again, the reason for this is not known, but female Red Deer may begin breeding in their second year, and it is reasonable to suppose that their susceptibility to mortality factors is related to this fact.

Before, we can compare Table 4–2 with Table 4–3, we will need further information. In 1957, the island of Rhum was established as a National Nature Reserve. Beginning in 1958, the deer population was culled each year in an attempt to stabilize it. From 1961 onward, the policy was to cull one-sixth of the adults counted in the spring. Assuming that Table 4–2 represents premanagement mortality rates reasonably well, at least for the first few age classes, we can see that in comparison with Table 4–3, mortality rates for the younger age classes were lower prior to 1957, except for the first age class, which was considerably higher. Lowe's data also showed that death by shooting in large part replaced natural deaths, rather than simply adding to them, thus demonstrating that the population had considerable "resiliency" in reacting to changes in mortality factors. More will be said about this phenomenon in the next chapter.

NATALITY

Survivorship data obviously are not sufficient for the estimation of changes in population size through time; to do this, additional data on the production of young are necessary. The simplest expression of natality is *crude birth rate,* which is the number of offspring produced per unit population size per unit time; but, as in the case of mortality, information on the production of young is most useful when expressed as an age specific rate. We define *age specific birth rate, m_x,* as the average number of female offspring produced per unit time by a female in the age class x. (Our discussion shall be limited to the female portion of a population, since this eliminates complexities resulting from unequal sex ratios at birth and differences in death rates between the sexes.)

Figure 4–4 shows how m_x could be defined using the concepts of calculus, as was done for q_x. The abscissa represents age and the ordinate represents the average total number of female offspring produced by a female from age 0 to age x. If Δx is equal to one unit of time, then the ratio $\Delta y / \Delta x$ is m_x as defined above. If the interval of time, and hence Δx, becomes smaller and smaller, Δy will also decrease, and the ratio $\Delta y / \Delta x$ will approach the slope of the line at x. The slope of the line is simply dy / dx, and so we could define m_x by the equation

$$m_x = \frac{dy}{dx}$$

Figure 4–4 is not the usual way of graphing m_x data; the usual way is shown in Fig. 4–5. More common yet is simply to include m_x data in an additional column in a life table. Tables 4–4 & 4–5 show m_x data obtained by Lowe for Red Deer. The meaning of the values labelled R_0 will be explained below.

Number of Offspring Produced by a Cohort. In order to calculate the total number of offspring produced by a cohort during its existence, we need to combine l_x and m_x information. To see how this is done, let us examine how many offspring are produced in a small unit of time Δt during which the cohort advances in age from x to $x + \Delta t$. If we make Δt sufficiently small, we can approximate the number of mothers alive during Δt as

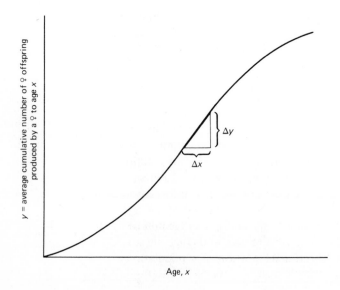

FIGURE 4–4. Total number of female offspring produced by a female by the time she has reached age x. The slope of this curve is the age-specific birth rate.

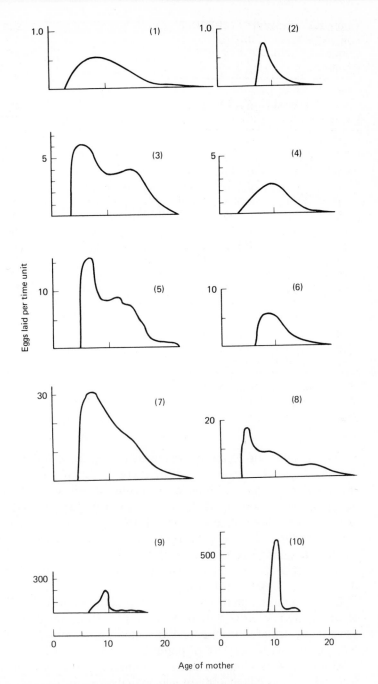

FIGURE 4–5. (1) *Microtus agrestis;* (2) *Niptus hololeucus;* (3) *Mezium affinie;* (4) *Daphnia pulex;* (5) *Calandra oryzae;* (6) *Pediculus humanus;* (7) *Tribolium castaneum;* (8) *Ptinus tectus;* (9) *Physa gyrina (Scio);* (10) *Physa gyrina (Argo).* All curves have been converted to the same time scale by using units = 0.1 of one generation (After Laughlin, 1965.)

TABLE 4-4. FECUNDITY SCHEDULE FOR THE HINDS (♀♀) IN THE STANDING POPULATION ON RHUM IN 1957.*

Age x (years)	% of foetuses ♀	♀♀/♀ m_x	$l_x m_x$
1	–	–	–
2	–	–	–
3	50.0	0.311	0.242
4	50.0	0.278	0.193
5	50.0	0.302	0.184
6	61.9	0.400	0.210
7	61.9	0.476	0.210
8	61.9	0.358	0.128
9	61.9	0.447	0.081
10	50.0	0.289	0.017
11	50.0	0.283	0.014
12	50.0	0.285	0.012
13	50.0	0.283	0.010
14	50.0	0.282	0.007
15	50.0	0.285	0.005
16	50.0	0.284	0.003

*From Lowe, 1969.

Note: $\Sigma l_x m_x (R_0) = 1.316$.

TABLE 4-5. FECUNDITY SCHEDULE FOR THE HIND COHORT AGED 1 YEAR IN 1957.*

Age x (years)	% of foetuses ♀	♀♀/♀ m_x	$l_x m_x$
1	–	–	–
2	–	–	–
3	50.0	0.311	0.292
4	50.0	0.278	0.209
5	50.0	0.302	0.153
6	61.9	0.400	0.122
7	61.9	0.476	0.089
8	61.9	0.358	0.047
9	61.9	0.447	0.011

*From Lowe, 1969.

Note: $\Sigma l_x m_x (R_0) = 0.923$.

l_x. The number of offspring produced during Δt will then be approximately $l_x m_x \Delta t$. For example, if there are 15 females (l_x) from a cohort still alive on day 97, and females of this age produce on the average 0.2 offspring per female per day (m_x), then during this one-day period, we would expect three offspring to be produced. The total number of offspring produced by the cohort during its existence would then be the sum of the offspring produced in all successive intervals, which is simply the area beneath a curve having the product $l_x m_x$ as the ordinate and x as the abscissa. This area is $l_x m_x dx$.

This integral is usually called the *net production rate, R_0.* Even though R_0 is clearly related to how rapidly the population might change in size, since it is calculated from l_x and m_x data, it is not equal to, and must not be confused with, the rate of population growth to be discussed below.

Although R_0 has been defined in terms of continuous functions, data from which it is calculated are usually gathered for discrete age classes. In these cases, R_0 is usually estimated by $\Sigma l_x m_x$. How satisfactory this estimate is depends on the length of the age classes used, how much l_x and m_x vary within an age class, and what use is being made of the estimate. On this last point, we must honestly admit that ecologists have not found R_0 to be especially useful, although it is often referred to and calculated from life-table data. Table 4–6 gives R_0 values for several species, together with values for length of one generation, T_c. This latter statistic is discussed below.

AGE DISTRIBUTION AND POPULATION GROWTH

Population growth (or decline) depends on both the distributions of birthrate and death rate with respect to age, plus one additional distribution, namely, the distribution showing the relative frequency of individuals in the population as a function of age. This is generally called the *age distribution,* and it is usually shown as the fraction of the population falling in successive age classes.

The dependence of population growth on the age distribution can easily be understood by the following reasoning. The number of offspring added to a population in a small interval of time is the sum of the number added by each age class; similarly, the number lost from the population is the sum of the losses from each age class. The net change in population size will be the sum of these gains and losses over all age classes. Therefore, a population of animals entering their peak reproductive period will grow rapidly in comparison with a population of old animals with low fertility and high death rates.

The most useful single index of population growth is the *instantaneous rate of change of population size per individual.* If population size, N, is graphed on time, t, as in Fig. 4–6, then the rate of change of N with respect to t is approximately the ratio $\Delta N / \Delta t$. By allowing Δt and ΔN to become very small, we see that $\Delta N / \Delta t$ approaches the slope of the line at time t, or dN/dt. The rate of change per individual, which we shall call k,* will then be given by

$$k = \frac{1}{N} \frac{dN}{dt} \tag{4–1}$$

Unfortunately, there is some confusion in terminology surrounding the value we have called k. The symbol r is sometimes used with the same meaning that we have given k, but it is more generally used for a special case of population growth. This special usage results mainly from the theoretical work of Lotka (Sharpe and Lotka, 1911; Lotka, 1925). Lotka was able to show that if the distributions of l_x and m_x are *stationary* (that is, unchanging through time), the age distribution will become stationary, and that the shape of the stationary age distribution is independent of the initial age distribution. This age distribution is called the *stable age distribution,* since, as Sharpe and Lotka (1911) showed, the population will return to the same age distribution if it is temporarily disturbed, so long as the l_m and m_x distributions are the same.

---

*This is not a generally agreed upon symbol; we have used it simply for convenience.

TABLE 4-6. VALUES OF R_0 AND T_c FOR A VARIETY OF SPECIES.*

Species		R_0	T_c (days)
Dugesia tigrina (planarian)		2.0	10.3
Physa gyrina (pond snail)		515.3	142.0
P. gyrina[a]		642.7	242.7
Daphnia pulex		21.4	10.6
D. pulex[a]		20.7	15.4
Pediculus humanus (human louse)		30.9	30.9
Nezara viridula (Hemiptera: Pentatomidae)		1.6	10.3
Phyllopertha horticola (Coleoptera: Rutelidae)		2.9	365.0
Lasioderma serricorne (Coleoptera: Anoliidae)		13.2	35.6
Calandra oryzae		113.6	58.1
Tribolium castaneum		275.0	80.0
Ptinus sexpunctatus	(Coleoptera: Ptinidae)[b]	4.1	227.5
P. fur	(Coleoptera: Ptinidae)[b]	10.8	182.3
P. tectus	(Coleoptera: Ptinidae)[b]	116.1	149.6
Niptus hololecus	(Coleoptera: Ptinidae)[b]	2.3	138.1
Eurostus hilleri	(Coleoptera: Ptinidae)[b]	3.1	115.6
Stethomezium squamosum	(Coleoptera: Ptinidae)[b]	41.0	190.6
Gibbium psylloides	(Coleoptera: Ptinidae)[b]	75.2	182.6
Trigonogenius globulus	(Coleoptera: Ptinidae)[b]	49.8	171.8
Mezium affine	(Coleoptera: Ptinidae)[b]	64.2	287.3
Microtus agrestis (vole)		5.9	170.4
M. orcadensis		12.2	332.4
Rattus norvegicus (pure stock)		16.5	344.1
R. norvegicus (cross albino stock)		16.8	236.5

*Compiled from several sources by Laughlin, 1965.

[a]Values of R_0 and T_c determined under different environmental conditions.

[b]Values for different species in this family determined under the same environmental conditions.

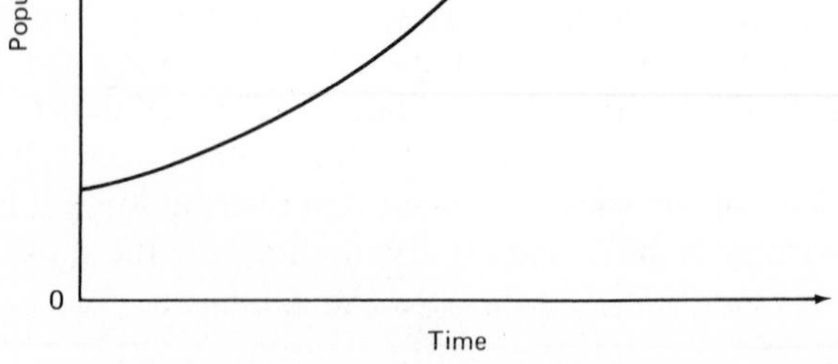

FIGURE 4–6. Population size plotted against time. The slope of this curve is the growth rate of the population. ΔN is the net increase in the population in the small interval of time Δt.

The mathematical proof that a growing population with stationary l_x and m_x distributions will develop a stable age distribution is beyond the scope of our discussion. An empirical demonstration of this process is provided, however, in Figs. 4–7 and 4–8. These graphs were generated from the life-table data given in Table 4–7. These data do not apply to any particular species; they were constructed only to provide a simple example. Inspection of the q_x column shows that a constant mortality rate of 15% per unit time has been assumed. Figure 4–7 shows the age distribution of the population every three units of time until the stable age distribution is reached. In this case, it was assumed that all individuals were in the youngest age class at time $t = 0$. This original cohort is clearly distinguishable as it progresses through time, becoming progressively smaller because of mortality and because its relative contribution to the total population is diluted as new animals are born. Figure 4–8 also shows a changing pattern of age distribution with time, but in this case all age classes were equal in size at time $t = 0$. Again, there is considerable change in the age distribution through time, but the distribution that eventually develops is identical to that in Fig. 4–7.

The length of time required for a stable age distribution to be reached depends both on the original age distribution and the l_x and m_x distributions. In general, the more rapidly the population is growing, the sooner the stable age distribution will be reached. We see in the example given that only a little more than twice the maximum length of life is required.

The shape of the stable age distribution that eventually develops can be predicted from the l_x and m_x distributions (Birch, 1948), but this is only occasionally done. We note, however, that the stable age distribution for a growing population will always be such that age classes become progressively smaller as age increases.

Under the conditions of stationary l_x, m_x, and age distributions, the rate of population growth per individual is a constant and the population grows exponentially. This constant rate of population growth is usually symbolized as r. Equation (4–1) then becomes

$$\frac{dN}{dt} = rN \qquad (4-2)$$

or, in integrated form,

$$N_t = N_o e^{rt} \qquad (4-3)$$

where N_t is the population size at some arbitrary time t, N_o is the initial population size, t is the length of time the population has been growing, and e is the base of the natural logarithms (2.718 . . .). The units of r (and k) may conveniently be thought of as *number added to population / unit time/individual in the population*, although this reduces to $1/t$.

The constant r is referred to by several different names, which further adds to the confusion already caused by using r with the meaning we have given to symbol k. Dublin and Lotka (1925) called r the "true rate

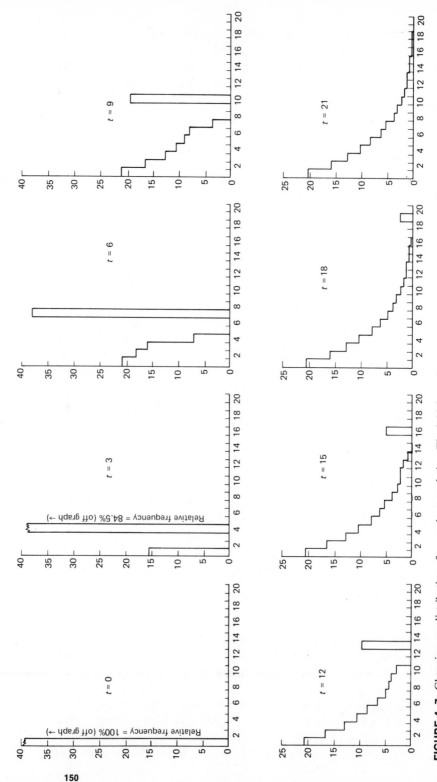

FIGURE 4–7. Changing age distribution of a growing population. The initial age distribution at time zero had all individuals in the first age class. The life table for the population is shown in Table 4–7. The abscissa in each graph represents age, the ordinate represents the fraction of the population that is in a given age class. Each age class is one time unit long.

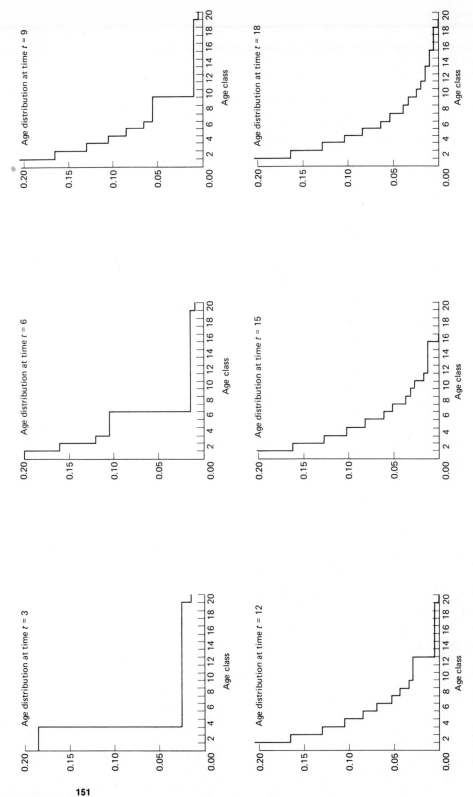

FIGURE 4–8. Changing age distribution of a growing population. The initial age distribution at time zero had all age classes equal in size, which would form a horizontal straight line. The life table for the population is shown in Table 4–7.

TABLE 4-7. HYPOTHETICAL LIFE-TABLE DATA USED IN CONSTRUCTING FIGS. 4-7 AND 4-8.

Age interval (x)	Fraction surviving to start of age interval	Age specific death rate per 1,000	Age specific birth rate per 1,000	Stationary age distribution
0–1	1.000	150.0	0.0	0.204
1–2	0.850	150.0	0.0	0.163
2–3	0.723	150.0	0.0	0.130
3–4	0.614	150.0	400.0	0.104
4–5	0.522	150.0	450.0	0.083
5–6	0.444	149.9	500.0	0.066
6–7	0.377	150.1	500.0	0.053
7–8	0.321	150.0	500.0	0.042
8–9	0.272	150.1	500.0	0.034
9–10	0.232	149.8	480.0	0.027
10–11	0.197	150.3	460.0	0.021
11–12	0.167	150.0	440.0	0.017
12–13	0.142	149.8	420.0	0.014
13–14	0.121	149.7	400.0	0.011
14–15	0.103	149.8	380.0	0.009
15–16	0.087	149.9	360.0	0.007
16–17	0.074	150.7	340.0	0.006
17–18	0.063	150.6	320.0	0.004
18–19	0.054	149.3	300.0	0.004
19–20	0.046	1000.0	280.0	0.002

Note:
$\Sigma l_x m_x = 1.7159.$
$\Sigma x l_x m_x = 12.490.$

of natural increase," which is not very suitable, since if a stable age distribution has not been reached, the value of r may differ considerably from the actual rate at which population is growing. Lotka and others have also referred to r as the "inherent" or "intrinsic" rate of increase, which has the disadvantage of implying that r is purely a characteristic of the species, rather than a consequence of the species characteristics interacting with environmental variables. Andrewartha and Birch (1954) have introduced the phrase "innate capacity for increase," which they symbolize as r_m. The innate capacity for increase, according to Andrewartha and Birch, is the value of r for a population growing in an environment in which "the quality of food, space, and animals of the same kind are kept at an optimum and other organisms of different kinds are excluded." Although population growth rate under these conditions may sometimes be of interest, the introduction of a new term and symbol, in our opinion, is confusing and unnecessary. We shall simply use the symbol r, rather than a name that may be misleading.

There is no simple formula for calculating r directly. Lotka (1925) has shown, however, that

$$\int_0^\infty e^{-rx} l_x m_x dx = 1 \qquad (4\text{-}4)$$

from which r can be calculated by numerical approximation techniques (Birch, 1948).

Curves A and B in Fig. 4–9 show population growth for the two hypothetical populations considered in Figs. 4–7 and 4–8. We see that even though these growth curves are based on the same life table (Table 4–7), they are significantly different; in the same length of time, the population begun with young animals (curve B) has reached a much larger size than the population begun with old animals.

Curves C and D are based on the same hypothetical life table (Table 4–7). Curve C was produced by assuming the initial age distribution to be identical to the stable age distribution. This curve is, then, idealized population growth as described by the exponential growth equation, Eq. (4–3). If we were to divide the slope of this curve at any point by the corresponding value of N, we would get the same value, namely r.

To construct curve D, all individuals were assumed to be in the fourth age class, that is, the age class in which reproduction begins according to our hypothetical life table. We see that there are regions of curve D that are steeper than regions of curve C having the same value of N, which shows that, at least for a short time, the actual population growth

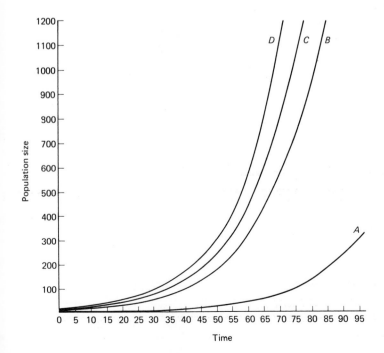

FIGURE 4–9. Population growth curves resulting from different initial age distributions. Age-specific mortality and fecundity rates are the same for all four curves. (A) All age classes initially equal in size; (B) all individuals initially in first age class; (C) initial age distribution identical to stable age distribution; (D) all individuals initially in fourth age class (the first age in which reproduction occurs).

rate, k, can be greater than r. This observation has particular significance since most animal species have a time during their life cycle when dispersal, or movement of some individuals outward from the population, takes place. The dispersing animals do not usually represent a random sample of the population but usually are the young ones that are soon to begin reproducing. Thus, initial population growth following colonization of an area may be enhanced.

The theory of population growth has been known to ecologists for many years, and values of r have been estimated for a variety of species, but only occasionally have these estimates been used to good advantage in demonstrating some interesting ecological phenomenon. One straightforward way in which estimates of r can be useful is in comparative studies involving one species under different conditions. Root and Olson (1969), for example, reared aphids (*Brevicoryne brassicae*) on several kinds of crucifers (broccoli and collards, which are two cultivated varieties of *Brassica oleracea;* Chinese cabbage, *Brassica pekinensis;* and yellow rocket, *Barbarea vulgaris*). Aphids were confined to small plastic "cages" clipped onto the leaves of host plants, which in all cases except one were placed in controlled environment chambers. The gathering of data was begun with a single newborn nymph per cage. Under suitable conditions, aphids reproduce parthenogenetically rather than sexually, and so it was sufficient to observe the number of offspring produced by each isolated aphid during its lifetime and to remove the newborn from the cage. Data for an l_x curve were obtained directly from observing the longevity of the original aphids.

The resulting data are shown in Figs. 4–10 and 4–11 and Table 4–8. The l_x data are essentially the same when collards, broccoli, and Chinese cabbage are the host plants, but they are distinctly different when yellow rocket is the host plant. The data for l_x on broccoli indoors and outdoors are not shown separately, since there was no observable difference in them. The differences among m_x curves for the aphids reared on different hosts are greater than the differences among l_x curves, with the curve for yellow rocket again being substantially lower than the others. In contrast to mortality rates, the m_x curves for aphids living on broccoli outdoors is higher than for those on broccoli in the controlled environment.

Values of r are given in Table 4–8. We see that r was smallest for aphids reared on yellow rocket, as we would expect from the shape of the l_x and m_x curves. The reason for this apparently was that 15 of the 48 nymphs placed on yellow rocket developed into alates (winged adults), which have lower fecundity and survival rates. No alates developed on the other hosts. The reason for this difference was not definitely established, but since conditions were approximately the same for aphids on different kinds of host except for those directly related to the plant characteristics, it is likely that plant factors were responsible. We note that yellow rocket is not a cultivated species but the other species are

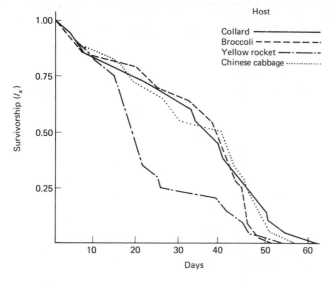

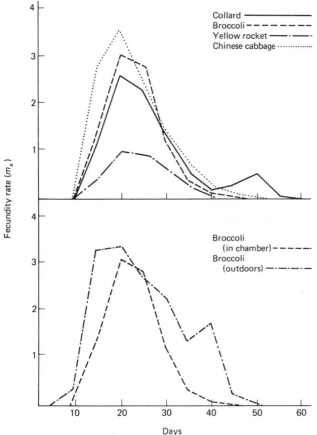

FIGURE 4–10. Age-specific survivorship curves for cabbage aphids reared on different hosts in a controlled environment room. (From Root and Olson, 1969.)

FIGURE 4–11. Age-specific fecundity curves for cabbage aphids reared on different hosts in a controlled environment room (above) and on the same host in constant and fluctuating environments (below). The m_x values are expressed as the mean number of nymphs produced per female on day x. (From Root and Olson, 1969.)

TABLE 4–8. THE INSTANTANEOUS RATE OF POPULATION INCREASE (r) AND
NET REPRODUCTIVE RATE (R_0) OF *BREVICORYNE BRASSICAE* COHORTS
REARED ON DIFFERENT HOST PLANTS AND IN DIFFERENT ENVIRONMENTS.*

Food plant	Environment	No. in cohort	Temp. ($^\circ$C) $\bar{X}$ (range)	r	R_0
Collard	Chamber	28	20 (19–21)	0.179	31.64
Yellow rocket	Chamber	48	20 (19–21)	0.094	5.58
Chinese cabbage	Chamber	29	20 (19–21)	0.223	44.04
Broccoli	Chamber	25	20 (19–21)	0.191	34.20
Broccoli	Outdoors	27	20 (5–34)	0.243	59.57

*From Root and Olson, 1969.

and they have been selected for characteristics that make them more
useful as a food crop for humans, some of which may also make them
more susceptible to aphids. Furthermore, if yellow rocket is not well
suited as a host for *B. brassicae,* then the development of alates on this
host could be adaptive in that it would greatly facilitate colonization of
more suitable hosts.

As Table 4–8 shows, the highest value for *r* was obtained from aphids
living on broccoli outdoors. This may seem surprising in view of the wide
range in temperature fluctuations experienced by these aphids. We note
that the midpoint of these fluctuations was about the same as the temperature within the controlled environment chamber, yet the m_x curve is
higher and peaks earlier. One possible explanation is that the higher
temperatures during the day accelerated developmental rates more than
enough to compensate for the cooler nights.

We must admit that the aphids in Root and Olson's study, even those
reared outdoors, were reared in somewhat unnatural environments. The
natural history of aphids, however, is such that the estimates of *r* may
have relevance in comparison to one another. As anyone who has gar-
dened knows, aphids seem to appear in large numbers rather suddenly on
suitable host plants. Careful observation will show, however, that almost
all of these are produced in a rapid phase of population growth from a
few colonists that are sparsely distributed. This early phase of population
growth may occur, therefore, under conditions similar to those devised
by Root and Olson, and thus the value of *r* seen under these conditions
may be quite a good indication of their behavior in nature.

GENERATION TIME

Thus far we have avoided saying anything about the length of a generation—and for good reason. Although "generation" is a common word in
everyday language and most of us have a rough idea of what a generation
is, for populations of continuously breeding organisms with overlapping
generations, the concept of "one generation" is quite abstract. Several

definitions have been proposed, and at this time there is no general agreement on which is preferable. These definitions do not generally give identical numerical results for the same set of data, although they may in certain special cases.

The definition that is perhaps the easiest to understand is suggested by the answer to the following question. Suppose that we are considering a population with stationary l_x, m_x, and age distributions growing at an instantaneous rate r. Suppose that we focus our attention on one cohort of females at the beginning of their life. We have already seen that under these conditions, the total number of female offspring produced by this cohort will be

$$\int_0^\infty l_x m_x dx$$

Given these suppositions, what is the average age of the mothers of newborn offspring produced by this cohort? This age is defined as one generation, T_c, and is given by the equation

$$T_c = \frac{\int_0^\infty x l_x m_x dx}{\int_0^\infty l_x m_x dx} \qquad (4\text{--}5)$$

T_c thus defined, is the mean of the $m_x l_x$ distribution. (The subscript c is used simply to distinguish this definition from others given below.)

One consequence of this definition may clarify its meaning. In reality, the females in a cohort produce offspring at various ages, as described by the m_x distribution. Suppose, instead, that all of the offspring were produced by the cohort when the mothers were of age T. The net effect would be that the population would grow at the same rate as before.

The numerical value of T_c, as defined above, is usually calculated by the approximation

$$T_c \cong \frac{\sum x l_x m_x}{\sum l_x m_x} \qquad (4\text{--}6)$$

Applying this approximation to the data in Table 4–7 gives $T_c \cong 7.279$. The total number of offspring that would be produced by cohort of 100 females with a life table as shown in Table 4–7 is 171.6. Thus, if upon reaching age 7.279, the females in this hypothetical population produced this many offspring and no more, the growth curve of the population would be the same as curve C shown in Fig. 4–9.

A second commonly used definition of generation time is

$$T = \frac{1}{r} \log_e R_0 \qquad (4\text{--}7)$$

which may give a somewhat different value from the first definition (Leslie, 1966). This definition has the advantage of being easy to apply in most cases, but if $r = 0$ (i.e., population size is not changing), then T is $0 \div 0$ and hence indeterminate. When defined in this way, the following interpretation of T can be made. Suppose we ask how much the population is multiplied by in time T if it is assumed that it is growing exponentially according to the equation $N_t = N_0 e^{rt}$. If we replace t by T and divide by N_0, we have

$$\frac{N_T}{N_0} = e^{rT} \qquad (4\text{–}8)$$

Taking natural logarithms of both sides and dividing by r gives

$$\frac{1}{r} \log_e \frac{N_T}{N_0} = T$$

Thus, we see that $N_T = R_0$; that is, if a generation is defined in this manner, then the population will be multiplied by R_0 in one generation.

Occasionally, r is estimated by first calculating the length of one generation according to Eq. (4–6) and then substituting the resulting value into Eq. (4–8), which is then solved for r. This procedure is generally unsatisfactory, since it may give an estimate of r that differs considerably from that which would be obtained from solving Eq. (4–4). For example, the value of r for the beetle *Calandra oryzae* estimated by using Eqs. (4–6) and (4–8) (Birch, 1948) gave an underestimate of about 25% (Caughley, 1967).

Leslie (1966) has proposed a third definition for generation time, namely,

$$\overline{T} = \int_0^\infty x e^{-rx} l_x m_x \, dx$$

Leslie states that if generation length is defined in this way, and if the l_x and m_x distributions are uncomplicated, the mean of the relative frequency distribution of the number of individuals in successive generations alive in the population at any given time will increase by $l / \overline{T}$ per time. Although this definition may eventually become widely used, this is not the case at the present time. For this reason, and since the interpretation of $\overline{T}$ is not all straightforward, we shall not discuss this definition further.

POPULATION GROWTH OF DISCONTINUOUSLY BREEDING SPECIES

Growth of real populations departs from the idealized case of stationary l_x and m_x distributions and stable age distribution discussed thus far. These departures are caused by a large array of variations in the physical environment, populations of other species, and changes in the size of the population itself. Departures are also caused by intrinsic characteristics of the species that have evolved in response to environmental conditions. In a majority of species breeding occurs seasonally, rather than throughout the year, and among many of these the generations do not overlap. Thus, a discussion of discontinuously breeding organisms could be considered more basic than continuously breeding organisms, but the latter kind has been given more attention by population biologists. This is partly due to the relative simplicity of the theory of population growth of continuously breeding organisms and partly because of the great interest in our own species, which breeds about as continuously as any to be found.

Unfortunately, a standard terminology and set of symbols have not been developed for seasonally breeding species. In the case of seasonal breeders with nonoverlapping generations (such as mayflies, for example), the statistic *I*, the *population trend index* (Balch and Bird, 1944), is often used. *I* is defined by the equation

$$I = \frac{N_{i+1}}{N_i}$$

where N_i and N_{i+1} are population sizes in two successive generations. Thus defined, the population trend index is analogous to R_0.

The length of one generation in this case is defined as the time between the start of successive breeding periods. If *I* remains constant for several generations, say *t*, we can write $N_t = I^t N_0$. This expression shows that, under this condition, population sizes at the start of successive generations will be on an exponential growth curve. Population size plotted against time will not, however, form a smooth exponential curve because after each breeding season, population size will decline according to whatever mortality rates are applicable.

The case of seasonally breeding populations with overlapping generations is more complex. Caughley (1967) considered this case in detail and suggested a useful approach. Caughley called a population having this kind of breeding pattern a *birth-pulse* population. For simplicity, we shall assume that breeding occurs at yearly intervals, although Caughley's ideas apply equally well for any constant interval between breeding seasons.

First, Caughley points out that the concept of an infinitesimal population growth rate, *r*, cannot be applied in a strict sense, since this type of reproductive pattern will produce a saw-toothed population growth curve, as does the case of seasonal reproduction with nonoverlapping generations. Similarly, the concept of a stable age distribution cannot be applied in its strict sense, since the age distribution of a birth-pulse population will be considerably altered every breeding season. We can, however, examine the age structure of the population each year (immediately following breeding, for example) and consider the population as having a stable age structure if it does not change from year to year. Caughley suggested that the statistic r be retained and interpreted as being the exponential growth rate of the hypothetical curve passing through points representing population size plotted at seasonal intervals. The value of r thus defined can be calculated from l_x and m_x by iteratively solving the equation

$$\sum e^{-rx} l_x m_x = 1$$

for r. (In this equation l_x and m_x are considered discrete variables spaced at yearly intervals.)

Finite birth and death rates are also redefined. For finite birth rate, Caughley suggests, "The finite rate at which a birth-pulse population with a stable age distribution defined at the same time each year would initially increase if deaths ceased." Mathematically, this may be expressed as

$$\text{Finite birth rate} = \frac{\sum l_x e^{-rx}}{\left(\sum l_x e^{-rx}\right) - 1}$$

Finite death rate is defined as "the finite rate at which a birth-pulse population would initially decline if births ceased" or

$$\text{Finite death rate} = 1 - \frac{\sum d_x e^{-rx}}{\sum l_x e^{-rx}}$$

where d_x is "the probability of death between x and $x + 1$."

Caughley also gives a new definition for R_0 for a birth-pulse population, but points out that this is not a very useful statistic, since its interpretation depends on length of a generation. As we have already said, generation length may be defined in several different ways, none of which is really very convenient or useful.

LIFE HISTORY PATTERNS AND POPULATION GROWTH

Just as we think of an organism's morphology as being a consequence of natural selection, we can safely assume that the pattern of its life history is a result of selection. Although this is by no means a new idea, evolutionists only recently have made notable progress toward developing quantitative models dealing with the specifics of how life history patterns and natural selection interact.

The diversity of life history patterns that exists in nature is enormous, as any student who has tediously memorized the life cycles of ferns, fungi, tapeworms, and crayfish in his introductory biology courses knows. The net sum of all of these details exhibited by any species is its ability to increase when conditions are favorable and survive when they are not. As we have already seen, this depends on the shapes of the l_x and m_x curves, which are, after all, abstract summarizations of the species' life history. Thus, we may conveniently think of natural selection as shaping the l_m and m_x curves of a species in a way that optimizes survival and population growth under the environmental conditions it typically encounters.

Cole (1954) was able to make a good start in this area by examining how r would be affected by various major changes in life history patterns. Cole simplified his analysis by considering only the consequences of various life history features without worrying about what particular kinds of environments they might be best suited to.

The arguments that Cole uses to develop his conclusions are beyond the level of mathematical sophistication that we have assumed our readers to possess. Cole's results, however, are straightforward. First, he considered the possible value of *iteroparity* (repeated reproduction). In his analysis, Cole made several simplifying assumptions: namely, that mortality during the reproductive period of life was negligible, that mothers in the iteroparic species began reproducing at one year of age and continued to produce the same number of offspring every year forever, and that litter size remained constant. Cole's analysis thus leads to the maximum gain that could be achieved by iteroparity. He concluded:

> *For an annual species, the absolute gain in intrinsic population growth which could be achieved by changing to the perennial reproduction habit would be exactly equivalent to adding one individual to the average litter size.* [Cole's italics.]

Cole also determined that the maximum possible percent gain in population growth rate that could be achieved by changing from *semelparous* (reproducing only once) to iteroparous reproduction. Figure 4–12 shows that gain plotted against age at first reproduction for several different litter sizes. This figure clearly shows that organisms that are slow maturing and produce small litters (such as many higher vertebrates) have much to gain through repeated reproduction, but rapidly maturing organisms that produce large litters have little to gain. Hence, selection among these organisms for traits allowing them to survive beyond one period of reproduction would be expected to be slight.

Figure 4–13 shows much the same information, but in a way that shows how the length of the prereproductive period and the number of annual births affect r. In the calculations, Cole assumed that the number of offspring per litter was one and the sex ratio 1:1, thus giving an average of 0.05 female offspring per litter. This pattern of reproduction is not unusual among higher vertebrates, and several specific examples are associated with the curves in Fig. 4–13. These curves show that the total

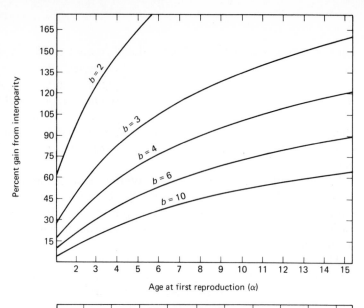

FIGURE 4–12. The effects of litter size, *b*, and age at maturity, *α*, on the gains attainable by repeated reproduction. The litter size is the number of female offspring per litter in the case of a sexual species. Percent gain from iteroparity is calculated by

$$\left(\frac{r_\infty - r_1}{r_1}\right) 100$$

where r_∞ is the value *r* would be if the species reproduced repeatedly and r_1 is the value *r* would be if the species reproduced only once. (From Cole, 1954.)

Percent gain from interoparity

Age at first reproduction (*α*)

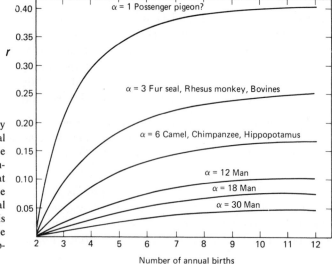

r

α = 1 Possenger pigeon?

α = 3 Fur seal, Rhesus monkey, Bovines

α = 6 Camel, Chimpanzee, Hippopotamus

α = 12 Man
α = 18 Man
α = 30 Man

Number of annual births

FIGURE 4–13. The effects of progeny number on the intrinsic rate of natural increase when the litter size is one ($b = \frac{1}{2}$. The ordinate scale shows the intrinsic rate of increase for species that produce an average of one-half female offspring per litter. For any given total progeny number, the intrinsic rate (*r*) is seen to be greatly affected by the age (*α*) at which the first offspring is produced. (From Cole, 1954.)

number of annual births for a species that matures quickly has a much greater effect on *r* than it does for a species that matures slowly. Thus, a rapidly maturing species, such as, perhaps, the passenger pigeon, would be more susceptible to environmental factors that reduce the longevity of adults than would a slowly maturing species, such as man. Furthermore, the three bottommost curves, which could apply to man, show once again that the age at which reproduction begins is very important in determining population growth rates. Indeed, we see that there is a larger percent increase in *r* if the number of annual births is held constant at 4 and age at first reproduction is reduced from 30 to 18 years than there is if the age at first reproduction is held constant at 18 and the total number of annual births is increased from 8 to 12.

Considering the seemingly great advantage in early reproduction and the comparatively small increase in population growth rate attainable through repeated reproduction, one might ask why perennial species are so common in most communities. We cannot yet provide a precise, quantitative answer to this question, but a qualitative explanation can be suggested.

We must keep in mind that Cole's analysis considered how changes in patterns of reproduction affect the potential value of r attainable by a species when an opportunity for exponential population growth arises. Selection favoring an increase in this rate has been termed "r selection" by MacArthur and Wilson (1967). Species associated with unstable or newly formed habitats presumably have undergone this kind of selection; such species are often called *opportunistic* or *colonizing* species.

Many species, however, utilize environmental resources that do not fluctuate greatly in their availability and perhaps do so in a more predictable manner. For these species, what is commonly called the "carrying capacity" of the environment—i.e., the population size that can be sustained for a long period of time—remains fairly constant. Under these conditions, there is no great advantage in being capable of rapid population growth, since the opportunity for such growth rarely exists. Selection, therefore, will tend to favor individuals that can most effectively utilize the resources available in maximizing their probability of surviving and producing a small number of offspring having a high probability of surviving. Since the hypothetical carrying capacity is often called K, MacArthur and Wilson (1967) have termed selection under these conditions "K selection."

Regardless of what kind of selection a population is subject to, it is reasonable to suppose that the resources, e.g., energy, time, available to a population are at times limited. By making this assumption together with several others, Gadgil and Bossert (1970) have been able to investigate the question of how optimal patterns of reproduction might vary in reponse to changes in the constraints placed on a population. The available resources must be allocated among the three basic life processes —maintenance, growth, and reproduction. Gadgil and Bossert point out that the allotments are not independent; energy that is used for maintenance is not available for reproduction, for example. The pattern of allocation is, in a sense, an abstract representation of the life history of the species, and we can reasonably expect the allocation to be modified by selection in the direction of an optimal combination. In particular, how much effort would one expect to be allocated to reproduction during successive age classes [reproductive effort by an individual of age i being defined as "the fraction of the total amount of resources of time and energy available to an individual at (age i) that is devoted to reproduction"]? With this question in mind, Gadgil and Bossert developed a model using three basic assumptions. First, increasing reproductive effort by any age class was assumed to decrease the survival rate of that age class; second, it was assumed that an increase in the reproductive effort of an age class resulted in a decrease in the growth of individuals of that age class; and third, the number of offspring a female could produce was assumed to increase as the size of the female increased.

Computer simulation techniques rather than conventional analysis were used by Gadgil and Bossert to obtain an idea of what constitutes an optimal pattern of reproduction. From the results of these simulations the following trends emerged:

1. For a species that reproduces once and then dies, the optimum age for reproduction tends to increase as the availability of resources becomes more limiting;

2. For a species that reproduces repeatedly, the amount of reproductive effort that is optimum for each age class increases as the availability of resources becomes more limiting;

3. If the fecundity of an organism increases slowly as it grows, the size at which it makes the greatest contribution to population growth can be less than its size when maximum fecundity is attained.

Hairston et al. (1970) also investigated the evolution of reproductive rates using computer simulation techniques. Their investigations, however, took a different approach than those of Gadgil and Bossert, in that Hairston et al. examined changes in gene frequencies associated with phenotypes having different reproductive rates.

The results of one of the simulations of Hairston et al. is shown in Fig. 4–14. In this simulation it was assumed that a single pair of alleles determined the pattern of reproduction. Homozygous AA individuals were assumed to produce 10 offspring at age 1 and then cease reproducing; aa individuals were assumed to produce 5 individuals at age 1 and 5 more at age 2; and Aa individuals were assumed to be intermediate. Survival rates were randomly selected from a normal distribution having a mean of 0.1 and a standard deviation of 0.04. This procedure was used so that the size of the population at any point in time did not affect the survival rate during any following interval of time. These particular values for the mean and standard deviation were chosen so that it was unlikely that the population would become excessively large or go to extinction. Figure 4–14 shows that the aa genotype was more abundant than the AA genotype for a large fraction of the time. More interestingly, the relative advantage held by each genotype appeared to shift depending on whether the population was declining (when aa was favored) or increasing (when AA was favored). This result agrees nicely

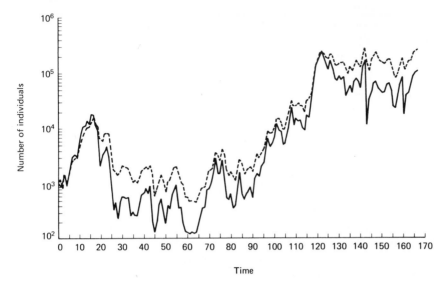

FIGURE 4–14. Changes in two hypothetical homozygous genotypes in a population. The solid line represents individuals that produce ten offspring at age one and then die; the broken line represents individuals that produce five young at age one and five more at age two. Heterozygous individuals are intermediate. (From Hairston, Tinkle, and Wilbur, 1970.)

with the more qualitative statements of MacArthur and Wilson (1967) regarding r and K selection.*

PATTERNS OF DISPERSION

Our discussion so far has centered on changes in populations through time but we have ignored how the members of populations are distributed through space, that is, their *pattern of dispersion*.† Common experience tells us that organisms are almost never regularly spaced in their habitat, like points in a grid, but beyond that, subjective observations are of little help. Two other possibilities suggest themselves: the members of a population may be randomly positioned throughout their habitat or they may tend to be aggregated. By aggregated we mean that in some areas the density of organisms is considerably higher than the overall mean density but in other areas the density is considerably lower; this pattern is more pronounced than could reasonably be expected by chance.

In fact, random spacing of individuals is almost as unusual in natural populations as patterns that tend toward uniform spacing. This should not

*Further investigations of r and K selection have been reported by Roughgarden (1971) and Charlesworth (1971), but these appeared too late to be included in the above discussion.
†*Dispersion* must not be confused with *dispersal*, which usually means the movement of organisms outward from a population.

be surprising since there are many possible reasons for expecting organisms to be aggregated. We shall list several possibilities, not simply for their own sake, but because one of the justifications for studying pattern is that it may aid our understanding of the underlying processes that account for pattern.

1. Social behavior may result in individuals being attracted to one another.
2. Organisms may respond to environmental heterogeneity so that they tend to settle in some areas and avoid others. (The environmental factors they are responding to in this way may not be apparent to us.)
3. Settling may be random, but environmental heterogeneity may lead to differential mortality with respect to location.
4. Offspring of low motility may develop from eggs laid in clusters, or seeds from a single parent plant may fall in a limited area.
5. Organisms may be moved passively but nonrandomly by wind or water currents so that they tend to accumulate in some areas.

This list is not exhaustive, and more importantly, we note that the existence of a particular spatial pattern is not, in itself, sufficient evidence for assuming that a particular process is responsible.

A second major reason for examining the dispersion of a population is so that relationships between density and various factors affecting mortality and natality can be more reasonably assessed. If, for example, mortality rates are known to be related to density, it is unreasonable to consider a population of randomly dispersed individuals as being equivalent to a population of the same size but in which the individuals are highly aggregated.

In almost all of our discussion we will assume that we are dealing with organisms that are *sessile,* or low in mobility. Almost all plants and animals have a period during their life history when they can move or be passively transported long distances. This period is usually short, though perhaps crucial, but we shall not consider it now because we are more concerned with their dispersion after they have "settled down."

Three distinct cases can be recognized for the purpose of facilitating our discussion of dispersion (Pielou, 1969). The first concerns organisms occurring only in certain discrete, objectively recognizable units within an area. We may know, for example, before we begin sampling that a certain kind of brittle star occurs only under rocks in the intertidal zone and almost never in the gravel between them. Or, the larval stages of a species of insect that interests us may reasonably be expected to be found only within the fruit on a particular kind of tree. These discrete units can be treated as sampling units, and we can examine the frequency with which various numbers of individuals occur in them.

In other instances we may have no a priori reason for expecting the species we are studying to occur in recognizable, discrete units of habitat. We may know only that it occurs scattered throughout an area that may vary from place to place but does so in a continuous manner. If we were studying the dispersion of collembolans in the leaf litter on a forest floor, we would probably find it quite impossible to designate some naturally occurring units for purposes of sampling as we did above. We would be forced to create sampling units by some means such as placing circles $1/10$ m^2 in area at random and determining the number of collembolans in each circle. Almost all studies of the dispersion of plant populations fall into this case.

The third case recognized by Pielou (1969) involves organisms that have a growth form making it impossible or completely impractical to distinguish individuals. Bermuda grass is an excellent example. Encrusting, colonial marine organisms such as ectoprocts, tunicates, and some corals are also good examples. Figure 4–15 shows two possible forms that this kind of pattern may take: one in which open areas are left scattered within a larger area covered by the organism; the other in which the organism forms spreading islands. Analysis of this third case has received little attention by ecologists in comparison to the first two.

Random Dispersion

If every sampling unit has the same probability of being occupied by an individual, and the presence of an individual does not change the probability of another individual being found there, the dispersion pattern will be random. In this case the probability of finding n individuals will be given by terms of the Poisson distribution, that is,

$$p_n = \frac{\lambda^n \, e^{-\lambda}}{n!}$$

where λ is the mean number of individuals per sampling unit and e is the base of the natural logarithms. Recalling that probabilities can be thought of as fractions between zero and one representing frequencies of occurrence, we see that if we multiply p_n for each n by the total number of sampling units, we will obtain the expected number of sampling units with $0, 1, 2, \ldots, n, \ldots$ individuals each.

The Poisson distribution has the property that the mean equals the variance, and therefore, the number of sampling units expected to contain $0, 1, 2, \ldots$ individuals each can be calculated from an estimate of either of these values. Moreover, since the expected ratio of the variance to the mean is one, this ratio obtained from data can be compared to the expected ratio of one in a simple statistical test to determine whether the data conform to a Poisson distribution. Thus, it is fairly easy to

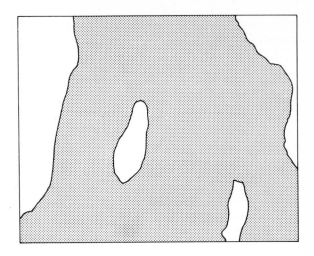

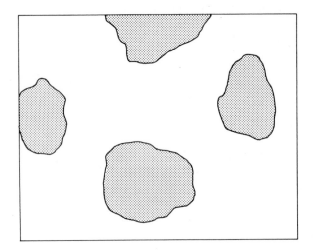

FIGURE 4–15. Two possible patterns of dispersion that could develop from the spread of colonial organisms.

determine whether the dispersion of a population can be assumed random.

We have pointed out that it is unusual for populations to have a random dispersion pattern. Demonstrating this statistically for a given population is not especially interesting. Understanding what is meant by random dispersion is important, however, since it serves as a convenient reference for discussing other possible dispersions. The variance to mean ratio, moreover, is sometimes used as an index of dispersion with the assumption being made that as a uniform dispersion is approached this ratio becomes progressively smaller than one, but as the degree of aggregation increases this ratio becomes progressively larger than one. A related statistic has been proposed by David and Moore (1954). This statistic, called the *index of clumping,* is given by

$$I = \frac{s^2}{\bar{x}} - 1$$

Since the expected variance to mean ratio for a randomly dispersed population is one, we see that in this case $I = 0$. In addition, the definition suggests that I increases as clumping increases.

Pielou (1969) points out that if deaths occur randomly among individuals in a population, the index of clumping, I, will decrease linearly as population density decreases. Pielou suggests ways this property could be used as a test of whether deaths among a cohort of sedentary organisms might be density dependent, but to our knowledge this has not yet been done.

The Negative Binomial Distribution

Since the Poisson distribution is inadequate for describing the dispersion of the great majority of populations, many attempts have been made to find more satisfactory distributions. (A good introduction into this literature may be found in Greig-Smith, 1964.) The distribution that appears to be the most widely applicable is the negative binomial. According to this distribution the probability of there being n individuals in a sampling unit is expressed as

$$\rho_n = \frac{k\,(k+1)\ldots(k+n-1)\,(k\lambda)^n}{n!\,Q^{k+n}} \tag{4-9}$$

where k is a fitted positive constant, λ is the mean number of organisms per sampling unit, and $Q = 1 + \lambda/k$.

This distribution thus requires only two parameters (λ and k) to be estimated. Although theoretically, k can be any positive number, in practice values tend to range from around 1 to about 8. It is trivially simple to estimate λ, but unfortunately the best estimate of k must be obtained by a somewhat tedious iterative method (Bliss and Fisher, 1953). This parameter has properties that make it useful in itself. Very small values of k are associated with a high degree of aggregation, and as $k \rightarrow \infty$ the negative binomial distribution approaches the Poisson. A second potentially useful property of k is that it does not change as a population decreases due to deaths occurring randomly (Pielou, 1969). Successive estimates of k through time could thus reveal whether or not the probability of dying was higher for individuals in areas of aggregation.

Table 4–9 shows data from a population of adult beet leaf-hoppers (Bliss, 1958). This population was examined twice, once in the spring

TABLE 4–9. FREQUENCY DISTRIBUTIONS OF ADULT BEET LEAF-HOPPERS PER BEET IN A $4\frac{1}{2}$ ACRE FIELD FROM TWO COUNTS IN 1937.*

| | May 20 | | | | | August 26–27 | | | | | |
u	f	$P\phi$	$NB\phi$	u	f	$P\phi$	$NB\phi$	u	f	$P\phi$	$NB\phi$
0	174	161.3	173.1	0	2	0.1	1.6	12	15	16.4	16.8
1	112	129.5	115.9	1	8	1.1	6.2	13	14	10.0	12.9
2	54	52.0	48.4	2	14	4.2	13.7	14	8	5.6	9.6
3	14	13.9	16.2	3	19	11.0	22.2	15	6	2.9	7.1
4	4	2.8	4.7	4	28	21.8	30.0	16	6	1.5	5.1
5	1	0.4	1.3	5	41	34.3	35.4	17	4	0.7	3.6
6	1	0.1	0.3	6	30	45.2	37.9	18	1	0.3	2.5
7+	0	0	0.1	7	38	50.9	37.5	19	3	0.1	1.7
				8	39	50.2	34.8	20	1	0.1	1.2
				9	32	44.0	30.8	21	2	0	0.8
				10	30	34.7	26.0	22	1	0	0.5
				11	18	24.9	21.2	23+	0	0	0.9
$P(x^2)$		0.13	0.74							<.001	.978

*From Bliss, 1958.

Note: On each date, the leaf-hoppers were counted on 10 sugar beets, selected at random, from among 3,000 plants in each of 36 plots. The number of beets (f) infested with $u = 0, 1, 2 \ldots$ leaf-hoppers has been fitted by a Poisson ($P\phi$) and a negative binomial ($NB\phi$) distribution.

and again in late summer. The spring population was made up of invading migrant adults. We see that their dispersion does not differ greatly from randomness, as is shown by a comparison with the Poisson distribution, but a negative binomial is a much better description. By late summer, after the population had grown considerably, a Poisson distribution had become totally inadequate but the agreement with a negative binomial is excellent.

Several different, biologically reasonable models that lead to a negative binomial distribution have been suggested, two of which we briefly describe. Suppose that a species usually occurs in clumps because of some biological phenomenon. We may imagine, for example, an insect that deposits its eggs in clusters in the fruit of some particular kind of tree, and that we are interested in describing the frequency distribution of number of larval insects found per fruit. Suppose that the females randomly select fruits from all those available in which to lay their eggs. (This implies that more than one cluster of eggs may be laid in a fruit.) Suppose that the number of larvae emerging per cluster follows a logarithmic distribution. (This means that the probability of their being n larvae from a cluster is given by

$$p_n = \frac{-\alpha^n \, n^{-1}}{\ln{(1 - \alpha)}}$$

where α is a fitted constant such that $0 < \alpha < 1$.) Under these assumptions the number of larvae per fruit will follow a negative binomial distribution.

We note that even though we have referred to the logarithmic and Poisson distributions in this model, neither of them is actually used in calculating the terms of the negative binomial. This is done by using Eq. (4–9).

A second, different model that leads to a negative binomial distribution can be developed by assuming that the sampling units differ in their "suitability." Let us again use a hypothetical insect as an example, but now suppose that eggs are laid individually rather than in clusters and that the female searches randomly for a fruit to oviposit in before laying each egg. If all fruits were equally susceptible to being oviposited in, the expected frequency distribution of eggs per fruit would follow a Poisson distribution. But suppose that the fruits differ in their susceptibility and that the relative frequency of fruits of differing susceptibility has some particular distribution. If this relative frequency distribution is a gamma distribution, the resulting frequency distribution of eggs per fruit will follow a negative binomial. (The gamma distribution that is described in almost all mathematical statistics texts is a family of unimodal curves; the chi-square distribution and the negative exponential are particular members of this family. The range of variation among members of this family makes it reasonable to suppose that many sets of biological data of the kind we are discussing could be described by members of this family.)

Several alternative sets of assumptions leading to the negative binomial have been proposed (Waters and Henson, 1959), and we may expect additional models to be suggested in the future. This illustrates an important point we made earlier, namely, the fact that a particular set of data conform to a prediction made by a model is not, in itself, conclusive evidence that the model is a valid explanation, since other models might have led to the same predictions.

Population Dispersion in a Continuum

Thus far we have assumed that the organisms being considered occur in discrete sampling units, which simplifies matters considerably. In many instances, however, there is no reasonable way to distinguish discrete sampling units (Pielou's case 2, above), and, therefore, other approaches must be used. The problems that arise mainly involve aspects of statistical techniques and are, therefore, not appropriate for a full discussion here. We shall briefly mention them, however, since data obtained with no awareness of technical problems may be misinterpreted and thereby distort our concepts of ecological processes.

Several approaches are used when discrete, naturally occurring sampling units are not available. The most common approach is to place a device, such as a wire hoop, wooden rectangle, etc., at random locations in the area being sampled and then make counts (and perhaps take other data) of the number of individuals within each selected area. Arbitrarily designated sampling units of this kind are called *quadrats*. Although quadrats are usually placed randomly, other methods are sometimes used for special purposes.

In considering population dispersion in a continuum, it is useful to recognize two different ways in which dispersions may differ. These are called *intensity* and *grain* by Peilou (1969). The difference between them can be visualized by imagining a jigsaw puzzle in which the pieces are different shades of grey. Suppose the darkness of each piece is proportional to population density and the size of each piece is proportional to the area covered by the population at that density. How much the shade of grey varied from place to place would then represent intensity, with high contrast being analogous to high intensity. The size of the pieces, on the other hand, would represent grain. Clearly, populations could differ in either or both of these two properties, and a single measure of dispersion could not adequately represent both. All such measures based on sampling using quadrats of one size are measures of intensity (Pielou, 1969).

Data obtained from quadrat sampling are commonly treated in the same way as data from discrete, naturally occurring sampling units; that is, they are compared to a Poisson distribution or to a negative binomial, and so forth. Major difficulties may arise in doing this, however, since the choice of quadrat size and shape can influence the result, as can the orientation of the quadrats.

That the choice of quadrat size can be important is shown by a simple example. Suppose that the species we are interested in occurs in aggregations that are not sufficiently distinct as to be self evident. Large quadrats may frequently encompass several areas of aggregation, but, in contrast, small quadrats may frequently cut across or fall entirely within areas of aggregation. In either case, misleading results may be obtained.

Table 4–10 shows variance to mean ratios and several other parameters for three species of plants in an old field in Michigan. We see that not only do these parameters change as quadrat size changes, but that they change differently for each species. The entire population of each species was included in this analysis so that the effects of changing quadrat size were not due to sampling error. We suspect that the effects of quadrat size would have been even more pronounced if random sampling had been used.

A variety of techniques for dealing with these problems has been proposed, e.g., using quadrats of several sizes for sampling a single population. We shall not pursue these suggestions here but good discussions can be found in Greig-Smith (1964) and Pielou (1969).

**TABLE 4-10. EFFECTS OF QUADRAT SIZE ON
VARIANCE TO MEAN RATIO FOR THREE
COMPONENT SPECIES IN AN OLD-FIELD
COMMUNITY, EDWIN S. GEORGE RESERVE,
LIVINGSTON COUNTY, MICHIGAN.***

Quadrat size (sq m)	$V/\bar{x}$		
	Lespedeza[a]	Liatris[b]	Solidago[c]
0.0625	1.102	1.038	1.007
0.125	1.217	1.082	1.024
0.25	1.478	1.164	1.038
0.5	1.843	1.341	1.085
1.	2.586	1.742	1.206
2.	3.864	2.424	1.310
4.	6.319	3.399	1.512
8.	9.809	5.874	1.983
16.	15.319	12.193	2.633

*From Evans, 1952.
Densities:

[a]*Lesedeza* 0.1014 plants/m^2.
[b]*Liatris* 0.0434 plants/m^2.
[c]*Solidago* 0.0565 plants/m^2.

Plotless Sampling

Since the effects of quadrat size and shape can be a serious problem, a class of techniques that avoids their use altogether has been developed. This class is referred to as *plotless sampling*. All these techniques are based on the same general principle, namely, using measurements of distance from either randomly selected points to the nearest individual or from one individual to another. Methods using the latter approach are called *nearest-neighbor* techniques. In some cases measurements are made from every individual to its nearest neighbor, and in other instances a random sample of individuals is chosen and measurements are made to the nearest neighbor of each of these. The difficulty in obtaining a truly random sample from which to take measurements is, however, a serious drawback with this latter approach. In all these methods the measurements are combined into statistics that reflect the degree of aggregation of the population and can be used in tests for randomness of dispersion. In most cases, however, good estimates of density must also be obtained. All the methods give approximately the same results for populations having random dispersions, but they may give different results for nonrandom populations.

Mean Crowding

Estimates of mean population density are a part of virtually all studies of population dynamics, yet we have seen that there is generally considerable variation in density from place to place within the area occupied by a population. Thus, a simple estimate of mean density for the entire area will be an underestimate for some places and an overestimate for others. The indices of aggregation that we have discussed, mean clumping, the parameter k in the negative binomial, and others that have been proposed reflect the tendency of organisms to be aggregated, but they are difficult to think of in intuitive terms. Moreover, by themselves they do not reflect population density. In response to this need Lloyd (1967) has introduced an index that he calls *mean crowding*. The definition of mean crowding is suggested by the answer to the question: "On the average, with how many other individuals does each member of a population share its quadrat?" The answer is given by

$$\overset{*}{m} = \frac{\sum\limits_{i=1}^{N} X_i}{N} \qquad (4\text{--}10)$$

where X_i is the number of individuals located in the same quadrat as the *ith* individual N is the total number of individuals, and $\overset{*}{m}$ is called *mean crowding*. This can be demonstrated by a simple example. Suppose we have 3 quadrats in which there are 4, 3, and 2 individuals. For each of the 4 organisms in the first quadrat, $X_i = 3$ with $i = 1, 2, 3, 4$. For the second quadrat, $X_i = 2$ with $i = 5, 6, 7$; and for the third quadrat, $X_i = 1$ with $i = 8, 9$. There are 9 individuals all together so

$$\overset{*}{m} = \frac{4 \times 3 + 3 \times 2 + 2 \times 1}{9} = 2.22$$

Mean density per quadrat, on the other hand, would be $9/3 = 3$.

One important advantage to $\overset{*}{m}$ is that, unlike mean density, it is not affected by empty quadrats (Lloyd, 1967). Thus, if there were two empty quadrats in the above example, $\overset{*}{m}$ would still be 2.22 but mean density would become 1.8. This may be especially advantageous in samples in which many empty quadrats occur, not because the population is sparse relative to its suitable habitat, but because we have failed to recognize unsuitable hatitat.

Equation (4–10) may be used directly if the quadrats cover the entire area of interest, but, unfortunately, it gives biased results if only a random sample is available (which is usually the case). Therefore, by making the reasonable assumption that the dispersion of the population approximates a negative binomial, Lloyd derives an equivalent statistic for data that came from a random sample. This statistic is called the *sample mean crowding, $\overset{*}{x}$,* and is given by

$$\overset{*}{x} = \bar{x} + \frac{\bar{x}}{k}$$

where $\bar{x}$ is the sample mean density. Thus we see that this simple statistic incorporates information on both mean density and pattern of dispersion. The expression is easily interpreted intuitively if we recall that as k becomes large, the negative binomial approaches the Poisson distribution. At the same time, $\bar{x}/k$ will approach zero and we see that mean crowding equals mean density for a randomly dispersed population. On the other hand, k would be small for a highly aggregated population and hence $\bar{x}/k$ would be large and $\overset{*}{x}$ would become larger than $\bar{x}$. This would agree with our intuition that tells us that the effects of crowding in an aggregated population would be greater than indicated by mean density.

REFERENCES

Andrewartha, H. G. and L. C. Birch. 1954. *The Distribution and Abundance of Animals.* Chicago: University of Chicago Press. 782 pp.

Balch, R. E., and F. T. Bird. 1944. A disease of the European spruce sawfly, *Gilpinia hercyniae* (Htg.), and its place in control. *Sci. Agr.,* **25**:65–80.

Birch, L. C. 1948. The intrinsic rate of natural increase of an insect population. *J. Anim. Ecol.,* **17**:15–26.

Bliss, C. I. 1958. The analysis of insect counts as negative binomial distributions. *Proc. 10th Int. Cong. Ent.,* **2**:1015–32.

———— , and R. A. Fisher. 1953. Fitting the negative binomial distribution to biological data and a note on the efficient fitting of the negative binomial. *Biometrics,* **9**:176–200.

Caughley, G. 1967. Parameters for seasonally breeding populations. *Ecology,* **48**:834–39.

Charlesworth, B. 1971. Selection in density-regulated populations. *Ecology,* **52**:469–74.

Cole, L. C. 1954. The population consequences of life history phenomena. *Quart. Rev. Biol.,* **29**:103–37.

David, F. N., and P. G. Moore. 1954. Notes on contagious distributions in plant populations. *Ann. Bot. Lond. N.S.,* **18**:47–53.

Deevey, E. S., Jr. 1947. Life tables for natural populations of animals. *Quart. Rev. Biol.,* **22**:283–314.

Dublin, L. I. and A. J. Lotka. 1925. On the true rate of natural increase as exemplified by the population of the United States, 1920. *J. Amer. Statist. Assoc.,* **20**:305–39.

Evans, F. C. 1952. The influence of size of quadrat on the distribution pattern of plant populations. *Contrib. Lab. Vert. Biol. Univ. Mich.,* No. 54.

Gadgil, M., and W. H. Bossert. 1970. Life historical consequences of natural selection. *Amer. Natur.,* **104**:1–24.

Greig-Smith, P. 1964. *Quantitative plant Ecology* (2nd ed.). London: Butterworths. 256 pp.

Hairston, N. G., D. W. Tinkle, and H. M. Wilbur. 1970. Natural selection and the parameters of population growth. *J. Wildl. Mgmt.,* **34**:681–90.

Laughlin, R. 1965. Capacity for increase: A useful population statistic. *J. Anim. Ecol.,* **34**:77–91.

Leslie, P. H. 1966. The intrinsic rate of increase and the overlap of successive generations in a population of guillemots (*Uria aalge* Pont.). *J. Anim. Ecol.,* **35**:291–301.

Lloyd, M. 1967. Mean crowding. *J. Anim. Ecol.,* **36**:1–30.

Lotka, A. J. 1925. *Elements of Physical Biology*. Baltimore: Williams & Wilkins.

Lowe, V. P. W. 1969. Population dynamics of the red deer (*Cervus elaphus* L.) on Rhum. *J. Anim. Ecol.,* **38**:425–57.

MacArthur, R. H., and E. O. Wilson. 1967. *The Theory of Island Biogeography*. Princeton, N. J.: Princeton University Press. 203 pp.

Pielou, E. C. 1969. *An Introduction to Mathematical Ecology*. New York: Wiley-Interscience. 286 pp.

Root, R. B., and A. M. Olson. 1969. Population increases of the cabbage aphid, *Brevicoryne brassicae,* on different host plants. *Can. Ent.,* **101**:768–73.

Roughgarden, J. 1971. Density-dependent natural selection. *Ecology,* **52**:453–68.

Sakagami, S. F., and H. Fukuda. 1968. Life tables for worker honeybees. *Res. Pop. Ecol.,* **10**:127–39.

Sharpe, F. R., and A. J. Lotka. 1911. A problem in age distribution. *Phil. Mag.,* **21**:435–38.

Southwood, T. R. E. 1966. *Ecological Methods*. London: Methuen. 391 pp.

Spinga, C. A. 1970. Population dynamics of the Uganda defassa waterbuck (*Kobus defassa Ugandae* Neumann) in the Queen Elizabeth Park, Uganda. *J. Animal Ecol.,* **39**:51–78.

Waters, W. E., and W. R. Hensen. 1959. Some sampling attributes of the negative binomial distribution with special reference to forest insects. *For. Sci.,* **5**:397–412.

Chapter 5

POPULATION LIMITATION

Populations are being continually buffeted by their changing environments. There always seems to be either too many individuals present for the resources at hand or fewer than could be supported. Sudden, unpredictable changes in the weather may cause massive mortality. An outbreak of a disease that usually occurs infrequently decimates the population, perhaps because a superabundance of food or a mild winter has permitted the population to build up to a density that promotes the spread of disease. In short, almost all populations, almost all of the time are not quite in tune with their environment.

As a population is buffeted by its environment, not only does its size change, but it also changes in many other respects — the frequency of pre-reproductive and postreproductive individuals, the sex ratio, the average size and state of health of the individuals all may change. Some of these changes will tend to bring the size of the population more into accord with prevailing environmental conditions, but other changes may have the opposite effect. Thus, a population together with its environment can be considered a complex system of interacting components. At present, there are several strongly divergent viewpoints among ecologists about the relative importance of various parts of this system and the interactions among these parts. It is becoming increasingly clear, however, that in order to understand how the abundance of organisms through space and time is determined, the population and its environment must be studied as a single system.

The diversity of ecological phenomena to be considered makes it difficult to avoid presenting a fragmented picture of population limitation. Therefore, we shall begin our discussion by pointing out in a qualitative manner the relationships among variables that affect population size and by discussing quantitative concepts that apply generally to these variables. After this general introduction, specific ecological phenomena, such as predation and competition, will be discussed individually in more detail. We hope that as this is done, the student will attempt to relate these phenomena to one another as components of unified systems. The chapter will conclude with an example of a study in which this was actually done.

Figures 5–1(a) and 5–1(b) show diagrammatically the relationships among the major factors influencing population size. The variable named at the tail of each arrow is potentially important in determining the size of

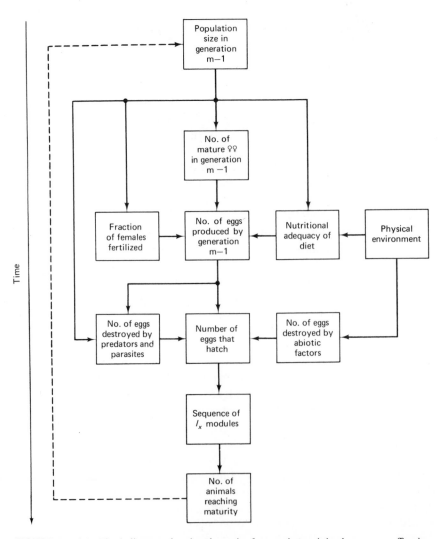

FIGURE 5–1 (a). Block diagram showing the major factors determining how many offspring are produced each generation.

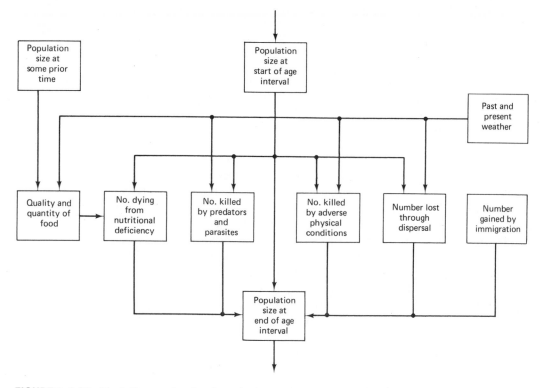

FIGURE 5–1 (b). Block diagram showing the major factors determining how many offspring survive to begin the next generation.

the variable at the head of the arrow. For the sake of simplicity the diagrams have been constructed for an egg-laying species having nonoverlapping generations. The arrangement of blocks is certainly not the only one possible, and modifications and additions must be made depending on the characteristics of the particular species involved. For example, if the species being considered were a wasp that could reproduce both with and without mating, considerable modification of Fig. 5–1(a) would be necessary. The student should find it instructive to choose a species with which he is familiar and attempt to construct a more detailed diagram than Figs. 5–1(a) and 5–1(b) for that species.

By arranging the variables that are important in limiting population size into a block diagram two important concepts become obvious. First, note that the diagram can conveniently be broken into two parts, one part [Fig. 5–1(a)] composed of variables determining how many new offspring are produced each generation, and a second part composed of variables determining how many of these offspring survive to begin the next generation. This latter part is consolidated into one block labeled "sequence of

l_x modules" in Fig. 5–1(a); Fig. 5–1(b) shows and l_x module in detail (strictly speaking, "number gained by immigration" should not be included here, but we have done so for convenience). As many l_x modules as are necessary to show accurately the dynamics of a population can be included in series by using smaller intervals of time for each module. The data for population size from all l_x modules in succession make up the l_x column of the species' life table. The importance of life tables should become evident as they are considered in the context of Figs. 5–1(a) and 5–1(b).

The utility of distinguishing factors causing variations in natality from factors responsible for variations in mortality has been demonstrated by Southwood (1967). Figure 5–2(a) shows that, in the population studied, the number of frit flies (*Oscinella frit*) reaching pupation is closely related to population size at the egg stage, but unrelated to mortality during the larval stages. In contrast, Fig. 5–2(b) shows comparable data for a population of the beetle *Phytodecta olivacea* in which mortality rather than natality is closely related to subsequent population size.

The second concept that becomes evident from Figs. 5–1(a) and 5–1(b) is that some of the factors influencing population size are not in turn influenced by the population, for example, weather. Other factors may both affect and be affected by the population; the number dying from starvation is an example. This distinction is important to make, since factors whose effect on population size is influenced by population size

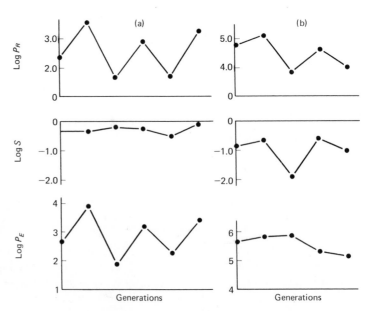

FIGURE 5–2. The visual correlation of the logarithm of resulting population (P_R) with log natality (P_E) and log survival (S). (a) Natality dominant, *Oscinella frit* (L.). (b) Mortality dominant, *Phytodecta olivacea* (Forster). Note that since survival rate, S, is between 0 and 1, log S is negative. (From Southwood, 1967.)

can induce population stability or instability in a way analogous to feed-back in an electrical circuit. If a factor acts in such a way that the pro-portion of the population affected by it is, at least in part, a function of population size, it is called *density dependent*. Figure 5–3 (Morris, 1963) illustrates the density dependent relationship between survival of first- and second-instar spruce budworm larvae and population density in a New Brunswick forest.

As the logical alternative to density dependence, we shall say that a factor is *density independent* if the fraction of the population affected is independent of population size. This does not mean that a density inde-pendent factor necessarily affects a constant fraction of the population, although this may be the case. More commonly, the fraction affected varies, but does so independently of population size, thus producing a more or less random scatter of points around a horizontal straight line when plotted as shown in Fig. 5–4.

It should not be inferred that density dependent factors are necessarily responsible for large changes in population size. The factor having the greatest effect on population size is often called the *key factor* (Morris, 1959, 1963). Figure 5–5 (Blank et al., 1967) shows the relationship be-tween mortality of partridge chicks, total mortality, and size of the par-tridge population on an estate in England. In this case, chick mortality is both the key factor and a density dependent factor. Density dependence of chick mortality is demonstrated by the fact that 51% of the variation in chick mortality was accounted for by variations in population size at time of hatching (not shown in Fig. 5–5). Variation in chick mortality

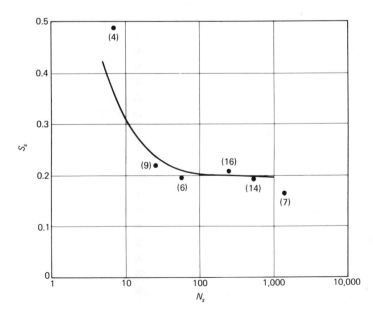

FIGURE 5–3. Relationship be-tween small-larval survival (S_s) and population growth (N_s). Points are means computed from the number of sets of data indi-cated. The fitted line is of the form: $S_s = a + \dfrac{b}{N_s}$. (From Morris, 1963.)

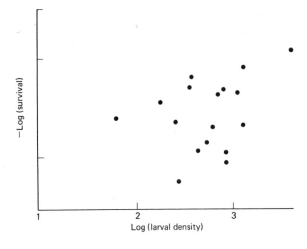

FIGURE 5-4. Plot of −log (survival) of winter moth larvae against log (larval density). (From Varley and Gradwell, 1968.)

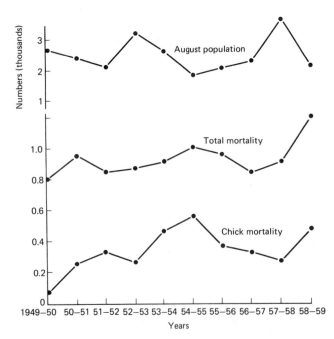

FIGURE 5-5. The relationship of the August population of the partridge population at Damerham, England, to −log (total mortality) and −log (chick mortality). (From Blank et al., 1967.)

accounted for 48% of the variation in total mortality each year and, hence, was the most important factor influencing population size.

The work of Klomp (1966) on the pine looper (a moth) illustrates a situation in which a strongly density dependent factor does not greatly influence population size. Figure 5–6 shows the relation between fecundity of the pine looper and larval density. These two variables are related because smaller moths are produced when larval densities are high and larval growth is reduced, and smaller moths produce fewer eggs. Figure 5–7 shows the negative logarithms of survival rates for each stage in the

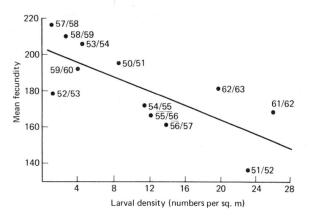

FIGURE 5-6. Correlation between larval density and mean adult fecundity. The numerals refer to years. Kendall's rank correlation method shows that the relation is significant ($P = 0.019$). The regression function is $y = 204 - 2.02x$. (From Klomp, 1966.)

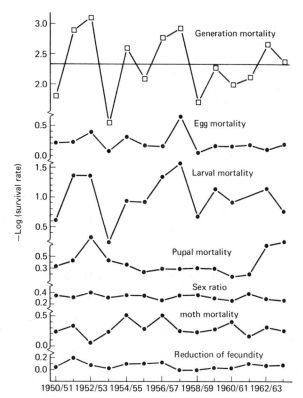

FIGURE 5-7. Comparison of fluctuations of age-internal mortalities, sex ratio, and fecundity, expressed as a fraction of its maximum potential, with the fluctuations of generation mortality. (From Klomp, 1966.)

pine looper's life cycle plotted for 14 years. These curves show that fecundity does not greatly influence the size of the population in the next generation, mainly because other factors, primarily larval survival, override the effects of variation in fecundity.

We have stressed the distinction between key factors and density dependent factors because knowledge of each tells us something different about a population. For example, if our main goal is to predict population size at some future time, then we must know something about the key factor. If, on the other hand, our interest is in understanding how more subtle interactions between variables can stabilize population size, then we must know something about the density dependent factors involved.

Figures 5–1(a) and 5–1(b) showed that there are many interrelationships that have the potential of acting as stabilizing forces on population size and some that can have a destabilizing effect as well. The relative importance of these interactions is not only the subject of debate among ecologists, but it is also of immense practical importance as well. The past decade has demonstrated the folly of trying to control pest species by indiscriminate use of pesticides, for example. The intelligent management of pest species requires an intimate knowledge of which interactions the species is most sensitive to and what the effects of modifying these interactions will be.

In other instances we may want to increase the abundance of species that we consider desirable. Examples are abundant: sport and commercial fishes, game birds, honey bees, all kinds of agricultural crops, and timber. Knowing what factors cause the greatest mortality in these populations is obviously important; less obvious is the importance of knowing what interactions may cause instability, perhaps by way of a positive feedback effect.

We must also point out the importance of knowing something about what determines the abundance of the many thousands of species that are not considered to be pests or as having a direct economic value. Although these species may be quite inconspicuous, they may play an important role in the ecosystem of which they are a part or they may simply delight us by their existence, an existence that may be endangered by our ignorance of what factors determine their abundance.

EFFECTS OF INTRASPECIFIC FACTORS ON POPULATION DYNAMICS

Little insight is required to realize that the individuals of a single population may affect one another in ways important to the dynamics of the population. The form of an interaction between individuals may be relatively straightforward, such as the shading out of one plant by another, or more subtle and complex, such as a reduction in fecundity of a subordinate female in a social hierarchy of animals. In Figs. 5–1(a) and 5–1(b) there is no block labeled "intraspecific interactions," but such interactions could affect many of the blocks shown. For example, social interactions may influence dispersal rates and would, therefore, affect "number lost through dispersal"; similarly, competition for food would affect "number dying from nutritional deficiency."

A variety of important interactions among both plants and animals has historically been placed under the heading of "competition." Intraspecific competition is defined as occurring when individuals of a species utilize an environmental resource that is in short supply. Let us consider how intraspecific competition can play a role in limiting population size.

Competition among animals can conveniently be divided into two categories. In some situations each individual attempts to use the resources it requires without particular regard for other individuals that are attempting to use the same resource. This kind of competition is called *scramble competition* and is well exemplified by a mass of fly maggots consuming a piece of meat that is insufficient for all of them. In other cases competitors may respond behaviorally to one another in such a way that some individuals retreat or give way in deference to other individuals. This kind of competition is called *contest competition,* since the behavioral interaction often appears to be in the form of a ritualized contest with the victor obtaining the sought-after resource after the contest, rather than a direct, physical conflict over the resource. Competition for territory such as is observed in many vertebrates and some invertebrates is a good example of contest competition. Of course, the concept of contest versus scramble competition is not applicable to plant populations. Let us defer our discussion of plant competition and first discuss competition among animals.

Although many ecologists believe intraspecific competition to be an important factor in population dynamics, there is not a great amount of direct evidence from natural populations supporting this belief. This lack of evidence is not surprising considering the complexity of the interactions among the components affecting naturally occurring populations. Consequently, several ecologists have extensively investigated competition in laboratory populations. A. J. Nicholson, one of the main proponents of the importance of competition, has, for example, studied scramble competition in laboratory populations of the Australian sheep-blowfly *Lucilia cuprina*. In one experiment Nicholson (1957) kept flies under uniform laboratory conditions in large cages and provided them with a constant amount of beef liver every other day. The larvae fed exclusively on the liver, and the adult females obtained from the liver the protein necessary for the formation of eggs. In addition, ample sugar and water were available to the adults. Figure 5–8 shows the changes in population size for a period of more than 450 days. The most impressive thing about this *Lucilia* population is that it fluctuated violently in spite of uniform culture conditions. In addition, note that the fluctuations were similar throughout the experiment and that there was no upward or downward trend in the average population size.

The elimination of extraneous environmental variables made it relatively easy to recognize what was happening in the blowfly population. When the adult population size was high, large numbers of eggs were laid, producing an excessively large larval population. Since the food resources were insufficient for such a large number of larvae, few of them obtained enough food to complete their development. Consequently,

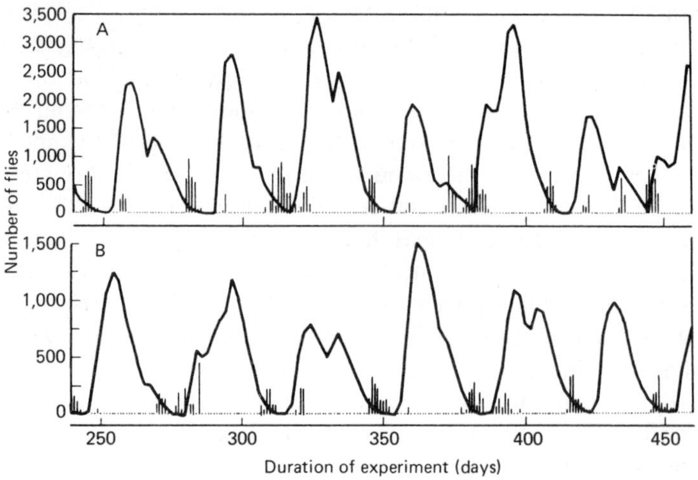

FIGURE 5-8. Heavy line, number of adult *Lucilia cuprina* in cage; vertical lines, numbers of adults eventually emerging from eggs laid on dates of plots. Both cultures were subject to identical conditions except that the daily quota of larval food was 50 g in A and 25 g in B. (From Nicholson, 1957.)

only a few adults were produced, which, in turn, laid many fewer eggs. Most of the larvae from these few eggs were able to mature, and the adult population again became excessive, thus again starting the cycle.

In another experiment Nicholson (1957) limited the amount of liver available to the adults, so that there was competition for food among the adults as well as the larvae. It might reasonably be predicted that the mean population size would be smaller than when the adults had ample food, but Fig. 5-9 shows that this was not the case. What apparently happened was that when the adult portion of the population increased because of the reduced mortality among the larvae, egg production by these adults was reduced by competition for protein, thus keeping the larval population from becoming excessively large. This example demonstrates how interactions among the factors affecting a population can affect its stability. To clarify this point the reader should trace the pathway of interactions for this example through the components of Figs. 5-1(a) and 5-1(b).

Although studies in the laboratory can demonstrate possible ways competition can affect populations, it is obviously necessary that competition be studied in natural populations as well. Following an experimental design suggested by Nicholson and others, Eisenberg (1966) studied populations of the pond snail *Lymnaea elodes* occurring in a small, permanent pond in Michigan. The design of Eisenberg's experiment was simple and consisted of deliberately increasing and decreasing densities of artificially isolated segments of a single population, while initially keeping other conditions as similar as possible. Eisenberg reasoned that if density dependent regulation was operating to a significant extent, the densities of the two altered populations should have converged toward that of unaltered control populations. If density dependent regulation was relatively unimportant, the altered populations should have

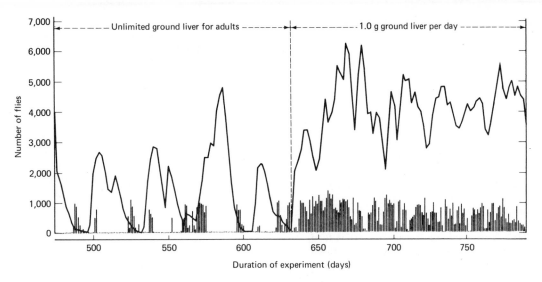

FIGURE 5–9. Effects produced by restricting daily quota of ground liver for adult *L. cuprina* to 1.0 g, after period of ample supply, in a population governed by larval competition for daily quotas of 50 g of meat. (From Nicholson, 1957.)

fluctuated in size about the new mean densities achieved by addition or removal of snails.

Eisenberg constructed 28 narrow rectangular enclosures at right angles to the shore line. After estimating the number of snails per enclosure, he added snails to four enclosures and removed snails from four enclosures, so that the altered densities were about 4.9 and 0.24 times the original densities. In two other enclosures, in which densities had not been altered, frozen spinach was introduced every few days to provide additional food. Of the remaining enclosures four were censused as completely as possible by total removal of all vegetation and living and dead snails in order to determine the actual density of snails in the habitat, four were used in predation experiments (which we shall not discuss here), and ten served as controls.

By using estimates of egg production per snail obtained for individuals from outside the enclosures, together with the estimated number of snails per enclosure, Eisenberg could estimate what the ratios of young snails in the experimental to the control enclosures would be if no regulation were taking place, that is, if egg production per adult female and survival of young were not significantly different between the experimental and control enclosures. For example, the ratio expected in early summer with no regulation was about 0.54 for the reduced density to control enclosures and about 3.72 for the increased density to control enclosures. Figure 5–10 shows the population sizes that were actually found for the young snails. Note that the positions of the experimental lines with re-

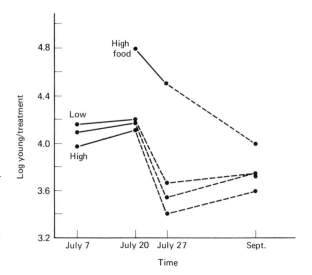

FIGURE 5–10. Relation of the numbers of young snails to treatments in the three sets in the basic design and the fed pens during July and in the fall. Numbers adjusted to a per pen basis. The control set is unlabeled. (High food = fed, low = reduced density, high = increased density. (From Eisenberg, 1966.)

spect to the control are the opposite of what would be predicted if no regulation had occurred. Analysis of size-frequency distributions showed that there was little or no growth of the young snails in any of the populations during the summer, and that, as a result, no significant size differences were shown among the experimental and control populations. Survival of adult snails also appeared to be unrelated to density. The effect of modifications of densities of adults on mortality of the young is uncertain. In the enclosure that had food added, however, mortality apparently increased considerably after the feeding was stopped.

If egg production in the various enclosures is examined, a clearer picture emerges. While no significant differences in the total number of eggs per enclosure were found among the altered density and control populations, Eisenberg found a significant inverse relationship between log number of eggs per egg mass and log number of adults per enclosure. Thus, the differences in clutch size approximately compensated for the differences in adult densities. In the enclosures receiving additional food, the mean clutch size was about 2.3 times that in the control enclosures.

To summarize, in Eisenberg's words, "The results from the basic experimental and fed pens indicate that in the altered pens of the basic design, the regulation of density was achieved through an adjustment in adult fecundity as influenced by the food supply." The results of these experiments also indicate that although we infrequently observe starving animals and although vegetation seems plentiful, we cannot assume that competition for food is an unimportant factor in population limitation.

In many species of animals scramble competition is rarely, if ever, observed. To have a limiting effect on populations, however, competition does not necessarily have to involve a scramble for resources. Complex social behavior appears to have evolved in many animals in the place of scramble competition. We say "appears" because the role of social behavior is often unclear; in some cases, it seems to be an alternative to scramble competition; in other cases, it seems to have other functions. Furthermore, even in cases of social interactions that are clearly competitive, we cannot be sure that the interaction limits population size.

We have already mentioned that in contest competition one of the competitors is able to "lay claim" to a mutually sought-after resource so that its rival is excluded from further use of the resource, and that the interaction is usually ritualized.

One of the more common manifestations of contest competition is territorial behavior. A territory is usually defined as an area that is defended by an individual or group of individuals against intrusion by other conspecific individuals. (The definition is sometimes extended to include instances in which members of closely related species are excluded, but we shall omit discussion of these.) Territorial behavior is most obvious and has been more extensively studied among birds than among other animals, although territorial behavior is known to exist within many taxonomic groups, including both vertebrates and invertebrates. The diversity of forms that territorial behavior takes is great; among various species of birds, for example, territories may be defended throughout the year or only during the breeding season. Both members of a pair may defend an area. In some cases only the immediate vicinity of the nest is defended; in others several acres may be defended. The common features among all of the many behavioral variations, however, are that there is an attachment to a particular area and also "hostility toward a certain category of other animals" when they enter this area (Tinbergen, 1957).

Behavioral attachment to a particular site may have several adaptive advantages. By limiting its activities to a certain area, an individual can become more familiar with it and hence more efficient in gathering food in it. Increased familiarity with one site would also decrease the amount of time required to take shelter from a predator. In many species of birds site attachment seems to play an important role in the formation of a bond between male and female during the breeding season. Other advantages of site attachment could be mentioned, but these are unnecessary for our discussion. The main point is that successful contenders acquire the advantages of site attachment that are denied to individuals without territories. If there are individuals in excess of the number of territories held, these excess animals will be prevented from breeding and possibly exposed to greater mortality risks; both of these possibilities could act to limit population size.

Although a simplistic view in which territoriality limits population size is attractive and has been assumed to be valid by many ecologists, there is more to be considered. This has been made clear by Brown (1969), who points out that the density of a population of birds may be classified into three levels:

1. At a level so low that all individuals are able to obtain territories in desirable habitats.
2. At a level such that some individuals set up territories in distinctly poorer habitats but no individuals are without territories, thus giving rise to what is called a "buffer effect."
3. At a level sufficiently high that some individuals are without territories and thus fail to breed.

Clearly at densities in the first level, territorial behavior cannot limit population size and may even promote population growth by improving the distribution of individuals with respect to their resources and allowing the other benefits of territoriality that we have already mentioned.

If a population grows to the extent that some females breed in marginal habitat (the second level) or do not breed at all (the third level), the effect will be a reduction in the rate of population growth, and we can say that territorial behavior has, in these instances, a regulatory effect. The importance of territorial behavior as a regulation of population density hinges, then, on how commonly populations reach the second or third level of densities. Unfortunately, data on this point are exceedingly scarce. In reviewing the literature, Brown (1969) found only six populations involving four species in two study areas for which the buffer effect seemed to be of significance. Even fewer data are available for judging the frequency with which populations reach the third level. The studies most commonly cited as demonstrating a surplus of potential territory holders involved the removal of territory-holding males followed by the observation of new males moving in to claim the vacated territory (Stewart and Aldrich, 1951; Hensley and Cope, 1951). Although these studies showed a surplus of males, they failed to demonstrate the existence of surplus females. Since it is the female portion of the population that is reproductively important, and we cannot assume a sex ratio of 1:1 in the species studied, we must conclude that these experiments did not demonstrate that territorial behavior was important in regulating the populations involved, although this may be the case.

Various attempts have been made to incorporate patterns of social behavior and reproductive success into a unified theory through hypotheses involving physiological mechanisms. The most detailed theory of behavioral-physiological system of population regulation has been suggested by the endocrinologist J. J. Christian (1963). Figure 5–11 shows a schematic representation of Christian's theory. The general idea of this system is that the intensity and form of social interactions are highly dependent on population density, and the function of the endocrine system is modulated by social interactions via neurosecretory pathways. Changes

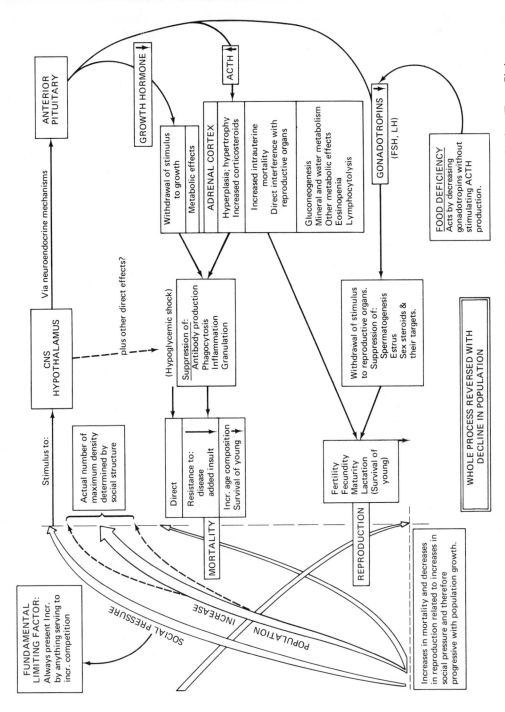

FIGURE 5–11. Schematic summary of the physiologic feedback regulation of population growth as it is envisioned today. (From Christian, 1961.)

in the endocrine system in turn modify such characteristics as fecundity, fertility, and disease resistance in such a way that population size tends to decrease when it is high and increase when it is low.

Many of the data now available bearing on Christian's hypothesis were obtained from small, confined laboratory populations of rodents (particularly voles, mice, and rats). Although these data are suggestive, they have been widely criticized, largely because of the unusually high densities and greatly simplified environmental conditions in which many of the experiments were performed.

In one of the more recent studies relating to this hypothesis, Christian and Davis (1966) censused a population of *Microtus pennsylvanicus* for several years by trapping. The adrenals of all trapped individuals were removed, weighed, and histologically examined. Since adrenal size is frequently correlated with adrenal activity, and adrenal activity is considered to be very important in Christian's "physiological feedback" hypothesis, a positive correlation between adrenal size and population size is predicted by the theory. Christian and Davis found a statistically significant positive correlation, suggesting that population size was being regulated to some extent by social interactions via the endocrine system. The immediate causes of changes in population density were not identified, however, and the functional relationship between changes in adrenal weight and factors producing changes in population size remains unknown. Similar criticisms can be made of virtually all field studies concerning the physiologic feedback hypothesis. Although this hypothesis still remains insufficiently tested under natural conditions, it is an interesting one and worthy of further study.

Limitation of Plant Populations by Competition

Plant populations can be limited by competition by increases in mortality or decreases in reproduction (either sexual or vegetative) due to insufficient light, water, or nutrients for all individuals.

Unlike animals, competition among plants cannot be modified by behavior, and hence contest competition does not occur. Another important difference between plants and animals is that terrestrial plants and higher aquatic plants are usually stationary once germination has occurred, but almost all animals are motile. Many animals are thus capable of making adjustments in spacing or emigrating in response to competition, but plants are not.

Much of our knowledge about competition among plants comes from experiments performed under greenhouse conditions or in cultivated fields. Moreover, almost all of these experiments span only a single growing season. Although these experimental simplifications are often necessary in order to identify and examine complex factors—and the results are often of basic interest to agronomists — extensive work remains to be done in natural settings, if the goals of the ecologist are to be attained.

The results of greenhouse experiments lead to several generalizations. First, in many species the mean weight per seed of seeds from plants grown at different densities and, presumably, different degrees of competition is relatively constant. On the other hand, the mean number of seeds produced per plant and the amount of vegetative growth both appear to be highly plastic. Figure 5–12 shows the results of experiments with several species of plants. Plants having highly determinate growth, such as sunflowers, do not follow the rule; neither would it be surprising if it were shown that plants relying largely on vegetative reproduction do not abide by the generalization. The generalization, however, does appear to apply in many cases.

The variation of vegetative growth shown in Fig. 5–12 also demonstrates another very important difference between plants and animals. That is, most animals have highly determinate development, and density of individuals is usually the most appropriate measure of density for competition studies (although there are exceptions, e.g., many fish species). For plants, however, biomass, leaf area, or some other measure per unit area may be more appropriate than number of individuals per unit area. This is especially true in vegetatively reproducing species in which it is very difficult to distinguish individuals.

That plants show considerable plasticity in vegetative growth but little variation in seed size when grown at different densities may be explained on the basis of selection pressures for successful seed dispersal and germination. Obviously, there is no advantage for an annual to maximize vegetative growth at the expense of reduced seed survival because of seeds of a suboptimal size being produced. We would expect, rather, that natural selection would result in plants sufficiently flexible in vegetative growth under a wide range of competitive situations to be able to devote a maximum amount of the biomass it produces to the formation of optimally sized seeds.

Some of the experimental work performed by J. L. Harper and his students serves to bridge the gap between studies of principal interest to agronomists and those of principal interest to ecologists. Figure 5–13 shows the results of an experiment in which seeds of *Trifolium repens* were sown in three densities (Harper, 1961). One set of replicates was watered throughout the experiment. Another set was watered for only the first 18 days, after which water was withheld in order to simulate drought conditions. At the end of seven weeks, the densities of the surviving seedlings were determined. In the replicates watered regularly, mortality was very low and not strongly density-dependent. But in the replicates deprived of water, density-dependent mortality is unques-

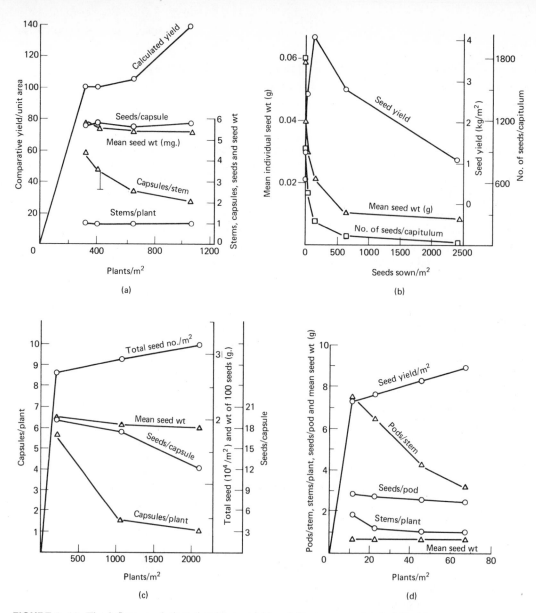

FIGURE 5-12. The influence of plant density on yield and the components of yield. (a) Linseed, *Linum usitatissimum*. (b) Sunflower, *Helianthus annuus*. (c) Conncockle, *Agrostemma githago*. (d) Field bean, *Vicia faba*. (From Harper, 1961.)

tionable; when the number of seeds sown was increased from about 18,000/m² to about 60,000/m², mortality of seedlings during the four-week drought nearly tripled.

In another experiment, Harper (1961) investigated the relationship between density of seeds sown and the microtopography of the soil. Seeds of the grasses *Bromus rigidus* and *B. madritensis* were sown on the surface of silty loam prepared in two ways. In one plot the soil was broken

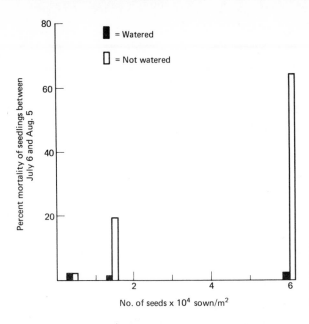

FIGURE 5–13. Percent *Trifolium repens* seedlings dying in seven weeks as a function of seed density. Unshaded bars represent mortality of seedlings that were not watered after the 18th day. (Redrawn from Harper, 1961.)

into lumps (ca. one-inch diameter); in another plot, the soil was first soaked with simulated rainfall and then allowed to dry and crack. The seeds were sprinkled on the prepared surfaces and the equivalent of one inch of rainfall applied. Figure 5–14 shows the results. Note that in the dried, cracked soil, the number of seedlings emerging increased as the number of seeds sown increased only to a limited extent and then levelled off. This means that the percent survival of seed decreased with increasing seed density. Harper interpreted these results to mean that there were only a small number of suitable sites for germination available in the dried, cracked soil. It seems reasonable that similar phenomena in nature might have strong regulatory effects on plant density.

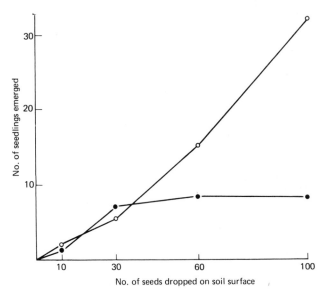

FIGURE 5–14. The relationship between the number of seeds sown and the number of seedlings emerging for a mixed population of *Bromus rigidus* and *B. madritensis* sown on the surface of a Yolo silty loam soil. The interaction of soil type with sowing density was significant at $P < 0.01$. ○, rough soil surface prepared by breaking large clods; ●, hard cracked surface prepared as above, but watered heavily and allowed to dry and crack. (From Harper, 1961.)

One of Harper's students, G. R. Sager (Harper, 1961) has provided data bearing on plant competition in a study of several species of plantains growing in fields. Sager mapped the positions of individuals of the species *Plantago lanceolata* at various times beginning with newly sprouted seedlings and then analyzed seedling survival in relation to proximity of an established plantain. A statistically significant relation was found indicating that the probability of a seedling's survival was contingent on the nearness of established plantains (Table 5–1), although the biological interaction responsible for it was not specifically identified.

POPULATION LIMITATION BY INTERSPECIFIC FACTORS

Thus far we have dealt with intraspecific factors limiting population growth, but there are also a variety of interspecific interactions that may act as limiting factors. Among these interactions interspecific competition has received much attention by ecologists, so much that we have devoted the next chapter to this topic. Aside from interspecific competition, almost all interspecific interactions involve the exploitation of members of one species by members of another, that is, the utilization of one species by another as a source of nutrients and energy. Methods of exploitation are commonly classified as predation, herbivory, or parasitism (with disease being considered a result of parasitism); but the distinctions between categories are not always clear, and, therefore, at times arbitrary decisions are made in separating them. Let us consider predation first.

To evaluate the effect of a population of predators on a population of prey several kinds of data are needed, namely:

1. The relation between prey density and the number of prey in each age class killed per predator per unit time.
2. The way in which the predator's population density responds to changes in the prey's population density.

TABLE 5-1. THE RELATIONSHIP BETWEEN THE CHANCE OF SURVIVAL OF SEEDLINGS OF *PLANTAGO LANCEOLATA* AND THE DISTANCE FROM NEAREST "NONSEEDLING" *P. LANCEOLATA.**

Fate of seedlings present in June 1957	Number of individuals	Mean distance (cm) to nearest nonseedling *P. lanceolata*
Absent in August 1957	113	7.09
Present in August 1957	81	8.93

*From Harper, 1961; original data from Sagar, 1960.

Note: The difference between the two values of mean distance was significant at $P < 0.01$.

Using the terminology proposed by the entomologist M. E. Solomon, we shall call the first relationship the *functional response* of the predator, and the second relationship, the *numerical response* of the predator.

C. S. Holling (1959) has demonstrated these two responses particularly well in his study of small mammal predation of the European pine sawfly. In Canadian forests the European pine sawfly, *Neodiprion sertifer* (which is a hymenopteran, not a fly) is a pest of conifers. The females lay their eggs in pine needles in the fall; the eggs overwinter and hatch in the spring. After the larvae have fed on the needles, they drop to the ground, crawl beneath the litter on the forest's floor, and pupate. The adults emerge in the fall and the cycle is begun again. Holling studied predation on sawfly pupae by three species of small mammals in pine plantations in Ontario. The plantations were well suited as study areas because the trees in each plantation were uniform in size, age, and spacing, there was little understory, and the substrate was flat and sandy. Thus, extraneous variables were minimized and sampling was simplified.

Three species of mammals preyed upon the sawfly pupae to a significant extent: the deermouse, *Peromyscus maniculatus;* and two shrews, *Sorex cinereus* and *Blarina brevicauda.* By examining the manner in which a cocoon had been opened, Holling could tell what mammal had opened it to eat the pupa inside. The densities of both living and preyed-upon pupae were estimated by stratified random sampling of the forest's floor. Areas of different sawfly densities were available because the plantations had been treated with several concentrations of spray containing a virus parasitic on the larval stages of the sawflies. Densities of *Peromyscus* and *Blarina* were estimated by a capture–mark–recapture technique; *Sorex* densities were estimated indirectly by using laboratory data on the rate of predation per individual. Thus, Holling was able to obtain the information necessary to plot both the numerical and functional responses of all three predators. These are shown in Fig. 5–15, together with the percent predation on the sawfly population by the mammal populations. These last curves are obtained by first obtaining the product of the number of days that the cocoons were preyed upon, the number eaten per predator per day, and the number of predators present at each prey density, and then dividing this product by the corresponding prey density and converting the quotient to a percent.

The numerical response curves have similar but not identical shapes for the three mammalian predators. This is not surprising, since the shapes of these curves depend on the factors that limit the population growth of each predator species, and it is reasonable to expect each predator species to be subjected to its own set of limiting factors. For example, for the two shrew species we suspect that *Sorex* was being limited at low cocoon densities by lack of food, but the *Blarina* was limited by something other than food, for although each *Blarina* ate many more pupae as they became available, the population size of *Blarina* did not increase.

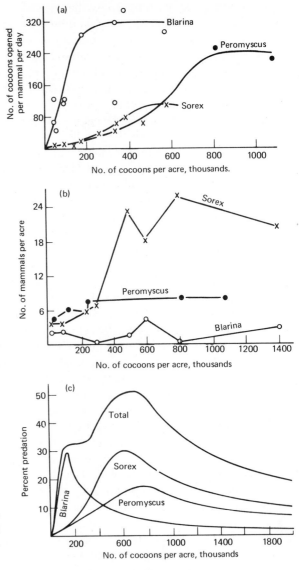

FIGURE 5–15. Functional (a) numerical (b) and combined (c) responses of small mammal predators to density of prey (sawfly cocoons). (After Holling, 1959.)

When the functional and numerical responses are combined (Fig. 5–15), the set of curves showing the total impact of the predators on the prey population is obtained. Here we see that all three predators together had a direct density-dependent effect on the prey population—but only up to a point. After a density of around 700,000 cocoons per acre was reached, the total effect of the predators became inversely density dependent. That is, increases in prey density were accompanied by decreases, not increases, in the percent of the population removed by predators. One might guess that after the peak of the percent predation curve is reached, the predators become ineffectual in regulating the prey population, but this is not necessarily the case. Figure 5–16 shows the way in which predation could have a regulatory effect even after the maximum point of percent predation is passed. In Fig. 5–16 the curved line represents percent of a prey population destroyed by a hypothetical predator. In the absence of predation the prey population will still suffer some mortality each generation, but assume that over the range of density shown, the prey population can increase from one generation to the next if there is no predation. Under these circumstances at each prey density there would be some rate of predation that would result in the prey population remaining constant from generation to generation. This amount of predation required to maintain a constant prey population size is shown as a horizontal straight line in Fig. 5–16, but it could (and probably would) be curved without affecting the argument. Now at prey densities below A and above C actual predation is less than "required" predation and the prey population can increase. Between A and C actual predation is greater than "required" predation and the prey population will decrease, even in the region between B and C, which is beyond the peak of predation. Points A and C are different, however, in that the prey population size tends to move toward A from either side, but if the prey population exceeds C, it escapes control by the predator and increases until it is checked by other factors.

In addition to providing us with an excellent demonstration of functional and numerical responses under natural conditions, Holling (1965, 1966) has devised an elegant model incorporating such variables as the predator's hunger level, the rates of movement of the predator and prey, the predator's perceptual field, and the length of time required to kill and consume a prey, to mention just a few. By extensive experimentation Holling has also obtained actual values for the parameters involved, and, using a digital computer, has successfully simulated the actual functional response of predators, e.g., mantids.

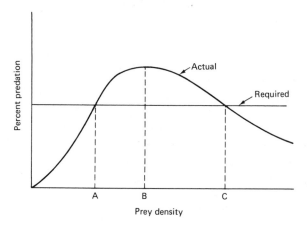

FIGURE 5–16. Relationship among actual percent predation, predation required to maintain a constant population density, and prey population density. Points A, B, and C are explained in the text. (Redrawn from Holling, 1959.)

Although Holling's work tells us a great deal about predator-prey relationships, it has not yet been extended to include changes that may be brought about in the prey population by predation. No satisfactory models have yet been developed to do this in general, but in fisheries sophisticated biological theories have been developed to relate rates of exploitation by fishing to changes in populations of fish. These models often involve considerations of the methods of fishing, imposed limits, and biological characteristics of particular species, and they are too specialized to consider here. We can examine to some advantage, however, a laboratory experiment conducted to investigate the relationships between rates of exploitation and changes in the population using the kinds of data usually gathered in commercial fisheries.

Silliman and Gutsell (1958) cultured four populations of guppies (*Lebistes reticulatus*) in aquaria for over three years. Two populations were not cropped, thus serving as controls. Guppies can reproduce about every three weeks, and Silliman and Gutsell argued, therefore, that cropping the populations every three weeks is similar to an annual fishing season in a fish that spawns annually. The fish were sieved so that fry were not removed, and a refuge for the small fish was available in each aquarium (guppies are notably cannibalistic).

The results of the fishing experiments are shown in Fig. 5–17. Three things of particular importance should be noted. First, increases in rate of exploitation resulted in an initial decline in total population size. But while the 25% and 50% exploitation rates were applied, the decline was, after a time lag, followed by a temporary population increase. Second, the temporary population increases at the 25% and 50% exploitation rates were due to peaks in the number of juveniles. Third, the populations probably could sustain themselves under a continuous 25% exploitation rate (although at a lower density than the controls), but probably would go to extinction under a 50% exploitation pressure. In fact, they did become extinct under a 75% rate of exploitation. In a similar set of experiments, Silliman (1968) exploited guppy populations that were being fed at different rates. Exploitation rates of 25%, 33%, and 50% per three-week interval were used for populations being fed in the ratios of 1:2:3 with respect to one another (thus giving a total of nine populations). The results of these experiments showed that, as one would expect, the populations receiving the most food produced the greatest yield and the populations receiving the least food produced the least yield. However, there was only a slight relationship between yield and exploitation rate; certainly, feeding rate was much more important in determining yield.

From these experiments and similar experiments performed with flour beetles, *Daphnia,* and blowflies, we may make several generalizations. It appears that populations have considerable resilience that allows them to adjust to various rates of predation. The immediate effect of predation is to decrease l_x for at least some values of x. Since $\sum l_x m_x$ must, on the average, equal one if the population is to survive, there must be additional

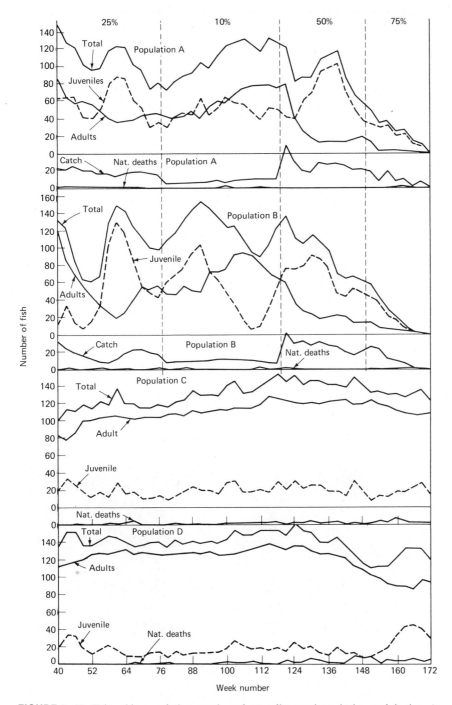

FIGURE 5–17. Triweekly population, catch, and mortality numbers during exploitation. A and B are exploited populations; B and C are controls. Percentages at the top are exploitation rates. (From Silliman and Gutsell, 1958.)

changes in the l_x curve or the m_x curve to compensate for the depression of l_x due to predation. A change in the l_x curve represents a change in the age structure of the population. From Fig. 5–18 we see that exploitation resulted in a considerable increase in the ratio of juveniles to adults, even, at times, to the extent that the ratio exceeded one. Coupled to this shift in age structure, though not obvious from the graphs, was an increase in the rate of recruitment of new adults when the population was under exploitation pressure. Thus, in the case of guppy populations, the resilience displayed apparently is a result of compensatory increases in l_x values for the younger fishes. Since the large guppies cannibalize the small and since a fixed, limited amount of food was provided, it appears that the compensatory adjustment was caused by a decrease in intraspecific predation on the young and a relaxation of competition for food.

The two conditions in Silliman and Gutsell's experiments differing most significantly from a natural situation were the constant exploitation rate and the relatively homogeneous environment. Both of these simplifications were justified in view of the purposes of the experiment, but it is well that we examine another set of experiments performed to determine the effect of fragmenting the predator and prey populations into partially isolated subpopulations and with the rate of predation not held constant by an experimenter.

C. B. Huffaker and his co-workers (1958, 1963) have very effectively used populations of herbivorous and predatory mites to investigate the importance of spatial complexity in predator-prey systems. Herbivorous mites (*Eotetranychus sexmaculatus*) were raised on oranges, the skins of which served as their food; and predatory mites (*Typhlodromus occi-*

FIGURE 5–18. Universe of 120 oranges used in studies of predator–prey interaction (prey: *Eotetranychus sexmaculatus;* predator: *Typhlodromus occidentalis*). Each orange has $\frac{1}{20}$ of its area exposed. Partial barriers of Vaseline form a complex maze of impediments between the oranges. Wooden dowels allow prey to disperse by climbing on a dowel, dropping on a silken strand, and being carried by an air current into a different area. (From Huffaker, 1958. Photograph by F. E. Skinner.)

dentalis) were allowed to feed on the herbivores. The oranges were arranged in patterns chosen by the experimenters and had predetermined amounts of skin exposed to the herbivores. Figure 5–18 shows a two-dimensional arrangement of the oranges.

Previous work had shown that when predator and prey were confined to a single orange, the predators destroyed the entire prey population and consequently eliminated themselves. Huffaker began his experiments with a four-orange universe in which half of each orange was exposed and in subsequent experiments increased the number of oranges and decreased the exposed area. In the simpler universes the population trends generally showed single, large peaks for both predator and prey, followed by the predator destroying most or all of the prey. In some cases a few prey escaped destruction when the predator population became extinct, and in some cases the prey population was totally destroyed. The results were the same—the end of the predator-prey system.

When the spatial complexity of the universe was considerably increased, however, the results were somewhat different. With a two-dimensional arrangement of 120 oranges, each orange having 1/20 of its area exposed, and with lines of Vaseline as partial barriers between the oranges, the system remained in existence for over seven months. During that time, there were three synchronized oscillations of both predator and prey, but once again the predator finally eliminated itself by destroying almost all of its prey.

For the final stage of spatial complexity, Huffaker et al. (1963) redesigned their experimental universe, utilizing three dimensions rather than two (Fig. 5–19). The mites now had to travel along the wires and dowels of the framework supporting the oranges to get from orange to orange. Under these conditions, the system proved to have considerable stability (Fig. 5–20). The system remained in existence for 490 days and was terminated not by overexploitation by the predator but by a viral disease that reduced the prey population to a level insufficient to maintain the predators.

Thus, it appears that the stability of a predator-prey system is highly dependent on the size and complexity of the space it occupies. Subpopulations on individual oranges were still going to extinction, but a sufficient number of subpopulations were out of phase with one another at any one time, so that extinct subpopulations could be reestablished. Considering the great spatial heterogeneity of many habitats, we suspect that this phenomenon is common in nature. Andrewartha and Birch (1954) give several good examples that support this belief.

We have considered important aspects of predator-prey relationships, but at this point the student might well ask whether or not there is evidence that predators are important in limiting prey populations in natural situations. As might be expected, some studies indicate that predators are very important; others suggest that they are not.

FIGURE 5–19. Universe of 252 oranges, each with $\frac{1}{20}$ of its area exposed. Each orange rests on a glass furniture coaster, and the coasters are arranged on grid-wire shelves supported on wooden dowels. The grid wires serve as a maze of impediments. (From Huffaker et al., 1963. Photograph by F. E. Skinner.)

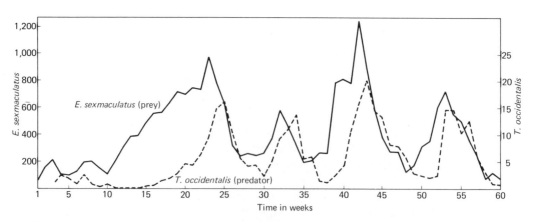

FIGURE 5–20. Predator-prey interaction with $\frac{1}{20}$ orange exposed on three-dimensional 252-orange dispersion. (From Huffaker, 1963.)

Economic entomologists working in biological control are understandably concerned with testing the effectiveness of predatory insects in controlling pest species. Some of the studies motivated by the desire to control insect pests clearly demonstrate the marked impact that predators and parasites may have on their prey. For example, the California red scale (*Aonidiella aurantii*) has been a serious pest of citrus trees in Southern California for many years and much work has been done to find out more about its natural enemies. In one particular series of experiments DeBach (1958) and his co-workers wanted to determine whether the natural enemies of *A. aurantii* had a significant depressing effect on its density. These experiments were conducted in a lemon grove which was not being sprayed with insecticides. Since the scale insects are sessile during their adult life and do not move much in the subadult stages, trees within the same grove can be treated differently without changes in scale density on one tree affecting the others.

The natural insect enemies of the scale insect were reduced in two ways. First, DDT-impregnated cloth sleeves were placed for several months over some branches on the tree and untreated sleeves open at the ends to allow predators and parasites to enter were placed on other branches of the same tree. The treated, closed sleeves completely eliminated the scale's natural insect predators and parasites. The second method was to spray entire trees lightly with DDT at monthly intervals. The sessile California red scale is not appreciably affected by DDT, but many of its enemies are.

On several other trees ant populations were eliminated. Ants are not enemies of *Aonidiella*; they indirectly benefit it. The ants collect honeydew from the citrus mealybug and the soft brown scale and in the process apparently interfere with the insects searching for, and feeding or ovipositing on the red scale.

The results of excluding ants and spraying lightly with DDT are shown in Fig. 5–21. When *A. aurantii*'s enemies were excluded by DDT impregnated sleeves, DeBach found that the enclosed branches were defoliated and many twigs had died. In one case the entire branch died within 16 months.

Even in the absence of quantitative data on particular species of the predators and parasites of the scale, we may safely conclude that they had a profound effect on the scale population. Moreover, as DeBach (1958) points out, we see that "there is no such thing as one average or normal population density equilibrium position for a given species except under one particular definite set of environmental conditions."

The reader might conclude at this point that predation is a generally important population limiting process. This may not be the case. We have selected examples to show that predation can be important, but there is evidence supporting the opinion that at least in some instances, losses due to predators have little effect on population density. Errington

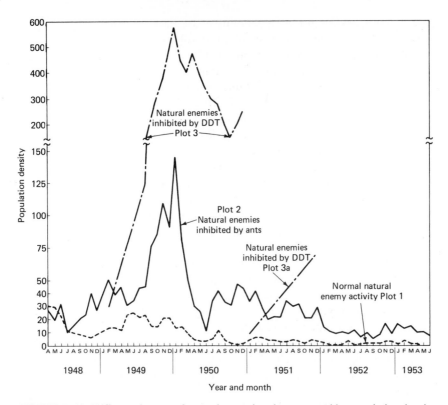

FIGURE 5–21. Different degrees of natural control as demonstrated by population density trends of the California red scale on lemon trees on adjacent plots under conditions of experimentally varied natural enemy effectiveness: (Plot 1) excellent natural control under conditions of normal enemy effectiveness; (Plot 2) fair natural control with natural enemy activity retarded by ant interference; (Plots 3 and 3a) very poor natural control with natural enemy activity greatly retarded by the toxic effects of DDT sprays. Neither ant interference nor DDT residue had any appreciable effect on red scale populations except indirectly through adverse effects on natural enemies. (From De Bach, 1958.)

(1963), for example, after years of work on muskrat populations, concluded that although predators, particularly mink, may take large numbers of muskrats, they have little effect on muskrat densities because muskrats preyed upon are primarily animals forced out of suitable habitat in the spring by dominant individuals. The number of animals excluded from good habitat increases as the population density decreases due to intensified intraspecific intolerance. The muskrats remaining in suitable habitat are seldom preyed upon, but those emigrating would probably die from other causes if they were not killed by predators. Thus, population limitation in this case should be attributed to social behavior rather than predation.

Errington (1946) reviewed the literature on predation among vertebrates and concluded that similar situations may exist in many species having territorial behavior. Among nonterritorial species, however, the same generalization cannot be made. Although Errington's conclusions appear well reasoned, there is one possible problem, which Errington recognized. That is, one of the components of a suitable habitat for a particular species is shelter from potential predators. So, in Errington's words, we must ask, "What proportions of the habitats that are marginal for various prey species might accommodate greater populations were it not for interspecific predation?"

Just as an animal population may be limited by exploitation pressure from other species, so may plant populations thus be limited. Both herbivorous animals and parasitic plants may use green plants for food and energy, but for want of space we will limit our discussion to plant-herbivore relationships.

The effects of herbivory on a plant population can be most clearly demonstrated in two ways: by excluding herbivores from areas in which they were previously present or by introducing them into areas where they were previously absent and then comparing the results to those noted for control areas.

The latter situation is well illustrated by efforts to control the Klamath weed, *Hypericum perforatum,* in California. *Hypericum* is a native of Europe but was accidentally introduced into northern California around 1900. By 1944, it occupied over two million acres of range land. Much attention was directed toward its control because it is harmful to livestock grazing on it.

In an effort to control *Hypericum* several species of herbivorous insects known to feed on it in Europe were introduced into California. One of these, the beetle *Chrysolina quadrigemina (C. gemellata)* has resulted in outstanding control; *Hypericum* has now been reduced to less than 1% of its former abundance in California (Holloway, 1964).

It can reasonably be argued that the effect of *Chrysolina* on *Hypericum* is not an acceptable example of plant population limitation by an herbivore since California is not the native home of either organism. It is probably true that *Chrysolina* is not the principal limiting agent in Europe, but another insect, the root borer, *Agrilus hyperici,* does appear to be an important agent in limiting *Hypericum,* at least in southern France (Wilson, 1943). Moreover, even a trained ecologist who had not made a detailed study of the *Hypericum-Chrysolina* interaction in California and did not know that both were introduced would probably consider *Chrysolina* to be unimportant as a limiting agent of *Hypericum* (Huffaker, 1957). It might also be erroneously concluded that *Hypericum* prefers a shady habitat, since *Chrysolina* is much less effective in controlling it in wooded areas. This suggests that many native plant species might be significantly limited by herbivorous insects or other animals in ways that are not at all obvious.

There have been several studies done on the role of herbivores in limiting plant populations by using the technique of excluding the herbivores from an area. This technique is commonly used for studying effects caused by large mammals, either domestic livestock or game animals such as deer, but let us consider two examples involving smaller animals.

As part of a study in the population dynamics of the vole *Microtus agrestis*, V. S. Summerhayes (1941) constructed exclosures in two areas of grassland in England. The vegetation within the exclosures and in control plots was sampled annually for eight years. Although the results were not nearly as dramatic as in the case of *Hypericum*, there was a decrease in abundance of several of the less important species of angiosperms and a pronounced decrease in abundance of mosses within the exclosures. The latter is illustrated in Fig. 5–22.

In a similar experiment, A. S. Watt (1957) excluded rabbits from grassland plots. Table 5–2 shows comparatively the difference between a plot ungrazed for two years and one ungrazed for 21 years. Note that two of three dominant species (as judged by dry weight biomass) are considerably more abundant in the area ungrazed for 21 years and that one species (*Astragalus danicus*) was entirely absent.

Although neither of these experiments conclusively demonstrated that any of the plant species concerned was being limited by herbivores

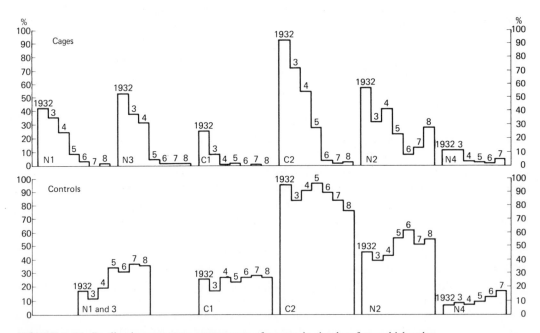

FIGURE 5–22. Decline in percentage occurrences of mosses in six plots from which voles have been excluded by cages. Experimental and control plots bearing the same designations were near one another. (From Summerhayes, 1941.)

TABLE 5–2. OVEN-DRY WEIGHT (g) OF SPECIES COMPOSING SWARDS FROM 250 cm x 25 cm IN GRASSLAND B, JULY, 1956.*

Species	Ungrazed for 2 years	Ungrazed for 21 years
Festuca ovina	15.556	59.920
Astragalus danicus	7.115	–
Koeleria gracilis	6.612	21.225
Carex spp.	5.747	10.963
Thymus drucei	5.283	–
Lotus corniculatus	4.986	0.070
Agrostis spp.	3.179	–
Galium verum	2.960	0.081
Asperula cynanchica	2.693	–
Heliototrichon pratense	2.691	9.995
Luzula campestris	0.693	0.425
Taraxacum spp.	0.198	0.136
Medicago lupulina } *Trifolium repens*	0.185	–
Anthoxanthum odoratum	0.166	–
Scabiosa columbaria	0.057	–
Linum catharticum	0.017	–
Senecio jacobaea	0.013	–
Sagina nodosa	0.001	–
Hieracium pilosella	–	0.033
Totals: Higher plants	58.152	102.848
Bryophytes	5.104	1.553
Lichens	2.312	–

*From Watt, 1957.

through their grazing activities, it appears to be a reasonable conclusion. The data also suggest that some of the plant species were more strongly limited by herbivores and that, in the absence of grazing pressure, these plants were dominants and were able to displace some of the other species.

THE ROLE OF PHYSICAL FACTORS IN LIMITING POPULATIONS

Earlier chapters discussed the physical environment and the adaptations organisms have evolved to cope with it, but we have not yet discussed the role of physical factors of the environment in limiting the abundance of organisms. Obviously, a species will not persist in areas where it is not adapted to the physical environment, and in this sense the distributions of populations are limited by physical factors. But assuming that a species can exist in a particular area, we may ask whether its population densities are closely related to physical factors as they vary through time or whether the physical environment is more like a stage setting with the biological interactions among the organisms determining their abundance. We feel that neither of these views is a sound generalization, but since a polarization of opinions has developed and has been important in stimulating much research, we will discuss the topic in its historical context.

There are many variables composing the physical environment, but we shall narrow our focus to the set that is commonly thought of as weather. Some ecologists attach little importance to the role of weather in limiting population density, arguing that weather is uninfluenced by changes in population density and, therefore, cannot interact with a population in a way that would tend to stabilize the population's size. These ecologists view weather as important in determining the potential productivity of an area over a long period of time, but they consider biotic interactions to be directly responsible for determining population densities on a year to year basis.

This point of view has been challenged by several ecologists, most notably by the Australians Andrewartha and Birch (1954). Their arguments were developed to apply specifically to animals, but they might, with some modification, be applied to plants as well. Two premises are at the core of their argument. First, they assume that most animal species are rare relative to their resources, and secondly, that extinction of localized populations is not an uncommon phenomenon. It is unfortunately true, as Andrewartha and Birch point out, that most population studies have dealt with relatively abundant species chosen primarily for the convenience of the persons working with them, and that because of this bias we may be seriously misled if we generalize from them. Assuming that Andrewartha and Birch's premises are valid, let us see what they infer from them.

The rarity of many species relative to their resources suggests that competition is correspondingly uncommon. When applied to species at different trophic levels, the same premise suggests that populations are seldom limited by exploitation pressure from other species.

If competition and exploitation are not of primary importance in limiting population densities, then what is? According to Andrewartha and Birch, weather often limits population density by causing r to become negative before other factors become regulatory. The population then declines until conditions again become favorable for an increase. Of course, since weather in many areas is notably erratic from year to year, populations would be expected to decline to extinction at least occasionally. According to Andrewartha and Birch's second premise, this is the case. However, many species are fragmented into several, perhaps many, semiisolated population living in areas differing in favorability, and fluctuations in population size are often not in synchrony among these units. Therefore, if extinction occurs in one area, the population may soon be reestablished by emigrants from more favorable areas. Unless data are collected with this in mind, an investigator might assume that he is censusing a single population, especially if the organisms are small. Because of the effects of averaging, such an error could lead him to the conclusion that the "population" under study was more stable than it really was.

Andrewartha and Birch have attempted to substantiate their theory with specific examples, most of which are insect populations in Australia such as *Thrips imaginis* (Davidson and Andrewartha, 1948) and the grasshopper *Austroicetes cruciata* (Andrewartha, 1957). In their analysis, they relied largely on correlations between fluctuations in population density and various components of weather. This method of analysis is open to criticism because correlations by themselves tell us little about causal mechanisms and may be quite misleading (Smith, 1961). It is possible for fluctuations in population density to reflect those of some component of weather, while the population is being regulated by density-dependent mechanisms. This is why we distinguished between key factors, which are often components of weather, and density-dependent factors in the first part of this chapter. Moreover, critical reevaluation of the ecology of the species used as examples by Andrewartha and Birch strongly suggests that they have overlooked important density-dependent processes (an excellent summary and criticism of these examples is provided by Clark, et al., 1967). Finally, if we grant that weather plays an important role in determining population size, then we should expect that an unusually long sequence of years of favorable weather would allow the population to build up to a high level. Predators and parasites that might otherwise have pronounced effects on the population are limited by factors particular to themselves and, therefore, may not respond sufficiently to limit the expanding host population. The population would expand to the point of overtaxing its resources and competition would ensue. Recent studies of economically important insects suggest that this sort of interaction between the effects of weather and regulation by biotic factors may be quite common.

The California oak moth *(Phryganidia californica)* demonstrates some of these generalizations quite well (Harville, 1955). *Phryganidia californica* ranges through the coastal counties of California from just above San Francisco Bay to the Mexican border (the southern limit of its range is not definitely known). The caterpillars feed on the leaves of both deciduous and live oaks and at times increase to enormous numbers and severely defoliate their host trees. There is no quiescent overwintering stage, and two successive generations are completed each year. In the San Francisco Bay region mortality among the caterpillars of the winter generation may be very high, especially during unusually cold winters.

Furthermore, winter-generation caterpillars emerging from eggs laid on deciduous oaks are doomed because of the loss of leaves from the trees. However, enough caterpillars survive on live oaks located in suitable microclimates to begin the summer generation.

During some winters, the weather is milder than usual and, consequently, winter mortality may be negligible. The number of caterpillars in the following generation may then be so great that the trees they feed upon are stripped of most of their leaves. The caterpillars move downward and outward through the trees in search of more food and eventually may reach the ground, where they disperse in search of other trees. The combination of the poorly oriented, trial-and-error search for additional trees to exploit and the patchy distribution of trees has a fortuitous consequence. Some trees are found by only a few caterpillars, which are therefore relieved of competition and have a good chance of surviving. Thus, although almost all of the caterpillars starve at this time, enough survive to begin the succeeding generation.

Parasites and predators appear relatively unimportant in limiting *Phryganidia*'s population size in central California. Rather, the interaction between weather and competition for food apparently is the principal factor involved. However, in southern California winter weather is much less likely to cause heavy mortality, and Harville has suggested that parasites and predators may be more important in this region of mild winter weather.

Intrapopulation Variation and Population Limitation

In a population that is being limited some individuals obviously are dying without reproducing. We also know that there is considerable genetic diversity within most populations. If the reasonable assumption is made that death is not a random event, then we have the components of natural selection. The consequence will be changes in gene frequencies in the gene pool of the population, but what part these changes play in population regulation is not yet known. At any rate, it is possible that important changes in the gene pool could simply involve fluctuations in the frequencies of alleles. A few suggestions have been made and some data have been collected along these lines, but undoubtedly a great deal remains to be discovered.

The western tent caterpillar *(Malacosoma pluviale)* is an excellent example of a species showing marked fluctuations in the frequency of genetically determined characteristics important to its population dynamics. Wellington (1960, 1964) has shown that tent caterpillars are polymorphic with respect to activity level and vitality. Within a population there may be individuals that are very active, individuals that are so sluggish that they usually do not survive, and intermediate individuals. Colonies of caterpillars construct silken tents varying in size from large to small depending on the proportion of active and sluggish caterpillars. Colonies made up largely of sluggish caterpillars are less likely to survive, and sluggish caterpillars may become adults that cannot fly well enough to reproduce successfully. On the other hand, highly active caterpillars are more prone to be lost from the population while dispersing, and active adults may be blown away from suitable habitats when they fly above the trees into fast-moving air.

These intrapopulation differences apparently interact to produce the following sequence of events. If all the tent caterpillars in a stand of trees have been destroyed by a long period of especially unfavorable weather, the stand eventually will be recolonized by active moths. Some of these active moths, however, will produce a few sluggish offspring. As the active forms are lost from the population by dispersal, the inactive forms increase in frequency. The inactive individuals are less viable, however, and eventually the population is again carried to extinction during a period of harsh weather. Figure 5–23 shows the kind of shift in frequency of active and sluggish forms that can take place in a population.

It is fairly obvious that shifts in frequencies of kinds of individuals in *Malacosoma* populations significantly influence their population dynamics. Indeed, were it not for the active forms, *Malacosoma pluviale* probably would not survive, but whether or not the existence of the sluggish forms is in some way advantageous is unclear. Considering their low vitality, it is difficult to understand why they have not been eliminated or modified through natural selection. More information about the ecology and genetics of *Malacosoma* is clearly needed to complete this interesting story.

Having seen that changes in the frequencies of different kinds of individuals within a population can strongly influence the population dynamics of the species, let us consider a genetical theory of population regulation proposed by Chitty (1960, 1964). Whether or not this theory

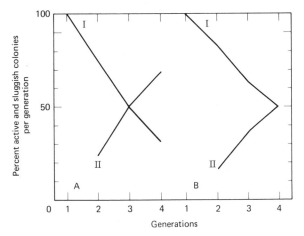

FIGURE 5–23. Shifts in the frequencies of active (I) and inactive (II) forms of caterpillars in two populations of *Malacosoma pluviale* in western Canada. Population A was from a farmland habitat and population B was from a forest. Numbers per generation in A were: 8, 25, 117, 208; in B: 4, 18, 80, 153. (From Wellington, 1964.)

is applicable to tent caterpillars is unknown, nor are there many relevant data from other species; but Chitty's theory is thought-provoking and worth considering.

Chitty developed his theory while studying the population fluctuations of voles, particularly those of *Microtus agrestis* in England. In his investigations he found little evidence for any of the hypotheses proposed to explain the population fluctuations of voles (e.g., predator-prey oscillations, herbivore-vegetation oscillations, epidemic diseases, Christian's endocrine-stress hypothesis) and consequently suggested the following hypothesis (Fig. 5–24): As the size of the population increases, its rate of increase declines because of decreases in reproduction and increases in death and emigration rates. But these three factors do not act equally on all the animals in the population; the less aggressive animals are more likely to be eliminated and thus the relative frequency of individuals that are genetically inclined to be aggressive increases. Selection for aggressiveness, however, is accompanied by selection against traits promoting viability. The rate of change in population size becomes negative and the

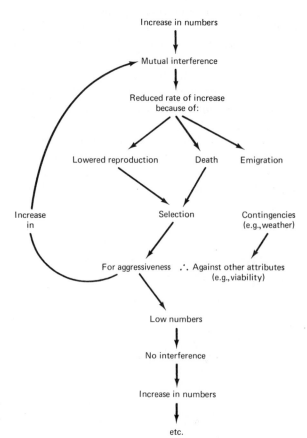

FIGURE 5–24. Chitty's hypothesis. The system is partly genetic and primarily behavioral. (From Krebs, 1964.)

population decreases in size until aggressiveness is no longer advantageous and traits favoring reproductive capacity and viability are strongly selected for. The population once again grows, and the cycle is begun anew.

Chitty's hypothesis may appear similar to Christian's, but there is a fundamental difference between them. Christian's hypothesis postulates no genetic changes in the population, but genetic changes are essential to Chitty's hypothesis. One of the main reasons Chitty introduced a genetic mechanism was to explain the observation that intraspecific "strife" during population peaks of voles appears to have little effect on the adults but the viability of their progeny is decreased. This decrease in the viability of the offspring persists even when they are reared under seemingly excellent laboratory conditions for which survival of voles of other generations is much higher.

MODELS OF POPULATION DYNAMICS

Considering how successful mathematics has been in representing complex processes in the physical sciences, it is not surprising that many attempts have been made to develop mathematical models of population processes. Indeed, many of these models were proposed by persons whose training was primarily in the physical sciences. These efforts at modelling population dynamics cover a period of roughly 50 years. To treat this topic thoroughly would take many pages filled with the symbols of higher mathematics, but to omit it altogether would be ignoring an area of population ecology that has stimulated much basic research and that is continuing to grow in importance.

There are several reasons why population ecologists engage in modelling. In applied areas such as insect pest control, fisheries biology, or management of game species, the goal may be to develop a tool that is useful in making forecasts or in evaluating prospective management strategies, e.g., Will an insect population reach an economically important level next summer because of mild conditions this winter? How much will fawn production be affected next spring if a doe season is allowed this fall? Models used in this way are usually closely tied to the characteristics of a particular species and make extensive use of empirically defined relationships among variables. On the other hand, we may not be interested in potential applications; instead, model development may be viewed as a way of organizing ideas in a rigorous, quantitative manner, finding out in the process what potentially important data are not available, identifying functional relationships that are insufficiently well understood, and, finally, testing our understanding of "how things work." This last point is especially important. In our discussion of population dispersion (Chapter 4) we pointed out that several different sets of assumptions may lead to the same predictions, and, therefore, an agreement between observation and prediction does not prove the validity of a model. On the other hand, lack of concordance generally indicates that either the model is inadequate because some important process has been inappropriately represented or omitted altogether, or the data used to develop or test the model are not representative. Careful examination of the assumptions made and the data used will often suggest reasons for the discrepancy, and these possibilities then lead to further model refinement and acquisition of data.

The Logistic Equation. One of the first models proposed (Verhulst, 1838; Pearl and Reed, 1920) for self-limiting population growth is represented by the *logistic equation*

$$\frac{dN}{dt} = rN\left(1 - \frac{N}{K}\right) \tag{5.1}$$

where dN/dt is the rate of change of population size, N, with respect to time; K is the equilibrium population size; and r is the instantaneous population growth rate per individual that would be attained under the same conditions but with the growth of the population having no self-limiting effects. That is, r is the instantaneous growth rate for a population increasing exponentially with a stable age distribution and stationary l_x and m_x distributions, as was discussed in Chapter 4.

Although this model is now generally considered inadequate for most purposes, it is worth examining because it illustrates an approach to modelling that is still important but not unduly complex mathematically. Furthermore, the logistic has played a historically important role in the development of population ecology by stimulating research.

We see from Eq. (5-1) that the logistic differs from the expression for exponential population growth by the presence of the factor $(1 - N/K)$. Examination of this factor shows that if $N = K$, then $(1 - N/K) = 0$ and hence the rate of change of population size, dN/dt, also becomes zero. Furthermore, if $N > K$, then $dN/dt < 0$ and the population decreases in size, but if $N < K$, then $dN/dt > 0$ and the population grows. Thus K is a stable equilibrium point for population size.

By rearranging Eq. (5-1) the logistic can be written as

$$\frac{1}{N}\frac{dN}{dt} = r - \frac{rN}{K}$$

The entire right side of this equation now represents the actually attained instantaneous rate of change of population size per individual (Chapter 4, p. 147). If $r - rN/K$ is plotted against N, a straight line will result, thus showing that the logistic assumes that population growth rate per individual decreases linearly as new individuals are added to the population. Figure 5–25 shows a logistic curve, together with data points for a laboratory culture of *Paramecium caudatum.*

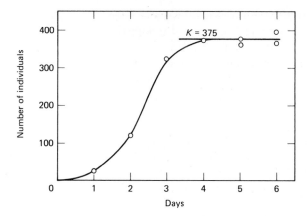

FIGURE 5–25. The growth of a laboratory population of *Paramecium caudatum* fitted to the logistic equation. Circles are observed counts; line is the fitted curve. (From Gause, 1934.)

It is true that growth curves of laboratory populations often start out with a sigmoid shape similar to what the logistic would predict. But this in itself is not sufficient evidence for the validity of the model. It is possible to find values of r and K for a set of observations of population growth so that the deviation between the fitted curve and the data points is minimized. These values for r and K, however, may be quite different from those that would be obtained from experiments designed to estimate the biological parameters that they are supposed to represent. Thus, fitting a logistic equation to a set of data points by choosing r and K in order to minimize the discrepancy is no more reasonable than fitting some other curve that gives a sigmoid shape but has no biological interpretation. This kind of curve fitting has legitimate uses, but is not a valid way of testing an ecological model. Smith (1952) provides a good discussion of this point.

It is not surprising that the logistic is inadequate in almost all instances in view of its basic simplifying assumption that the population growth rate per individual decreases linearly as new individuals are added to the population. This assumption has several biologically unreasonable corollaries. It implies that the absolute decrease in the population growth rate per individual caused by the addition of a new individual to a sparse population just beginning to grow is the same as the decrease caused by the addition of an individual to a dense population. If the species is sexually reproducing, females may not always find mates, so that the addition of new individuals could increase rather than decrease the population growth rate. In other cases, a population of intermediate density may modify its physical or chemical environment in a beneficial way. (Allee et al,. 1949). Another serious shortcoming is that all individuals in the population have the same effect on its growth rate; there is no allowance for changes in the impact an individual has on the population as it develops and matures. These and other difficulties with the logistic model are discussed thoroughly by Smith (1952).

Many other models of single-species population dynamics have been developed, several of which are variations or extensions of the logistic, but as we have already pointed out, it is not our intent to carry out a general review. Instead, we shall briefly examine another classic model in population ecology, but here the model deals with the interaction between two species.

Lotka-Volterra Predator-Prey Equations. This model was proposed independently by two different persons, Lotka (1925) and Volterra (1928) and hence bears the name of both. Mathematically, this model consists of a pair of simultaneous equations, one representing the rate of change of a predator population, the other representing the rate of change of the population it is preying upon. These equations are

$$\frac{dN_1}{dt} = (\alpha_1 - \beta_1 N_2) \, N_1$$

$$\frac{dN_2}{dt} = (-\alpha_2 + \beta_2 N_1) \, N_2$$

where N_1 = prey population size, N_2 = predator population size, α_1 = growth rate per individual of the prey population in the absence of the predator, $-\alpha_2$ = rate of decline per individual of the predator population in the absence of its prey, β_1 = decrease in prey population growth rate per individual predator present, and β_2 = increase in predator population per individual prey present.

If both N_1 and N_2 are plotted on the same graph against time, the predator and prey populations appear to oscillate out of phase with one another (Fig. 5–26). Although these curves bear a resemblance to those sometimes observed for real predators and their prey, critical examination of the similarity usually shows it to be superficial. Some of the reasons for the inadequacy of this model are the same as those listed for the logistic. In addition, Smith (1952) has pointed out several other unrealistic assumptions, which, in summary, are:

1. Neither species enters its own average growth function; hence, neither ever inhibits its own growth. . . .
2. Prey density has no effect upon the likelihood of a prey being eaten; there is no safety in numbers.
3. Predator density has no effect upon the likelihood of a predator catching a prey; there is no competition for food.
4. The predators have an unlimited rate of increase.

Even though we have pointed out that the shortcomings of the logistic and the Lotka-Volterra predator-prey models lie in their unrealistic assumptions, it would be a mistake to conclude that an adequate model must not make simplifying assumptions. If the details of a model were

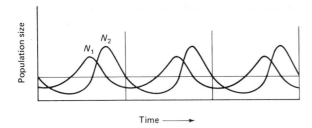

FIGURE 5–26. Curves for prey (N_1) and predator (N_2) population sizes as predicted by the Lotka-Volterra equations. (From Volterra, 1931.)

in a one-to-one correspondence with those of reality, the model would be impossibly complex and of no use to us. Indeed, the successful model is one that includes enough detail to serve the purposes for which it was intended but not so much that it becomes unwieldy.

A Posteriori Population Models. In the models discussed above the approach was to begin with a set of assumptions about how some particular population process might behave. In each case the model was developed without reference to a particular population and afterward data were obtained to test the model's adequacy. A very different approach to modelling population dynamics is often taken by ecologists interested in a particular species. This approach is to gather a large quantity of data on variables that are thought to be important and then, by using various statistical techniques, to develop a model. Thus, this approach is highly empirical rather than analytical.

An excellent example of an empirical approach to modelling is Morris' (1963) study of spruce budworm *(Choristoneura fumiferana)* in New Brunswick. In addition to providing an example of this approach, the spruce budworm study is a good closing example for this chapter because it is sufficiently comprehensive to show how many factors act to influence population density.

The spruce budworm is a lepidopteran endemic to North America. In New Brunswick the larvae feed principally on balsam fir *(Abies balsamea)* and often become serious economic pests. The life cycle of *Choristoneura* takes one year in eastern North America, although individuals occasionally take two years. Masses of approximately 20 eggs are deposited on the needles in midsummer, and each female can lay about 200 eggs in all. The larvae soon emerge and move toward the branch tip. Turblent winds may cause them to drop by silk threads and may result in their being blown away, either to another part of the same stand of trees or out of the stand. The first-instar larvae spin hibernacula in which they overwinter. Early in the following May the larvae emerge as second instars. They again move to the tips of the branches to feed and many disperse on silk threads. The preferred food is newly opening staminate flowers, but vegetative buds and mature needles are also eaten. Fifth- and sixth-instar larvae are able to utilize old needles but this causes a smaller pupa and reduced adult fecundity. Pupation takes place in the host tree's foliage in mid-July, the adults emerge in 8 to 12 days, and a new cycle is begun.

During the course of their 15-year study, Morris was able to develop life tables for the spruce budworm in several areas under various conditions. Table 5–3 shows a life table for one area during an outbreak. A life table is only a beginning, however, and what is ultimately desired is a knowledge of what environmental factors have the greatest influence in modifying the life table through space and time and how these factors

TABLE 5-3. LIFE TABLE FOR THE SPRUCE BUDWORM BASED ON MEAN VALUES OBTAINED IN AREA 1 DURING THE OUTBREAK.*

x	N_x	M_xF	M_x	$100M/N$	S_x
Age interval	No.[a] alive at beginning of x	Factor responsible for M_x	No.[a] dying during x	M_x as percentage of N_x	Survival rate within x
Eggs (to Instar I)	200	Parasites	18.0	9	
		Predators	12.0	6	
		Other	8.0	4	
		Total	38.0	19	0.81
Instar I	162	Fall and spring dispersal, etc.	132.8	82	0.18
Instar III	29.2	Parasites	11.7	40	
		Disease	6.7	23	
		Other[b]	6.7	23	
		Total	25.1	86	0.14
Pupae	4.10	Parasites	0.53	13	
		Predators	0.16	4	
		Other[b]	0.70	17	
		Total	1.39	34	0.66
Moths	2.71	Sex (46.5% females)	0.19	7	0.93
Females x 2	2.52	Reduction in fecundity	0.50	20	0.80
'Normal' females x 2	2.02	(No adult mortality or dispersal)			
Generation	—	—	197.98	98.99	0.010
'Normal' females x 2	2.02	Adult mortality ± dispersal	0.99	49	0.51
Actual females[c] x 2	1.03	—			
Generation	—	—	198.97	99.49	0.005

*From Morris, 1963.

[a]Number per 10 square feet of foliage.

[b]Minus interaction among all factors.

[c]Actual egg density of 103 eggs in generation $n + 1$ divided by 200.

operate. Ideally, one would wish to perform many experiments in which selected factors were systematically varied, but this would be impossible under realistic conditions for an insect such as the spruce budworm. Instead, data were gathered from several areas over many years, with the variation of the variables in space and time analogous to variation provided by an experimenter. Through statistical analyses the variables having the greatest effect on the budworm's population dynamics could be identified and synthesized into a quantitative model by utilizing an approach developed by Watt (1961).

The first step in the analysis is to identify the critical stages in the life cycle, that is, the stages most influential in producing variability in population density in space and time. This was done by using the statistical technique of multiple regression analysis with survival rate of an entire generation as the dependent variable and survival rates for each life cycle stage, fecundity, sex ratio, and gain or loss through emigration as the independent variables. Figure 5–27 (Morris, 1963) shows a simple linear regression in which only survival of large larvae (instars 3–6) is considered an independent variable. Survival of the large larvae can be seen to be very important in determining S_G; in fact, in the area from which the data were gathered, 68% of the variation in S_G could be accounted for by the variation in survival of the large larvae.

Once the more important stages of the life cycle (in terms of their effects on generation survival) have been identified, one can then proceed to construct submodels for survival of those stages. For example, the model arrived at for the survival of the small larvae (S_s) of *Choristoneura* in the unsprayed area was

$$S_s = 0.448 - 0.0245 \ F_d + \frac{1.1123}{N_s} - 0.0533 \ \log D_{cc}$$

where F_d = average tree diameter in the particular stand being considered, N_s = density of small larvae, and D_{cc} = an index of the extent of cumulative defoliation in the stand. The stands examined could roughly be separated into two groups, one containing trees of approximately four inches in diameter, the other containing trees of about eight inches in diameter. It is unlikely that trunk diameter *per se* is important to the spruce budworm. Trunk diameter, however, is probably correlated with a variety of other factors that were not measured but are important to the *Choristoneura,* such as survival of dispersing larvae, intensity of insolation to which the larvae are exposed, and various other microclimate factors.

We also see that survival of small larvae decreases as N_s and D_{cc} increase. Both N_s and D_{cc} may be related to the abundance of food available per larva and hence competition, which was not measured. Or,

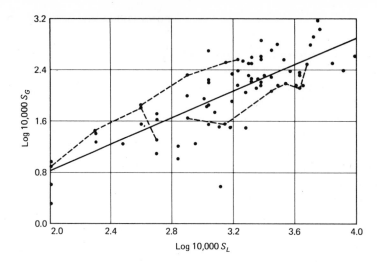

FIGURE 5–27. The relation of generation survival (S_G) to the survival of large larvae (S_L). The broken lines above the regression line connect the points contributed by a plot in an immature forest where the spruce budworm population had not yet reached an outbreak level. The broken line below the regression line connects points contributed by plots in a mature forest having had a severe outbreak lasting several years.

the extent of small larval dispersal may be dependent on larval density. At any rate, the model for small larvae survival accounts for 49% of the variation in S_s.

Table 5–4 summarizes the submodels that were developed during the course of the budworm study. When all of the submodels were combined into one model, 41% of the variation in generation survival was accounted for.* A intuitive idea of what this means can be gotten from Fig. 5–28. The horizontal axis represents S_G as calculated by substituting values for the independent variables into the model to obtain predicted values of S_G. The vertical axis represents S_G as actually measured in the field. If the model accounted for 100% of the variation, the points would all lie on the diagonal line. If the model accounted for none of the variance, the points would appear to be scattered at random.

Whether a model that accounts for 41% of the variation in generation survival is successful depends on one's point of view, and, at any rate, it is an unimportant question for our purposes. The main points are that Morris has demonstrated the importance of quantitative data on many aspects of the ecology of a species in studying its population dynamics, and he has also shown one way that these data can be integrated.

There are, however, disadvantages in this approach. Perhaps the most important is that the models developed in this way lack generality, that is, they are closely tied to particular species and tell us little about the dynamics of a broad class of species. To reiterate an important point we made at the beginning of this section: the form that a population model takes will be strongly influenced by the purposes it is intended to serve.

*The observant student may wonder why 68% of the variance in S_G could be accounted for by variations in the survival of large larvae but only 41% is accounted for by the entire model. This is because in obtaining the figure of 68%, direct estimates of survival rates for large larvae were used, but in obtaining the 41% figure, no direct estimates of survival rates were used; only estimates of the variables thought to be important in determining them were used.

TABLE 5-4. MODELS FITTED TO EACH AGE INTERVAL AND PROPORTION OF VARIATION IN SURVIVAL IN THE AGE INTERVAL EXPLAINED BY THE MODELS.[a]

Age interval	Model	Proportion of variation in survival in age interval explained by model
Egg	None	—
Small larva	$0.448 - 0.0245\,F_d + \dfrac{1.1123}{N_s} - 0.0533\,\log D_{cc}$	0.49
Large larva	$\{S_{par}\}\,\{0.096 + N_L\,e - (4.87 + 0.009N_L)\}\,\{1.5797 - 0.0396\,Z\}$ $\{-3.85 + 6.30\,T/H - 1.97\,T^2/H^2\}\,\{0.67 + 0.009\,I\ *\}$	0.56**
Pupa	$2.488 - \dfrac{120.35}{T_{max}}$	0.60
2P♀	$0.879 - \dfrac{28.52}{T_{mP}}$	0.44
P_F	$100.72 - 0.16\,N_L + \dfrac{89.35}{D_y}$	0.85
Adult	None	—

[a]From Morris, 1963.

*This model contains two nonsignificant factors.

**Proportion of variation in log (survival of large larvae) explained by model.

Abbreviations:

F_d = mean diameter of trees

D_{cc} = index of cumulative defoliation of balsam fir caused by many years of feeding by budworms

N_s = density of small larvae

S_{par} = fraction of small larvae surviving parasites

N_L = density of first instar larvae

Z = number of days separating the time of emergence of budworm larvae from hibernacula and the opening of shoots or flowers on balsam fir

T = daily mean temperature averaged over last 14 days of larval development

H = daily mean humidity averaged over last 14 days of larval development

I = ratio of egg density in any given year to the egg density in the preceding year

T_{max} = maximum daily temperature averaged over period of pupal development

T_{mP} = mean temperature during period of pupal development

D_y = index of degree of defoliation of balsam fir due to current generation of budworms

P♀ = fraction of adults that are female

P_F = fraction of maximum fecundity that is actually achieved

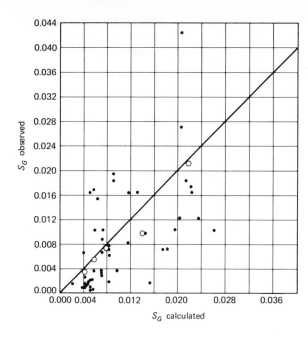

FIGURE 5–28. Observed generation survival as a function of generation survival calculated from the model. The dots represent raw data; the circles are arithmetic means of 10, 10, 10, 10, and 6 sets of data, respectively, from left to right. (From Morris, 1963.)

REFERENCES

Alle, W. C., A. E. Emerson, O. Park, T. Park, and K. P. Schmidt. 1949. *Principles of Animal Ecology*. Philadelphia: Saunders. 837 pp.

Andrewartha, H. G. 1957. The use of conceptual models in population ecology. *Cold Spring Harbor Symp. Quant. Biol.,* **22**:219–36.

——— , **and L. C. Birch.** 1954. *The Distribution and Abundance of Animals*. Chicago: University of Chicago Press. 782 pp.

Blank, T. H., T. R. E. Southwood, and D. J. Cross. 1967. The ecology of the partridge. I. Outline of processes with reference to chick mortality and nest density. *J. Animal Ecol.,* **36**:549–56.

Brown, J. 1969. Territorial behavior and population regulation in birds. *Wilson Bull.,* **81**:293–329.

Chitty, D. 1960. Population processes in the vole and their relevance to general theory. *Canad. J. Zool.,* **38**:99–113.

——— . 1964. Animal numbers and behavior. In J. R. Dymond (ed.), *Fish and Wildlife: A Memorial to W. J. K. Harkness*. Ontario: Longmans.

Christian, J. J. 1963. Endocrine adaptive mechanisms and the physiologic regulation of population growth. In W. V. Mayer and R. G. Van Gelder (eds.), *Physiological Mammology*, Vol. 1. New York: Academic Press.

——— , **and D. E. Davis.** 1966. Adrenal glands in female voles *(Microtus pennsylvanicus,* as related to production and population size. *J. Mamm.,* **47**:1–18.

Clark, L. R., P. W. Geier, R. D. Hughes, and R. F. Morris. 1967. *The Ecology of Insect Populations in Theory and Practice*. London: Methuen. 232 pp.

Davidson, J., and H. G. Andrewartha. 1948. Annual trends in a natural population of *Thrips imaginis* (Thysanoptera). *J. Anim. Ecol.,* **17**:200–222.

DeBach, P. 1958. The role of weather and entomophagous species in the natural control of insect populations. *J. Econ. Entomol.,* **51**:474–84.

Eisenberg, R. M. 1966. The regulation of density in a natural population of the pond snail, *Lymnaea elodes. Ecology,* **47**:889–906.

Errington, P. L. 1946. Predation and vertebrate populations. Quart. Rev. Biol., **21**:144–77.

———. 1963. *Muskrat Populations.* Ames: Iowa State University Press. 665 pp.

Gause, G. F. 1934. *The Struggle for Existence.* New York: Hafner. 163 pp.

Harper, J. L. 1961. Approaches to the study of plant competition. *Soc. Exp. Biol. Symp. No 15.,* pp. 1–39.

Harville, J. P. 1955. Ecology and population dynamics of the California oak moth *Phryganidia californica* Packard (Lepidoptera: Dioptidae). *Microentomology,* **20**:83–166.

Hensley, M. M. and J. B. Cope. 1951. Further data on removal and repopulation of the breeding birds in a spruce-fir forest community. *Auk,* **68**:483–93.

Holling, C. S. 1959. The components of predation as revealed by a study of small mammal predation of the European pine sawfly. *Canad. Entomol.,* **91**:293–320.

———. 1965. The functional response of predators to prey density and its role in mimicry and population regulation. *Mem. Canad. Entomol. Soc. No. 45.*

———. 1966. The function response of invertebrate predators to prey density. *Mem. Canad. Entomol. Soc. No. 48.*

Holloway, J. K. 1964. Projects in biological control of weeds. In P. DeBach (ed.), *Biological Control of Insect, Pests and Weeds.* New York: Reinhold Pub. Corp.

Huffaker, C. B. 1957. Fundamentals of biological control of weeds. *Hilgardia.,* **27**:101–57.

———. 1958. Experimental studies on predation: dispersion factors and predator-prey oscillations. *Hilgardia,* **27**:343–83.

———, **K. P. Shea, and S. G. Herman.** 1963. Experimental studies on predation: Complex dispersion and levels of food in an acarine predator-prey interaction. *Hilgardia,* **34**:305–30.

Krebs, C. J. The lemming cycle at Baker Lake, Northwest Territory, during 1959-1962. *Arctic Inst. N. Amer. Tech. Paper No. 15.*

Lotka, A. J. 1925. *Elements of Physical Biology.* Baltimore: Williams & Wilkins.

Morris, R. F. 1959. Single-factor analysis in population dynamics. *Ecology,* **40**:580–88.

——— ed. 1963. The dynamics of epidemic spruce budworm populations. *Mem. Canad. Entomol. Soc. No. 31.*

Nicholson, A. J. 1957. The self-adjustment of populations to change. *Cold Spring Harbor Symposia on Quant. Biol.,* **22**:153–73.

Pearl, R. and L. J. Reed. 1920. On the rate of growth of the population of the United States since 1790 and its mathematical representation. *Proc. Nat. Acad. Sci. Wash.,* **6**:275–88.

Pielou, E. C. 1969. An introduction to mathematical ecology, New York: Wiley-Interscience. 286 pp.

Ricker, W. E. 1954. Effects of compensatory mortality upon population abundance. *J. Wildl. Management,* **18**:45–51.

Silliman, R. P. 1968. Interaction of food level and exploitation in experimental fish populations. *U.S. Fish. Wildl. Serv. Fishery Bull.,* **66**:425–39.

————, and J. S. Gutsell. 1958. Experimental exploitation of fish populations. *U.S. Fish. Wildl. Serv. Fishery Bull.,* **58**:214–52.

Smith, F. E. 1952. Experimental methods in population dynamics: A critique. *Ecology,* **33**:441–50.

————. 1961. Density dependence in the Australian thrips. *Ecology,* **42**:403–07.

Southwood, T. R. E. 1967. The interpretation of population change. *J. Anim. Ecol.,* **36**:519–29.

Stewart, R. E., and J. W. Aldrich. 1951. Removal and repopulation of breeding birds in a spruce-fir forest community. *Auk,* **68**:471–82.

Summerhayes, V. S. 1941. The effect of voles *(Microtus agrestis)* on vegetation. *J. Ecol.,* **29**:14–48.

Tinbergen, N. 1957. The functions of territory. *Bird Study,* **4**:14–27.

Turnbull, A. L. 1962. Quantitative studies of the food of *Linyphia triangularis* Clerck (Aranea: Linyphidae). *Canad. Entomol.,* **94**:1233–49.

Varley, G. C., and G. R. Gradwell. 1968. Population models for the winter moth. In T. R. E. Southwood (ed.), *Insect Abundance. Sym. Royal Entomol. Soc. London, No. 4.* Oxford: Blackwell Sci. Publ.

Verhulst, P. F. 1838. Notice sur la loi que la population suit dans son accroisissement. *Corresp. math. phys., A. Quetelet,* **10**:113–21.

Volterra, V. 1928. Variations and fluctuations of the number of individuals in animal species living together. *J. du Conseil intern. pour l'explor. de la mer III,* Vol. I. Reprinted in R. N. Chapman, *Animal Ecology.* New York: McGraw-Hill, 1931.

Watts, A. S. 1957. The effect of excluding rabbits from grassland B (Mesobrometum) in Breckland. *J. Ecol.,* **45**:861–78.

Watt, K. E. F. 1961. Mathematical models for use in insect pest control. *Canad. Entomol,* **93**, Suppl. 19.

Wellington, W. G. 1960. Qualitative changes in natural populations during changes in abundance. *Canad. J. Zool.,* **38**:289–314.

————. 1964. Qualitative changes in populations in unstable environments. *Canad. Entomol.,* **96**:436–51.

Wilson, F. 1943. The entomological control of John's wort *(Hypericum perforatum L.)* with special reference to the weed in southern France. *Australian Council of Scientific and Industrial Research Bull., No. 169.* 87 pp.

Chapter 6

RESOURCE RELATIONSHIPS OF SPECIES

In earlier chapters we examined the relationships of individual organisms to conditions of their physical environment. These relationships, however, do not completely determine the distribution and abundance of plant and animal species. In a sense, the adaptations of individuals for meeting challenges of the physical environment set potential limits to the distribution of the species. The distributional limits actually realized, and the abundance achieved, are determined to a great extent by the results of interactions with other species which act as food sources, competitors, predators, parasites, or disease agents. Thus, a fuller understanding of the factors determining the distribution and abundance of species must be sought through an examination of processes operating at the population, or even the ecosystem, level of organization.

In this chapter we shall concentrate on interactions between populations of species that are similar in their patterns of utilization of environmental resources and that are potential competitors for these resources. First, we shall discuss the nature of environmental resources, the kinds of environmental factors that constitute resources, and the unique features of the relationship between organisms and the resources they utilize. Then we shall examine patterns of competitive interaction among species having similar resource requirements. We shall study the relation of these interactions to, and their modifications by, other biotic and abiotic factors of the environment, and we shall consider the evolutionary consequences of these interactions for the species involved.

ENVIRONMENTAL RESOURCES

Environmental resources consist of the various biotic and abiotic components of the environment that are utilized by organisms in such a way as to make them temporarily or permanently unavailable to other organisms. Environmental resources vary in type from those that are present in a fixed or limited quantity, such as nest sites for animals, to those which are made available at a continuous, but limited rate, such as light energy for plants. Likewise, environmental resources may be made unavailable to other organisms in various ways. Some resources, such as energy-rich food materials, may be taken from the environment and converted or degraded into an unusable form. Others, such as specific chemical nutrients, may be taken from the environment and held, in living body tissues, during the lifetime of the organism. Still others, such as nest sites, may be occupied for shorter periods, but in a way that prevents other individuals from simultaneously using them.

The specific nature of the environmental resources required by a particular organism varies greatly, depending on the mode of nutrition and on many individual features of the life history of the organism. For microorganisms, environmental resources comprise the various organic and inorganic compounds required for active metabolism. For green plants they consist of water, carbon dioxide, specific chemical nutrients, and light energy. Since, for plants, the occupation of some minimum area of substrate is frequently the guarantee of an adequate supply of these materials, space may also be regarded as constituting a resource for these organisms. For animals, food materials, water, and specialized environmental features such as nest sites, oviposition sites, hibernation sites, and other shelters, together with specific materials used in nests and similar structures, constitute environmental resources.

From the above, it is apparent that the common feature of environmental resources does not lie in their physical nature, but rather in their unique relationship to the organisms that utilize them. The two important features in this relationship are the limited quantity or rate of supply of resource materials, and the unsharable nature of individual resource units. These features may lead to situations in which competition, here defined as a combined demand for resources by organisms in excess of the available supply, becomes an important determining factor of the distribution and abundance of ecologically similar species.

COMPETITIVE EXCLUSION PRINCIPLE

Studies of interactions between species having similar resource requirements have led to the recognition of an important generalization known as the *competitive exclusion principle* (Hardin, 1960). This principle states that two or more resource-limited species having identical patterns of resource utilization cannot continue to coexist in a stable environment. Under this principle it is assumed that if two or more species having the same resource requirements come together, one will prove better adapted for utilizing the available resources under the existing environmental conditions, and it will outcompete and eliminate the others. It is important to note, however, that this principle is restricted to resource limited species, i.e., to species for which the direct limiting factor for population density is the supply of available resources, and to stable environments in which an advantage possessed by one species will be maintained long enough for competitive elimination of other species to occur.

The roots of the competitive exclusion principle may be traced to Darwin (1859) who suggested that the struggle for existence was most severe among species of similar structure and behavior, and who gave examples of cases of apparent replacement of individual species by stronger competitors. This idea was further developed by Grinnell (1904), a vertebrate ecologist working in California, and by Gause (1934), a Russian population ecologist. Gause conducted an extensive series of studies dealing with competition in single-species and mixed-species populations of various yeasts and protozoans. The best known of these studies involved the closely related ciliate protozoans *Paramecium caudatum* and *P. aurelia*. In simple laboratory cultures where both species were forced to utilize the same food resources, Gause found that *P. caudatum* was eventually eliminated from mixed-species populations, although it was able to survive well under identical environmental conditions in the absence of *P. aurelia*. In recognition of Gause's important contributions to the development of the above principle, it is often referred to as *Gause's law*. Current interest among ecologists in the problems of interspecific competition and coexistence of similar species dates primarily from these experimental studies by Gause.

The validity of the competitive exclusion principle is supported by studies of interactions between ecologically similar species in both the laboratory and field. To illustrate the nature of the available evidence, we shall examine in detail two major studies, one dealing with interspecific competition under laboratory conditions, the other with competitive replacement in nature.

Most laboratory studies have utilized small invertebrates having short life cycles, such as protozoans, microcrustaceans, and insects. Some of the most extensive studies have dealt with competition between the flour beetles *Tribolium confusum* and *T. castaneum* (Park, 1962). These species are ideally suited to laboratory studies of competition since they are small, approximately 4–5 mm in length as adults, and are able to carry out their entire life cycle in a vial of dry, finely milled flour. The animals can be separated from the flour, and the various life cycle stages sorted out by putting the flour matrix through a series of sieves of graded mesh size. Since populations can be maintained in small vials, the behavior of many replicate populations under identical environmental conditions can be studied by placing a series of vials side by side in a controlled environment chamber.

Park (1962) studied the growth of single- and mixed-species populations of these beetles at a series of temperatures and relative humidities. The temperatures used were 24°, 29°, and 34°C. At each temperature experiments were conducted at relative humidities of 30% and 70%. Single-species populations of both *T. confusum* and *T. castaneum* survived and grew under the entire range of conditions, although maximum population densities differed for the two species and varied with the environmental conditions. The maximum population density for *T. confusum* was achieved at 34°C and 70% relative humidity, but that for *T. castaneum* occurred at 29°C and 70% relative humidity (Table 6–1).

In the mixed-species populations, however, one of the two species was eliminated in all cases. Results of these experiments involving many replicate populations at each temperature–humidity combination are summarized in Fig. 6–1. Under a given regime, one or the other species was clearly the better competitor, although not necessarily being the winner in all replicate vials. This latter observation suggests that factors such as the genetic constitution, physiological condition, or age of individuals used to initiate the mixed-species cultures also influenced the outcome of competition.

The species that typically survived, however, varied in a consistent manner with temperature and relative humidity (Fig. 6–1). At temperatures of 24°C, regardless of humidity level, and at higher temperatures and low humidity, *T. confusum* was the usual survivor. At temperatures of 29° and 34°C and high humidity, *T. castaneum* survived most frequently.

The results of these experiments demonstrate the validity of the competitive exclusion principle for situations in which species are forced to

TABLE 6–1. RANKINGS OF MAXIMUM DENSITIES ACHIEVED BY *TRIBOLIUM CONFUSUM* AND *T. CASTANEUM* IN SINGLE-SPECIES CULTURES UNDER VARIOUS TEMPERATURE AND HUMIDITY REGIMES.*

Regime	Rank of *T. confusum*	Rank of *T. castaneum*
34°C, 70% r.h.	1	3
34°C, 30% r.h.	6	5
29°C, 70% r.h.	2	1
29°C, 30% r.h.	3	4
24°C, 70% r.h.	4	2
24°C, 30% r.h.	5	6

*Adapted from Park, 1962.

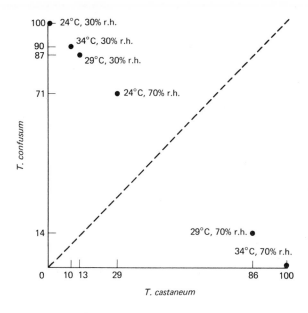

FIGURE 6–1. Percentages of interspecific competition contests between the flour beetles *Tribolium confusum* and *T. castaneum* won by each species at different temperatures and relative humidities.

utilize the same resources in a stable, structurally simple environment. They also illustrate the fact that relationships with the physical environment may determine the outcome of competition. In this example the competitive advantage can be shifted from one species to the other by changing conditions of temperature and relative humidity.

On the basis of further studies of the competitive interaction between these two species, Park (1962) suggested that the specific factors responsible for the elimination of one of the competing species fall into two categories, *exploitation* and *interference*. Exploitation refers simply to the utilization of common resources, with different efficiencies, by the competing species. The effects of this aspect of competition are felt through modification of the available supply of resources relative to the number of individuals competing for them. Interference, on the other hand, refers to direct detrimental effects of one species on the growth, survival, and reproduction of the other. In the case of *Tribolium* species, these effects include the predation of eggs and larvae of one species by adults of the other and physical disturbance of individuals of one species by individuals of the other. The latter may result in inhibition of egg production, retardation of larval development, and reduction of longevity. From these studies Park concluded that competition occurs, in a sense, if species interact through either exploitation or interference, and that the competitive interaction will be most intense when both processes are involved (Fig. 6–2).

In the case of these flour beetles, as in the case of virtually all laboratory and field studies of competition, evidence for the importance of exploitation as a competitive process is almost entirely circumstantial.

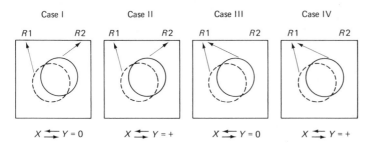

This fact may appear surprising, since exploitation seems to represent competition in its purest form. However, the data needed to demonstrate differences in the efficiency of utilization of, for example, food resources generally require much more sophisticated techniques than are needed to demonstrate direct effects of one species on another. The direct observation of mechanisms leading to the decline and extinction of one species have thus, in almost all cases, involved interference effects. This situation raises the interesting question of whether or not competition, in the sense of an interaction closely related to resource utilization, is actually occurring in many of these situations. The importance of interference effects, and the lack of direct evidence for exploitation effects, have led Sokoloff and Lerner (1967), for example, to suggest that for *Tribolium* species mutual predation, rather than competition, is the important interaction. Ryan et al. (1970) have demonstrated for *Tribolium confusum* and *T. castaneum,* however, that adults and larvae of one species selectively prey upon pupae of the other species in preference to those of their own. Dawson (1968) has suggested, furthermore, that the behavioral activities producing interference may show intraspecific genetic variation and thus may be subject to selection pressures based on the overall resource utilization relationship. Many interference mechanisms may thus be selected as "weapons" in the competitive interaction because of the advantages in resource exploitation for individuals possessing them. Nevertheless, care must be taken in reaching this conclusion, and the paucity of evidence for exploitation effects in studies of ecologically similar species is a major weakness of most studies of competition.

Support for the competitive exclusion principle has also been furnished by a variety of other laboratory studies. For organisms as diverse as yeasts and ciliate protozoans (Gause, 1934), hydras (Slobodkin, 1961), cladocerans (Frank, 1957), grain weevils (Utida, 1953), fruit flies

(Merrell, 1941; Miller, 1964; Moore, 1952), and duckweeds (Clatworthy and Harper, 1962), the experimental confinement of two species under constant conditions where both are forced to utilize the same food resources has shown that one species is invariably eliminated. These studies have also demonstrated the influence of a variety of factors, both abiotic and biotic, on the course and outcome of competition. For both yeasts and ciliate protozoans Gause (1934) found that the competitive advantage of one species was modified by changes in the chemical conditions of the culture medium resulting from accumulation of metabolic wastes of the organisms themselves or of bacteria serving as their food. The addition of a sporozoan parasite to mixed-species populations of flour beetles was found by Park to shift the competitive advantage from one species to the other. Merrell (1951) showed that aging of the food medium shifted the competitive advantage from *Drosophila melanogaster* to *D. funebralis*. Moore (1952), working with *Drosophila melanogaster* and *D. simulans,* found that the outcome of competition was related to the nature and location of the food surfaces on which oviposition could occur.

In contrast to the abundance of experimental studies illustrating competitive exclusion, detailed field observations of the replacement of one species by an ecologically similar form have rarely been obtained (Miller, 1967). Still less frequent are observations of replacement in situations for which experimental analysis of the possible mechanisms of replacement can be conducted. These characteristics are approximated closely, however, in a study by Bovbjerg (1970) on the relationships of two congeneric species of crayfish in stream and pond habitats in northern Iowa and southern Minnesota.

The two species *Orconectes virilis* and *O. immunis* are widely distributed in North America east of the Rocky Mountains, occurring from the Ohio River drainage northward into southern Canada. These species are morphologically similar, and appear to have identical food requirements, being generalized omnivores and scavengers. Typically, *O. virilis* is a stream or lake margin form and *O. immunis* inhabits ponds, but the local distribution in the area studied by Bovbjerg was much more complex, Both species occur in the upper portion of the Little Sioux River (Fig. 6–3). The headwater region of this river is intermittent and consists of a series of isolated, muddy ponds during dry summer periods. Here, *O. immunis* is the only species present. Farther south, the stream is permanent, and is mostly gravel- or rock-bottomed. In this area, *O. virilis* occurs alone. Between these two regions is a section in which muddy pools and rocky riffles alternate; here both species occur. Although both may be caught in the same seine-haul, generally *O. virilis* is found in the riffle sections and *O. immunis* is restricted to muddy pools.

Several additional observations complicate this pattern. Certain mud-bottomed streams in nearby locations harbor permanent populations of *O. virilis,* in the absence of *O. immunis*. Furthermore, after two summers

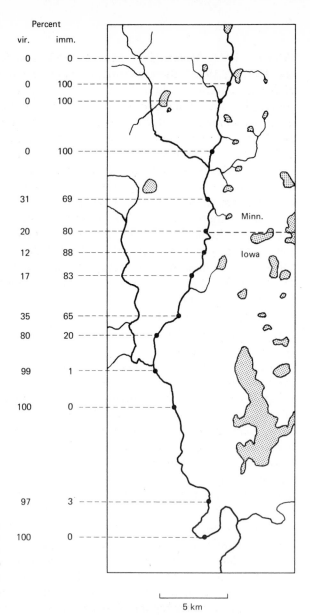

Percent

vir.	imm.	
0	0	
0	100	
0	100	
0	100	
31	69	
20	80	
12	88	
17	83	
35	65	
80	20	
99	1	
100	0	
97	3	
100	0	

Minn.

Iowa

5 km

FIGURE 6–3. Distribution of the crayfish, *Orconectes virilis* and *O. immunis* in the headwater region of the Little Sioux River in Minnesota and Iowa. The larger lakes and ponds are stippled. Numbers at the left give percentages of the two species in collections at various sites. (From Bovbjerg, 1970.)

of intense drought and unusual stagnation of the Little Sioux River, followed by a period of severe spring flooding in 1969, many of the rocky sections of the river, which were normally occupied by *O. virilis,* possessed high populations of *O. immunis*. These populations persisted until the end of the summer of 1969, by which time they had almost completely been replaced by *O. virilis.*

Finally, it was noted that in an oxbow lake on the flood plain of the Little Sioux River, permanently occupied by *O. immunis,* spring floods frequently introduced large numbers of *O. virilis*. During the summer, this pond normally dries up and the individuals of *O. virilis* die, but those of *O. immunis* survive by constructing elaborate burrows in the mud of the pond bottom. These burrows extend to depths at which high moisture levels occur.

These observations raise several questions about the factors restricting the distributions of the two species and leading to their occasional replacement in portions of the Little Sioux River. Bovbjerg investigated these through an extensive series of experiments. First, he examined responses of both species to conditions of drying and stagnation of pond habitats. Faced with simulated drying of a pond, *O. immunis* constructed burrows, topped by a protective "chimney" reaching above the mud surface. Under the same conditions *O. virilis* showed very weak patterns of burrowing and suffered high mortality. In addition, when tolerances of the two species to low oxygen tensions in water (less than 1 ppm) were compared, the mortality rate for *O. virilis* was much higher (Fig. 6–4).

Thus, it appears that competition is of little importance in the exclusion of *O. virilis* from temporary pond situations. Regardless of whether or not mechanisms of exploitation or interference occur between the two species when they come together in such a situation, *O. virilis* is eventually eliminated by virtue of its intolerance of oxygen depletion and ponddrying. In other words, even if *O. immunis* were absent, *O. virilis* would be unable to occupy this area permanently.

A second series of experiments was conducted to examine factors of the stream environment. Preferences of the two species for rock, gravel, and muck substrates were determined in experimental aquaria. Surprisingly, both species showed strong, nearly equal preferences for rock

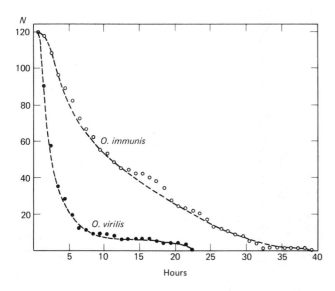

FIGURE 6–4. Survival of the crayfish *Orconectes virilis* and *O. immunis* in water depleted of oxygen to less than 1 ppm (240 animals). (From Bovbjerg, 1970.)

substrates when the preference test was conducted in the absence of the alternate species (Fig. 6–5). When both species were present, the preference for rock by *O. virilis* was enhanced, and *O. immunis* switched to a strong preference for the muck substrate.

These results suggested the occurrence of behavioral interaction between individuals of the two species. Upon subsequent investigation it was found that in agressive encounters *O. virilis* showed strong behavioral dominance over *O. immunis*. This dominance was strongest on rocky substrates, but was also true, to a lesser degree, on muddy substrates (Fig. 6–6).

These observations, together with considerable evidence that *O. immunis* can survive in, and indeed prefers, rocky stream habitats, suggest that it is excluded from such habitats by aggressive behavior of *O. virilis*. Clearly, this mechanism is one of interference rather than exploitation. Assuming, however, that any resource were in short supply, the operation of this mechanism would act as a guarantee of exclusive use of the resource by the dominant species. Bovbjerg concluded that, for these species in the stream habitat, natural crevices in which cray-

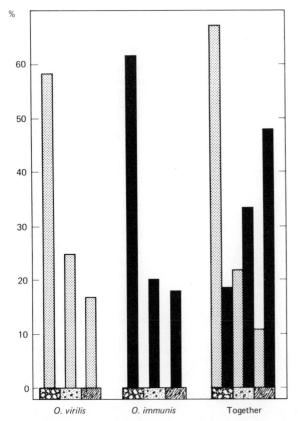

FIGURE 6–5. Substrate preferences (rock, gravel, muck) for the crayfish species, *Orconectes virilis* and *O. immunis,* when alone and when together, expressed as a percentage of total positions recorded (5,600 observations). (From Bovbjerg, 1970.)

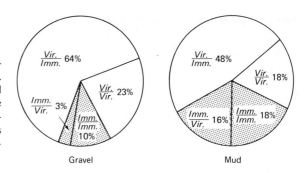

FIGURE 6–6. Interspecific aggression between the crayfish *Orconectes virilis* and *O. immunis* recorded as percentages of the total number of tension contacts. Each sector of the circle represents one of the four possible outcomes of a contact; the numerator indicates dominant species, the denominator the subservient. (From Bovbjerg, 1970.)

fish may hide constitute one of the most important environmental resources. These locations cannot readily be constructed in rocky areas and are necessary for protection of individuals during molting, egg-carrying, predator attacks, or floods.

Thus, these two species possess similar resource requirements and seem likely to be limited by the availability of such resources. Although both species potentially are able to occupy a variety of habitats, including rocky streams, only one species is usually found. Exclusion of the other, in this case of *O. immunis* from streams, occurs as a result of an interaction that does not directly involve the resources important to the species but functions to assure individuals of the dominant species an adequate supply of resources.

Other examples of competitive replacement in nature are given by Connell (1961), DeBach (1966), DeBach and Sundby (1963), and Lack (1947, pp. 140–42). Although the number of well-documented cases of replacement of species over wide geographical areas is small, such occurrences must be very common on the local level. The process of biotic succession, for example, involves the replacement of entire communities of plant and animal species by other communities better adapted to the modified environmental conditions, and, presumably, more efficient in their utilization of environmental resources.

In the statement of the competitive exclusion principle given earlier, application of the principle was restricted to resource-limited species. This restriction is necessary since, if populations are limited by factors other than the supply of resources, this limitation must be at a level below that set by resources. If this is the case, there may be no effective competition for resources, and forms with similar or identical patterns of resource utilization may be able to coexist.

Hairston et al. (1960), in a short paper on population control and competition, have argued that terrestrial grazing herbivores as a group may exhibit this situation. Their argument is based largely on the observation that vegetation is rarely depleted by the grazing action of herbivores, although observations from disturbed or artificial situations indicate that these animals are capable of causing such depletion. The serious overgrazing of the Kaibab Plateau in Arizona by deer in the early 1900s, as an apparent result of predator removal and protection of the deer herd from hunting, is an example of such depletion. The unusual occurrence of situations of this kind suggests that some other factor, such as predation, normally holds populations of grazing herbivores well below the limit set by the available plant food. This conclusion, in turn, suggests that competition between herbivore species may often be negligible, and that species with similar patterns of food resource utilization may be able to coexist. In this regard, it is interesting to note that Utida (1953), working with species of grain weevils which normally show strict competitive exclusion in simple cultures, found that the introduction of a parasitic wasp caused reduction of populations of both competing species below levels set by food resources and permitted their indefinite coexistence. The discussion by Hairston et al. (1960) has, however, stimulated considerable controversy among ecologists. Several of the important papers embodying this controversy are reprinted in Hazen (1970).

MATHEMATICAL MODELS OF INTERSPECIFIC COMPETITION

In his studies of population growth and interspecific competition between species of *Paramecium* and other microorganisms, Gause (1934) combined techniques of experimentation and mathematical modelling. The models developed by Gause provide the theoretical basis for many of our current ideas about interspecific competition, as well as the point of departure of most current attempts at modelling competitor systems. We will examine the basic model formulated by Gause, and explore some of the implications that may be derived from it.

Let us assume the presence of two species, the maximum abundances of which are limited, by resource availability, at the levels K_1 and K_2, respectively. We may also assume that each of these species, when alone, shows a pattern of population growth that can be described satisfactorily by the logistic equation (see p. 216). Growth of independent populations of these species could thus be represented by the equations:

$$\frac{dN_1}{dt} = r_1 N_1 \left(\frac{K_1 - N_1}{K_1} \right) \qquad (6\text{--}1)$$

$$\frac{dN_2}{dt} = r_2 N_2 \left(\frac{K_2 - N_2}{K_2} \right) \qquad (6\text{--}2)$$

where

N_1, N_2 = Population sizes of species 1 and 2
r_1, r_2 = Intrinsic rate of natural increase of species 1 and 2
t = Time

We may consider the right side of these equations to consist of two expressions: (1) rN, giving the potential growth of the population considering the number of individuals present and their inherent capacity of increase and (2) $(K - N)/K$, the degree of realization of this potential increase. According to the logistic equation, as N approaches K, the value of this degree of realization expression declines toward 0.

Now, if these two species occur together and show some degree of competitive interaction, the degree of realization of potential growth by one species is influenced not only by its own population density but also by that of its competitor. This influence would reduce the degree of realization of potential growth by a factor related to the population size of the competitor. Such effects may be represented by the following modifications of Eq. (6–1) and Eq. (6–2):

$$\frac{dN_1}{dt} = r_1 N_1 \left(\frac{K_1 - N_1 - \alpha N_2}{K_1} \right) \tag{6-3}$$

$$\frac{dN_2}{dt} = r_2 N_2 \left(\frac{K_2 - N_2 - \beta N_1}{K_2} \right) \tag{6-4}$$

In these equations the coefficients α and β serve to convert the population densities of the competitor species into units equivalent in value to individuals of the species itself. For example, this coefficient may take into account the fact that individuals of the two species differ in body size and food intake and thus that their K levels differ. They will also reflect the intensity of competition, including both overlap in resource use and intensity of interference effects.

These two equations enable prediction of the outcome of competition between two species. According to the precise values of α and β the outcome differs. The different outcomes depend on the relative intensity of intraspecific and interspecific competition for each species. Thus, we may first state the conditions under which intraspecific and interspecific competition would be equal. In this situation the following would be true:

$$\alpha = \frac{K_1}{K_2} \tag{6-5}$$

$$\beta = \frac{K_2}{K_1} \tag{6-6}$$

Thus, if species 1 had a K-value of 100 and species 2 a K-value of 200, α would be equal to 0.5. In effect, this means that one individual of species 2 is equivalent to 1/2 individual of species 1 in competitive influence. Since species 2 reaches a maximum population density twice that of species 1, this is equivalent to saying that when the difference in resource use per individual is taken into account, the effect of species 2 on species 1 is just as intense as the intraspecific effect of species 1 on itself.

Based on this relationship, four cases may be postulated:

1. $\alpha > \dfrac{K_1}{K_2}$ $\beta > \dfrac{K_2}{K_1}$ Interspecific competition more intense than intraspecific for both species

2. $\alpha < \dfrac{K_1}{K_2}$ $\beta < \dfrac{K_2}{K_1}$ Intraspecific competition more intense than interspecific for both species

3. $\alpha < \dfrac{K_1}{K_2}$ $\beta > \dfrac{K_2}{K_1}$ Intraspecific competition more intense than interspecific for species 1, the opposite for species 2

4. $\alpha > \dfrac{K_1}{K_2}$ $\beta < \dfrac{K_2}{K_1}$ Interspecific competition more intense than intraspecific for species 1, the opposite for species 2

These four cases are shown graphically in Fig. 6–7(a) and the consequences can be explored in Fig. 6–7(b). In these diagrams the horizontal and vertical axes correspond to abundance of species 1 and 2, respectively. Thus, a point within the graph represents a combination of population densities of the two species.

For each of these species we may show equilibrium population sizes possible at various abundances of its competitor. Thus, for species 1, maximum populations can range from K_1, when species 2 is absent, to 0, when species 2 has an abundance equal to K_1/α. Likewise, species 2 can vary from K_2, when species 1 is absent, to 0, when species 1 has an abundance equal to K_2/β. These relationships can be shown as lines of maximum abundance for each species on the four case graphs in Fig. 6–7(a).

From the relationship of these lines the outcome can be derived for any initial combination of abundances of the two species [Fig. 6–7(b)]. For example, any point below and to the left of both lines represents abundances of the two species below those possible. Thus both can increase in abundance, causing the point to move upward and to the right. These increases may continue until the point encounters a line arranged such that an increase in one species is still possible, but only at the expense of a decrease in the second. If this occurs, the second will eventually be reduced to extinction by continued increase of the first.

Only in case #2 does a situation prevail where the combined populations gravitate to a point where both are limited. This point corresponds to the intersection of the two lines. For case #1, one species or the other survives, depending upon the initial ratio of their abundances. In this case, coexistence of the species at abundances corresponding to the intersection, s, should in theory be possible. However, any accidental deviation from this ratio will initiate a sequence leading to elimination of one of the species. In cases 3 and 4, one particular member of the species pair survives, regardless of the initial composition.

This model suggests that the basic requirement for coexistence is that the intensity of intraspecific competition exceed that of interspecific

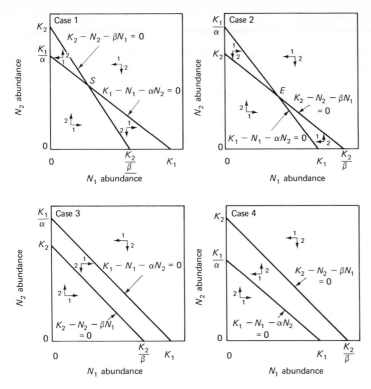

FIGURE 6–7(a). Competitive interaction between pairs of species. In each case the lines $K_1, K_1/\alpha$, and $K_2, K_2/\beta$ represent the saturation values for species N_1 and N_2 respectively. As indicated by the short arrows, neither species can increase above its saturation line. Below its saturation line each species will increase. There are four cases:

Case 1: Unstable equilibrium is possible at point S, but it is to be expected that one and only one species will survive. Which species survives depends on whether the initial mixture of species lies to the right or left of a line from O through S.

Case 2: Stable equilibrium occurs at point E. Initial concentrations of the two species are irrelevant.

Case 3: Species N_1 will always win in competition since the region $K/\alpha, K_2, K_2/\beta, K_1$ is below the saturation level of N_1 but above the saturation level of N_2.

Case 4: Species N_2 will always win in competition since the region $K_2, K_1/\alpha, K_1, K_2/\beta$ is below the saturation level of N_2 but above the saturation level of N_1. (From Slobodkin, 1963.)

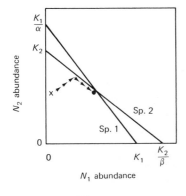

FIGURE 6–7(b). Illustration of the hypothetical course of populations of species 1 and 2 from an initial starting point, x, to an equilibrium state in a Case 2 relationship.

competition for all of the species involved. This is most easily achieved if the species differ in the specific nature of the resources required. However, coexistence of species may still occur even though both are dependent on the same resource if the competitors differ in their strategy of resource exploitation (Frederick E. Smith, personal communication). A resource, such as an insect food species for insectivorous birds, may be considered to consist of units that have a certain average vulnerability to being taken. This average vulnerability changes with abundance of the resource, since at low resource abundances, the resource will exist only in the locations most favorable to it. Consequently, exploitation of a resource by a species may modify the vulnerability of the resource. If two species employ different strategies of resource exploitation, the vulnerability of the resource may be reduced for one species, but it still may be high for the other. Thus, changing patterns of vulnerability of a single resource to exploitation in various ways may permit coexistence of several competitors. In terms of the above model, this amounts to the suggestion that the coefficients, α and β, may themselves be variables, rather than constants.

These basic models may be elaborated upon in various ways. For example, Slobodkin (1963) has shown how nonselective predation may influence a competitive relationship. If a predator takes fraction m of individuals, regardless of species, per unit time, such an action can be incorporated into the competition model as follows:

$$\frac{dN_1}{dt} = r_1 N_1 \left(\frac{K_1 - N_1 - \alpha N_2}{K_1} \right) - mN_1$$

$$\frac{dN_2}{dt} = r_2 N_2 \left(\frac{K_2 - N_2 - \beta N_1}{K_2} \right) - mN_2$$

The influence of this interaction on the outcome of competition varies with the case and with specific values of r. However, as shown in Fig. 6–8, it may modify case 3 to permit coexistence of both species. This would occur when species 1 always wins in the absence of the predator but species 2 has the higher value of r.

ECOLOGICAL ISOLATING MECHANISMS

An important corollary of the competitive exclusion principle states that resource limited species that continue to coexist must possess mechanisms for the prevention or reduction of active interspecific competition. These mechanisms are termed *ecological isolating mechanisms* and are said to confer *ecological isolation*. The evolutionary origin of ecological isolating mechanisms is an important step in the process of speciation. The simple production of two reproductively isolated segments of a parental species population does not guarantee their survival as new

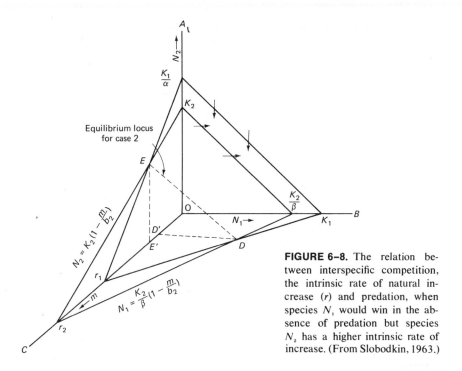

FIGURE 6–8. The relation between interspecific competition, the intrinsic rate of natural increase (*r*) and predation, when species N_1 would win in the absence of predation but species N_2 has a higher intrinsic rate of increase. (From Slobodkin, 1963.)

species. If two or more incipient species come into contact or sympatry, it is likely that all but one will be eliminated through competitive processes unless mechanisms have concurrently evolved to provide each species with unique features of resource utilization that prevent its replacement in all environmental situations.

Spatial Mechanisms of Ecological Isolation

Allopatry. Two or more species may utilize similar habitats and environmental resources but may occupy separate geographical areas. This distribution pattern may reflect the fact that each species is best adapted to the biotic and abiotic conditions of its own geographical area and is most efficient in utilizing environmental resources within this area. On the other hand, it may simply reflect the fact that the species involved are separated by geographical barriers that prevent contact and competitive challenge between the species. In the former case, the pattern of ecological isolation is evidence of the failure of the species to differentiate to a degree adequate to permit actual coexistence. An example of this pattern of ecological isolation is furnished by the Black-capped Chickadee, *Parus atricapillus,* and the Carolina Chickadee, *P. carolinensis,* in eastern North America (Brewer, 1961). These two species are similar in size, behavior, and feeding ecology. The Black-capped Chickadee occurs throughout forest areas of Canada and northern United States and is replaced by the Carolina Chickadee in similar habitats in the central and southern United States.

Habitat Isolation.   Two or more ecologically similar species may occur in the same geographical area, but may occupy different habitats. These species may utilize similar kinds of environmental resources, but within each habitat one species is superior to all others in adaptation to environmental conditions and in efficiency of resource utilization. An excellent example is furnished by the jays (Corvidae) of the western United States. In the Colorado Rockies, four species of jays often occur in the same geographical areas. These species are restricted to different vegetation types, however. The Scrub Jay, *Aphelocoma coerulescens,* occurs in broad-leafed shrub vegetation, the Piñon Jay, *Gymnorhinus cyanocephala,* in pinyon-juniper woodlands, the Stellar's Jay, *Cyanocitta stelleri,* in the lower elevation coniferous forests, and the Gray Jay, *Perisoreus canadensis,* in spruce-fir forests at high elevations. This habitat separation is maintained in spite of the fact that in many areas the different vegetation types lie within a few minutes flying distance of each other.

Temporal Mechanisms of Ecological Isolation

Diurnal.   Species may occur together and utilize similar environmental resources but at different times of day. To the extent that this separation in time leads to differentiation of the specific resource utilized, it may function as a mechanism of ecological isolation. For example, the food resource consisting of flying insects is fed upon by various species of swifts and swallows during daylight hours, by nighthawks during the twilight hours, and by bats at night. Since the species composition of the flying insect population varies diurnally, it is likely that each of these bird groups is feeding upon different insect species. This separation of feeding activities is probably based on differences in mechanisms of food capture that give each species the advantage at particular light intensities. Pianka (1970), in studies of sympatric species of lizards of the Australian genus *Ctenotus,* has found that certain species pairs, otherwise similar in their feeding behavior, differ in the time of day at which their principal feeding activities occur.

Seasonal.   Different species may also utilize the same areas and the same general kinds of resources but concentrate their activities at different seasons. This pattern of isolation reflects differential adaptation of the species to seasonally varying environmental conditions. An example of seasonal isolation is furnished by the brown shrimp, *Penaeus aztecus,* and the white shrimp, *P. sertiferus,* in the Gulf of Mexico (Aldrich et al.,

1967). Postlarval shrimp of both species enter bays and estuaries along the Gulf Coast of Texas and Louisiana, where they carry out most of their growth in size. The brown shrimp enter these nursery areas in the late fall, hibernate in burrows in the bottom mud, and emerge to carry out feeding and growth during the following spring. The white shrimp enter these areas in summer, when the brown shrimp have moved back into deeper water, and they undergo development during the summer and fall. This example presents an almost complete separation of seasonal activity cycles, but even the partial staggering of life-cycle activities of similar species furnishes a degree of ecological isolation of this kind.

Differentiation of Mechanisms of Resource Exploitation

Morphological. Coexistence of species may also occur if resource requirements of the species are differentiated. This may result from the evolution of differences in morphological characteristics related to resource exploitation. In the case of vertebrates this often involves the morphology of the feeding apparatus. Examples of such differentiation are especially common in birds, which are highly adapted for specific modes of feeding in features such as beak structure, body size, and leg and foot structure. The Hairy Woodpecker, *Dendrocopus villosus,* and the Downy Woodpecker, *D. pubescens,* for example, are two species occurring together in many of the same habitats over much of North America. The Hairy Woodpecker is larger in body size and has a beak proportionately larger than that of the Downy Woodpecker. These species differ markedly in their specific feeding patterns, and they apparently utilize quite different kinds of insect food.

Physiological. In plants and most microorganisms, exploitation of resources largely involves the uptake of specific chemical nutrients by physiological processes. Differentiation of these processes may lead to differences in patterns of resource utilization among species, thus allowing their coexistence. An example of this pattern of isolation is furnished by various kinds of nitrogen-fixing microorganisms. These nitrogen-fixers, primarily species of bacteria and blue-green algae, are able to utilize molecular nitrogen as a raw material for the synthesis of nitrogen-containing organic compounds. Many other microorganisms, in contrast, require ammonium or nitrate ions for such syntheses. These materials are made available largely through decomposition of organic matter by bacteria and fungi and through weathering of parental rock materials. These differences in mechanisms of nitrogen uptake and metabolism effectively prevent competition between these two groups of microorganisms when nitrogen is a potentially limiting nutrient.

Behavioral. Differences in behavioral mechanisms of resource exploitation may also lead to differences in the resources utilized and thus permit coexistence. Carnivores and insectivores show many specific behavior patterns related to the capture of food organisms. Among insectivorous songbirds, the vireos and flycatchers are two groups which co-occur widely, and which are often found feeding in close proximity in the forest canopy. Members of these two groups, however, possess very different patterns of feeding. Flycatchers occupy exposed perches from which they dart out to capture flying insects. Vireos, in contrast, forage by hopping from branch to branch in the denser portions of the tree canopy where they glean small insects from the leaves and twigs. By virtue of these different feeding patterns birds of the two groups encounter different kinds of insect food, and they are thus able to coexist without competing for this resource.

We must, of course, recognize that a variety of evolutionary processes may lead to divergence among species in their use of resources. Thus, simple observation of differences in one or more of the above categories is not adequate for the conclusion that competition has been the primary selective process responsible.

DETAILED STUDIES OF PATTERNS OF ECOLOGICAL ISOLATION

To illustrate how ecological isolating mechanisms determine patterns of distribution and coexistence of species, we will examine three situations in detail. These were selected because they show strongly contrasting patterns of ecological isolation and because they introduce several new problems related to resource utilization by ecologically similar species. Two of these involve related species of birds, the first dealing with the Galapagos Finches, studied intensively by Lack (1947, 1969) and Bowman (1961), and the second with five species of warblers of the genus *Dendroica,* studied by MacArthur (1958). The third situation, studied by Hanes (1965), involves the ecological relations of two congeneric species of chaparral shrubs in California.

The Galapagos Finches are a group of 14 species of sparrowlike birds that constitute the subfamily Geospizinae of the family Fringillidae. This group is restricted to the Galapagos Islands, a group of volcanic islands located on the equator about 600 miles west of Ecuador (Fig. 6–9), and to Cocos Island, located about 600 miles northeast of the Galapagos Islands. In the Galapagos Islands the group is represented by 13 species (Fig. 6–10), usually grouped into three genera (Lack, 1947), and apparently derived from a single ancestral form which colonized the islands from the mainland of South America. These species constitute over half of the resident land bird species of the islands. Cocos Island possesses a single species, usually assigned to a fourth genus (Fig. 6–10). The Galapagos Finches were first observed by Charles Darwin, during the voyage of the *Beagle,* and his observations on the differentiation of species on

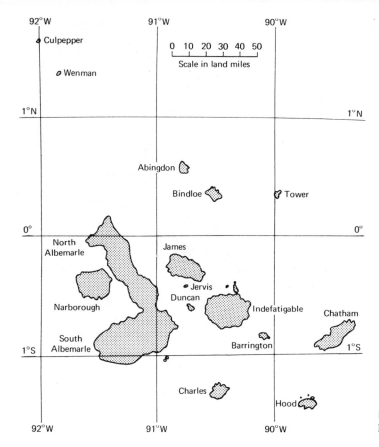

FIGURE 6–9. The Galapagos Islands. (From Lack, 1947.)

the various islands contributed importantly to his development of the theory of natural selection.

We will be concerned only with the relationships of the species occurring in the Galapagos Islands. In this group there are 13 major islands separated by distances varying from a few miles to about 90 miles. The islands vary greatly in size, with the largest, Albemarle, being about 80 miles long. Most of the smaller islands are low in elevation. Albemarle, however, reaches an elevation of over 4,000 feet, and several of the other large islands reach elevations of 2,000–3,000 feet. The topography of the islands is rugged, especially in areas of more recent volcanic activity. At lower elevations the vegetation consists of semiarid thorn scrub, in which large species of prickly-pear cacti of the genus *Opuntia* are common. On the larger islands this vegetation type shows a gradual transition, with increasing altitude, to a humid forest.

The Galapagos Finches provide an excellent illustration of the evolutionary processes of speciation and adaptive radiation. The semi-isolated nature of the islands, and of the populations occupying them, has favored

FIGURE 6–10. The Galapagos finches. (From Lack, 1953.)

the independent evolution of different island populations and allowed their divergence in many characteristics. New species have arisen where this divergence has continued to the point that interbreeding between the different populations is no longer possible. This process has been accompanied by a complex pattern of dispersal of newly evolved species throughout the island group. More detailed accounts of factors important in this process are given by Lack (1947) and Hamilton and Rubinoff (1963).

Adaptive radiation is clearly shown in the feeding patterns of the Galapagos Finches, which have diverged to a degree comparable to that shown among members of several different families of birds in continental areas. This divergence has been possible because of the impoverished nature of the Galapagos bird fauna. Many of the major families of land birds characterized by specialized feeding habits are absent from the Galapagos.

In this discussion, however, our primary interest lies in the evolutionary development of mechanisms of ecological isolation. Table 6–2 gives the distribution of species on the major islands, together with information on their habitat occurrence and feeding habits. From this table it can be seen that the pattern of distribution and co-occurrence of the various species is quite complex. The larger islands possess from seven to eleven species of finches. The coexistence of these species, all of recent evolutionary origin, suggests the operation of well-defined systems of ecological isolation.

Several of the different types of ecological isolating mechanisms outlined earlier are utilized by this group. Only a single clear case of geographical isolation or allopatry occurs, however. In the genus *Geospiza,* the species *conorostris* and *scandens,* which utilize the same general foods and occupy the same habitat type, are found only on different islands. The former occurs only on Hood, Tower, and Culpepper Islands, and the latter occurs on most of the remaining islands.

Several cases of habitat isolation occur. At the generic level, the various *Geospiza* species are, for the most part, restricted to the lower elevation arid and transitional habitats, while the *Camarhynchus* species are found in the higher elevation transitional and humid vegetation zones. At the species level, two clear cases are shown. In the genus *Geospiza,* the species *difficilis* and *fuliginosa* are separated by habitat on the large central islands, with *difficilis* being restricted to the higher elevation humid forest areas and *fuliginosa* to the lower elevation arid habitats. On the outer, low-lying islands of Culpepper, Wenman, and Tower, however, *difficilis* occurs alone and occupies the existing arid habitats. Thus, it appears that *difficilis* is potentially able to occupy arid habitats but is competitively excluded from them on the large central islands by *fuliginosa.* The presence of *difficilis* in arid habitats on the outer islands suggests that these are islands which *fuliginosa* has not succeeded in colonizing.

TABLE 6-2. ISLAND OCCURRENCE, HABITAT DISTRIBUTION, FOOD TYPE, AND FEEDING MICROHABITAT OF THE GALAPAGOS FINCHES.*

	Feeding microhabitat	Culpepper	Wenman	Tower	Abington	Bindloe	James	Indefatigable	Albemarle	Narborough	Barrington	Chatham	Hood	Charles
Geospiza magnirostris	1		AT SL	AT SL	AT SL	AT SL	AT SL	AT SL	AT SL	AT SL	AT SL			AT SL
G. fortis	1				AT SM	AT SM	AT SM	AT SM	AT SM	AT SM	AT SM	AT SM		AT SM
G. fuliginosa	1				AT SS	AT SS	AT SS	AT SS	AT SS	AT SS	AT SS	AT SS	AT SS	AT SS
G. difficilis	1(3)	AT SS,C	AT SS,C	AT SS	HT SS		HT SS	HT SS	HT SS	HT SS				HT SS
G. scandens	3				AT C	AT C	AT C	AT C	AT C		AT C	AT C		AT C
G. conirostris	1(3)	AT SL		AT C									AT SL,C	
Camarhynchus psittacula	2				HT IL	HT IL	HT IL	HT IL	HT IL	HT IL	HT IL			HT IL
C. pauper	2													HT IM
C. parvulus	2				HT IS		HT IS	HT IS	HT IS	HT IS	HT IS	HT IS		HT IS
C. pallidus	5						HT I	HT I	HT I	HT I		HT I		
C. heliobates	5								M I	M I				
C. crassirostris	6				HT B	HT B	HT B	HT B	HT B	HT B		HT B		HT B
Certhidea olivacea	4	AHT IS	AHT IS	AHT IS	AHT IS	AHT IS	AHT IS	AHT IS	AHT IS	AHT IS	AHT IS	AHT IS	AHT IS	AHT IS
Species on island		3	3	4	9	7	10	10	11	10	7	7	3	10

*Data from Lack, 1947, 1969.

Habitat: AT: Arid and Transitional
        HT: Humid and Transitional
        M: Mangrove
        AHT: Arid, Humid, Transitional

Food: S: Seeds (SS, small seeds; SM, medium seeds; SL, large seeds)
      C: Cactus
      B: Buds, leaves, fruit
      I: Insects (IS, small insects, etc.)

Feeding Microhabitat: 1. Ground
    2. Trees and shrubs (excavation of wood)
    3. Opuntia
    4. Tree and shrub foliage
    5. Tree trunk and branches (surface)
    6. Trees and shrubs (fleshy parts)

In the genus *Camarhynchus,* a second example is furnished by the species *heliobates* and *pallidus,* both of which feed on insects captured by foraging over the trunks and branches of trees. The species *heliobates* is restricted to mangrove swamps along the island coasts, while *pallidus* is widely distributed in forest areas of the island interior. In addition, *pallidus,* the "woodpecker finch," employs the unique behavior pattern of using a cactus spine to probe in bark crevices and flush out small insects.

These spatial mechanisms, however, account for only part of the over-all pattern of distribution and coexistence of the Galapagos Finches. For example, in the genus *Geospiza,* the species *magnirostris, fortis,* and *fuliginosa* coexist in the same habitat on many of the large islands and utilize the same general kinds of food (Table 6–2). The primary mechanism of ecological isolation for these species appears to be their morphological specialization for different foods. These species differ strikingly in the size and structure of the beak (Fig. 6–11). That of

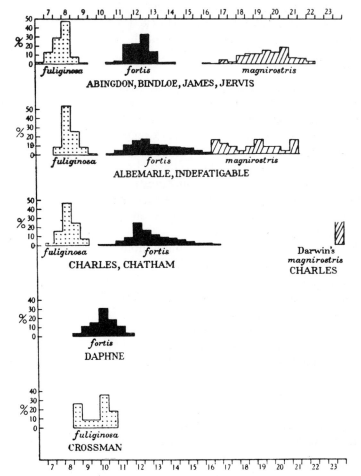

FIGURE 6–11. Histograms of beak-depth (mm) for *Geospiza magnirostris, G. fortis,* and *G. fuliginosa* on various islands of the Galapagos group. (From Lack, 1947.)

magnirostris is massive in structure and up to 20 mm or more in depth. The beak of *fortis* is intermediate in size and thickness, and that of *fuliginosa* is small and relatively slender. So strong is this differentiation that samples of individuals of the three species from a given island show little or no overlap (Fig. 6–11). The significance of this differentiation has been studied in detail by Bowman (1961) through field observations of feeding behavior and analysis of stomach contents. Bowman has shown that, although there is considerable overlap in the kinds of seeds taken, the diets of the three species differ markedly in the proportions of various seed types. Large, thick-walled seeds predominate in the diet of *magnirostris*. The diet of *fuliginosa* consists primarily of smaller, thin-walled seeds and includes a greater diversity of seed species. An intermediate type of diet is shown by *fortis*. Thus, morphological differentiation of the feeding apparatus, correlated with the utilization of different specific food resources, is the principal basis for ecological isolation in these species.

Ecological isolation through specialization in food resource utilization is also shown by *Geospiza scandens*. This species shows a nearly complete dependence on flowers and fruit of prickly-pear cacti (Lack, 1947; Bowman, 1961). Although two other *Geospiza* species occasionally rely heavily on this food resource, they do so only on islands from which *scandens* is absent (Lack, 1947).

Members of the genus *Camarhynchus* are, for the most part, isolated from the *Geospiza* species through differences in types of foods utilized, as well as the habitat differences previously noted. With one exception, the *Camarhynchus* species are primarily insectivorous. This exception is the species *crassirostris,* which feeds almost entirely on fleshy fruits, seeds, leaves, and flowers (Bowman, 1961). This diet is also distinct from that of any species of the genus *Geospiza* in its concentration on soft, cellulose-rich foods rather than seeds (Bowman, 1961). As an adaptation for this diet, *crassirostris* possesses the largest gizzard and the longest intestinal tract of any of the Galapagos Finches.

Within the genus *Camarhynchus* there is a beak-size differentiation comparable to that shown by the three *Geospiza* species. The *Camarhynchus* species involved are *psittacula, pauper,* and *parvulus.* All three coexist on Charles Island, and *psittacula* and *parvulus* occur together on many of the larger islands. In beak size and thickness *psittacula* is largest, *pauper* intermediate, and *parvulus* smallest (Fig. 6–12). From field observations of feeding behavior and analysis of stomach contents of *psittacula* and *parvulus* on Indefatigable Island Bowman (1961) concluded that the large-beaked form, *psittacula,* was able to excavate soft plant tissues and rotting wood to a greater depth and thus encountered larger insect larvae than did the small-beaked form.

Finally, *Certhidea olivacea,* the warbler-finch, is distinct from all other species in its feeding location and behavior. This species forages

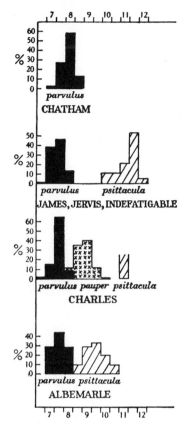

FIGURE 6–12. Histograms of beak-depth (mm) for *Camarhynchus psittacula, C. pauper,* and *C. parvulus* on various islands of the Galapagos group. (From Lack, 1947.)

through the foliage of trees and shrubs and gleans small insects in the manner of a warbler or vireo. It is morphologically adapted to feeding on small insects because of its small body size and slender beak (Fig. 6–10). The ecological isolating mechanism of this species is thus a combination of morphological and behavioral characteristics.

Among these finches, each species, through a combination of spatial, morphological, or behavioral features, thus possesses a unique pattern of utilization of food resources. Spatial mechanisms, especially those involving occupation of different habitats, and differentiation of the morphology of the feeding apparatus are especially important mechanisms in the ecological isolation pattern for this group.

The second study that we will examine in detail deals with ecological isolating mechanisms among five species of wood warblers (Parulidae) that breed together in many coniferous forest areas of the northeastern United States (MacArthur, 1958). These five species present a strong challenge to the hypothesis that differentiation of patterns of resource utilization is a requirement for coexistence, since they not only occur together in the same habitat but also show little differentiation in body size or morphology. These five species are members of the genus *Dendroica:* the Cape May Warbler, *D. tigrina;* the Myrtle Warbler, *D. coronata;* the Black-throated Green Warbler, *D. virens;* the Blackburnian Warbler, *D. fusca;* and the Bay-breasted Warbler, *D. castanea.*

Much of the field work in this study was devoted to obtaining standardized observations on the specific locations in which individuals of each species fed and on the particular behavior patterns used in feeding. For observations of feeding location, trees in the study area were divided into six vertical zones, each approximately ten feet in height. The lowest zone, Zone 6, corresponded to the area beneath the canopy of the tree. Each of the five remaining zones within the tree canopy was subdivided into sections relative to distance from the trunk of the tree. The innermost section corresponded to the area of bare or lichen-covered branches near the trunk, the middle section to the region of branches with old needles, and the outermost section to the area of branches with young needles and buds. For each species MacArthur obtained data on the number of individuals seen and the total time spent foraging in each zone and section. In addition, records of the frequency of occurrence of specific behavior patterns used in feeding were kept. These records included observations of the frequency with which individuals of each species (1) moved radially along branches, thus crossing the various canopy sections, (2) moved circumferentially around the tree, thus staying in the same section, or (3) moved vertically between branches at different heights, but within the same or adjacent zones. Still other observations were obtained on the length of intervals between flights to new feeding locations and on the frequency of behavior patterns such as hovering at the undersides or tips of branches, hawking flying insects, and carrying out long flights to distant feeding locations.

The data obtained (Fig. 6–13) show a number of important differences in the patterns of feeding activity of the different species. The precise location of feeding activities in and beneath the tree canopy differed for each species. On the basis of data on the time spent foraging in the various zones and sections, overlap between species in feeding areas varied from 24.5% for the Cape May and Bay-breasted Warblers to 73.2% for the Cape May and Blackburnian Warblers (Fig. 6–13). In addition, the frequency of specific patterns of feeding behavior differed for the five species. Data on the directional pattern of movements during feeding indicate that the Cape May Warbler uses primarily vertical movements, the Blackburnian and Bay-breasted Warblers radial movements. None of the species combinations having over 60% similarity in feeding locations, however, was alike in these movement patterns. Also, the species that were most similar in feeding location showed differences in frequency of use of the specific behavior patterns of hovering, hawking, and long flights.

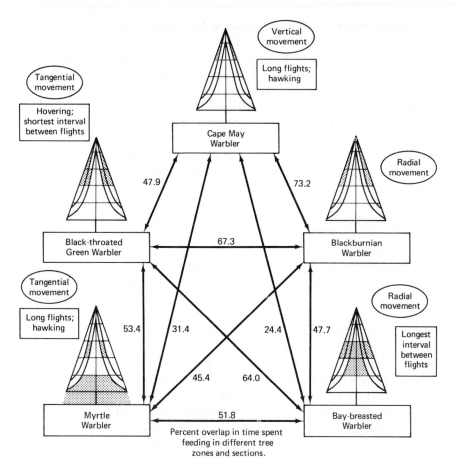

FIGURE 6-13. Ecological isolation among five species of warblers (Parulidae) in coniferous forests of the northeastern United States. Shaded portions of tree diagrams show zones and sections in which 50% of total feeding time is spent. Arrows between species give percent overlap in feeding areas on the basis of time spent in various zones and sections of tree. Principal direction of movement while feeding (radial, tangential, vertical), frequently used specialized feeding patterns (long flights, hawking, hovering), and other behavioral patterns (length of interval between flights to new feeding areas) are indicated adjacent to tree diagrams.

These behavioral differences, combined with those of feeding location, suggest that, as a result of a combination of rather subtle differences in behavior, the patterns of exploitation of insect food resources by the five warbler species may be well differentiated. Although no analyses of stomach contents were obtained in the area in which these observations were made, data from other areas suggest that there are consistent differences among them in specific insect foods taken (MacArthur, 1958). Thus, by feeding in different locations and in different manners, the five warbler species apparently encounter different kinds of food insects and achieve a relatively high degree of ecological isolation.

In addition to these differences, the Cape May Warbler shows a slight degree of differentiation in beak morphology. This species is unique in possessing a semitubular tongue and a beak somewhat more slender at the tip than those of other warbler species. These specializations are correlated with the fact that this species is to some extent a nectar feeder, especially during periods of rainy weather when insects are inactive.

Data on nesting dates for the different species also suggest an element of temporal isolation by the staggering of the dates of clutch completion. Although the median date of clutch completion varies by only about 10 days for the five species, this results in the fact that the critical period of reproduction, that of feeding the young in the nest, does not occur simultaneously for all species.

Ecological isolation in these warblers is thus achieved by mechanisms quite different from those utilized by the Galapagos Finches. Here the mechanisms of primary importance are behavioral, involving differences in the precise location and manner of foraging, rather than spatial and morphological as in the former case.

The study by Hanes (1965) provides an excellent example of some of the mechanisms of ecological isolation important in higher plants. Hanes studied the ecological relations of Chamise, *Adenostoma fasciculatum,* and Red Shank, *A. sparsifolium,* two species of shrubs occurring together in certain areas of the coastal chaparral of California and northern Baja California. Coastal chaparral is a dense, shrubby vegetation type dominated by small-leaved, evergreen species and occurs in a climatic region characterized by mild, moist to wet winters and hot, extremely dry summers.

These shrub species differ in several important features of the crown and root system (Fig. 6–14). At maturity Chamise is a spreading shrub from two to eight feet in height, with small, needle-like leaves and shredding gray bark. Red Shank is similar, but it reaches a height of 6–20 feet and possesses finely dissected leaves and reddish-brown bark which shreds and peels away from the stem, thus giving the plant its name. Chamise has the more extensive root system, relative to the size of the above-ground portions of the plant. It possesses a tap root and a system of major roots that penetrates deeply into the fractured parent material below the surface soil. Red Shank, on the other hand, lacks a tap root and has a less extensive root system that is concentrated within the top soil layer. Red Shank also possesses a series of very thick buttress roots, which apparently support the plant crown, and a system of small lateral roots which form a dense mat within the layer of top soil. The ecological significance of these different root systems presumably lies in the differentiation of patterns of water and nutrient uptake from the soil.

Important differences exist between the two species in seasonal

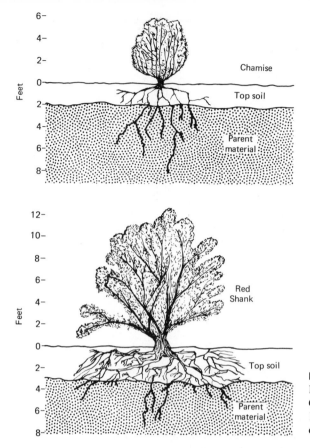

FIGURE 6–14. Root system of Chamise and Red Shank compared. The root system of Chamise is extensive in proportion to its top. Red Shank has a lateral root system with masses of small fibrous roots. (From Hanes, 1965.)

patterns of activity (Fig. 6–15). Both species are relatively inactive during late fall. Vegetative growth in both begins in the early winter, and in Chamise reaches a peak during April and May. Flowering in Chamise occurs from April through June, and seed is set through July. In Red Shank, vegetative growth gradually increases through the spring, reaching a peak in late spring, but with continued active growth throughout summer and early fall. Flowering occurs in August, and seed is set in October. Laboratory and field studies of growth and photosynthetic activity of the two species indicate that Red Shank exhibits not only greater drought tolerance, but also an unusual capacity to remain active during drought stress conditions, which characterize the habitat of these species during the summer. Thus, through differential adaptations to the physical environment, a considerable component of temporal isolation operates between these species. This temporal isolation, together with the differentiation of root systems in extent and depth of penetration, results in differential patterns of resource utilization adequate to allow these two species to coexist.

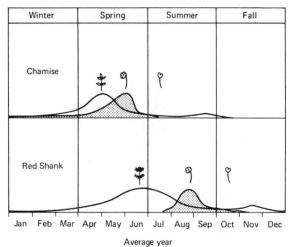

FIGURE 6-15. Generalized phenology of Chamise and Red Shank. The precise activities of each species vary from year to year and with changes in altitude, latitude, proximity to the sea, aspect, and microenvironment. (From Hanes, 1965.)

The preceding examples have illustrated a variety of ecological isolating mechanisms related primarily to utilization of food resources. It should be kept in mind that competition and ecological isolation may operate in relation to environmental resources other than those constituting food or nutrients. Broadhead and Wapshere (1966) have conducted a detailed study of life cycles and patterns of resource utilization of two related psocids (Insecta, Psocoptera) which coexist on larch trees in England and feed on the algal-fungal film present on the bark. Analysis of the feeding ecology of these species reveals no important differences, but examination of the details of reproduction reveals differences in the sites required for oviposition. From a detailed consideration of these differences the authors conclude that the two species are separately limited by intraspecific competition for oviposition sites and that this limitation is at a level that results in little or no interspecific competition for food.

This discussion of the mechanisms of ecological isolation in various groups of animals and plants has shown that differentiation of the patterns of utilization of resources, both food resources and other specific requirements, is necessary for the coexistence of resource limited species.

From the earlier studies of competitive exclusion in laboratory and field situations, and the preceding discussion of mechanisms of ecological isolation, a basic principle relating to the requirement for coexistence emerges. This principle is simply that, to coexist, each of a group of species, when very abundant, must inhibit its own further increase more than it inhibits that of any of the other species (MacArthur, 1958). This condition should prevail when each species population is limited by intra-specific competition for different resources. As we shall see later, such a condition develops, not from any "altruism" of the species involved, but instead from evolutionary specialization forced upon the species by the selective force of interspecific competition.

TRANSIENT COEXISTENCE OF COMPETITOR SPECIES

Transient situations involving the co-occurrence of direct competitors may exist during the replacement of one resource-limited species by another. Differential adaptations of the species involved to conditions presented by environmental disturbance, together with different dispersal abilities, may make such transient coexistence frequent, however, and prevent the extinction of the poorer competitor over broad geographical areas.

One situation of this type involves so-called "fugitive species." A fugitive species is a form adapted to taking advantage of unusual environmental conditions or newly created areas of habitat. In a given location, under stable environmental conditions, such a species will gradually be outcompeted by other forms and will decline in numbers or disappear. Where areas of new habitat appear, however, the fugitive species, because of its more efficient dispersal mechanism, may be the first species to invade and colonize. Or, given a short period of unusually favorable environmental conditions, it may be able to increase its population more rapidly than can its competitors, thus offsetting a competitive disadvantage maintained over a longer period of stable conditions.

Hutchinson (1959) has given an example of a fugitive species relationship in the European insects *Corixa punctata* and *C. dentipes* (Order Hemiptera), which are aquatic forms commonly known as backswimmers. These species occur throughout much of northwestern Europe and are similar in size and structure. Newly formed ponds are first colonized by *C. dentipes,* while older pond habitats are occupied by *C. punctata.* Thus, *C. dentipes* acts as a fugitive species and survives throughout its overall range by virtue of its greater ability for dispersal into and colonization of newly created ponds or ponds of temporary duration.

Examples of such relationships are probably quite common. In a sense, any species adapted to early stages in biotic succession is employing this strategy. These species are eventually outcompeted and eliminated by members of later successional stages. Their survival is based on the fact that natural selection has favored efficient dispersal mechanisms and the ability to rapidly colonize unoccupied habitats, at the expense, perhaps, of efficiency in the utilization of environmental resources.

MacAthur (1958) in his analysis of competitive relationships among wood warblers describes a related situation. The Cape May and Bay-breasted Warblers, two of the five species discussed earlier, were noted to have clutch sizes larger than those of the remaining species. Furthermore, the Bay-breasted Warbler was able to increase its clutch size still more under conditions of high food availability. Thus, under conditions such as those prevailing during outbreaks of coniferous forest insect pests, primarily the spruce budworm, these species are favored. In this situation the Cape May and Bay-breasted Warblers are able to produce larger clutches and raise more young than can the other warbler species, thus compensating for a gradual decline in numbers that may have occurred due to interspecific competition from the other species over a period of years. Observations by Kendeigh (1947) during spruce budworm outbreaks in Ontario indicate that these species, upon arrival in breeding areas in the spring, selectively establish territories in areas of high spruce budworm populations. Survival of these warbler species is thus favored not only by their reproductive response, but also by their tendency to concentrate in areas of unusually favorable food conditions.

PERMANENT COEXISTENCE OF COMPETITOR SPECIES

These examples of fugitive species call attention to the fact that many natural environments show a high degree of instability. In our earlier discussion of competitive exclusion processes the occurrence of relatively constant conditions favoring one species was suggested as a condition necessary for competitive replacement. Under fluctuating environmental conditions competitive replacement of a species may be prevented, or at least slowed, by changes in the environment that shift the competitive advantage from one species to another. In fact, a number of ecologists (Andrewartha, 1961; Ross, 1962) feel that the instability of natural environments and the impact of physical conditions of the environment on populations are so great that interactions of the sort required for competitive replacement are rarely of importance in nature.

Hutchinson (1953) has suggested that the role of regular environmental fluctuations in permitting the coexistence of ecologically similar forms depends on the relationship between the period of environmental fluctuation and the generation time of the organism (Fig. 6–16). If the generation time of the organism is very short relative to the period of environmental fluctuation, competition may occur over several generations within one phase of the environmental cycle and thus be able to run its course before environmental conditions change. In other words, conditions may favor one species long enough to eliminate the other. If the generation time is long relative to the period of environmental fluctuation, the species involved must be adapted to the entire range of environmental conditions throughout all stages in the life cycle. Thus, although temporary conditions may favor one species more than the other, the single species best adapted to the overall conditions of the environmental cycle will very

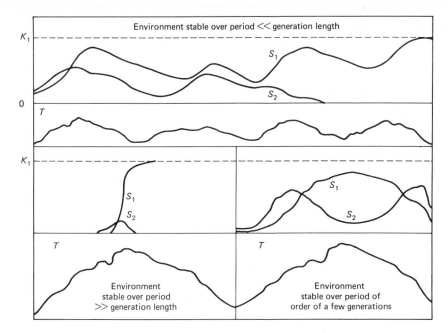

FIGURE 6–16. Ideal course of competition between two species as regulated by the relation between generation length and the period over which the environment may be taken as stable. (From Hutchinson, 1953.)

likely eliminate its competitors. However, if the relationship between environmental period and generation time is intermediate, the environment may alternately favor different species for periods of a few generations. Thus, one species might not possess the competitive advantage long enough to eliminate its competitors, and coexistence could result. Hutchinson (1953) feels that this relationship may be particularly important for species of freshwater phytoplankton, which show little differentiation in resource requirements but coexist in a very homogeneous environment. These forms often show a seasonal progression in dominance, suggesting that the environment is continually shifting the competitive advantage among species during the seasonal cycle.

These observations indicate that, although competitive relationships and ecological isolating mechanisms are of major importance for resource limited species, the spatial and temporal variability of natural environments allows a variety of strategies for survival of species in close association with competitors. Broadhead and Wapshere (1966) have also pointed out that continued coexistence of direct competitors may occur in areas of intermediate habitat in which populations of the competitor species are maintained by dispersal of individuals from areas where each predominates. In addition, they suggest that predators or parasites, if they concentrate on the more abundant species in a competitive situation, may prevent it from increasing to the point of completely replacing its competitors.

HABITAT DISTRIBUTION AND THE BALANCE OF INTRASPECIFIC AND INTERSPECIFIC COMPETITION

Some of the strongest evidence for the importance of competition and mechanisms of ecological isolation comes from observation of patterns of evolutionary change related to resource utilization by species. These evolutionary changes can best be understood by considering the influence of selective forces related to the balance between intraspecific and interspecific competition.

The contrasting effects of relatively high interspecific and intraspecific competition on habitat occupation or resource utilization by species are illustrated in Fig. 6–17. Intense interspecific competition, represented by a situation in which several similar species occur together, favors specialization of each species in the patterns of resource use and habitat occupation. This specialization results from selection favoring characteristics that enable individuals of each species to exploit particular resources or habitats more efficiently than can their interspecific competitors. Intense intraspecific competition, on the other hand, favors a broadening of the range of resources and habitats utilized by a single species. In this situation, represented in Fig. 6-17 by the presence of a single species in a given area, selection favors characteristics that enable individuals to utilize any resources or habitats not already being heavily exploited by other individuals of the same species.

Lack (1947) gives a number of examples of situations illustrating different relative intensities of interspecific and intraspecific competition and their effects on habitat occupation. One of these involves the two species of chaffinches in the Canary Islands. One, the European Chaffinch, *Frin-*

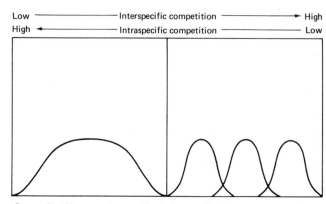

FIGURE 6–17. Influence of the relative intensities of interspecific and intraspecific competition on the range of food resources or habitats utilized by species.

gilla colebs, is conspecific with a European mainland form; the other, the Blue Chaffinch, *F. teydea,* is endemic to the Canary Islands. On the island of Palma, where only the European Chaffinch occurs, this species occupies both the pine forest zone at higher elevations and the chestnut-laurel forest at lower elevations. On the islands of Gran Canaria and Tenerife, however, both species are present. Here the European Chaffinch is restricted to the chestnut-laurel forest zone and the Blue Chaffinch to the pine forest zone. This situation is similar to that of the Galapagos Finches *Geospiza difficilis* and *G. fuliginosa,* as described earlier, in which *difficilis,* although apparently able to occupy arid habitats, was restricted to high elevation humid forests on the larger islands because of the presence of *fuliginosa.*

A further example of this relationship, involving freshwater planarians, is provided by Beauchamp and Ullyott (1932). Two species, *Planaria montenegrina* and *P. gonocephala,* occur in southern Italy and the Balkan Peninsula. When *montenegrina* occurs alone in a drainage system, it is distributed from the stream springhead where temperatures are as low as 6.6°C, to lower stream sections where temperatures reach 16–17°C (Fig. 6–18). In drainages occupied only by *gonocephala,* this species also occurs from the springhead, at temperatures of about 8.5°C, to lower parts of the stream where temperatures reach 20°C or more. When both species occur in the same drainage system, however, *montenegrina* is restricted to areas having temperatures below 13–14°C, and *gonocephala* to areas having temperatures above this value.

EVOLUTIONARY CONSEQUENCES OF COMPETITIVE RELATIONSHIPS

In addition to these effects on habitat distribution, different competitive regimes may cause evolutionary changes in characteristics of species

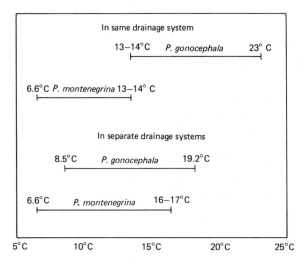

FIGURE 6–18. Ranges of water temperatures occupied in nature by the flatworm species *Planaria gonocephala* and *P. montenegrina* in stream systems in which they occur singly and together.

related to their utilization of resources or habitats. In cases of intense interspecific competition selection may lead to divergence in the characteristics of the species involved. The phenomenon of increased difference in characteristics of species in areas of contact or sympatry, as compared to areas of allopatry, is termed *character displacement* (Brown and Wilson, 1956; Hutchinson, 1959). An example of this phenomenon, involving two species of Galapagos Finches, is given in Table 6–3. The difference between beak measurements of the two species is greater when populations from one of the large central islands, where both occur, are compared than when measurements from populations from small, isolated islands with only one species or the other are compared (Lack, 1947). Character displacement thus represents a stage in the perfection of ecological isolating mechanisms by species that have come into contact with each other relatively recently.

Conditions of low interspecific competition and, consequently, relatively high intraspecific competition, often lead to a phenomenon termed *ecological release,* which refers to the expansion of the activities of a species population, resulting in the utilization of an increased range of resources and habitats. This phenomenon is illustrated in the bird fauna of the Bermuda Islands (Crowell, 1962). The presence of an impoverished North American bird fauna in these islands has allowed those species present to maintain higher population densities than are shown by the same species in mainland areas. This difference is apparently due to the utilization of food resources and habitats made available by the absence of various mainland species. Crowell (1962) was also able to detect changes in specific characteristics of at least one species under these conditions. In the White-eyed Vireo, *Vireo griseus,* the diversity of specific feeding patterns was significantly greater in the Bermuda populations than for birds observed in mainland areas.

TABLE 6–4. DISTRIBUTION OF VARIOUS SEX AND AGE GROUPS BY HEIGHT IN VEGETATION AND BRANCH DIAMETER FOR *ANOLIS CONSPERSUS* ON GRAND CAYMAN ISLAND.*

Group	Perch height in feet			Perch diameter in inches		
	1–2	3–5	>5	>3.0	0.5–3.0	<0.5
Adult males	17	50	35	68	32	2
Subadult males	25	56	18	48	49	2
Adult females	37	50	13	35	58	7
Juveniles	60	29	11	16	41	43

*From Schoener, 1967.

Several workers have subsequently attempted to determine if reduction in interspecific competition in such situations leads to increased variability in morphological characteristics of the feeding apparatus. Since island faunas in general show reduced species diversity, and thus, in many instances, reduced interspecific competition, almost all studies have involved comparison of island and continental populations. Van Valen (1965) compared beak measurements of various bird species from populations on the Azores, Canary Islands, and the island of Curaçao (near Venezuela) with measurements for the same species in mainland areas. These comparisons, involving males and females of six species, showed significantly greater variability in the island populations in all but one case. This exception, interestingly enough, was the European Chaffinch in the Canary Islands. For this species, as described earlier, the mainland represents the area of low interspecific competition, since only the European Chaffinch occurs there, and the islands, where both the Blue Chaffinch and European Chaffinch occur, the area of high interspecific competition. The increased variability shown by these island populations suggests that natural selection has favored intraspecific diversification of the feeding apparatus as a means of reducing intraspecific competition for food.

Grant (1967) conducted an analysis of measurements for bird species occurring both on the Tres Marias Islands, located along the Pacific Coast of Mexico, and on the Mexican mainland. From this study he concluded that the pattern of variability in island and mainland populations is more complex than suggested above. He also suggested that the lower diversity of habitats and food resources on many islands may lead to reduced variability in species characteristics, even though fewer species are present in these areas. Analyses for 30 species occurring both on the Tres Marias and on the Mexican mainland showed no consistent variability trend. One abundant island species, however, *Turdus rufo-palliatus,* showed significantly greater variability in beak morphology in the islands than on the mainland, where it occurs in sympatry with a strong competitor, *T. assimilis* (Grant, 1967). These observations serve to emphasize the difficulties involved in obtaining comparisons between populations differing only in the intensity of interspecific competition.

Other studies have suggested that continued selection under conditions of reduced interspecific competition can lead to major patterns of differentiation of resource utilization by individuals within single species populations. One such pattern concerns the specialization of different sex and age groups within the population for utilization of different habitats or resources.

One of the most interesting examples of this phenomenon occurs in species of lizards of the genus *Anolis* on various islands of the West Indies

(Schoener, 1967, 1968). On islands occupied by this genus the number of species ranges from one, on many small, isolated islands, to 22, on Cuba. On islands with only a single species, a pronounced sexual dimorphism in body size is frequently shown. This dimorphism is absent in populations of the same or closely related species on other islands having several *Anolis* species. This differentiation represents a parallel trend that has arisen in at least six evolutionary lines within the genus.

Schoener (1967) studied the ecology of *Anolis conspersus,* the only species found on Grand Cayman Island. A pronounced sexual dimorphism in body size is shown in this population, with males reaching lengths 1.3–1.5 times those of females. In this study data were obtained on the height and diameter of branches on which individuals of various sex and age classes were found in the field. The stomach contents of a large sample of animals were also examined and the numbers and sizes of food items of various types determined. These sex and age groups showed important differences in their patterns of distribution in the vegetation (Table 6–4). Adult males were found most frequently at the highest levels and on branches of greatest diameter. The subadult males, adult females, and juveniles were distributed progressively at lower levels and on smaller branches.

Analysis of stomach contents showed that within each sex the average size of food items taken varied with body size of the lizard. The average size of food items was greater for males than for females of the same size, however (Fig. 6–19). When the percentage of the total food volume made up of items of various sizes is considered, this difference is even more apparent (Fig. 6–20). The bulk of the diet of adult males consisted of items of quite large size. The size of the items contributing most to the overall food volume was progressively smaller for the subadult males, adult females, and juveniles.

These observations do not prove that, given the same range of food items to choose from, individuals of the different sex and age groups actively select food items of different size. They do, however, indicate

TABLE 6-3. BILL LENGTHS (CULMEN) FOR *GEOSPIZA FORTIS* AND *G. FULIGINOSA* IN AREAS OF SYMPATRY AND ALLOPATRY IN THE GALAPAGOS ISLANDS.*

	Islands		
Species	Indefatigable (Sympatric)	Daphne (Allopatric)	Crossman
Geospiza fortis	12.0 mm	10.5 mm	—
Geospiza fuliginosa	8.4 mm	—	9.3 mm
Difference	3.6 mm	1.2 mm	

*Data from Lack, 1947.

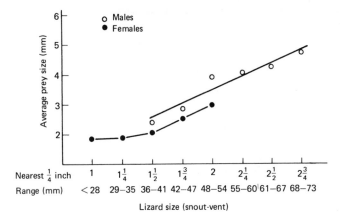

FIGURE 6–19. Relationship of body size of male and female lizards, *Anolis conspersus,* to the average size of prey insects taken. (From Schoener, 1967.)

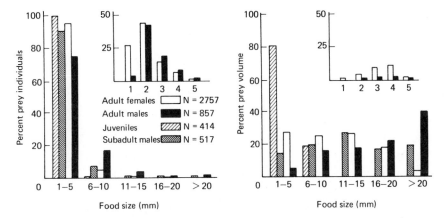

FIGURE 6–20. Percentage of prey individuals (left) and prey food volume (right) in five size categories for four age and sex classes of the lizard *Anolis conspersus.* A detailed breakdown of the first food size category is given in the inserted graphs, which have the same axes as the main graphs. (From Schoener, 1967.)

that the various sex and age groups are differentiated in resource utilization by mechanisms related to body size and habitat distribution.

In direct contrast to this situation, Schoener (1968) has demonstrated that on the island of South Bimini, in the Bahamas, the four *Anolis* species present do not show such intraspecific specialization. Instead, the different species divide the habitat according to perch height and diameter, with size classes of different species being distributed in such a way that interspecific groups overlapping most in habitat differ most in food prey size.

Selander (1966) has demonstrated a similar pattern of intraspecific differentiation in the woodpecker *Centurus striatus* under conditions of low interspecific competition. This species is the only woodpecker found on the island of Hispaniola. In contrast to related continental forms that show specific habitat preferences *C. striatus* occupies all kinds of woodland and forest vegetation occurring on the island. It achieves population densities four to five times those of related species on the North American mainland. It also exhibits the greatest degree of sexual dimorphism in beak structure of any woodpecker species in the Western Hemisphere. Correlated with this dimorphism are differences between the sexes in manner and height of foraging. Males use probing techniques more frequently than do females, and females feed by gleaning small insects from the surfaces of leaves and twigs more often than do males. The foraging activities of males are also concentrated at greater heights than are those of females. These differences in feeding behavior are not shown by related woodpecker species on the North American mainland.

EXPERIMENTAL STUDIES OF INTERSPECIFIC
COMPETITION AND EVOLUTIONARY CHANGE

A number of experimental studies have recently been directed toward analysis of the genetic basis of competitive ability and measurement of the rate of change of competitive ability in species under experimental conditions. Lerner and Ho (1961), working with the flour beetles *Tribolium confusum* and *T. castaneum,* found that genotypic strains differing in competitive ability could be obtained. Furthermore, they demonstrated that either species could be given the competitive advantage under certain environmental conditions by varying the specific genetic strains matched in competition.

Pimentel et al. (1965) conducted a series of experiments designed to compare the effects of intraspecific and interspecific competition on the competitive abilities of fly species in experimental systems. These experiments were based on the hypothesis that in a situation in which one member of a competing pair of species is abundant and the other rare selection should favor improved ability of the rare species to counter interspecific competition and improved ability of the abundant species to counter intraspecific competition. Improvement in the interspecific competitive ability of the rare species should thus occur while this characteristic remains unaffected in the abundant species. Eventually, the rare species should improve in competitive ability to such a degree that it is able to reverse the abundance relationship. Pimental and his coworkers further hypothesize that if competitive differences between two species are slight, this feedback process may allow their coexistence under a system of alternating species dominance and changing competitive ability.

To test this hypothesis a series of experiments using the housefly, *Musca domestica,* and the blowfly, *Phoenicia sericata,* was conducted.

These species have life cycles approximately two weeks in length, and they can be cultured in laboratory cages in which the larvae are fed a medium of agar fortified with milk, brewer's yeast, and liver. Competition between these species in a simple population cage ultimately leads to the elimination of one of the species, although many weeks may be required for this to occur.

The experimental design used by these workers involved comparing the competitive abilities of wild flies of each species with those of experimental flies which had experienced competition with each other over many generations in a population cage consisting of 16 semi-isolated compartments. The structure of the cage was designed to inhibit dispersal of the two species somewhat and thus lead to differences in the competitive regime in different compartments that would tend to prevent the rapid elimination of one species by the other. This mixed-species system was maintained for 38 weeks. During this time, the housefly, which was the more abundant species, averaged 200–300 adults per compartment, and the blowfly averaged about 25 adults per compartment.

The competitive abilities of the two species were tested by placing small groups of both species in a single-compartment competition cage and observing the populations until one species was eliminated. Results of various tests are given in Table 6–5. At the start of the experiment tests with wild houseflies and wild blowflies resulted in two victories for each species. Tests with wild-stock houseflies and blowflies, kept in separate cultures, showed four victories for the housefly and one for the blowfly at the end of the 38-week period. These differences seemed to reflect only chance variation in the initial conditions of the test populations.

At the end of the 38-week period the competitive abilities of the experimental houseflies and blowflies were also tested. In tests between the experimental houseflies (the more abundant species in the experimental

TABLE 6–5. RESULTS OF COMPETITION TESTS BETWEEN HOUSEFLY AND BLOWFLY POPULATIONS INVOLVING INDIVIDUALS OF WILD STOCK AND INDIVIDUALS FROM A MIXED-SPECIES POPULATION CAGE MAINTAINED FOR 38 WEEKS WITH HIGH DENSITIES OF HOUSEFLIES AND LOW DENSITIES OF BLOWFLIES.*

| | Week | Contests won by | |
		Housefly	Blowfly
Wild housefly X wild blowfly	0	2	2
Wild housefly X wild blowfly	38	4	1
Exp. housefly X wild blowfly	38	3	2
Wild housefly X exp. blowfly	38	0	5
Exp. housefly X exp. blowfly	38	0	5

*Data from Pimentel *et al.,* 1965.

system) and wild-stock blowflies there were three victories for the housefly and two for the blowfly. This outcome suggested that little change in
the interspecific competitive ability of the housefly had occurred. In tests
matching the experimental blowfly with either wild-stock or experimental
houseflies all ten contests were won by blowflies. The competitive ability
of the blowfly had thus improved considerably during the 38-week period
in the mixed-population cage.

Especially interesting results were obtained in two of the competition
tests with experimental houseflies and experimental blowflies. In these
tests both species survived for over 500 days, with alternating periods of
dominance by each species. Because of the unusual behavior of these
test populations, individuals were periodically taken from them, and they
themselves tested. At the end of periods of housefly dominance in these
populations, the blowfly was found to possess the competitive advantage.
At the end of periods of blowfly dominance, the housefly had regained
the competitive advantage.

These experiments clearly demonstrate that significant changes in
competitive ability may occur in populations over periods of a few generations, and that ability of a species to respond genetically in a manner
affecting its competitive ability may be very important for its survival in
the face of competition.

COMPETITION AND THE EVOLUTION OF MAJOR ADAPTIVE SYSTEMS

So far we have seen that interspecific and intraspecific competition may
lead to a variety of evolutionary changes relating to the utilization of
resources and habitats by a species and to the competitive ability of the
species. In most cases these changes have involved single characteristics
of species which are of direct importance in competition between closely
related forms. The evolutionary significance of competition and resource
relationships, however, extends far beyond this level. It is likely that such
relationships have played a major role in the evolutionary origin of major
systems of adaptation such as territoriality, reproductive systems, life
cycle patterns, and seasonal migration.

Cox (1968) has considered the possible roles of interspecific and intraspecific competition in the evolutionary origin of migration in land birds.
Until recently, the major selective agents for migratory behavior in birds
have generally been considered to relate to the physical environment and
to include factors such as climatic conditions and changes in regional
climatic patterns. It is now apparent, however, that both interspecific and
intraspecific competition can act as selective agents for migratory behavior.

Migration, wherever it occurs, is correlated with environments showing
seasonally differentiated conditions. Let us assume, initially, that such an

environment is occupied by a permanent resident species, which experiences optimal conditions year-round within its range but is limited in its distribution by conditions unfavorable at one season or another beyond its range limits. In this case, it is possible that seasonally favorable areas will exist in areas adjacent to the range boundaries. For resident bird species of the temperate zone, for example, it is likely that areas to the north of the permanent resident area in the summer and to the south in the winter would offer favorable conditions during these seasons. In such a situation migratory behavior would be favored if individuals entering these seasonally favorable areas showed greater survival or reproduction due to an overall reduction in the intensity of competition. The selective agent for incipient migratory behavior in these cases might be either interspecific or intraspecific competition, or a combination of both. If the selective agent for migratory behavior is intraspecific competition alone, growth of a migratory segment of the species population would continue until the seasonally favorable areas are occupied to the extent that benefits from reduced competition no longer exist. A species population consisting of both resident and migratory segments would thus be produced. Intraspecific competition alone, however, should not bring about the elimination of the resident segments of the original population.

If interspecific competition is the primary selective force, or if it comes into play through contact of the species with strong competitors, selection may lead to the production of populations that are entirely migratory. Two of the ways in which this change may occur are shown diagrammatically in Fig. 6–21. These situations essentially show specialization of the competing species either for areas occupied only through migration or for areas occupied only as permanent residents. The evolutionary origin of disjunct migration patterns thus represents, in many cases, a complex, behavioral-spatial solution to the problem of interspecific competition.

To test the hypothesis that interspecific competition has been of major importance in the evolution of migration, Cox (1968) examined the degree to which interspecific differentiation of the feeding apparatus is shown by members of migrant and resident bird groups. In this analysis it was assumed that if problems of interspecific competition and ecological isolation were not important in the evolution of migration, migrant and resident species groups should not differ in such characteristics. Figure 6–22, however, shows that variation in beak size among species in mainly resident groups is greater than that in mainly migrant groups. Migration thus appears to have evolved most frequently in groups that have been unable, or that have not been forced, to differentiate in morphology as a means of attaining ecological isolation. It, therefore, appears that migration functions as a complex behavioral and spatial mechanism of ecological isolation which is, at least in part, an alternative to morphological isolating mechanisms. Among land birds migration is best developed among members of the order Passeriformes, which is the most recently evolved bird order. It is possible that the failure of members of this order to emphasize morphological isolating mechanisms reflects the fact that a diverse fauna of nonpasserines morphologically specialized for various food resources already existed when the passerines appeared on the evolutionary scene.

It seems likely that similar analyses will reveal that resource relationships have also played a major role in the evolution of other important adaptive systems for both plants and animals.

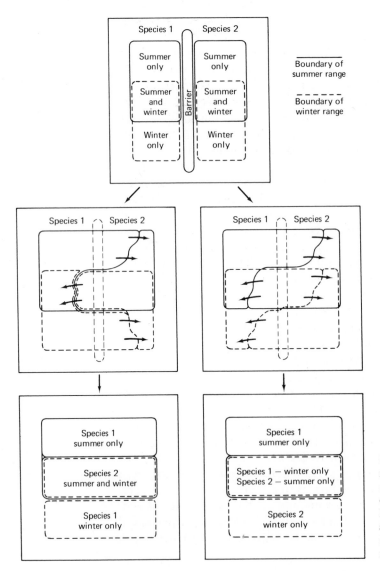

FIGURE 6–21. Diagrammatic representation of two possible sequences by which interspecific competition between closely related partial migrants may bring about conversion of one or both species into complete or disjunct migrants. (From Cox, 1968.)

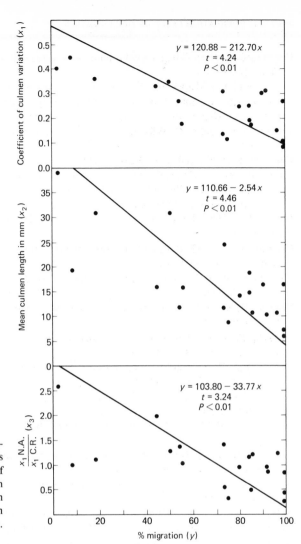

FIGURE 6-22. Relationship of percent migration in various North American land bird groups to the coefficient of culmen (upper mandible of bill) variation and mean culmen length of North American group members, and to the culmen variation coefficient ratio for North American (N.A.) and Costa Rican (C.R.) group members. (From Cox, 1968.)

REFERENCES

Aldrich, D. V., C. E. Wood, and K. N. Baxter. 1967. Burrowing as a temperature response in postlarval shrimp (Abstract). *Bull. Ecol. Soc. Amer.,* **48**(2):80.

Andrewartha, H. G. 1961. *Introduction to the Study of Animal Populations.* Chicago: University of Chicago Press. 281 pp.

Beauchamp, R. S. A., and P. Ullyott. 1932. Competitive relationships between certain species of fresh-water triclads. *J. of Ecol.,* **20**:200–208.

Bovbjerg, R. V. 1970. Ecological isolation and competitive exclusion in two crayfish (*Orconectes virilis* and *Orconectes immunis*). *Ecology,* **51**:225–36.

Bowman, R. I. 1961. *Morphological Differentiation and Adaptation in the Galapagos Finches.* University of California Pub. Zool., Vol. **58**. 302 pp., 22 pl.

Brewer, R. 1961. Comparative notes on the life history of the Carolina Chickadee. *Wilson Bulletin,* **73**:(4):348–73.

Broadhead, E. 1958. The psocid fauna of larch trees in northern England. *J. Anim. Ecol.,* **27**:217–63.

―――― , **and A. J. Wapshere.** 1966. *Mesopsocus* populations on larch in England – the distribution and dynamics of two closely related coexisting species of Psocoptera sharing the same food source. *Ecol. Monog.,* **36**:327–88.

Brown, W. L., and E. O. Wilson. 1956. Character displacement. *Systematic Zool.,* **5**:49–64.

Clatworthy, J. N., and J. L. Harper. 1962. The comparative biology of closely related species living in the same area. *J. Exp. Bot.,* **13**:307–24.

Connell, J. H. 1961. The influence of interspecific competition and other factors on the distribution of the barnacle *Chthamalus stellatus. Ecology,* **42**:710–23.

Cox, G. W. 1968. The role of competition in the evolution of migration. *Evolution,* **22**(1):180–92.

Crowell, K. L. 1962. Reduced interspecific competition among the birds of Bermuda. *Ecology,* **43**(1):75–88.

Darwin, C. R. 1859. *The Origin of Species by Means of Natural Selection.* London: Murray.

Dawson, P. S. 1968. Xenocide, suicide, and cannabalism in flour beetles. *Amer. Nat.,* **102**:97–105.

DeBach, P. 1966. The competitive displacement and coexistence principles. *Annual Rev. Ent.,* **11**:183–212.

―――― , **and R. A. Sundby.** 1963. Competitive displacement between ecological homologues. *Hilgardia,* **34**(5):105–66.

Frank, P. W. 1957. Coactions in laboratory populations of two species of *Daphnia. Ecology,* **38**:510–19.

Gause, G. F. 1934. *The Struggle for Existence.* Baltimore: Williams & Wilkins. 163 pp.

Grant, P. R. 1967. Bill length variability in birds of the Tres Marias Islands, Mexico. *Can. J. of Zool.,* **45**:805–15.

Grinnell, J. 1904. The origin and distribution of the Chestnut-beaked Chickadee. *Auk,* **21**:364–82.

Hairston, N. G., F. E. Smith, and L. B. Slobodkin. 1960. Community structure, population control, and competition. *Amer. Nat.,* **94**:421–25.

Hamilton, T. H., and I. Rubinoff. 1963. Isolation, endemism, and multiplication of species in the Darwin Finches. *Evolution,* **17**:388–403.

Hanes, T. L. 1965. Ecological studies on two closely related chaparral shrubs in southern California. *Ecol. Monog.,* **35**:213–35.

Hardin, G. 1960. The competitive exclusion principle. *Science,* **131**:1292–98.

Hazen, W. E. 1970. *Readings in Population and Community Ecology,* 2nd ed. Philadelphia: Saunders. 421 pp.

Hutchinson, G. E. 1953. The concept of pattern in ecology. *Proc. Nat. Acad. Sci.,* **105**:1–12.

―――― . 1959. Homage to Santa Rosalia *or* why are there so many kinds of animals? *Amer. Nat.,* **93**:145–59.

Kendeigh, S. C. 1947. *Bird Population Studies in the Coniferous Forest Biome During a Spruce Budworm Outbreak.* Biol. Bull. No. 1, Ontario Dept. Lands and Forests, Division of Research. 100 pp.

Lack, D. 1947. *Darwin's Finches.* Cambridge, England: Cambridge University Press. 204 pp.

_____. 1953. *Darwin's Finches,* Scientific American, **188** (4): 66–71.

_____. 1969. Subspecies and sympatry in Darwin's Finches. *Evolution,* **23**:252–63.

Lerner, J. M., and F. K. Ho. 1961. Genotype and competitive ability of *Tribolium* species. *Amer. Nat.,* **45**:329–43.

MacArthur, R. H. 1958. Population ecology of some warblers of northeastern coniferous forests. *Ecology,* **39**:599–619.

Merrell, D. J. 1951. Interspecific competition between *Drosophila funebris* and *Drosophila melanogaster. Amer. Nat.,* **85**:159–69.

Miller, R. S. 1964. Larval competition in *Drosophila melanogaster* and *D. simulans. Ecology,* **45**(1):132–48.

_____. 1967. Pattern and process in competition. *Adv. Ecol. Res.,* **4**:1–74.

Moore, J. A. 1952. Competition between *Drosophila melanogaster* and *Drosophila simulans.* I. Population cage experiments. *Evolution,* **6**:407–20.

Park, T. 1955. Experimental competition in beetles, with some general implications. In J. B. Cragg and N. W. Pirie (eds.), *The Numbers of Man and Animals.* London: Oliver and Boyd. 152 pp.

_____. 1962. Beetles, competition, and populations. *Science,* **138**:1369–75.

Pianka, E. R. 1970. Sympatry of desert lizards (*Ctenotus*) in Western Australia. *Ecology,* **50**:1012–30.

Pimentel, D., E. H. Feinberg, P. W. Wood, and J. T. Hayes. 1965. Selection, spatial distribution, and the coexistence of competing fly species. *Amer. Nat.,* **94**:97–109.

Ross, H. H. 1957. Principles of natural coexistence indicated by leafhopper populations. *Evolution,* **11**:113–129.

_____. 1962. *A Synthesis of Evolutionary Theory.* Englewood Cliffs, N. J.: Prentice-Hall. 387 pp.

Ryan, M. F., T. Park, and D. B. Mertz. 1970. Flour beetles: Responses to extracts of their own pupae. *Science,* **170**:178–80.

Schoener, T. W. 1965. The evolution of bill size differences among sympatric congeneric species of birds. *Evolution,* **19**:189–213.

_____. 1967. The ecological significance of sexual dimorphism in size in the lizard *Anolis conspersus. Science,* **155**:474–77.

_____. 1968. The *Anolis* lizards of Bimini: Resource partitioning in a complex fauna. *Ecology,* **49**:704–26.

Selander, R. K. 1966. Sexual dimorphism and differential niche utilization in birds. *Condor,* **68**:113–51.

Slobodkin, L. B. 1961. Preliminary ideas for a predictive theory of ecology. *Amer. Nat.,* **95**(3):147–53.

_____. 1963. *Growth and Regulation of Animal Populations.* New York: Holt, Rinehart & Winston. 184 pp.

Sokoloff, A., and I. M. Lerner. 1967. Laboratory ecology and mutual predation of *Tribolium* species. *Amer. Nat.,* **101**:261–76.

Utida, S. 1953. Interspecific competition between two species of bean weevil. *Ecology,* **34**(2):301–7.

Van Valen, L. 1965. Morphological variation and width of ecological niche. *Amer. Nat.,* **94**:377–90.

Part IV
THE COMMUNITY LEVEL OF ORGANIZATION

Chapter 7
PATTERNS OF
COMMUNITY CHANGE

Ecosystems vary in their degree of stability. In terrestrial environments from which most of our ideas of community change have come, forest communities may be relatively stable for hundreds of years. Selander (1950) states that the major plant communities of the mountains in Scandinavia have remained essentially unchanged for the past 2,600 years because there has been no major change in climate. True stability is probably illusory, however, and the degree to which communities appear to be static depends somewhat on the focus of the investigation. Although some aquatic communities, especially small lakes and ponds, may have short periods of stability, they are usually considered to be transient and leading to terrestrial communities (Lindemann, 1942).

After a short period in which ecologists regarded communities as static entities, emphasis in community and ecosystem studies focussed on dynamic processes. Regardless of the different philosophical positions taken about the nature of the community, all ecologists agree that some communities are more transient than others.

Almost all ecologists who have studied community dynamics have concerned themselves with relatively short term changes. A forest is cut, perhaps the land is farmed for a short time, and then the farm is finally abandoned. Plants and animals from the surroundings immediately start to invade the area until eventually, given time measured at most in hundreds of years, a vegetation similar to the preagricultural one covers the area. This process of change is referred to as *ecological succession*.

At the same time we understand that other forces are at work. Long term geological changes such as mountain building, continental drift and glaciation have changed the landscape. Major and minor climatic fluctuations have occurred. The plants and animals themselves are evolving—new species are arising, old ones are becoming extinct. Thus, in our study of ecological change, as defined above, we must make some assumptions about nature.

We assume, for example, that the climate has been relatively stable for some hundreds of thousands of years. Our concept of relative stability in this case includes our knowledge that as recently as about 100 years ago the earth was emerging from a "little ice age" and that since then most of the mountain glaciers of the world have retreated from their terminal moraines (Matthes, 1942). When we speak of relative climatic stability, we mean simply that so far as we can determine there have been no substantial shifts in any vegetation boundary within the historical past, except for those attributed to man's activities.

We distinguish these minor fluctuations in climate from those that occurred in conjunction with continental glaciation during the Pleistocene and from the worldwide warming that occurred during the period from 7,000 to 4,000 years ago (Antevs, 1948). Both of the latter are known to have changed distributions of plant species. Before continuing our discussion of ecological succession we shall consider some of the kinds of change associated with climatic shifts.

Lamarche and Mooney (1967) have described a change in the position of the upper altitudinal margin of the bristlecone pine forest that occurred during the warming period noted above. In the White Mountains of California upper timberline now occurs at 3,430 to 3,810 m. At altitudes some 120 to 150 m above this there are the standing and fallen remains of dead pine trees. Aging the outermost (and youngest) wood of these trees by radiocarbon methods shows that they died some 2,000 to 4,000 years ago. The arid climate of the White Mountains has prevented their decay. Similar evidence collected in a number of high-altitude arid mountain ranges around the Great Basin supports a conclusion that several thousand years ago timberline occurred at higher altitudes than it does now.

Another example, also from the arid western United States, comes from the work of Wells and Berger (1967). Their data consist of the readily identifiable plant remains that were accumulated by wood rats (*Neotoma*) during the construction of nests and middens that have since been abandoned, presumably because of climatic and accompanying vegetation changes. The deposits are particularly interesting because some have been carbon-14 dated to a time corresponding to the end of the last glacial stage (12,000 years ago). Wells and Berger show that during the time the nests were occupied, wood rats were cutting and storing plants whose lower altitudinal limits are now 600 m above the fossil midden sites. The most ubiquitous and important of these species is *Juniperus osteosperma,* the one-seeded juniper. When this species is the only woody plant in the middens, it suggests that the vegetation of the site was a dry phase of the Piñon Pine-Juniper Woodland now occurring on many of the mountains of the desert and Great Basin areas.

These kinds of changes, occurring in a relatively short time under the impact of changing climates, are probably rather common. We take special note of them because of our tendency to think of vegetation boundaries as static. It is probable, however, that we are not sufficiently attuned to nature to notice smaller changes that may be taking place every day.

We have ample evidence of even more extensive shifts in vegetation distribution correlated with major climatic changes in the past as evidenced by plant and animal remains that have been buried and preserved or fossilized. The field of *palynology,* for example, is concerned with analysis of preserved pollen grains and spores. These microfossils may be extracted from the sediments in which they occur, identified, and compared to the modern pollen "rain" occurring beneath the vegetation growing in the areas today. By this method numerous studies have shown such major changes as the replacement of coniferous forest by grassland or of grassland by desert shrubs (Martin and Mehringer, 1965). These changes (Figs. 7–1 and 7–2) took place in the southwestern United States during and following the last major glaciation.

There are several lessons to be learned from these examples:

1. Shifts in the distribution of plant species occur with changes in climate.
2. Apparently not all of the species of the community are necessarily involved in the movement.
3. The changes at the margins of the community suggest that other changes may be taking place inside the community.
4. Successional changes during periods of climatic stability should be separated from changes in community composition caused by climatic change.

The emphasis in this chapter is on ecological succession in biotic communities. Community dynamics have been an extremely important aspect of ecological research and thought for many years. The problems involved here are important not only in terms of their fundamental significance to the field of ecology but also because of the practical significance they have had in the management of the landscape.

SUCCESSION AND THE CLIMAX

The observation that the development of plant and animal communities in one area includes the replacement of some species by others through time, or of some communities by others, is called *biotic* or *ecological succession.* In North America the idea was discussed first by Cowles (1899), and later, in much greater detail, by Clements (1916).

Clements studied naturally occurring biotic communities in many parts of North America. He was impressed that vegetation was strongly correlated with climate and eventually came to believe that climate, especially temperature and precipitation, exercises primary control over

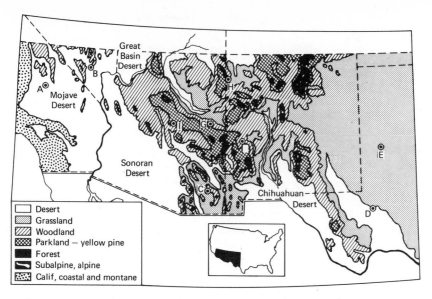

FIGURE 7–1. Modern vegetation of the Southwest. (From Martin and Mehringer, 1965.)

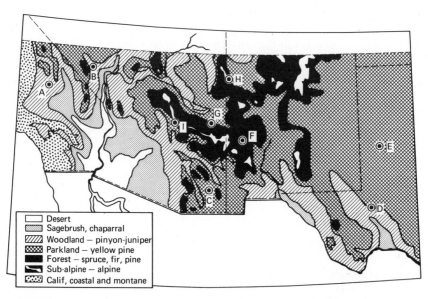

FIGURE 7–2. Full-glacial vegetation of the Southwest 17,000–23,000 years ago, based mainly on carbon-dated fossil-pollen spectra from (A) Searles Lake, (B) Tule Springs, (C) Pluvial Lake Cochise, (D) Crane Lake, (E) Rich Lake, (F) San Augustin Plains, (G) Laguna Salada, (H) Dead Man Lake, (I) Potato Lake. (From Martin and Mehringer, 1965.)

the composition and distribution of communities. Much of his early work was accomplished during the time when considerable natural vegetation remained on this continent. He recognized that similar kinds of communities occurred in similar climates, and this led to his classification of the major plant formations in North America.

Clements also became identified with what is called the dynamic approach to ecology. He saw changes taking place in communities under the impact of grazing and agriculture, and he understood that change is also a natural aspect of community ecology. Furthermore, within any climatic zone with its characteristic association Clements determined that reestablishment of the typical community followed similar pathways regardless of the nature of the initial disturbance. Finally, he proposed his climax theory of the community, which states: Within any region characterized by a more or less homogeneous climate the communities culminating successional processes are called the *climax* communities. Although animals are obviously important parts of both successional and climax communities, much of the ecological theory that is concerned with community dynamics emphasizes the autotrophic organisms.

Kinds of Succession

The two recognized basic kinds of succession are *primary* and *secondary*. Although they are not always clearly distinguishable from one another, primary succession begins on terrestrial or aquatic bedrock surfaces that are altered relatively little or not at all. If organisms have been growing on a site and have modified it by their activity, e.g., have contributed their remains to soil formation, succession on such places is secondary. Examples of primary succession are the sequences (*seres*) of plants that occur on bedrock, sand dunes, and glaciated surfaces. Secondary succession occurs on sites that have been burned, farmed, or otherwise denuded without removing all traces of previous organic activity.

Climax Theory

An axiom of Clementsian ecology is that whether succession is primary or secondary, and whether the initial communities in the sere are wet (*hydrosere*) or dry (*xerosere*), the outcome of the succession will converge on a self-reproducing community occupying a *mesic* (intermediate on the moisture gradient) habitat and existing in dynamic equilibrium with the regional climate (Fig. 7–3). In Clements' view only one such community could exist in any bioclimatic zone, and this constituted the climatic climax community, which is the basis of the *monoclimax hypothesis*. In order to understand Clements' ideas on this point, it is necessary to imagine that time is unimportant. Mountains erode away, ponds and lakes fill with sediment, and finally the landscape assumes a gentle rolling topography with relatively uniform and "medium" conditions throughout. When this ideal is achieved, the vegetation responds directly to the regional climate, because neither topography, exposure, soils, nor drainage

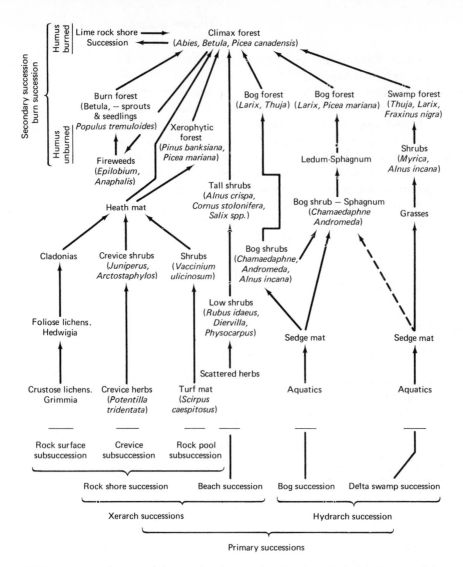

FIGURE 7–3. A diagram of the trends of succession for the principal habitats on Isle Royale, Lake Superior. This is one of the early complete condensations of a successional story for an entire region. On this pattern similar diagrams have been worked out for many sections of the country. Note that the system shows at a glance the kinds of habitats in which succession originates, the interrelationship of trends, and the major dominants in each of the stages of succession. Study of the diagram should help to clarify concepts of succession and climax. It must be remembered that not all trends progress with equal speed. (After Cooper, 1913.)

are so extreme in their expressions as to override the effects of the climate.

To account for those situations of apparent stability in which communities are in equilibrium with some nonclimatic feature of their environment such as fire, grazing, soil conditions, and the like, Clements proposed an elaborate nonclimax terminology. If, for example, an area does not develop its expected climax composition because of continued disturbance caused by the grazing of cattle, the term *disclimax* (disturbance climax) is applied. Similarly, the community might be delayed in the stage immediately preceding the climax by its topographic position. This is a *subclimax* in Clementsian terminology. A community whose composition reflects cooler and moister conditions than the average is called *postclimax,* and that whose composition reflects warmer and drier conditions is called *preclimax.* The logic behind the use of the post- and preclimax terms is based on Clements' view of latitudinal zonation of climatic areas. Thus, as one moves from south to north, one encounters cooler, wetter conditions *after* warmer, drier ones. This system of classification of communities on a dynamic basis was attractive in that it accounted for all communities and placed them in a context of gradual and inevitable progress toward the climatic climax.

In the years since Clements expressed his views on the nature of the climax, other points of view on the permanence of plant communities have emerged. One opinion now widely held by most plant ecologists is that any community that is self-maintaining deserves to be classified as a climax community even though it may be in equilibrium with something other than climate. This has been called the *polyclimax* hypothesis. Thus, Daubenmire (1952) recognized edaphic, topographic, zootic, and fire climaxes in addition to a climatic climax in the vegetation of northwestern Idaho and adjacent Washington. In the case of fire, for example, it can be argued that the plants of a fire climax evolved in an environment in which fire is a powerful selective agency. The coastal chaparral communities of coastal southern California are good examples of this point. Ample evidence exists that fires occur frequently over much of the area (Figure 7–4). The shrubby plants of which the chaparral is composed are evergreen and because of thick waxy coatings and the production of volatile substances are extremely flammable. After a fire many of the plants sprout within a few weeks from heavy and fire-resistant root crowns. Seeds of some of the shrubs germinate only after a fire cracks the nearly impermeable outer layers. Clearly, the plants are adapted to fire situations and to discuss these communities in a context in which fire is excluded makes little sense. Yet, in the monoclimax approach it would be necessary to "guess" the nature of a vegetation that probably has never existed in areas where fires are characteristic parts of the environment.

Despite their differences in approach to vegetation dynamics, it should be understood that both the adherents of the monoclimax and the polyclimax interpretations believe in the objective reality of communities. Succession also is a feature common to both views. The major difference is very likely one of perspective and scale. The monoclimax proponent directs his attention to the patterns that occur more or less continuously over great areas of gently rolling or "average" relief. He tends to regard as exceptional those communities that do not conform to his expectations about what the vegetation should look like at maturity. Ecologists who

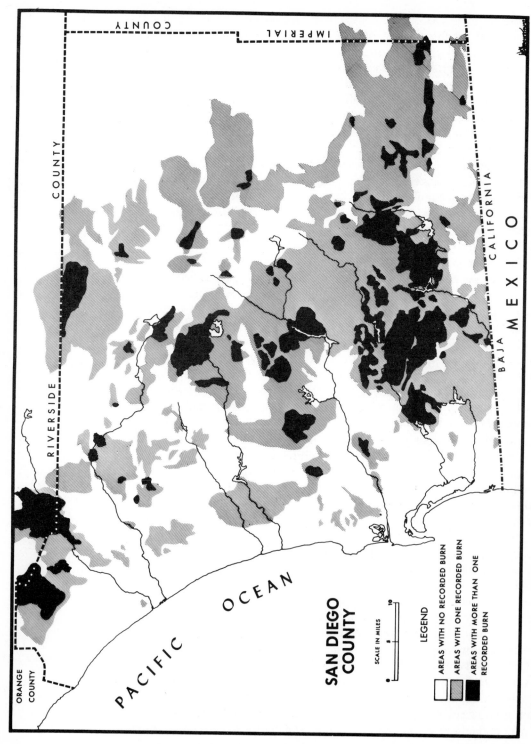

FIGURE 7–4. Vegetation and fire map of San Diego County, California, showing the occurrence of major fires over the past 60 years.

feel that the polyclimax interpretation of vegetation better represents nature tend to view it as a mosaic, with some pieces in equilibrium with the climate, others under the control of special soil or other factors that modify or delay indefinitely the culmination of succession, and still others in the process of changing following expected successional sequences.

Not all ecologists agree that communities exist other than as abstract and somewhat arbitrary groupings of species. The work of Whittaker (1953, 1956), for example, emphasizes continuous variation along environmental gradients. Whittaker thinks in terms of *climax patterns* in the vegetation, by which he means that the continuous distribution of species that he reports is reasonably stable, i.e., it is not successional. To workers of his persuasion the emphasis on discontinuous communities is misplaced; the emphasis should be placed on which populations in any area replace other populations and "then maintain themselves" (Whittaker, 1953).

Differences among ecologists over the interpretation of vegetation played a prominent role in writings in the field for several decades. That much of this discussion has been replaced by other concerns should not be interpreted as meaning that the issues are settled (e.g., McIntosh, 1967). We discuss these ideas further in Chapter 8.

Directional Change

Progressive Change. Clements considered most of the successional changes occurring in communities to be progressive; i.e., they follow ordered and more or less predictable sequences from simple to more complex communities. The climatic climax was said to be the most highly developed community that could develop under the prevailing climate. As the climax is approached there is an increasing tendency toward self-maintenance, uniformity within and between stands, and soil maturity. Whittaker (1953) concludes that progressive change in communities includes an increase in the biomass per unit area in productivity, in the complexity and diversity of species and life forms, and in the relative stability and regularity of populations. These points will be discussed in more detail below.

Retrogressive Change. A retrogression would imply a directional change that is, in some respects, in the reverse direction from normal succession. Clements denied that such reverse order sequences could exist. In fact, such changes are rather common under circumstances where disturbance occurs, either naturally such as in the arctic and alpine parts of the world by soil freezing and thawing, or, widely, under the impact of man's activities.

Woodwell (1970) describes the effects of seven years of chronic gamma irradiation on an oak-pine forest at the Brookhaven National Laboratory in New York. Zones of vegetation modification occurred outward from the source of irradiation, with the greatest changes in the composition of the forest in the areas of highest radiation exposures. Only mosses and lichens survived these exposures which were greater than 200 R per day. Outward from this zone Woodwell identified a sedge zone, a shrub zone, an oak zone, and, finally, a zone in which no differences in species composition could be found between undisturbed forest and an area that was receiving less that 2 R per day. Woodwell's work is also described in Chapter 11.

The changes described by Woodwell are retrogressive; i.e., they are similar in reverse to the successional sequences that give rise to the oak-pine forest. Woodwell summarizes the disturbances in the forest as changes in structure, reductions in diversity and primary production, increases in respiration, and a loss of nutrients.

What is rather surprising here is that the kinds of retrogressive changes occurring under the influence of irradiation are very much like those occurring under a variety of other kinds of disturbance, such as fire, air pollution, and herbicides. Gorham and Gordon (1960, 1963), for example, show species reductions along a gradient of effects associated with sulfur pollution from a smelter. Similarly, Tschirley (1969) showed that the highly diverse tropical forests in Viet Nam are replaced by bamboo thickets under the impact of repeated spraying with herbicides. Apparently, when communities are disturbed, by whatever cause, the structure of the system breaks down and what may have once been a diverse, relatively stable system becomes monotonous and unstable. It is a commentary on man's understanding of natural systems to note that in order to maintain monotony in ecosystems huge amounts of energy must be invested thus reducing what man hopes to gain from them. The implications of retrogressive changes by man's influences are up to now only dimly perceived, but in a sense they lie at the very center of our concern about environmental pollution.

Cyclic Change. In every community that has been studied intensively, even those that have reached apparent stability, changes continue to occur. Although many of these probably should be regarded as fluctuations within the normal composition of the community, they do, nonetheless, constitute departures from the static concept of the climax community. In some cases these departures are sufficiently repetitive that they may be referred to as cyclic changes or as fluctuations around an average. The use of the term "cycle" in this context should not be construed as implying that the periods with which these changes occur is regular.

One of the best descriptions of the relationship between the community and the environment, and one which includes examples of the three kinds of change described above is by Drury (1956) who studied the succes-

sional and climax communities on the flood plain of the Upper Kuskokwim River in interior Alaska. These alluvial lowlands are covered by
forests, bogs, and a myriad of small ponds and lakes. The river meanders
across this area of several thousand square miles, and within the meander
zone the surfaces and the vegetation on them are young. The areas adjacent to the meander zone are older, although it is assumed that they were
once disturbed by the river.

Perennially frozen ground (permafrost) is commonly found, although
not continuously, in much of interior Alaska, being especially prevalent
where surface insulation provided by vegetation does not allow the complete elimination of the winter's frost by summer melting. In the meander
zone, however, permafrost is mostly absent. The presence of permafrost
in soils and its position relative to the soil surface is extremely important
in the ecology of the area.

Several different river terraces, differing in height above the river, can
be recognized. These river terraces, and the vegetation on them, can be
arranged chronologically in terms of their relationship to the river. At any
river bend, the river is cutting against the outside of the bend and deposit-
ing on the inside (Fig. 7–5). The river migrates and deposits sand and silt,
irregularly, on these surfaces. When the river floods, it flows across the
new surfaces and deposits alluvial material on them. Active meanders
may occasionally be cut off and become oxbow lakes by the formation
of a new channel.

Since permafrost is not associated with the zone of active meanders,
the alluvial soil is deep and well-drained, particularly after the spring
run-off has occurred. On these surfaces succession occurs as indicated
in Fig. 7–6. Horsetails (*Equisetum*) and willows (*Salix*) initiate the succession on the recently exposed mud and sand bars. With time and additional deposition dense stands of willows and alders provide a nearly
complete cover. At the first flood-plain surface at the top of the banks,
an open forest of balsam poplar (*Populus balsamifera*) with a rich herb

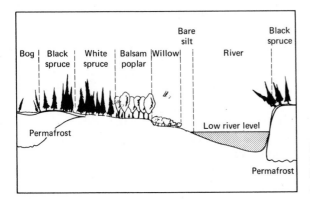

FIGURE 7–5. Diagrammatic cross section of
typical distribution of vegetation and permafrost across a meander of a river in interior
Alaska. (From Viereck, 1970.)

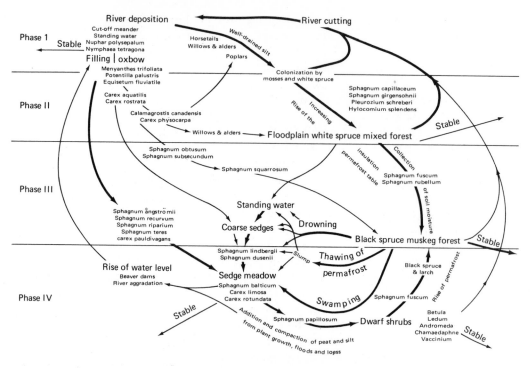

FIGURE 7–6. Flow diagram of vegetation and physiographic processes on the floodplain. This diagram emphasizes the coordination of changes of vegetation and of underlying alluvial sediments. The size of letters is an indication of the relative conspicuousness of area occupied by the various vegetation types, and the breadth of arrows indicates how widespread the changes are. As indicated, several of the vegetation types maintain stability until their site conditions are altered by physiographic processes. This emphasizes the point that the changes indicated are not unidirectional with the age of the floodplain. (From Drury, 1956.)

flora replaces the willow-alder thicket. With time the balsam poplar forest is replaced by a white spruce forest mixed with balsam poplar, and, in older parts of the drainage, with paper birch (*Betula papyrifera*). As the forest matures, a complete carpet of mosses develops on the forest floor. Dwarf shrubs such as mountain cranberry (*Vaccinium vitis-idaea*) and the blueberry (*Vaccinium uliginosum*) along with several herbs become common.

This pure or mixed white spruce forest is the most productive and highly developed vegetation type in the Alaskan subarctic. Lutz (1956) regards it as the climatic climax forest growing on deep mesic soils. Quite unexpectedly, however, this forest is replaced by a less mesophytic forest in a series of regressive changes. Thus, as Drury (1956) comments, "This forest can hardly be considered climax."

Some of the details of the next events are shown in Fig. 7–6. Basically the forest floor becomes progressively moister because of the insulating effects of the dense moss carpet. Concomitantly, the moss carpet is invaded by sphagnum mosses that eventually replace the other species. Sphagnum mosses have the ability to hold large quantities of water and are well known for their insulating properties. Under their influence the soil becomes progressively wetter and colder as the permafrost level rises. White spruce, which has a low tolerance for a high water table, is then replaced by black spruce (*Picea mariana*) and heath plants (*Ledum,* dwarf birch, etc.)

The fate of the black spruce forest is dependent somewhat on its topographic position. In the higher, drier portions of the area black spruce is stable. In the lower areas, where swamping becomes pervasive, the forest is drowned out and conversion to sedge meadow, or, in extreme cases, open water occurs. Commonly, in this area sedge meadows are converted to dwarf shrub heaths by the accumulation of organic and mineral debris. Under the most favorable conditions for each of these types, however, some period of stability may ensue.

Another interesting feature of this vegetation is seen in Fig. 7–6. River cutting may initiate a new cycle of events in all but the oldest and highest portions of the flood plain. Likewise, black spruce forest may give way to open water and the initiation of a new hydrosere leading back to itself.

White spruce-mixed forest is also said to be climax on uplands in interior Alaska (Lutz, 1956), but as shown by Sjörs (1963) the boreal forest generally is subject to swamping and replacement by vegetation adapted to wetter conditions, and this probably applies to the interior Alaska uplands.

Finally, Pruitt (1958) has demonstrated another cycle in Alaskan white spruce-mixed forest in which forest openings are created among the mature trees. The opening may be initiated by accident (wind, lightning) or by disease. It grows by the unequal distribution of snow accumulating on side branches of large trees. The branches on the side of the tree facing the opening grow larger presumably because of increased light and lack of crowding on that side. Snow, which in interior Alaska may accumulate over a period of months in a virtually wind-free atmosphere, bends the crown of the tree toward the opening and, if enough weight has accumulated, breaks off the upper one-third to one-half of the tree. Succession beginning with willows and alders proceeds until the opening is eventually filled with smaller spruces, birches, and poplars.

The special conditions that favor vegetation type conversion and the cyclic processes described here are related to the extreme environmental conditions in which the presence or absence of soil ice or, in the latter case, snow figures prominently. The changes described, however, are different only in degree from those that occur in all vegetation. In general, it appears that:

1. Progressive, regressive, and cyclic changes occur within the vegetation.
2. Several well-delineated vegetation types are sufficiently long-lasting that they may individually be called "climax."
3. The idea of climax in vegetation as related to stable soils has little relevance in areas of high soil instability.
4. The processes by which the white spruce-mixed forest of the flood plain is replaced by black spruce which, in turn, is swamped out and replaced by bogs are *autogenic,* i.e., caused by the plants themselves.
5. The changes described occur under what appears to be a stable climate.
6. The classification of this vegetation into associations or ecosystems must include not only the end points of the several pathways described but also the intermediate points as well.

Other examples of cyclic change in community composition have been described by Watt (1947) in grassland and beech woodlands and by Billings and Mooney (1959) in alpine tundra vegetation. Kershaw (1964) documents several cases of so-called "marginal" effects in plant communities. Plants growing at the margins of the population frequently show enhanced growth when compared to individuals of the same species growing behind the margin. Attempts to show that in specific cases the growth of the plants at the margin had depleted the soil of minerals so that the plants behind the margin were inhibited proved fruitless. Instead, Kershaw concluded from the work of Donald (1961) that competition for light may be the cause. Indeed, Donald demonstrated that when taking the plant as a whole the photosynthesis of older plants whose lower leaves were severely shaded was at or sometimes below the compensation point. Plants growing under these conditions are susceptible to being replaced by plants having a more favorable photosynthesis/respiration ratio, either by younger plants of the same species or by different ones.

Physical Changes during Succession

Why does one plant replace another during succession? Often we hide behind our ignorance and ascribe the replacement to competition. What is the nature of the competitive effect? "An organism may operate to alter a physical condition in the environment that will enable another organism to enter. The reaction of the new organism will be to the altered physical condition and not to the organism that produces the altered condition" (Mason, 1947). Relatively few studies of succession have addressed themselves to the nature of the physical changes that occur during the process. Instead, most studies of plant succession have been those in which the "story" of succession is told by piecing together sequences of plant communities which, in the opinion of the investigator, provide an abstract view of the way these communities replace one another in time.

Ideally, it would be desirable if each community could be studied consistently over hundreds of years. Although this ideal has never been achieved, Crocker and Major (1955) summarize a group of observations and studies, including their own, which began in the 1890's at Glacier Bay, Alaska and which approach this standard.

The area around Glacier Bay has been the object of several studies of plant succession (Cooper, 1923, 1931, 1939; Lawrence, 1953; and Crocker and Major, 1955). Before that, miscellaneous observations concerned with the position of the front of the glacier had been made by John Muir in the late nineteenth century. Like most mountain glacier complexes in the world the ice in the several glaciers at Glacier Bay has retreated more or less steadily for at least 200 years or so (Figure 7–7). On sites that have been free of ice for several hundred years, a mixed spruce-hemlock climax forest grows on well-drained acidic soils. In the wettest habitats the forest is swamped out and replaced by a muskeg bog forest.

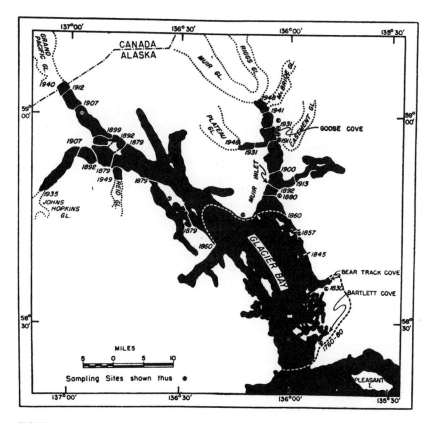

FIGURE 7–7. History of ice recession at Glacier Bay, Alaska.

The vegetational changes leading to mature forest climax require about 200 to 250 years from the initial seral stages on bare morainal gravel. The early work of Cooper and Lawrence at Glacier Bay described the sequences of plants that successively occupy these sites, and Crocker and Major correlate the changes in plant cover with those that occur in the soil.

Bare glacial soils at Glacier Bay are alkaline (pH 8.0 to 8.4), primarily because marble, which is high in calcium, is present in the bedrock of the area. Soil nitrogen and organic carbon are essentially lacking. The pioneer plant stages consist of mosses, herbaceous plants, and low shrubs; in a few years, mostly fewer than ten, taller willow species replace the low herb mat. Soil acidity immediately begins to decrease, depending on the degree to which the soils are covered by plants. There is relatively little change so long as soils remain bare. Organic carbon and soil nitrogen increase only slightly during the first 25 years or so.

The most rapid changes in these parameters occur following invasion of the sites by alder [*Alnus crispa* (Ait.) Pursh.]. Alder reproduces vegetatively as well as by seed, so that following the initial establishment of alder seedlings, the individual plants grow laterally through rhizomes to produce dense thickets that eventually dominate their habitats. The litter produced by alder has a marked acidifying influence on the soils, and the pH falls rapidly, reaching equilibrium values at around pH 5.0 in 50 years. Naturally, these changes occur most rapidly in the highly organic layers of the forest floor but are realized in the upper layers of mineral soil relatively soon (Fig. 7–8).

Alder roots have bacterial nodules, and their ability to fix atmospheric nitrogen very quickly increases the amount of nitrogen in the soil by a factor of about five (Fig. 7–9) in about 50 to 75 years. The nitrogen is contributed to the soils by the annual fall of leaves, which also increases organic carbon markedly (Fig. 7–10).

During this period, forest tree species, especially Sitka spruce (*Picea*

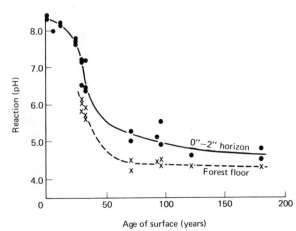

FIGURE 7–8. pH of litter residues and horizons of the mineral soil with increasing age of surface. (From Crocker and Major, 1955; courtesy of *J. Ecol.*)

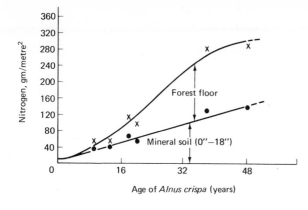

FIGURE 7–9. Accumulation of nitrogen in the mineral soil and forest floor under *Alnus crispa*. (From Crocker and Major, 1955; courtesy of *J. Ecol.*)

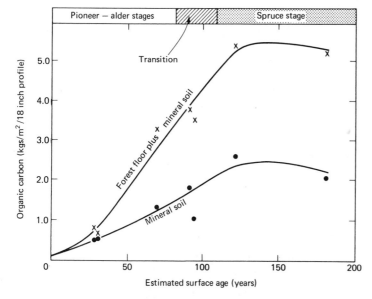

FIGURE 7–10. Organic carbon accumulation in mineral soil and forest floor. (From Crocker and Major, 1955; courtesy of *J. Ecol.*)

sitchensis), invade the areas and eventually replace the alders. Because their own leaves are of about the same degree of acidity as alder, no further changes in pH occur. Since spruce leaves yield somewhat less organic carbon annually than do alder leaves, a slight decrease in organic carbon occurs. More striking is the reduction in nitrogen content (Fig. 7–11). Since spruce roots have no nodules, there is a slow decline in soil nitrogen with the increasing age of the coniferous forest.

Sitka spruce and two species of hemlock, *Tsuga heterophylla* Sarg. and *T. mertensiana* Carr., eventually compose the climax forest community, and the soil characteristics described above presumably stabilize in this forest.

Although these data are among the most complete available that relate changes in the environment to plant succession, it must be admitted that the details of the changes in soil chemical conditions related here are largely unknown. Particularly, all of these changes almost certainly involve soil microorganisms. For example, there is abundant evidence of

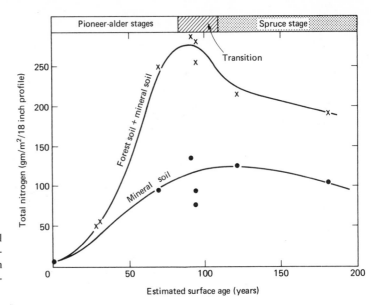

FIGURE 7–11. Change of total nitrogen content of soils on surfaces of varying age. (From Crocker and Major, 1955; courtesy of *J. Ecol.*)

the importance of mycorrhizae in the nutrition of coniferous forest trees (Wilde and Lafond, 1967). Yet the processes that allow forest mycorrhizae to become established during succession are almost unknown.

SUCCESSION IN HETEROTROPHIC ORGANISMS

Each stage in a sere is characterized by a specific group of plant and animal species. Our discussions of succession thus far have emphasized the plants, primarily because much of what is known about community change has been detailed from plant communities. Also, it should be obvious that changes involving the plants in a community are certain to be reflected among the animals. Hutchinson (1959), in discussing the causes for the great diversity of animals in nature, concluded that it is "due largely to the diversity provided by terrestrial plants." Johnston and Odum (1956) show how passerine birds are related to various stages in a secondary upland sere in the southeastern United States, and Margalef (1967) describes successional changes among the phyto- and zooplankton in coastal waters.

One interesting series of studies that deserves special attention involves recent attempts to study changes in the animal component of simple ecosystems without modifying the plant component. Wilson and Simberloff (1969) and Simberloff and Wilson (1969, 1970) devised techniques of defaunating small mangrove islands in Florida Bay and then studying the colonization that took place in the two years following the removal of the animals.

The islands studied were small enough to be covered completely by a tent and fumigated. Methyl bromide, the fumigant, was used in concentrations that proved lethal to arthropods but not to most of the plants. Other advantages to using small islands were their simplicity (one species of mangrove was the only autotroph on most of them), their isolation from other small islands and the mainland, and their diverse animal faunas. Also, because the islands differed in their distances to the nearest source areas for immigrants, the effect of these distances on the rate of recolonization and the number of species at equilibrium could be studied.

Each of the six islands was minutely censused before and immediately after fumigation, and a list of species was compiled. Table 7–1 is representative of the kinds and numbers of arthropod species found on these

TABLE 7–1. ARTHROPOD SPECIES FOUND ON E7 JUST PRIOR TO AND FOLLOWING FUMIGATION.*

Insects
Embioptera:	gen. sp.[a]
Orthoptera:	*Latiblattella* n. sp.
	Cycloptilum sp.
	Tafalisca lurida[a]
Coleoptera:	*Pseudoacalles* sp.[b]
	Leptostylus sp.
	Styloleptus biustus[b]
	Tricorynus sp.
Psocoptera:	*Archipsocus panama*
Hemiptera:	*Pseudococcus* sp.
Lepidoptera:	*Nemapogon* sp.[a]
Hymenoptera:	*Casinaria texana*
	Camponotus floridanus[a]
	Camponotus planatus[a]
	Paracryptocerus varians[a]
	Pseudomyrmex elongatus[a]
	Xenomyrmex floridanus[a]

Arachnids
Araneae:	*Ariadna arthuri*
	Eustala sp.
	Leucauge venusta[a]
Acarina:	*Galumna* sp.

Other
Chilopoda:	*Orphnaeus brasilianus*[a]
Isopoda:	*Rhyscotus* sp.[a]

*From Simberloff, 1970.

Notes: [a]Discovered dead after fumigation.
[b]Discovered mostly dead after fumigation.

small islands. The faunas on the six islands included approximately 20 to 40 species each of the nearly 100 species that occurred on all of the islands collectively. Some larger animals such as water snakes, birds, and mammals occasionally visited the islands, but they were not included in the censuses.

Following defaunation, each island was monitored frequently for approximately one year. After a second year four of the six islands were censused again. Figure 7–12 illustrates the pattern of recolonization seen on the four islands. On most of the islands the number of species present, either before defaunation or at some later time, is inversely related to the distance of the island from the nearest source of immigrants.

At the end of the first year, five of the six islands (three of the four shown in Fig. 7–12) had reached species levels that were about the same as in the pre-fumigation period. Before reaching this point, however, each of the islands had gone through a brief interval during which species numbers were higher than before defaunation. The authors interpret these peaks as "noninteractive species equilibria." In other words, during the early recolonization period a population that arrives on the island finds itself in a situation in which it is essentially isolated from its own or other species populations. If it is adapted to the conditions of the island, it is

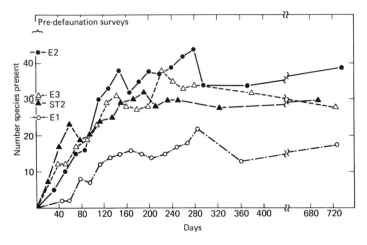

FIGURE 7–12. The colonization curves of four small mangrove islands in the lower Florida Keys whose entire faunas, consisting almost solely of arthropods, were exterminated by methyl bromide fumigation. The figures shown are the estimated numbers of species present, which are the actual numbers seen plus a small fraction not seen but inferred to be present by the criteria utilized by Simberloff and Wilson (1969) and Simberloff (1969). The number of species is an inverse function of the distance of the island to the nearest source of immigrants. This effect was evident in the predefaunation censuses and was preserved when the faunas regained equilibrium after defaunation. Thus, the near island E2 has the most species, the distant island E1 the fewest, and the intermediate islands E3 and ST2 intermediate numbers of species.

free to grow in an essentially nonlimited environment. The same is true of other early colonizers. Eventually, of course, resources start to become limiting, a process that is accelerated if the resource requirements of different species overlap somewhat. Now species interact in a variety of ways, some species become extinct and new "interactive species equilibria" are established at somewhat lower levels than those that had reached temporary equilibria under conditions of lower population numbers.

Two other points are worth making. Although the number of species remained at about the same level from year one to year two, a significant turnover of species occurred, amounting to, in some cases, more than 50%. The other point relates to longer term effects. Assuming that the prefumigation species composition was at equilibrium (which is not known), does the recolonized island show a tendency to be inhabited by the same species as before? Wilson (1969) suggests that as turnover occurs longer-lived species tend to accumulate in the system, until finally a particular set of species occurs together that has the greatest probability of persisting. He calls this an *assortative equilibrium*. The data collected in the study described here are too incomplete to draw conclusions about final equilibrium compositions, but as is seen in Table 7–2 after two years the fauna more closely resembles the prefumigation fauna than after only one year. As these islands are studied in the future additional information will be gathered relevant to these questions.

In this chapter we have concentrated on succession as a process of change in ecological systems. That species are replaced during succession can be understood in terms of the specific tolerances of the organisms involved. The presence of one species sets the stage for the possibility of others coming later either because the first conditions the habitat for them or because it provides resources in terms of food, shelter, or pro-

TABLE 7–2. PERCENTAGES OF SPECIES THAT WERE PRESENT AT BOTH OF TWO GIVEN CENSUSES ON FOUR OF THE EXPERIMENTAL ISLANDS.*

Name of experimental island	Censuses just before defaunation and one year later			Censuses just before defaunation and two years later			Censuses one and two years after defaunation		
	No. spp. in common	Total no. in both censuses	Percent in common	No. spp. in common	Total no. in both censuses	Percent in common	No. spp. in common	Total no. in both censuses	Percent in common
E1	2	29	6.9	5	26	19.2	7	18	38.9
E2	10	54	18.5	13	51	25.5	16	34	37.2
E3	8	40	20.0	7	35	20.0	16	31	51.6
ST2	11	37	29.7	17	31	54.8	12	34	35.3

*From Simberloff, 1970.

tection. Eventually, in most systems a group of organisms occurs that is self-maintaining. One of the large and still mostly unanswered questions in ecology is concerned with this group. Is the climax assemblage of organisms a group phenomenon or are its members selected on the basis of their individual survival capacities?

However this question is answered, it appears that those species of plants and animals that make up climax systems maintain their dominance and control over their systems by rather precise adjustment to the environment. Man is now learning, perhaps too late, that these relationships can be easily disturbed. Man often depends on succession to erase the scars associated with his activity. We may now be reaching the point where the changes we are causing exceed the capacity of the recovery mechanism.

REFERENCES

Albertson, F. W., F. W. Tomanek, and A. Riegel. 1957. Ecology of drought cycles and grazing intensity on grasslands of the central Great Plains. *Ecol. Monog.*, **27**:27–44.

Antevs, E. 1948. *The Great Basin, with Emphasis on Glacial and Post-Glacial Times. Climatic Changes and Pre-White Man.* University of Utah Bull. No. 3, pp. 168–91.

Billings, W. D., and H. A. Mooney. 1959. An apparent frost hummock-sorted polygon cycle in the alpine tundra of Wyoming. *Ecology,* **40**:16–20.

Clements, F. E. 1916. *Plant Succession, an Analysis of the Development of Vegetation.* Carnegie Institute of Washington Publ. No. 242. 512 pp.

Cooper, W. S. 1913. The climax forest of Isle Royale, Lake Superior, and its development. *Bot. Gaz.,* **55**:1–44, 115–40, 189–235.

_____ . 1923. The recent ecological history of Glacier Bay, Alaska. *Ecology,* **6**:197.

_____ . 1931. A third expedition to Glacier Bay, Alaska. *Ecology,* **12**:61–95.

_____ . 1939. A fourth expedition to Glacier Bay, Alaska. *Ecology,* **20**:130–59.

Cowles, H. C. 1899. The ecological relations of the vegetation on sand dunes of Lake Michigan. *Bot. Gaz.,* **27**:95–117, 167–202, 281–308, 361–91.

Crocker, R. L., and J. Major. 1955. Soil development in relation to vegetation and surface age at Glacier Bay, Alaska. *J. Ecol.,* **43**:427–48.

Daubenmire, R. F. 1952. Forest vegetation of northern Idaho and adjacent Washington, and its bearing on concepts of vegetation classification. *Ecol. Monog.,* **22**:301–30.

Donald, C. M. 1961. Competition for light in Australian crop plants. *S. E. B. Symposia,* **51**:282–313.

Drury, W. H., Jr. 1956. Bog flats and physiographic processes in the Upper Kuskokwim River region, Alaska. *Contributions from the Gray Herbarium,* **178**, pp. 1–130.

Gorham, E., and A. G. Gordon. 1960. Some effects of smelter pollution northeast of Falconbridge, Ontario. *Can. J. Bot.,* **38**:307–12.

_____ . 1963. Some effects of smelter pollution upon aquatic vegetation near Sudbury, Ontario. *Can. J. Bot.,* **41**:371–78.

Hutchinson, G. E. 1959. Homage to Santa Rosalia or why are there so many kinds of animals? *Amer. Nat.*, **93**:145–49.

Johnston, D. W., and E. P. Odum. 1956. Breeding bird populations in relation to plant succession on the Piedmont of Georgia. *Ecology*, **37**:50–62.

Kershaw, K. A. 1964. *Quantitative and Dynamic ecology*. London: Edward Arnold. 183 pp.

Lamarche, V. C., Jr., and H. A. Mooney. 1967. Altithermal timberline advance in western United States. *Nature*, **213**:980–82.

Lawrence, D. B. 1953. *Development of Vegetation and Soil in Southeastern Alaska, with Special Reference to the Accumulation of Nitrogen*. Final Report ONR, Project NR, pp. 160–83.

Lindemann, R. L. 1942. The trophic-dynamic aspect of ecology. *Ecology*, **23**:399–418.

Lutz, H. J. 1956. *Ecological Effects of Forest Fires in the Interior of Alaska*. U.S. Dept. Agric. Tech. Bull. No. 1133. 121 pp.

Margalef, R. 1967. The food web in the pelagic environment. *Helgolander Wiss. Meeresunters*, **15**:548–59.

————. 1968. *Perspectives in Ecological Theory*. Chicago: University of Chicago Press. 111 pp.

Martin, P. and P. J. Mehringer, Jr. 1965. Pleistocene pollen analysis and biogeography of the Southwest. In H. E. Wright and D. Frey (eds.), *The Quaternary of the United States*, pp. 433–51. Princeton, N.J.: University Press.

Mason, H. L. 1947. Evolution of certain floristic associations in western North America. *Ecol. Monog.*, **17**:201–10.

Matthes, F. E. 1942. Glaciers. In O. E. Meinzer, (ed.), *Hydrology*, pp. 149–219. New York: McGraw-Hill.

McIntosh, R. P. 1967. The continuum concept of vegetation. *Bot. Review*, **33**:130–87.

Pruitt, W. O. 1958. Qali, a taiga snow formation of ecological importance. *Ecology*, **39**(1):169–72.

Selander, S. 1950. Floristic phytogeography of southwestern Lule Lappmark. *Acta Phytogeogr. Suecica*, **27**:1–200.

Simberloff, D. S., and E. O. Wilson. 1969. Experimental zoogeography of islands: The colonization of empty islands. *Ecology*, **50**(2):278–96.

————. 1970. Experimental zoogeography of islands: A two-year record of colonization. *Ecology*, **51**(5):934–37.

Sjörs, Hugo. 1963. *Bogs and Fens on Attawapishat River, Northern Ontario*. National Museum of Canada Bull. No. 186, Contributions to Botany, 1960–1961, pp. 45–133.

Tschirley, F. H. 1969. Defoliation in Viet Nam. *Science*, **163**:779–86.

Viereck, L. A. 1970. Forest succession and soil development adjacent to the Chena River in interior Alaska. *Arctic and Alpine Research*, **2**:1–26.

Watt, A. S. 1947. Pattern and process in the plant community. *J. Ecol.*, **35**:1–22.

Wells, P. V., and R. Berger, 1967. Late Pleistocene history of coniferous woodland in the Mohave Desert. *Science*, **155**:1640–47.

Whittaker, R. H. 1953. A consideration of climax theory: The climax as a population and pattern. *Ecol. Monogr.*, **23**:41–78.

————. 1956. Vegetation of the Great Smoky Mountains. *Ecol. Monogr.*, **26**:1–80.

Wilde, S. A., and A. Lafond. 1967. Symbiotrophy of lignophytes and fungi: Its terminological and conceptual deficiencies. *Bot. Rev.*, **33**:99–104.

Wilson, E. O. 1969. The species equilibrium. In G. M. Woodwell and H. Smith (eds.), *Diversity and Stability in Ecological Systems*, pp. 38–47. Upton, N.Y.: Brookhaven Symposia in Biology, No. 22.

———, **and D. S. Simberloff.** 1969. Experimental zoogeography of islands: defaunation and monitoring techniques. *Ecology*, **50**(2):267–78.

Williams, G. C. 1966. *Adaptation and Natural Selection*. Princeton, N.J.: Princeton University Press. 307 pp.

Woodwell, G. M. 1970. Effects of pollution on the structure and physiology of ecosystems. *Science*, **168**:429–33.

Chapter 8

THE STRUCTURE AND ORGANIZATION OF COMMUNITIES

INTRODUCTION

In nature populations of individual species are integrated into complex biotic assemblages within which individual organisms interact with each other and with their physical and chemical environments. This integrated whole constitutes the ecosystem. The living organisms, although inseparable functionally from the abiotic environment, have received special attention and recognition as a unit termed the biotic community. The biotic community, however, does not represent a system or level of organization separate from that of the ecosystem (Rowe, 1951; Shultz, 1967). Rather, it represents a group of ecosystem components that may be separated for purposes of discussion. In this discussion we shall not only concentrate on the biotic components of the ecosystem, but we shall also consider ecosystem processes, both biotic and abiotic, as they influence certain characteristics of these components.

Biotic communities may be considered to possess both a structure and an organization. Community structure is a descriptive concept, referring to species composition and to the quantity and distribution of organic matter within the community. Description of species composition requires consideration of not only the kinds and variety of species present, but also of their patterns of abundance, spatial dispersion, and temporal occurrence and activity. A statement of the kinds of species present should include, ideally, a description of their patterns of adaptation and response at the organismal and population levels, as well as of their taxonomic relationships. Description of the biomass structure of communities involves a statement of the quantity and pattern of spatial distribution of organic tissues within the community environment.

Community organization, on the other hand, is a functional concept and concerns the relationships among community members that are important in determining their presence and role in the community. Some of these relationships are competition, mutualism, interference, and various food-chain interactions. The degree to which these relationships are important in determining the structure of communities is a measure of the degree of community organization. Thus, a community of species largely independent of each other and having a structure largely determined by relationships with the physical environment possesses little organization. A community in which the presence and quantitative importance of particular species is largely influenced by biotic relationships possesses a high degree of organization.

Historically, much effort has been directed toward gaining an understanding of patterns of community composition and structure and of correlating these characteristics with those of the physical and chemical environment occupied. More recently, however, attention has been directed toward study of the causal relationships behind these correlations and toward examination of the extent and nature of biotic interrelationships.

Our discussion of the structure and organization of communities will center on two major questions. The first concerns the determinants of structure and organization of a community occupying a given area of habitat. We shall approach this question by considering first the factors determining the set of species potentially available for inclusion within a community. Then we shall examine the processes by which the actual group of member species is sorted out. Finally, we shall consider the relationships that quantitatively define the position of each of these species in the overall structure of the community.

The second major question concerns the pattern of change occurring in community structure and organization as environmental conditions change through space or time. Our examination will consider the degree to which the community behaves, or does not behave, as a unit adapted to a certain range of environmental conditions, and which consequently appears and disappears as a unit as gradual change occurs in environmental conditions.

DETERMINANTS OF COMMUNITY STRUCTURE

The structure and organization of biotic communities result from a complex set of trial and error interactions. These interactions not only sort out a particular group of species from those potentially available for inclusion, but also determine the location, abundance, and productivity of the member species within the community. Furthermore, the interactions that determine the structure and organization of communities occur continuously. The assemblage of species available for inclusion in the community is continually changing as a result of arrival of new species and extinction of species which have been present. Also, the ecological characteristics of each species are continually changing as a result of evolutionary processes. The community itself is thus a changing entity reflecting the varying results of the interactions involved in community organization. This pattern may be further complicated by patterns of gradual modification of regional climate or by disturbance factors that initiate directional changes in community characteristics, e.g., biotic succession.

In suggesting that this is a trial and error process we do not mean to say that community structure is not governed by basic ecological principles. Rather, we are suggesting that basic ecological principles relating to the interaction and coexistence of different species are continually being tested by new combinations of species. Our knowledge of these principles is weak, however, and this area remains one of the least explored regions of modern ecology.

We can recognize several of the major processes that operate to define the structure and organization of communities. General descriptions, and in some cases preliminary statements of theory, can be presented for them. These processes relate to (1) dispersal, (2) the physical environment, and (3) the biotic environment. It should be borne in mind that a satisfactory community theory must integrate all of these processes, rather than separating them, as we are doing for convenience of discussion.

Dispersal

The species available for inclusion in a community developing in a given area of habitat consist of those reaching the locality by dispersal. In some cases, dispersal mechanisms may provide a large group of species of very diverse ecological characteristics; in other cases, only a few ecologically specialized forms may reach an area. The resulting structure of communities reflects such differences.

Several of the determinants of the number and ecological diversity of species reaching an area by dispersal are biogeographic in nature. These include the existence of geographical or ecological barriers to dispersal, the distance over which dispersal occurs, the influence of air or water currents in promoting or hindering dispersal, and the size of source and invasion areas.

Barriers to dispersal are relative and operate with different intensity for different kinds of organisms. In addition, dispersal barriers are all, in the final analysis, ecological. Barriers to dispersal represent essentially unfavorable areas of habitat separating areas of favorable habitat. Some, such as oceans to terrestrial organisms, are completely unfavorable. Others, such as patches of coniferous forest to deciduous forest birds, may only be unfavorable in the sense that other species existing within them are better adapted for life in the intervening areas. These extremes are occasionally distinguished as geographic and habitat barriers respectively, but all gradations exist between these extremes. The degree of difference between ecological characteristics of the barrier area and those of the areas separated by it thus determines in part the nature and frequency of species crossings.

The difference between geographical and habitat barriers is significant, however, in one important way. A "habitat" island, as opposed to an area of land surrounded by water, not only receives species successfully crossing the intervening area of different habitats, but it also receives immigrants from the area of different habitat itself (MacArthur and Wilson, 1967).

The frequency of crossing a barrier is inversely related to the width of the barrier and directly related to the size of the source and recipient areas. Furthermore, the existence of currents of air or water may increase or decrease the probability of barrier crossing, depending on the situation.

MacArthur and Wilson (1967) have developed several models of dispersal between islands. These models incorporate several of the above factors and utilize the following notation (see Fig. 8–1):

W_s = Diameter of the source island

W_r = Diameter of the recipient island (perpendicular to axis of dispersal)

d = Distance between islands

Λ = Mean dispersal distance of the organism

One of the first questions that may be asked concerns the rate of departure of propagules from the source island. This will vary with the characteristics of the organism affecting the distance a propagule will travel, on the average. These characteristics determine the value of Λ. If Λ is large, relative to W_s, the width or diameter of the source island, then departures may in fact occur from anywhere within the source island,

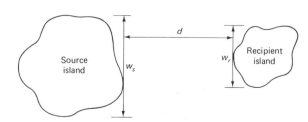

FIGURE 8–1. Definition of terms used in discussions of patterns of dispersal (see text).

and the rate will be related consequently to the area of the source island (that is, assuming that the island is circular in shape, the square of the island diameter).

$$\text{Departures per unit time} = \alpha W_s^2$$

In this formula, the coefficient, α, has a value determined by ecological characteristics such as mobility, population density, etc.

Alternatively, if Λ is much smaller than W_s, departures can occur only from the marginal areas of the source island. Thus, their rate will be related more closely to the circumference of the islands or, as a result of the constant relationship of diameter and circumference, to the island diameter:

$$\text{Departures per unit time} = \beta W_s$$

In this formula the coefficient, β, has a value determined in a manner similar to that of the coefficient, α.

Next, we may ask how the number of propagules may be expected to drop off with distance from the source island. Several possibilities exist. First, if Λ is much greater than d, there may be essentially no decrease in number of propagules over the distance involved. This pattern might be approximated by strong-flying birds in an area of small, closely grouped islands. This pattern may be termed a *uniform* distribution.

Second, it is possible that, per unit distance travelled, the propagules may have a certain constant probability of being inactivated. For example, an airborne seed being carried over water may have a certain probability, per unit distance, of falling into the water—the probability not, however, changing with actual distance from the source island. Under this assumption, the fraction of propagules reaching a given distance from the source island would be given by the expression:

$$e^{-d_s/\Lambda}$$

where

$$d_s = \text{Distance from source island}$$

This dispersal pattern may be termed *exponential*.

Lastly, it may be assumed that the distance reached by propagules is determined by some factor or characteristic having a certain mean duration of effectiveness, with the exact values for individual propagules varying about this mean according to a normal curve. Thus, if a group of insects disperse by active flying from a central point and continue flying until exhausted, with distance flown until exhaustion varying normally around some mean, such a situation would prevail. In this case the proportion of insects reaching some given distance, d_s, from the source would be given by the expression:

$$1 - 2 \int_0^{d_s} \frac{-1}{\sigma \sqrt{\pi}} \, e^{-1/2(x/\sigma)^2} \, dx$$

where

d = Distance from the source point in units of σ

σ = Standard deviation of dispersal distance of organism

This expression is a modification of the expression giving rise to the normal curve and represents the summation (indicated by the integral sign) of the fractions of propagules reaching the distance, d, or farther, from the source. This pattern of dispersal may be termed *normal*.

Finally, the fraction of the total propagules that move in the correct direction to intercept a particular area at some distance from the source point is determined by the ratio of the angle subtended by the recipient area to 360°; this may be calculated as follows (see Fig. 8–1):

$$\frac{2 \tan^{-1} \dfrac{W_r}{2d_s}}{360°}$$

This value would, of course, be subject to considerable modification by factors such as directional navigation tendencies of the propagules or the action of air or water currents moving the propagules.

From these expressions, however, a rough estimate of the number of propagules reaching a given area from a source could be determined by combining expressions for (1) number of propagules leaving the source island, (2) fraction of propagules surviving to the distance of the recipient areas, and (3) the ratio of the angle of the recipient area to 360°.

Let us assume that the number of propagules leaving a source island is related to its area and that decline in numbers with distance is exponential. The number reaching a recipient island of diameter W_r would be the product of

1. $\quad \alpha W_s^2 \quad$ = Number of propagules leaving source island

2. $\quad e^{-d_s/\Lambda} \quad$ = Fraction of propagules reaching distance equal to that of recipient

3. $\quad \dfrac{2 \tan^{-1} \dfrac{W_r}{2d_s}}{360°}$ = Ratio of angle subtended by recipient area to 360°

Two factors of a primarily ecological nature also influence the pattern of dispersal: abundance in the source area and adaptations of the species for dispersal. In the previous expressions these characteristics determined the value of the coefficients α and β, associated with the estimates of departures from the source island and the estimate of Λ, of mean dispersal distance.

The problem associated with examination of these relationships is the fact that dispersal itself is very difficult to observe. Dispersal, followed by successful colonization of a new area, is a very visible occurrence; dispersal without successful establishment is rarely recorded.

Much evidence indicates, however, that high dispersal ability is associated with a number of ecological characteristics of both plants and animals. High dispersal abilities, coupled with high reproductive potentials, characterize species occupying habitats of transitory duration or early successional status. Plants of open field habitats, for example, characteristically possess very light seeds (Table 8–1). The relationship of high dispersal ability with occupation of impermanent habitat type has been extensively documented for terrestrial arthropods by Southwood (1962). This can be illustrated by an analysis of various species of British water beetles according to whether they are able to fly, unable to fly, or show individual variability in flight ability (Table 8–2). This analysis shows the

TABLE 8-1. SEED WEIGHT FOR BRITISH PLANT SPECIES CHARACTERISTIC OF DIFFERENT HABITATS.*

Habitat	Number of species	Average seed weight (mg)
Open	98	1.315
Semienclosed	22	2.214
Forest edge and scrub	32	4.438
Closed woodlands	27	13.686
Woodlands (shrubs)	24	85.435
Woodlands (trees)	20	653.000

*From Salisbury, 1942.

TABLE 8-2. THE FLYING CAPABILITIES OF SPECIES OF BRITISH WATER BEETLES IN RELATION TO HABITAT OCCUPIED.*

	Rivers, springs	Lakes, tarns, canals	Brackish water, peat pools, bog ponds	Ponds, ditches	Artificial ponds, gravel pits, cattle troughs
Able to fly	3	10	30	27	24
Variable	6	13	14	15	8
Unable to fly	13	7	7	2	0

*From Southwood, 1962.

increased frequency of flying species in temporary pond habitats. Cox (1971) has similarly shown that in temperate zone bird communities, the frequency of migratory species is greatest in communities of early successional status and low species diversity. Generally speaking, species having high dispersal abilities meet the requirements for fugitive species, as described by Hutchinson (1951). These species survive by virtue of their ability to colonize new areas of habitat rapidly before the appearance of forms having greater competitive ability (see p. 259).

More detailed studies have been conducted on the characteristics of species that are effective colonizers of island situations. Wilson (1961) has suggested that a general distributional and evolutionary pattern, termed the *taxon cycle,* characterizes the faunas of island groups. This cycle may be divided into four stages (Ricklefs and Cox, 1972):

I. Invasion and spread of a new taxon through a major portion of an island group with little or no taxonomic differentiation.
II. Taxonomic differentiation of various island populations to subspecies or species level.
III. Contraction of distribution and development of distributional discontinuities through extinction of island populations.
IV. Restriction of taxon to a single island.

In this cycle stage I species have been shown to be characterized by several features. Wilson (1961) found, for ants in Melanesia, a concentration of these species in marginal habitats such as open forest, savanna, and coastal areas. Individual species, on the average, occurred in a wider variety of habitat types, showed greater plasticity of nest site requirements, and exhibited larger colony sizes. Similar observations for birds have been obtained by Greenslade (1968) in the Solomon Islands, and Ricklefs (1970) in the West Indies. Ricklefs and Cox (1972) have also shown that stage I species in the West Indies show migratory movements and flocking behavior more frequently than do species of later stages. Williams (1969) has characterized the two *Anolis* lizards, *A. carolinensis* and *A. sagrei,* which are the most frequent colonizers of the West Indian islands, in a similar fashion. Both are species of open forest or savanna areas and possess very generalized requirements. Williams characterizes these forms as versatile, and he suggests that the key to their successful colonization is the fact that they are able to use whatever environments and resources exist in areas into which dispersal brings them.

Challenges of the Physical Environment

Species that reach an area in numbers adequate to allow colonization must meet the combined challenge of the new physical and biotic conditions to survive. The physical component of this challenge will be discussed in this section and the biotic component in the next.

Conditions of the physical environment in a particular site act as a filter for newly arriving species. This filter specifies required tolerances of physical conditions for the species to become established. These tolerances are related ultimately to the extremes of external physical factors impinging upon the site. They are, however, determined precisely by the extent to which these extremes are modified by the existing community and by the degree of direct exposure of the species to conditions of the physical environment. The overall biotic structure, and the activities of species already present, may modify environmental conditions considerably within the spatial limits of the community. These modifying influences may be of major importance in permitting or prohibiting the establishment of new species. Likewise, characteristics of the newly arrived species, such as size and pattern of seasonal activity, determine whether individuals of the species will be exposed to the full extremes of the physical environment or to a much more restricted set of conditions.

These relationships may be exemplified by considering conditions that might be encountered by different species in their colonization of a temperate grassland community. In this area wind, temperature, and intensity of solar radiation represent important conditions of the physical environment. These conditions vary on a daily and seasonal basis and in response to regional weather patterns. During a given period, the greatest extremes of these factors will occur in the upper part of the vegetational canopy. At lower layers of the vegetation lesser extremes will occur, as a result of various modifying influences of the vegetation itself.

Because of their large size and perennial growth habit, tree or shrub species must be adapted to nearly the full extremes of these factors throughout the year in order to successfully invade a grassland area. Small annual plants, on the other hand, may be exposed only to considerably modified conditions during a limited portion of the year. This modified and restricted exposure may either favor the species, if the extreme conditions are detrimental, or prove detrimental to it, if extreme conditions such as intense solar radiation are required.

Biotic Influences on the Community Environment. The community may thus be regarded as possessing an overall macroclimate, which is modified internally—through the presence and activities of the community members—into a complex of microclimates. This effect is termed *ecological dominance*. Such effects are obviously well developed in certain communities, e.g., coral reefs and forests, in which biotic processes result in the growth of an extensive and complex physical structure. However, they may be seen even in communities in which phytoplankton or decomposers are the predominant organisms. Metabolic processes of such organisms may effect basic physical (absorption of light by dense phytoplankton populations) or chemical (depletion of oxygen in water) changes in the medium.

We will examine in greater detail some of the influences shown in a forest community and discuss their importance in control of community composition. In Chapter 3 we examined factors of the physical environment and considered in detail how a forest canopy influences these factors. We will briefly summarize these influences here.

The two factors most strongly modified by the presence of the forest canopy are solar radiation and wind. Almost all of the other microclimate differences noted within forests are directly or indirectly the result of their modification. A considerable portion of the incoming solar radiation is reflected from the canopy. For closed-canopy hardwood forests this amounts to about 16–37%; for coniferous forests approximately 10–14% (Ovington, 1965). Much of the remaining radiant energy is absorbed by leaves, so that the amount actually reaching the forest floor may be reduced to 1–6% or less of that reaching the outer canopy (Ovington, 1965). This effect is greater than that typically seen for any other environmental factor. The extent of reduction is obviously related to the density of trees and to the degree of closure of the forest canopy. The nature of the species making up the canopy, however, is also important. Species that are highly shade tolerant and predominate in climax or near-climax communities produce the greatest reduction in radiation reaching the forest floor (Kittredge, 1948).

Wind is strongly modified by the forest canopy. Wind velocities exceeding 1–2 mph are seldom recorded beneath well-developed canopies. On the average, reduction in wind velocity amounts to 40–80% of that recorded outside the canopy (Kittredge, 1948).

Reduced radiation and wind intensity beneath forest canopies are correlated with markedly different temperature conditions within the forest. Both the daily and seasonal extremes of temperature are less within the forest than they are at stations at similar heights outside. In temperate zone forests, for the most part, the mean daily temperature is lower in summer and higher in winter than for stations outside the forest. The mean annual temperature is generally lower within the forest since the reduction in the summer is usually greater than the increase during the winter period (Kittredge, 1948).

Precipitation is also modified. Some of the precipitation is intercepted by the canopy and ultimately lost by evaporation. During very light showers this may amount to nearly 100% of the total rainfall; in heavier rainstorms it is usually in the range of 10–40% (Kittredge, 1948).

Humidity is closely related to temperature. Absolute humidity, the amount of water per unit volume of air, tends to be only slightly higher beneath vegetational canopies. Because of the lower maximum temperatures, however, relative humidity is increased and the vapor pressure deficit, the difference between absolute humidity and saturation capacity, is reduced. Consequently, evaporation from moist surfaces is reduced considerably.

Within the soil, conditions are modified even more, especially if a well-developed litter layer is present. The temperature range is strongly reduced and evaporation decreased by such a layer.

These microclimatic influences are of importance in determining the composition of the subordinate strata of the forest community. In southeastern Ohio, Wolfe et al. (1949) studied macroclimates and microclimates in a small valley over a period of six years. During this study measurements were made of the temperature microclimates associated with various herbaceous plant species of the forest floor. These species, all perennials, possess active above-ground shoots for part of the year and dormant vegetative parts beneath the litter layer for the remainder of the year. Microclimatic measurements were made in corresponding locations, correlated with the annual cycle of the species. The measurements obtained were compared to the range of temperatures recorded at the Lancaster, Ohio, Weather Bureau Station.

Data for the Spring Beauty, *Claytonia virginica,* are shown in Fig. 8–2. This species overlasts the summer as a button-shaped corm beneath the leaf litter layer. The plants begin to grow in October and for the first 20 weeks are entirely or largely beneath the litter layer where the temperatures encountered rarely drop below freezing. At the same time, air temperatures as low as −18°F may be recorded at a weather bureau station. During the early part of this growth period, removal of the leaf litter

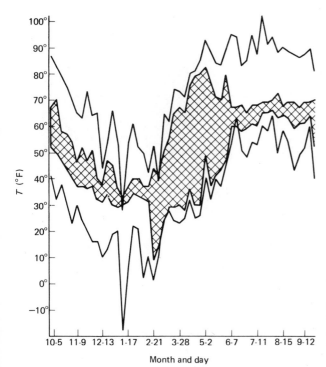

FIGURE 8–2. The temperature microclimate (cross-hatched) of the Spring Beauty, *Claytonia virginica,* in Neotoma Valley, central Ohio, compared to the regional macroclimate. (From Wolfe, *et al.,* 1949.)

causes the plants to die from exposure to temperatures only slightly below freezing. Upon emergence of the leaves in February and March, however, tolerance of low temperature has increased. During this period, low temperatures of 9°F are tolerated without injury. From March–May the plants flower, fruit, and the shoots die back. Through the summer the living portions of the plants again consist of the corms lying beneath the leaf litter. Thus, for this species the total range of conditions encountered varies from about 9–82°F, although the macroclimate range was from −18 to 102°F. A second species, Squirrel Corn, *Dicentra canadensis,* which does not emerge from beneath the leaf litter until mid-April, encounters a still more restricted range of temperatures (Fig. 8–3).

Despite the fact that the phenomenon of ecological dominance was one of the first relationships to be recognized as being of importance in community organization, it has received little careful study. If we consider relationships between dominant and subordinate members of the plant community, such as canopy tree species and the species occurring in the understory vegetation, certain general statements can be made, however. Several major studies of forest vegetation, both in the eastern and western United States (e.g., Daubenmire, 1968; Whittaker, 1956), have shown that, in general, there is considerable independence in the habitat

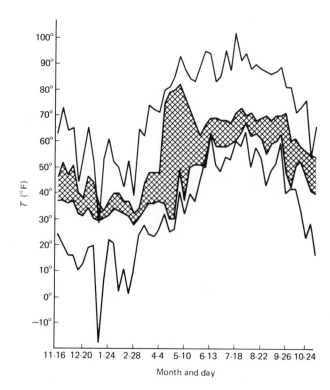

FIGURE 8–3. The temperature microclimate (cross-hatched) of Squirrel Corn, *Dicentra canadensis,* in Neotoma Valley, central Ohio, compared to the regional microclimate. (From Wolfe, *et al.,* 1949.)

distributions of canopy and understory species. In the Idaho–Washington region of the northern Rocky Mountains, for example, similar ground-cover vegetation may occur under forest canopies of completely different species composition. At the same time, forest stands similar in canopy composition may have quite different kinds of ground cover (Dauben-mire, 1968). Nevertheless, if the overstory is removed, certain species disappear and others invade at the ground level, although many of the original forest floor species may persist for some time. Furthermore, during succession, a climax equilibrium tends to appear earlier in the understory vegetation than in the forest canopy. These observations, in combination, suggest that the understory vegetation shows greater sensitivity to soil and microclimatic conditions than does the canopy composition, and that the effects of dominant canopy species set only rather broad limits to the species composition of understory portions of the community.

Species Sorting by Structural-Functional Characteristics. To facilitate the analysis and comparison of composition of communities, various systems of grouping species by similarities of overall pattern of adaptation have been devised. These systems have usually been developed for particular groups of plants or animals and have been defined on the basis of different expressions of some major feature shown by all members of the group. The most widely used classification for plants has been the system of life forms developed for flowering plants by Raunkiaer (1937). This system (described in Chap. 2), based on the presence of a perennating bud and its location below, at or above the ground surface, recognizes five major categories: Phanaerophytes (Ph), chamaephytes (Ch), hemicryptophytes (H), cryptophytes (Cr), and therophytes (Th).

Although defined on the basis of structure, in reality this classification incorporates a number of adaptive characteristics, both structural and physiological. For example, plants in the different categories above differ in size and consequently in depth of root penetration in the soil. As a result, important differences in the physiology of water and nutrient uptake are incorporated into this classification.

This fact has led recent workers (e.g., Knight and Loucks, 1969) to propose systems incorporating a variety of both structural and functional characteristics. However, since the approach is still in its early stages, we will examine the life form systems utilized by previous workers— recognizing, as fully as possible, the combination of structural and functional features included in them. In this discussion we will consider two examples, one dealing with life form analysis of plant communities and the other with animals. Our objective will be to consider the manner in which species are sorted for inclusion in communities by their patterns of adaptation and response to conditions of the physical environment and to the basic patterns of resource availability determined by these conditions.

Cantlon (1953) conducted a study of vegetation and microclimates on north and south slopes of Cushetunk Mountain in central New Jersey. This study provides data illustrating the role of environmental conditions in sorting species according to both life forms and patterns of physiological response to environmental conditions. It further illustrates the influence of dominant species in modifying conditions of the physical environment and the effects of this modification on community composition.

Cushetunk Mountain rises to an elevation of about 600 feet above the surrounding area and has north and south-facing slopes of about 20°. The basic environmental differences between the two slopes result from two relationships: angle of incidence of solar radiation and summer-winter differences in the direction of prevailing winds. Figure 8–4 illustrates the seasonal pattern of incident radiation on the two slopes. The low angle of incidence of radiation on the north slope during midwinter means that a unit of incoming solar radiation is spread over a larger area of ground surface. Combined with the interception of incoming radiant energy by the trees themselves, it further means that virtually no radiation really strikes the ground surface. At the same time the south slope is receiving energy at an angle of incidence nearly equal to that received on the north slope in midsummer. During the summer, solar radiation strikes the south slope at an angle of about 90°. Coupled with this difference, weather patterns change seasonally so that during the winter frontal storms strike the north slope with greatest severity and during the summer prevailing winds strike the south slope.

As a result, winter conditions are more severe on the north slope.

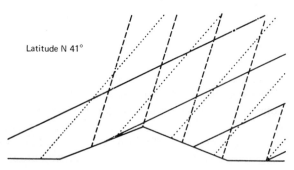

Latitude N 41°

Angle of insolation

Time of year	20° north slope	20° south slope	Level
Dec. 22, winter solstice	$5\frac{1}{2}°$	$45\frac{1}{2}°$	$25\frac{1}{2}°$
Mar. 21 and Sept. 22, equinox	29°	69°	49°
June 22, summer solstice	$52\frac{1}{2}°$	$92\frac{1}{2}°$	$72\frac{1}{2}°$

FIGURE 8–4. The angle of midday insolation for the seasons on 20° north and south facing slopes at the latitude of Cushetunk Mountain, New Jersey (N 41°). The solid line is the angle of insolation on December 22, the dotted line on March 21 and September 22, and the broken line on June 22. (From Cantlon, 1953.)

Here mean air temperatures at a height of 5 cm average 3.0–8.8°F lower during the winter months than for similar south slope situations. Soil temperatures are also lower than for the south slope and the soil is frozen to a depth of at least 4 cm for several weeks. Snow persists for a longer period than for the south slope. In contrast, on the south slope daily maximum air temperatures at 5 cm average 7.9–17.8°F higher than for the north slope and the soil rarely freezes to a 4 cm depth.

During the summer the north slope is cooler and more moist than the south slope. Mean temperatures at a height of 5 cm average 4.1–6.6°F less than for the south slope. Largely because of these cooler temperatures the vapor pressure deficit (saturation capacity of air—absolute humidity) is lower on the north slope. On the south slope the higher potential evaporation rate is correlated with generally higher wind velocities because of the prevailing wind patterns and the more open tree layer present there. Thus, during midsummer, when the water requirements of plants are highest, the available soil moisture is lowest on the south slope.

Throughout the year the north and south slopes differ in the profile of daytime air temperature with height above the ground. On the north slope this profile shows an inversion, with lowest temperatures at the surface. On the south slope, the reverse is true: temperatures are highest at the surface and decrease with height.

These environmental differences are reflected in the vegetation of the two slopes. Both are forested, but important differences in the characteristics of both forest canopy and undergrowth occur.

Table 8–3 contrasts the forest canopy characteristics of the two slopes. Among the trees over four inches in diameter the number of species and total basal area are greater on the north slope. Thus, the upper forest canopy is more open on the south slope. This is reflected in the occurrence of a greater number of species and a greater basal area of lower

TABLE 8–3. NUMBER OF TREE SPECIES AND TOTAL BASAL AREA (ft² /3,500 m²) FOR THE CANOPY AND UNDERSTORY OF FOREST COMMUNITIES ON THE NORTH AND SOUTH SLOPES OF CUSHETUNK MOUNTAIN, NEW JERSEY.*

Stratum	North slope		South slope	
	Number of species	Basal area	Number of species	Basal area
Canopy (> 4 in. d.b.h.)	14	85.405	11	64.278
Understory (1 in. – 4 in. d.b.h.)	17	4.257	19	8.051
Totals	25	89.662	27	72.329

*Data from Cantlon, 1953.

Note: d.b.h. = diameter at breast height.

canopy trees on this slope. The upper canopy and lower canopy thus show a reciprocal relationship of variety and basal area of tree species. The lower canopy trees include both species not reaching large size and young individuals of species that are part of the upper canopy.

Both slopes possess a well-defined shrub stratum composed of young trees and shrubs (Table 8–4). The young tree component is best developed on the south slope, the shrub component on the north. However, although the two slopes do not differ appreciably in the number of species of trees and shrubs, the total density of the shrub stratum is greatest on the north slope.

All of the woody species discussed above fall into the phanaerophyte life form class of Raunkiaer. Raunkiaer developed this system on the basis of the height of the perennating bud above the ground surface and its consequent exposure to direct action of severe environmental factors during the most unfavorable season. The greater representation of this group on the north slope, with its more severe winter conditions, thus appears to be a contradiction of the importance of this adaptive characteristic. For these large woody plants, however, summer conditions of temperature and drought may actually represent a greater challenge from the physical environment. The species of these strata are directly exposed to the full intensities of solar radiation and wind, and consequently the greatest evaporation stress. Since they also exhibit root systems of deepest penetration in the soil, they possess the greatest potential for high metabolic activity throughout the summer, providing adequate deep soil moisture is available. The greater representation of phanaerophytes on the north slope suggests more favorable conditions of moisture throughout the growing season.

The differences in microclimate beneath the forest canopy, particularly the inverted temperature profile on the north slope, are in part the result of the tree and shrub strata themselves. Differences in the ground-level vegetation on the two slopes are thus determined not only by the

TABLE 8-4. NUMBER OF SPECIES AND INDIVIDUALS (PER 3500 m²) IN THE SHRUB STRATUM ON NORTH AND SOUTH SLOPES OF CUSHETUNK MOUNTAIN, NEW JERSEY.*

	North slope		South slope	
Group	Number of species	Number of individuals	Number of species	Number of individuals
Young trees	24	5,262	24	9,251
Shrubs and woody vines	20	19,595	19	9,993
Totals	44	24,857	43	19,244

*Data from Cantlon, 1953.

overall environmental differences of the two slopes but also by the differential influence of the forest canopies on light intensity, temperature,
and evaporation potential. Furthermore, the ground vegetation, being
shallowly rooted, is more subject to these seasonal extremes of temperature and moisture.

Table 8–5 summarizes the characteristics of the ground layer. Seedlings of tree and shrub species, together, are more abundant on the north
slope and constitute a much greater portion of the total ground layer
than on the south. Conversely, members of other life form classes show
greater abundance on the south slope. Within the latter groups, several
important differences may be seen. Chamaephytes, ferns, and cryptophytes show greater abundance on the north slope. Cryptophytes, among
perennials, possess overwintering buds with the greater degree of protection because they are deeply buried. Their success on the north slope
may thus reflect an adaptation to the more severe freezing of surface
soils during the winter. Chamaephytes and ferns, with buds at or above
the soil surface may, on the other hand, be favored by the more favorable
summer conditions of moisture and moderate temperatures.

Therophytes and hemicryptophytes reach their greatest development
on the south slope. Therophytes, overlasting unfavorable periods by seeds,
are most highly protected against conditions of severe drought, as well
as low winter temperatures. In addition, these species are most plastic
in their seasonal activity cycle, being able to grow and flower whenever
periods of favorable temperature and moisture conditions occur during

TABLE 8-5. OVERALL COMPOSITION OF GROUND-LAYER VEGETATION (TREES
AND SHRUBS < 1 FT. IN HEIGHT, ALL HERBACEOUS PLANTS) ON NORTH AND
SOUTH SLOPES OF CUSHETUNK MOUNTAIN, NEW JERSEY.*

	North slope			South slope		
		Individuals			Individuals	
Group	Species	Density	Percent	Species	Density	Percent
Phanaerophytes:						
Trees	18	406 }	37.84	19	488 }	14.08
Shrubs	20	1,193 }		19	477 }	
Chamaephytes	1	50	1.18	1	5	0.07
Ferns	6	188	4.45	3	10	0.14
Hemicryptophytes	34	1,932	45.72	49	4,876	71.14
Cryptophytes	12	457	10.81	15	507	7.40
Therophytes	0	0	0.00	7	491	7.16
Totals	91	4,226	100.00	113	6,854	99.99

*Data from Cantlon, 1953.

Note: Densities are numbers of individuals per 140 m².

the year. One of these species, *Polygonum convolvulus,* may flower at any time between May and November, for example. Two of the species represented on the south slope usually germinate in midwinter, carry out significant growth during mild periods through late winter and spring, and begin to flower as early as April and May.

As a group, the hemicryptophytes, with perennial buds at or on the soil surface, vary considerably in their seasonal patterns of activity. This group may be subdivided into three categories: (1) protohemicrypto-phytes (Hp), which possess no basal rosette of leaves, but produce a leafy flowering stalk; (2) rosette plants (Hr), which possess a perennial rosette of leaves at ground level and produce leafless flowering stalks; and (3) semirosette plants (Hs), which have a basal leaf rosette and leafy flowering stalks. The abundance of each of these groups is shown in Table 8–6.

Protohemicryptophytes are represented in about the same relative degree on both slopes. Rosette plants, however, are more important on the north slope. These plants, among hemicryptophytes, are the ones most closely tied to the protected conditions of the soil surface. Thus they are favored during the winter by a protective snow cover and during the summer by the cooler moist conditions at this level. Semirosette plants, which include most grasses and sedges, must produce an erect leafy flowering stalk. This is facilitated by an ability to capitalize on short periods of conditions favorable for vegetative growth. On the south slope, where they are dominant, Cantlon (1953) observed that many of these species show metabolic activity during warm winter periods.

Thus, although species of a variety of life forms and activity patterns are present on both slopes, it is apparent that the physical environment exerts a strong sorting action that influences the composition of communities developing under even locally different conditions.

TABLE 8–6. NUMBER OF SPECIES AND ABUNDANCE OF VARIOUS GROUPS OF HEMICRYPTOPHYTES ON NORTH AND SOUTH SLOPES OF CUSHETUNK MOUNTAIN, NEW JERSEY.*

| | North slope | | | South slope | | |
| | | Individuals | | | Individuals | |
Group	Species	Density	Percent	Species	Density	Percent
Protohemicryptophytes	15	831	43.12	18	2,035	41.95
Semirosette plants	14	618	32.07	25	2,697	55.60
Rosette plants	5	478	24.80	6	119	2.45
Totals	34	1,927	99.99	49	4,851	100.00

*Data from Cantlon, 1953.

Note: Densities are numbers of individuals per 140 m².

The concept of life forms may be extended to animals, although few such analyses have been attempted. Salt (1953) devised a system for grouping bird species into ecological categories on the basis of foraging location and general food type (Fig. 8–5). The relative abundance of bird species in these categories reflects primarily the general vegetational structure of the habitat and the abundance and general type of plant and animal food available.

This system involves the initial grouping of species into four categories of major feeding location: air, foliage, timber, and ground. The air feeders are those that capture flying insects on the wing. Two groups may be recognized within this category: (1) *air-soaring,* including swifts, swallows, nighthawks, and other species that may feed at higher levels and that fly continuously while foraging, and (2) *air-perching,* including flycatchers and similar species that perch in exposed locations and fly out to capture passing insects.

The foliage feeders may be subdivided into three groups, depending on the major food type: (1) *foliage-nectar,* including hummingbirds and other nectar feeders; (2) *foliage-insect,* consisting of birds feeding on insects and berries; and (3) *foliage-seed,* applying to birds feeding on dry nuts, seeds, and other hard materials. The timber category includes species that forage over tree trunks and branches and includes two

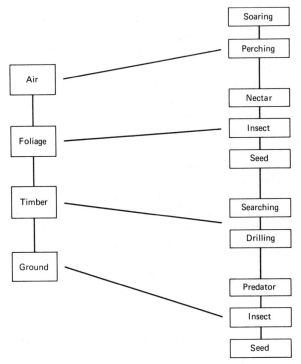

FIGURE 8–5. Ecological groupings used by Salt (1953) in analysis of the composition of three California bird communities.

groups: (1) *timber-searching,* consisting of species that search only over bark surfaces, and (2) *timber-drilling,* including woodpeckers that drill into bark and wood for their food.

Ground feeders can also be subdivided into groups of (1) *ground-insects,* (2) *ground-seed,* and (3) *ground-predator,* the last including species of hawks, owls, and other large predators capturing their prey largely on the ground.

Salt (1953) used these categories to compare the bird faunas of three localities in California throughout the year. These areas were (1) Boca Spring, Nevada County, (2) Yosemite Valley, and (3) Glen Oaks Canyon, Los Angeles County. Boca Spring, located at 6,000 feet on the east slope of the Sierra Nevada, is a wet spring meadow surrounded by an open Jeffrey Pine forest having an undergrowth of sagebrush-desert species. The climate (Fig. 8–6) of the area is characterized by a fairly cold winter, with appreciable snow cover, and a warm, dry summer. Yosemite Valley (4,000 ft) possesses a mixed forest of conifers and deciduous species. The region has a moderately cold winter, with light snow cover, and a dry, mild summer (Fig. 8–6). Glen Oaks Canyon (2,000 ft) is an area of riparian forest with live oaks, sycamore, and cottonwood in the foothills of the San Gabriel Mountains near Los Angeles. The climate (Fig. 8–6) shows a mild, moist winter and a long, dry summer.

Salt (1953) collected information on the principal species occurring in these three areas and prepared avifaunal spectra (Fig. 8–6) showing the number of species falling into each of the ecological groups defined above. Boca Spring exhibited a strong seasonal differentiation in this spectrum. During January and February, the low temperatures and relatively heavy snowpack exclude almost all groups except the foliage-seed, timber-searching, and timber-drilling groups. Throughout the remainder of the year ground foraging is of major importance and is about equally divided between insects and seeds. The foliage-nectar group is essentially absent, except during migration, reflecting the low density of nectar-producing flowering plants.

The Yosemite fauna shows important seasonal change but general dominance of insect-eating forms, especially the foliage-insect group. Many of these forms actually switch to berry-feeding during the colder months. The foliage-nectar group is represented during the summer months, as are the aerial feeders. In contrast to Boca Spring, ground feeders are active year-round, presumably because of the lighter snow cover at Yosemite.

The Glen Oaks Canyon spectrum shows the least seasonal change in ecological composition. The only major groups of limited seasonal occurrence are the aerial feeders. The timber-searching category is represented only in summer because of the shift in feeding activity of the species involved into the foliage-seed category during the winter. The largely evergreen nature of the vegetation and the absence of freezing conditions during the winter permit year-round occurrence of the foliage and ground-feeding groups, with little change in relative abundance.

The above examples of life form analyses show that a variety of morphological, physiological, and behavioral adaptations to the physical environment play a role in the sorting of species for inclusion in communities. They further emphasize the variety of ways in which organisms may successfully counter challenges of the environment, as well as the unsatisfactory nature of systems of analysis, such as the Raunkiaer life form system, based on only a few characteristics of the species involved.

Challenges of the Biotic Environment

In addition to successfully meeting challenges of the physical environment, potential community members face a variety of biotic challenges. Presence of other species having certain ecological characteristics may be necessary for their survival. Other species may act as competitors, predators, or parasites, or they may exert antagonistic effects in other ways.

Some of these interactions have been discussed adequately in earlier chapters. Here we will consider two main topics: (1) the extent to which resource relationships govern community composition, and (2) the types of species-specific interactions that are important in community organization.

Resource Partitioning. Earlier, we considered the problem of competition between ecologically similar species. This was discussed in terms of the mechanisms of ecological isolation between coexisting species. At the level of the community this may be viewed more broadly as the manner in which resources are divided, or partitioned, among community members. The question that may be raised here concerns whether or not a requirement exists for clear-cut differentiation of resource use among the species comprising major trophic groups within communities.

Ecologists have only begun to evaluate the extent and importance of resource partitioning among the member species of communities. Cody (1968) has examined this relationship for a series of grassland bird communities in both North and South America. These communities were selected for study because of their simplicity of vegetational structure and because the number of bird species present was small (from two to four passerine birds).

Cody's studies were concerned with the mechanisms of resource partitioning within grassland areas. For the bird species present at the same season, three major mechanisms seem possible. Different species may show:

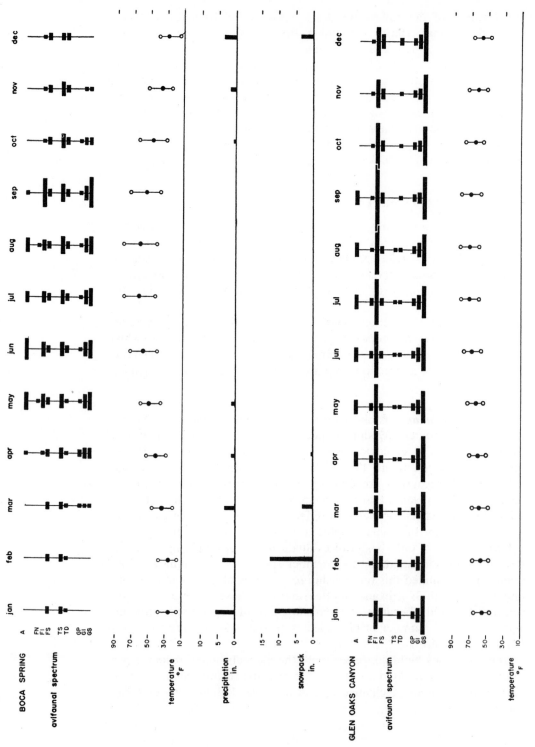

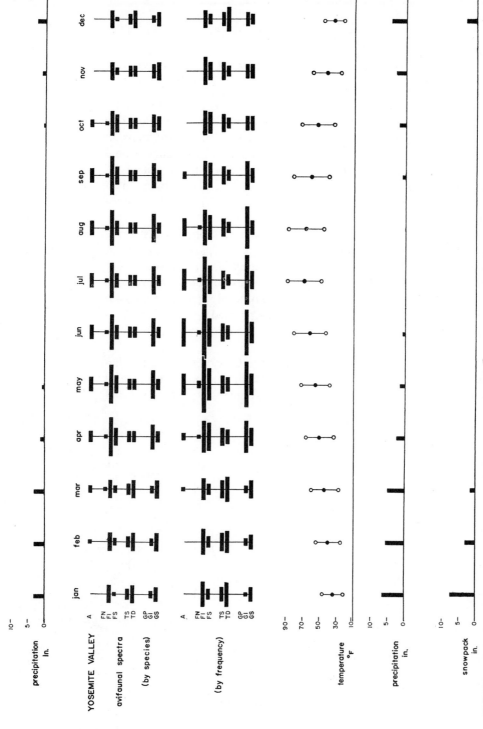

FIGURE 8–6. Avifaunal spectra and environmental conditions for Boca Spring, Yosemite Valley, and Glen Oaks Canyon. (From Salt, 1953.)

1. Direct specialization in morphology and behavior for different food materials.
2. Vertical separation through occupation of different strata within the vegetational profile.
3. Horizontal separation by occupation of habitat patches of differing characteristics.

If we consider two particular species we can see that, conceivably, adequate differentiation may be achieved by a complete difference in any one of these ways, or by a combination of differences in two, or perhaps all three, ways. This fact may be illustrated by using a three-dimensional graph in which each axis corresponds to one of the possible mechanisms of resource partitioning (Fig. 8–7). In this model any combination of differences adding up to a constant total (e.g., 100%) would fall on a plane, designated ABC in the illustration.

This model raises several questions that Cody attempted to answer by collecting data on grassland bird communities. For a particular group of species occupying an area of grassland, is differentiation of resource use achieved primarily through the use of one of these mechanisms or by a combination of two or all three? What degree of total separation tends to occur among species? How are the mechanisms used related to characteristics of the habitat? Does the degree of separation among species tend to be constant?

This last question was of particular interest to Cody. He reasoned that if considerations of resource partitioning were of major importance in determining which species could coexist in a community, the degree of

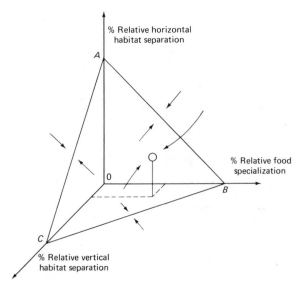

FIGURE 8–7. Graphical representation of the three mechanisms of resource partitioning examined by Cody (1968) in his analysis of grassland bird communities. Any pair of species, or a community average of such comparisons, can be represented as a point in these three dimensions. The plane ABC is the locus of pairs of species, or communities, the sum of whose ecological differences is constant. Arrows show the probable direction of selective forces.

difference between species should tend to be constant. In other words, for a given community, a quantitative index of differentiation should be similar for different pair combinations of the species present. For communities of similar type, in different geographical areas, the average degree of difference between species should also be similar. In such cases, points representing indices of differentiation between species should fall on a plane similar to ABC in Fig. 8–7. A point falling beneath this plane would indicate a pair of species having very low differentiation. In these cases it might be expected that either one member of the pair would be eliminated by competition or that evolutionary divergence would occur as a result of selection for increased difference in patterns of resource utilization. Conversely, a point falling above this plane would indicate a species pair having differences greater than necessary. In this situation the opportunity for invasion of the community by a new species would seem to exist.

To answer the questions posed earlier, Cody collected quantitative data on the three mechanisms of resource partitioning by songbird species in six grassland areas in the United States and Canada and from comparable areas in Chile. To obtain a measure of degree of feeding specialization, data were obtained for two aspects of beak morphology and three patterns of foraging behavior. By a rather involved series of calculations these data were combined to give a measure of the average percentage of food specialization by community members. In a similar manner observations on the height of foraging activities in the vegetational profile and on the occurrence of species at individual points within the study areas were analyzed to provide measures of the average percentages of vertical and horizontal habitat separation, respectively.

Data on the degree of separation by these three mechanisms are summarized in Table 8–7. These data suggest that the sum of the three ecological differences represented by the axes of Fig. 8–7 does tend to be constant. For the ten areas examined, the calculated sums lie closely around a total of 135.3%. Although in theory a 100% difference in any one of these characteristics should provide adequate separation, it appears that in practice differentiation is achieved by a combination of several differences totaling about one third more. These values can be plotted as in the original graphical model (Fig. 8–8), in which it is apparent that they tend to lie on or near a plane ABC having an intercept of 135.3 on each of the axes. It can also be seen in this diagram that the mechanisms most used by grassland birds involve food specialization, and those least used correspond to vertical and horizontal separation in space.

From these results Cody was able to design a single model combining the prediction of mechanisms of resource partitioning and the census characteristics of the bird communities involved (Fig. 8–9). In this model bird species are represented as a series of cells arranged in a central arc. The three major mechanisms of resource partitioning are represented as

TABLE 8–7. MEAN ECOLOGICAL DIFFERENCE BETWEEN SPECIES OF GRASSLAND BIRDS IN TEN COMMUNITIES STUDIED BY CODY (1968).

Study area	Horizontal habitat separation	Vertical habitat separation	Relative food speciali- zation	Sum of ecological differences	Deviation of sum from mean sum
North America					
Saskatchewan	36.5	7.7	83.0	127.2	−8.1
Minnesota	93.3	12.7	48.0	154.0	+18.7
Colorado	21.2	14.3	85.2	120.7	−14.6
Kansas	36.5	22.0	81.7	140.2	+4.19
Maryland (*Spartina*)	25.2	40.0	74.5	139.7	+4.4
Maryland (*Juncus*)	9.5	92.0	33.5	135.0	−0.3
Chile					
Phalaris field	37.5	16.7	79.6	133.8	−1.5
Irrigated field	40.0	11.1	79.2	130.3	−5.0
Cerro castillo	27.8	27.7	71.5	127.0	−8.3
Punta delgado	28.5	41.0	75.9	145.4	+10.1
Mean values	35.6	28.5	71.2	135.3	

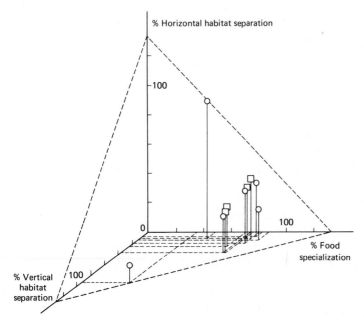

FIGURE 8–8. Three-dimensional graph of average ecological differences between species pair combinations for ten grassland bird communities studied by Cody (1968). The points lie approximately on a plane whose intercepts on the three axes do not differ significantly. The axes correspond to those in Fig. 8–6. Squares designate Chilean study areas; circles, North American.

FIGURE 8–9. The relation of methods of resource division in grassland bird communities studied by Cody (1968) to grass height. The three coordinates, H, V, and F, correspond to degree of horizontal habitat separation, vertical habitat separation, and food specialization, respectively. Adjacent species, represented by the sections (S_1, S_2, etc.) of the curved central unit, can be projected perpendicularly onto the three axes, and the relative overlap between species in the three methods of resource division predicted (as illustrated for S_{10}, S_{11}, and S_{12}). The H axis is scaled in terms of vegetation height in feet. A line (corresponding to vegetation height, h', at a specific location in a grassland area) or a band (corresponding to the range of heights with a mean of h' existing in a census plot) may then be projected from the H axis through the species curve to predict the species occuring in point and plot censuses, respectively.

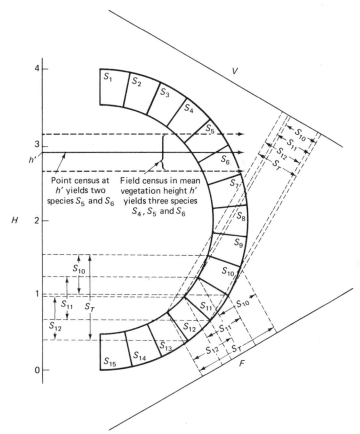

coordinate axes on three sides of the species arc. Projections of the species cells perpendicularly onto these axes, as illustrated in the diagram, indicate the degree of expected overlap in mechanisms of resource partitioning.

The number of species occurring at a given point or plot in a grassland of particular characteristics also may be determined on the basis of the vegetation height or range of heights. To obtain this prediction, a line (corresponding to a single habitat point) or a band (corresponding to an area of habitat having a certain range of vegetation height) is projected perpendicularly from the vegetation height axis through the species-cell diagram. The cells through which this projection passes indicate the species expected to occur in nature.

The degree of correspondence between predictions from this model and the actual number of species occurring at points within the study areas examined by Cody is indicated in Fig. 8–10. This correspondence is high. It may be noted that the prediction of low species numbers in

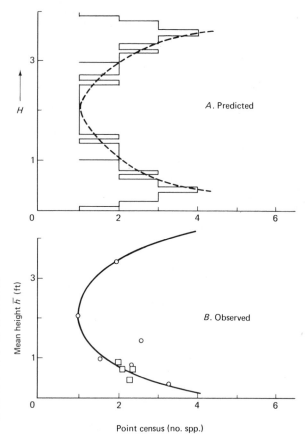

FIGURE 8–10. Comparison of predicted and observed numbers of species in grassland bird censuses conducted by Cody (1968). Curve *A* shows the relation between vegetation height *H*, and predicted number of species in point censuses, as given by the model in Fig. 8–8. Curve *B* shows the observed relationship for the ten grassland bird communities studied. Squares designate Chilean study areas, circles North American. The symbol *h* refers to the mean height of vegetation in the various census areas.

grasslands of intermediate height is related to the arc-like arrangement of the species cells in the model (Fig. 8–9). This feature was incorporated into the model as a result of observations suggesting that bird species could employ different feeding patterns in low vegetation and could occupy different strata in tall vegetation, but they could not employ either of these mechanisms in vegetation of intermediate height.

Although this model deals with a very simple community and incorporates only a limited number of characteristics of the species and habitat, its success indicates that mechanisms of resource partitioning play an important role in the organization of communities. It further suggests that, at least for species with relatively stable population sizes, such as vertebrates, the total degree of differentiation among the various species present in a given community tends to be a constant.

The role of resource partitioning in determination of the structure of plant communities has been examined by Whittaker (1965), who used quite a different approach. This analysis is based on extensive data on variation of community composition along environmental gradients in the Great Smoky Mountains. Whittaker has also been able to obtain estimates of net annual production by member species for many of these communities.

These data permit the examination of patterns of abundance and productivity of the various species along major gradients of moisture and altitude. For example, smoothed curves showing the relative abundance of each of the major species in habitats ranging from mesic to xeric may be prepared (Fig. 8–11). Curves of abundance vary considerably in form and show broad overlap for the 28 species shown in this figure. Peaks of these abundance curves, however, tend to be scattered along the moisture gradient rather than organized into groups of species having strongly coinciding peaks. This separation is increased if altitudinal preference is also included. Almost all species having similar peaks along the moisture gradient are distinctly different in their distribution with altitude. These observations suggest that evolutionary processes have favored divergence of the ecological optima of the various species as a mechanism for reduction of competition for resources. Thus, each species is characterized by a set of environmental conditions in which its ability to exploit and utilize resources is superior. For the species shown in Fig. 8–11, it further appears that the basic mechanism of resource partitioning is comparable to that of occupation of habitat patches differing in environmental characteristics – one of the major mechanisms recognized by Cody (1968).

The data obtained by Whittaker (1965) may also be examined from the standpoint of the relationship between diversity of species and their productivity within individual communities. This relationship may be shown graphically by plotting the net annual production against the rank of the species in the sequence from most to least productive (Fig. 8–12). The examples shown in Fig. 8–12 indicate that communities in the Great Smoky Mountains vary widely both in number of species and form of the productivity-diversity curve. Several patterns can be seen in these curves, however. The three cove forests (numbers 15, 18, and 23 in Fig. 8–12) are rich in species (40, 43, and 39 respectively) and show S-shaped curves which can be closely fitted by a single line. In contrast, the Fraser Fir forest (number 33), with only 7 species, exhibits a productivity-diversity relationship fitted by a straight line. One of the questions intriguing many ecologists is whether or not such diversity patterns can be explained on the basis of basic ecological relationships among the various species.

Various workers have suggested that particular mathematical series provide satisfactory fits for data on the abundance or importance of species when such data are ordered in a manner similar to the productivity-diversity curves of Whittaker (Fisher et al., 1943; Preston, 1948). The weakness of almost all of these analyses is two-fold: First, almost all such models do not adequately fit data from communities varying in

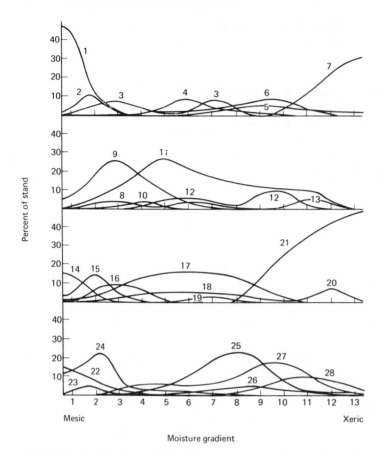

Percent of stand

1   2   3   4   5   6   7   8   9   10   11   12   13

Mesic   Xeric

Moisture gradient

FIGURE 8–11. Plant populations along an environmental gradient. The gradient is the topographic moisture gradient from mesic (moist) ravines (at left) to xeric (dry) southwest-facing slopes (at right), between elevations of 460 and 760 m in the Great Smoky Mountains. Populations of major tree species are plotted by percentages of the total numbers of tree stems over 1 cm in diameter 1.4 m above the ground. All the species illustrated are part of the same vegetation gradient, but they are separated into four panels for the sake of clarity. Although, with 28 species and 13 steps of the gradient, some species must have their modes in the same step, the modes of species populations appear to be scattered along the gradient. Pairs of species having their modes in the same step of the moisture gradient may be shown to be differently distributed in relation to the elevation gradient. Some species are bimodal, with two ecotypes having different population centers. Plant communities intergrade continuously from cove forests (transect steps 1–4), through oak forests (steps 6–8), to pine forests (steps 10–13). The species are as follows: 1. *Halesia monticola;* 2. *Acer saccharum;* 3. *Hamamelis virginiana;* 4. *Carya tomentosa;* 5. *Nyssa sylvatica;* 6. *Pinus strobus;* 7. *P. rigida;* 8. *Querus borealis;* 9. *Tsuga canadensis;* 10. *Fagus grandifolia;* 11. *Acer rubrum;* 12. *Qu. alba;* 13. *P. echinata;* 14. *Aesculus octandra;* 15. *Betula allegheniensis;* 16. *B. lenta;* 17. *Cornus florida;* 18. *Carya glabra;* 19. *C. ovalis;* 20. *Qu. marilandica;* 21. *P. virginiana;* 22. *Tilia heterophylla;* 23. *Cladrastis lutea;* 24. *Liriodendron tulipifera;* 25. *Qu. prinus;* 26. *Qu. velutina;* 27. *Oxydendrum arboreum;* 28. *Qu. coccinea.* (From Whittaker, 1965.)

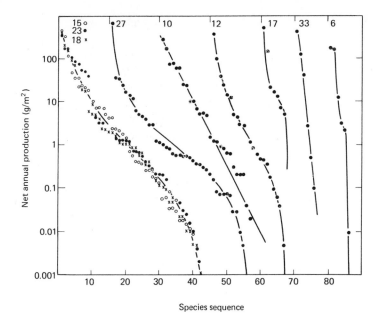

FIGURE 8–12. Dominance-diversity curves for vascular plant communities in the Great Smoky Mountains. Points represent species, plotted by net annual above-ground production (on the ordinate) against the species' number in the sequence of species from most to least productive (on the abscissa). In each curve the highest point represents the most productive species (species number 1 in the sequence) and the lowest point the least productive species. For the sake of graphic clarity, however, the curves have been arbitrarily spaced out, their origins being separated by 10 or 15 units along the abscissa. Positions of their origins on the abscissa are indicated by the vertical ticks along the top border of the figure. (From Whittaker, 1965.)

structure to the extent shown in Fig. 8–12. Second, the ecological meaning of these models is usually obscure.

More recently, attempts have been made to derive mathematical models of community structure from specific assumptions or relationships among community members. MacArthur (1957) has developed a series of models based on assumptions about the nature and use of environmental resources by community members. The three models presented differ in specific assumptions concerning the discrete or continuous nature of environmental resources and the occurrence or absence of overlap in resources utilized by different species. All of these models, however, assumed that allocation of resources among the species present is a random process. Almost all data on composition of plant and animal communities are poorly fitted by any of these models (Whittaker, 1965). Furthermore, models that give the same predictions may be developed on the basis of quite different assumptions (Cohen, 1968). Agreement of observed community composition with predictions of any of these models thus has very little meaning.

Whittaker (1965), without proposing formal mathematical models, has suggested how the various productivity-diversity curves (Fig. 8–12) seen in the plant communities of the Great Smoky Mountains may arise from patterns of resource partitioning by community members. To illustrate these, we may assume available resources to be represented, diagramatically, as a square that may be partitioned in various ways among member species of the community (Fig. 8–13). Here it is assumed that sizes of rectangles resulting from a partitioning procedure correspond to the productivity of the species in the community. Thus, productivity-diversity curves corresponding to different patterns of partitioning the "resource square" can be derived (Fig. 8–13).

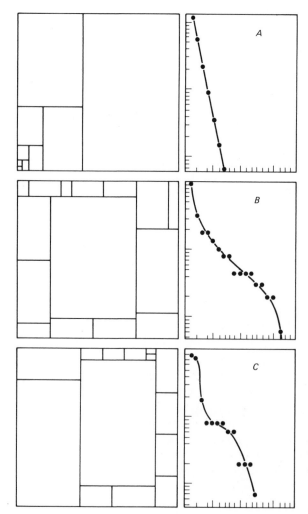

FIGURE 8–13. Models for species and resource relations which may underlie dominance-diversity curves. The squares in each case represent resource space which is divided among the species of the community, represented by rectangles. Sizes of the rectangles for species represent their share of environmental resources, as expressed in their population density, productivity, or other "importance" measurement. In the curves to the right of each model, species are plotted (on the ordinate) on a logarithmic scale by areas of their rectangles against species number in the sequence of species from most to least important (on the abscissa). (From Whittaker, 1965.)

Perhaps the simplest technique of partitioning is the assumption that in a community one species is best adapted to existing conditions and is able to exploit some fraction, k, of total resources available. The second best-adapted species exploits fraction k or the remaining resources, and so on. This pattern of partitioning corresponds to a simple geometric series. When production is plotted on a $\log_{10}$ scale, the productivity-diversity curve produced by this pattern is a straight line, (Fig. 8–13). If random variation of k values around some mean is assumed, this pattern of resource allocation corresponds to that apparently occurring in many communities. Curves of this kind are often seen in communities of rigorous environments and low species diversity. In the Great Smoky Mountains the high altitude Fraser Fir forest (Fig. 8–12, number 33) shows such a curve. In the same area, however, a low elevation pine forest (Fig. 8–12, number 10), with a greater number of species, also showed this basic pattern.

If, however, it is assumed that a greater number of species is available and that no single species has an overwhelming competitive advantage, a somewhat different pattern may emerge (Fig. 8–13, section B). For this model it is assumed that some relationship sets an upper limit to the extent of resource exploitation; in the graphical model this is a restriction that the length-width ratio of the partitioned areas must be between 1.5 and 2.0. In this model the best-adapted species may be assumed to occupy the central portion of the resource square, and the remaining species successively occupying the largest possible areas remaining.

The curve produced by such a division, when plotted as before, will be sigmoid in shape (Fig. 8–13, section B). The steep upper portion of this curve may be taken to correspond to the most important community dominants, the flatter central portion to the subordinate, or understory, species, and the lower, steep portion to the rare species restricted to specialized microhabitats within the community environment. This pattern corresponds to that shown by the richer cove forests in the Great Smoky Mountains (Fig. 8–12, numbers 15, 18, 23).

Other variations may be derived assuming, for example, the presence of two species of nearly equal competitive ability (Fig. 8–13, section C), with division of remaining resources occurring as in the previous case. The pattern produced here is similar to that found by Whittaker in one of the heath communities (Fig. 8–12, number 6).

The general conclusion suggested by this analysis is that the variety and productivity of species in a community are not governed rigidly by a simple relationship. Rather, the nature of the environment, the characteristics of the species, especially the dominants, and the evolutionary time available for their mutual adjustment may determine the conditions for division of environmental resources.

Species-Specific Interrelationships. Community structure is influenced by interspecific relationships which, because of their evolutionary history, tend to involve two or more particular species. These interactions may be termed *species-specific*. However, they may vary greatly in their degree of specificity. For example, a certain interaction may occur only between species x and species y, thus being highly specific. In another case an interaction may occur more frequently between species a and species b than between a and c, a third species similar to b and equally available to a.

These interactions may also vary from being obligatory to being facultative. In an obligatory interaction the species involved cannot exist in the absence of each other. In a facultative interaction the species may exist alone. Finally, we may note that such relationships may be positive or negative, favoring the co-occurrence of the species on one hand, or their nonoccurrence together on the other.

The significance of these interactions to community organization results from the fact that the presence of one member of such a group determines, to some degree, the probability of others also being present. The greater the number of specific and obligatory interactions involving potential community members, the greater is the degree of determination of community structure that results.

Species interactions have been classified in various ways (Burkholder, 1952; Odum, 1959). The simplest grouping, however, may be to recognize interactions between members of the same trophic level and those between members of different trophic levels.

Interactions between Members of the Same Trophic Level. A large number of studies, particularly over the past 20 years, have demonstrated that many plant species produce chemical substances that are released into the environment in various ways and act to inhibit germination and growth of other plants (Whittaker, 1970a). This phenomenon, termed *allelopathy,* may be considered to be a form of competition by interference (see Chap. 6). Only a small fraction of the chemical substances involved have been identified. These include a large variety of phenolic compounds (having one or more benzene rings in their chemical structure), terpenoid compounds (complex organic compounds based on 5-carbon, branched-chain hydrocarbon units), alkaloids, and a variety of other substances. These substances are released from the producer plants in various ways. Water soluble substances are apparently carried from leaves to the soil by rainwash. Examples are the chemical inhibitors produced by Chamise, *Adenostoma fasciculatum,* and various manzanitas, *Arctostaphylos* spp., of the California hard chaparral. Other shrub species characteristic of the California coastal sage scrub, including California sagebrush, *Artemesia californica,* and various true sages, *Salvia* spp., release volatile terpenes which are absorbed by soil particles and subsequently inhibit germination and growth of other plants (Muller et al., 1968). In still other cases chemical substances may be released from the root systems of living plants, from seeds and fruits, or from dead plant tissues (Whittaker, 1970a).

McPherson and Muller (1969) have examined the mechanism of allelopathy in Chamise, *Adenostoma fasciculatum,* and have discussed its significance in determining the composition of the plant communities of which it is a part. Chamise is the most widespread and abundant constituent of hard chaparral, a vegetation type consisting of evergreen, small-leaved shrubs. This type occurs in Mediterranean climates (summer-dry) along the Pacific Coasts of North America and Chile, portions of southern Australia and South Africa, and the Mediterranean area of Europe. Chamise is highly adapted to recovering from grassland fires and possesses an enlarged root crown that lies just beneath the soil surface. Following cutting or burning of the above-ground portions of the shrub, this root crown quickly gives rise to a thick cluster of new shoots. Chamise also germinates abundantly following fire and in areas of frequent brushfires it often forms nearly pure stands.

One of the features of Chamise stands studied by McPherson and Muller (1969) was the virtual absence of herbaceous undergrowth. They also observed that light intensities and soil moisture levels in these stands were not much lower than in nearby areas where abundant herb growth occurred. It was further noted that following either fire or removal of Chamise by cutting abundant germination and growth of a large variety of herbaceous species occurred in response to the first autumn rains. The immediacy of this response demonstrated that these plants were germinating from seeds already present in the soil of the Chamise stand. Thus, it was apparent that some mechanism directly associated with the presence of Chamise was preventing the development of herbaceous vegetation beneath the canopy.

Several possible mechanisms were investigated. For example, the abundant growth of herbaceous plants following chaparral fire has often been attributed to the fertilizing effects of ash from the burned shrubs. Experiments involving application of commercial fertilizers or ash derived from burned Chamise crowns to the ground beneath a Chamise canopy, however, produced no increased growth of herbaceous species.

The possibility that various small animals, active in the shelter of the shrubs, might destroy the young herbaceous plants was also examined by means of exclosures established within stands. Bartholomew (1970) has shown that such activities can be of major importance in the creation of bare zones surrounding chaparral shrubs. Results of experiments conducted by McPherson and Muller (1969) showed a five-to-sevenfold increase in number of herb seedlings within the exclosures. However, in clearings resulting from cutting or burning, herbaceous seedlings developed in numbers 15–50 times greater than in the exclosures. Furthermore, even within the exclosures, the germinating seedlings were stunted and, generally, failed to mature. Thus, it was apparent that although some small animal effects occurred, these also were inadequate to account for the major inhibition of the herbaceous vegetation.

Still other suggestions were examined. The abundant germination of herbs and shrubs following fire has been attributed to stimulation of seed germination by heat, but experiments in which shrub and herb seeds were oven-heated showed no significant increase in germination. When soil samples taken from beneath Chamise stands were similarly heated, increased seedling germination did occur. This last observation suggests that heating, rather than affecting the seeds themselves, affects soil conditions in some manner favorable to seed germination.

Finally, McPherson and Muller (1969) tested the possibility that a chemical inhibitor is produced in the Chamise crown. Through a careful series of experiments they were able to show that some water soluble material is released onto the surface of the leaves as a result of the metabolism of the leaf tissues. This material accumulates during dry periods, especially the period from May–September when little or no rain falls in coastal southern California. When rain does fall, this substance is carried to the soil where it inhibits the germination and growth of many species of plants, including those of Chamise itself.

Since similar processes involving different chemical substances and modes of release have been documented for other shrub members of the chaparral community (Muller et al., 1968), it thus appears that the vegetation cycle following chaparral fire is largely controlled by allelopathic processes. Chaparral fire removes the source of many chemical inhibitors and very likely destroys some of those that have accumulated in the soil. Thus, germination of large numbers of shrubs and herbs occurs, as well as extensive resprouting of shrubs which possess protected root crowns specialized for fire recovery. The herbaceous species appearing during the first season are primarily native species, apparently highly adapted to the fire cycle by virtue of having seeds capable of lying dormant beneath chaparral for many years.

In subsequent growing seasons, before an extensive shrub canopy redevelops, other herbaceous plants also appear, the result of seed dispersal into the area after the fire. Gradually, as the shrubs regenerate or grow, allelopathic mechanisms assume greater importance, and the herbaceous vegetation declines in abundance.

Composition of these communities, in terms of the balance of herbaceous and shrub components, is thus highly controlled by certain shrub species. In addition, allelopathy may further favor the development of single-species stands, if inhibition extends to shrub species and is greater for other species than for the producer species.

Whittaker (1970a) has suggested that allelopathic processes are virtually universal in occurrence in natural communities and that such mechanisms may be highly important in controlling community development during such sequences as succession and fire recovery. Further, such interactions may be the basis of interspecific associations among various plant species in mature communities. Since allelopathic effects have been observed on soil fungi and bacteria, it is also likely that allelopathic processes may influence basic patterns of nutrient cycling.

Interactions between Members of Different Trophic Groups. Species-specific relationships between members of different trophic levels are probably derived, evolutionarily, from situations in which one member initially exploited the other as a required resource, for example, as food. Many interactions of this sort can be cited: pollination and seed dispersal relationships of plants and insects, mutualistic relationships between algae and organisms such as fungi (in lichens), flatworms, and corals, and various forms of parasitism. We will examine in detail two general examples involving herbivore-plant interactions.

It has long been observed that many species of butterflies with brightly-colored wings appear to be distasteful or unattractive to potential predators. Furthermore, some of these, such as the Monarch Butterfly, *Danaus plexippus,* are apparently mimicked by other species. Much theory, based largely upon the assumption of unpalatability of certain groups of butterflies, has developed regarding the evolution of warning coloration and mimicry.

This area has recently been investigated experimentally (Brower and Brower, 1964) and many of the inferences of palatability and unpalatability of butterfly species demonstrated by experiments with specific vertebrate predators. It has also become apparent that these distasteful species are concentrated in several subfamilies and tribes of butterflies (Brower and Brower, 1964), which, in turn, tend to possess highly specialized food habits and preferences for plant hosts rich in toxic chemical substances (Ehrlich and Raven, 1967). These toxic chemicals include the terpenoids, alkaloids, and others, many of which belong to the same chemical groups noted in the discussion of allelopathy.

These observations have suggested several theories. First, the functional significance of many of the exotic and toxic chemicals produced by plants may be protection against herbivore grazing. For example, members of the coffee family, a very large and diverse family of tropical flowering plants, are rarely fed upon by butterfly larvae. These plants are also rich in alkaloid substances, such as caffeine and quinine.

Second, it is suggested that the specialization of certain butterfly groups for specific groups of food plants is due to their successful counter-response to the chemical defenses of the plant. The butterfly family Pieridae, the whites and yellows, shows a strong preference for plants of the caper and mustard families, many of which possess mustard oils.

These substances, which very likely discourage grazing by certain herbivorous insects, act as stimulants to feeding action of these butterflies (Ehrlich and Raven, 1964).

Third, it has been suggested that the distastefulness of certain butterfly species is a result of their utilization of plant poisons as predator deterrents (Brower and Brower, 1964). In a number of cases the presence of chemically similar poisons on both herbivores and their primary food plants has been demonstrated. More interestingly, Brower et al. (1967) have examined the mechanism of unpalatability in the Monarch Butterfly, *Danaus plexippus*. The Monarch feeds almost exclusively on milkweed plants of the genus *Asclepias* which contain chemical substances that are strong cardiac poisons in vertebrates. In the plant tissues these substances are usually present in chemical combinations with carbohydrate units, forming a complex termed a *cardiac glycoside*. The combination of a toxic chemical with such a glycoside is a common mechanism for storage of the toxin in inactive form. A number of herbivorous insects of various orders are specialized milkweed-feeders. In some of these the presence of cardiac glycosides similar to those occurring in their milkweed food has been demonstrated (Brower and Brower, 1964). Brower et al. (1967), working with the Monarch Butterfly, successfully reared adults on cabbage. When these individuals were offered experimentally to predators, they were found to be perfectly palatable. Thus, evidence exists for the conclusion that unpalatability in herbivorous insects is based on the utilization of toxic or unpleasant chemical substances derived from their plant food.

It should not be concluded from this example, however, that the specialized milkweed-feeding herbivores have made no adjustments to their food plants other than to the tolerance and utilization of the chemical toxin. The fact that these substances continue to be produced by the milkweed suggests that (1) they do provide a useful function in reducing herbivore grazing, (2) very likely the action of milkweed herbivore specialists is less severe than that of others, and (3) these milkweed specialists provide some other valuable service for the plant.

As a second example, we may examine the relationship between various ant species and their plant hosts, termed *myrmecophytes,* in tropical America. Similar situations occur in other tropical areas. In tropical America these relationships involve trees of the genera *Acacia* and *Cecropia* especially. These trees are characteristic of disturbed, early successional communities. The symbiotic ants are members of the genera *Pseudomyrmex,* for the Acacias, and *Azteca,* for the Cecropia (Jantzen, 1966, 1967, 1969).

The Acacia and Cecropia hosts provide both living sites and food for the symbiotic ant species. The Acacias possess greatly enlarged stipular thorns that are filled with soft parenchyma tissue when they first develop. The *Pseudomyrmex* ants cut an opening in the wall of these thorns and excavate the soft tissues, creating a hollow cavity up to 7 cm³ in size. Food is provided by enlarged nectaries located on the petioles of Acacia leaves and by the production of protein-rich structures, known as *Beltian bodies,* at the leaflet tips. These form the basic food for the ants. *Azteca* ants occupy hollow internodal portions of the young branches of the Cecropias. The ants gain access to these cavities by cutting through regularly located thin portions of the branch wall which represent an apparent adaptation of the plant to allow entry by ants. The Cecropia provides a specialized food source consisting of small structures rich in lipids, proteins, and carbohydrates, termed *Mullerian bodies,* which grow from the petiole base. Both the Acacias and Cecropias, although occurring commonly in woodlands which are largely deciduous, retain their leaves year-round, thus assuring a continuous food supply for the ants.

The ants, likewise, provide valuable services to their hosts. The *Pseudomyrmex* ants occupying swollen-thorn Acacias attack other insects, and in most cases they drive them off the Acacia foliage. They also attack and kill growing tips of other plants that either enter the Acacia canopy and touch the Acacia foliage or spring up in an area from 10 to 150 cm around the base of the Acacia tree. This may act as a "firebreak," protecting the Acacias from ground fires. This action by ants is essential to survival of the Acacia in most situations. Without the ant colonies the Acacia is heavily damaged by insect herbivores and usually dies within 6–12 months. For the Cecropia, the action of *Azteca* ants is largely in attacking and killing vines, branches of other tree species, and epiphytes that tend to invade the Cecropia canopy. When Cecropias grow in the absence of *Azteca* ants or when colonies fail to become established, the incidence of such foreign species within the Cecropia canopy is high. For a species of early successional status, as the Cecropia is, this action would be effective in prolonging the reproductive life of the tree.

Thus, in the case of the Acacias, although not the Cecropias, the interaction is obligatory for both the ants and their plant host. It is also interesting that a specialized fauna of herbivorous insects, including at least nine species, is restricted to the swollen-thorn Acacias. These species have apparently evolved some adjustment to prevent their being excluded by the *Pseudomyrmex* ants.

Coevolution. Any species interaction, such as competition or feeding exploitation, leads to selection for particular characteristics in the participating species, as we have seen. This selection tends to produce a relationship of maximal adjustment of the interacting species to each other. This evolutionary adjustment has been variously referred to as *coevolution* (Ehrlich and Raven 1967), *genetic feedback* (Pimental, 1968), and *counter-adaptation* (Ricklefs and Cox, 1972).

In the long run, these evolutionary adjustments should tend to mitigate the detrimental aspects of the relationship and enhance the beneficial aspects, and an increase in the specificity and obligatory nature of the interaction will likely result.

Thus, beginning with a plant-herbivore feeding interaction we may expect evolutionary changes to lead to the development of chemical or morphological defenses of the plant to grazing action. At the same time, specialization of the herbivore to surmount these defenses may be expected. Ultimately, both species may evolve close specialization to each other, so that in addition to the food energy movement from the plant to the herbivore, the herbivore provides some valuable service, e.g., protection, pollination, or dispersal to the plant.

The frequency of these relationships in communities will thus to some degree be dependent on the environmental stability, through geological time, of the region in which such interactions begin to evolve. It is virtually axiomatic that the frequency of highly evolved interspecific relationships is greatest in the tropics. However, although many of these interactions were described by the early naturalists, interest in their detailed examination in terms of adaptive significance and evolution has not appeared until recently. Consequently, considerable difference of opinion exists, especially among ecologists having experience limited to temperate-zone areas, on the importance of these relationships in community structure. Evaluation of the frequency, functional importance, and influence on community structure of such interactions will constitute an active area of ecology in the future.

Species Diversity

The subject of species diversity has proved to be both one of the most attractive and difficult topics dealt with by modern ecology. The concept of species diversity relates simply to the "richness" of a community or geographical area in species. At the simplest level of examination, species diversity corresponds to the number of species present. Since, however, the relative abundance of the species present is of major importance to the functional characteristics of the community, many ecologists have preferred to use measures of diversity that also incorporate a measure of the equitablility of abundance or importance of the various species.

This characteristic may be illustrated by considering two communities possessing the same number of species but having differing patterns of relative abundance. Let us assume that each community possesses five species differing in their ecological requirements and activities. If in one of these the five species were of equal abundance, the pattern of community appearance and function would be quite complex. If in second community 96% of the individuals were of one of the species and only

4% of the remaining four species, the pattern and function would very likely appear quite different.

Two indices of diversity that are sensitive to both aspects of community structure have been widely used by ecologists. Both indices increase in value either as the number of species increases or as the distribution of individuals among species increases in equitability. One, developed by Simpson (1949), is calculated by the formula:

$$\text{Diversity index } \frac{N\,(N-1)}{\sum \left[\,(n_i)\,(n_i-1)\,\right]}$$

where

N = Total number of individuals of all species

n_i = Number of individuals of the ith species

The value given by this calculation expresses the number of samples of random pairs of individuals which must be drawn from a population to have at least a 50% chance of obtaining a pair with both individuals of the same species.

The second index, more widely used but more difficult to calculate, is derived from information theory. This index is calculated by the formula:

$$\text{Diversity index } (H) = -\sum p_i \log_2 p_i$$

where

p_i = Decimal fraction of individuals of the ith species

This index, symbolized by H or termed the *Shannon-Wiener index,* expresses the degree of uncertainty of predicting the species to which a given individual, drawn at random from the community, will belong. In the calculation of this index, the minus sign is present simply to produce a numerical index having a positive value. A fuller discussion of these indices, together with a table to facilitate calculation of the latter index, is given by Cox (1967).

Interest in species diversity has centered around attempts to account for differences in species diversity both on a broad geographical scale and at the local level. In some cases the interest has centered on the biotic richness of different, major geographical areas. Since many of the differences at this level are accountable for by differences in the degree of physiographic and climatic differentiation, and thus in the variety of different community types represented, these approaches are of little interest to ecologists concerned with the structure of individual communities.

We will thus limit our consideration to the problems of the factors determining the diversity shown by specific communities within which the species involved possess potential interactions. For such cases it may be noted that some portion of the causal explanation for diversity patterns may be evolutionary in nature. That is, it may primarily involve questions of length or continuity of the evolutionary history of the environment and group of species in question. These problems will be dealt with more fully in a later chapter. Now we will concern ourselves with the possible mechanisms that maintain a given level of species diversity in communities or which operate on the time-scale of biotic succession to produce changes in species diversity.

A variety of different mechanisms, most of which are not mutually exclusive, have been proposed to account for variations in species diversity (Pianka, 1966, 1967). Those relevant to our discussion include:

1. The ecological time hypothesis.
2. The environmental stability hypothesis.
3. The spatial heterogeneity hypothesis.
4. The productivity hypothesis.
5. The predation hypothesis.

We will briefly consider the ideas involved in each of these hypotheses and then examine a series of studies that have attempted to evaluate them.

The ecological time hypothesis relates to the dispersal of species into new areas of suitable habitat. This hypothesis simply suggests that low species diversity in certain situations may be the result of failure of species to cross particular barriers by dispersal and become established in favorable habitat situations. This mechanism would be assumed to be of greatest importance for communities occupying habitats surrounded by sharp or extensive barriers, such as island or high mountain areas.

The environmental stability hypothesis assumes that relative constancy or predictability of favorable conditions (for the group of organisms in question) of the physical environment increases species diversity by guaranteeing the availability of supplies of critical resources. More species are able to occur since species with relatively specialized resource or habitat requirements are not eliminated by severe or erratic environmental conditions.

The spatial heterogeneity hypothesis suggests that spatial variation of environmental conditions within the community habitat increases diversity by providing environmental patches for which different species may specialize. In most habitats such heterogeneity can occur horizontally, through microtopographic differentiation of the substrate, or vertically, as in the case of vegetational stratification.

The productivity hypothesis assumes simply that a higher rate of energy flow per unit area of habitat will permit more species to exist through increased specialization in food resource use. In a highly productive environment the amount of energy reaching higher trophic levels may permit the existence of species which otherwise would not be able to gather enough food energy within the maximum area over which they are adapted to forage. At lower trophic levels the higher rate of energy flow would permit more species of lower mobility and greater specialization of habitat requirements to exist.

The predation hypothesis suggests that existence of many species of predators, in turn determined by other factors, maintains prey species populations at lower levels relative to available resources. Thus, competition among prey species is reduced, and more prey species are able to maintain populations without competitive elimination.

Since these hypotheses are not mutually exclusive, a number of theories merging two or more of them have been suggested. This fact also suggests that the real question, in attempting to explain patterns of species diversity at the local level, may concern the relative contribution to diversity provided by each.

Pianka (1967) has examined communities of lizards in the deserts of western North America and attempted to evaluate the importance of the several hypotheses. He first obtained data on the occurrence of lizard species in study areas several square miles in size. These areas were located throughout the western deserts from southern Idaho to southern California and Arizona (Table 8–8). The numbers of lizard species recorded in these areas varied from four (Idaho and Utah) to ten (Arizona). Data on vegetational features, abundance of lizards and other animals, and climate were also collected for these same areas.

Consideration of the ecological time hypothesis in light of these data leads to the conclusion that for most species, adequate time has been

TABLE 8–8. NUMBER OF LIZARD SPECIES AND VEGETATIONAL CHARACTERISTICS OF STUDY AREAS EXAMINED BY PIANKA (1967).

State	Location	Number of lizard species	Plant volume (cu m/plot) x	s	Total plant coverage (%) x	s	Plant species diversity	Plant volume diversity
Utah	West central	4	–	–	20.2	–	1.32	–
Utah	West central	5	–	–	21.6	–	1.37	–
Utah	West central	4	–	–	8.0	–	1.55	–
Idaho	Southwestern	4	–	–	$\sim$20	–	1.76	–
Nevada	Northwestern	5	16.1	0.9	6.0	1.2	0.80	0.08
Nevada	West central	5	49.4	1.4	14.8	2.0	1.46	0.13
Nevada	Southwestern	6	74.3	1.6	10.8	2.2	1.24	0.36
Nevada	Southwestern	6	55.5	1.5	9.1	2.1	0.91	0.25
Nevada	Southern tip	6	54.1	2.4	10.3	3.4	1.24	0.41
Baja California	Northeastern	6	–	–	$\sim$6	–	0.57	0.02
California	Central Mojave Desert	7	–	–	$\sim$10	–	1.09	0.54
California	Western Mojave Desert	8	142.7	2.3	13.3	3.3	1.73	0.44
California	Southern Mojave Desert	9	90.9	2.5	9.1	3.6	1.14	0.82
Arizona	Southwestern	9	77.4	4.1	9.3	5.7	1.23	0.62
Arizona	South central	10	73.1	6.4	10.5	9.0	1.36	0.53

available for dispersal to the geographical limits set by their ecological tolerances. Even the Great Basin Desert, much of which was under water during Pleistocene pluvial periods, could have been completely repopulated by species with dispersal rates as low as 60 m per year. However, it is possible that the genus *Uma,* the fringe-toed sand lizards, may be limited in this manner. The members of this genus are restricted to open sand dune areas. At present the members of this genus are allopatric in distribution and are confined to several disjunct areas in southern California and Mexico. Sand dune areas are present in more northern parts of the western deserts, and it is uncertain whether the absence of this genus from them is due to ecological restrictions or inability of the lizards to cross the intervening barriers of nonsandy desert.

Examination of the environmental stability hypothesis requires analysis of a number of possible relationships. Precipitation is clearly one of the more important environmental factors in desert areas. However, in the western deserts the annual variability of precipitation is inversely related to total rainfall. Variability of annual rainfall is highest in southern desert areas where lizard species diversity is highest. Predictability of the time and amount of future precipitation also varies in different desert regions. It is low in the Great Basin Desert and higher in the Mojave and Sonoran Deserts. In the Mojave Desert, however, this predictability largely involves certainty of little or no precipitation during the summer when temperatures permit lizard activity. Since both the Mojave and Sonoran Deserts possess high lizard species diversities, predictability of future rainfall, per se, seems of little importance to these animals.

Species diversity of lizards, however, does show strong correlations with mean July temperature and length of the frost-free period (Fig. 8–14). These measurements, themselves strongly correlated, may be important largely because they permit greater diversification among species as a result of the longer potential season of activity. Thus temporal separation, either seasonally or diurnally, or greater feeding specialization, may become possible. It is interesting to note that the lizard species having the most specialized feeding and activity patterns are restricted to this region. These are:

1. The nocturnal gecko, *Coleonyx variegatus.*
2. The herbivorous desert iguana, *Dipsosaurus dorsalis.*
3. The Gila monster, *Heleoderma suspectum,* a secondary carnivore feeding largely on other lizards and on eggs of birds.

In examining the spatial heterogeneity hypothesis, Pianka sampled the perennial vegetation on sample plots in each of the study areas. Density and coverage were measured for each plant species, and the volume of each individual plant was also recorded. From these data Pianka could determine, for all plant species combined, total plant volume and total plant coverage, together with estimates of the variability in those values

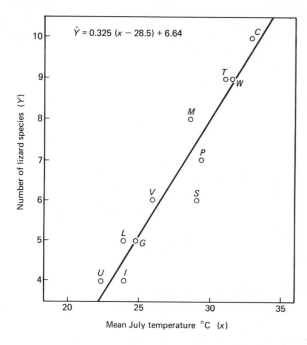

FIGURE 8-14. Relationship between number of lizard species, Y, and mean July temperature (°C), X, for 11 study areas in the southwestern deserts of the United States. (From Pianka, 1967.)

among sample plots (Table 8-8). Examination of these data suggests little correlation of number of lizard species with amount of vegetation, expressed either as total volume or as percentage of ground covered. However, the number of lizard species is significantly correlated with the measures of variability *(s)* in plant volume and percent cover among sample plots. Thus, spatial heterogeneity, in terms of horizontal variation in vegetational features, is correlated with lizard species diversity.

Pianka also examined the relationship between lizard species and two other aspects of habitat heterogeneity. Using the Shannon-Wiener diversity index, he made calculations of the plant species diversity and plant volume diversity in each study area (Table 8-8). These indices were calculated from data on the proportion of individuals belonging to each species and the proportion of individuals falling into various size (volume) classes, respectively. Plant species diversity showed no correlation with lizard species diversity. Plant volume diversity, however, showed a stronger correlation with lizard species diversity than either of the measures of horizontal variation in vegetational characteristics (Fig. 8-15). Thus, environments of high vegetational heterogeneity, both from place to place and in sizes of individual plants, possess the greatest diversity of lizard species. Presumably this heterogeneity permits specialization of species for different patterns of resource exploitation.

In testing the productivity hypothesis, Pianka examined the correlation between number of lizard species and various precipitation characteristics

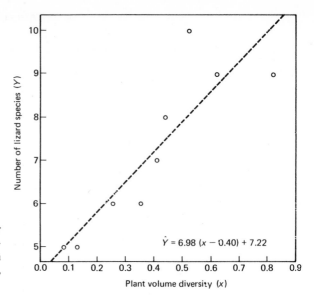

FIGURE 8-15. Relationship between number of lizard species, Y, and plant volume diversity, X, for nine study areas in the southwestern deserts of the United States. (From Pianka, 1967.)

$$\hat{Y} = 6.98\,(x - 0.40) + 7.22$$

of the study areas. This analysis assumed that primary production is closely related to moisture availability in the desert environment. No significant correlation with lizard species diversity was found, however.

Finally, Pianka inferred that if the predation hypotheses were important, evidence of increased intensity of predation should be shown in the study areas having greater numbers of lizard species. For lizards, one indicator of predation—in this case only attempted—is the frequency of individuals having broken and regenerated tails. For all four of the lizard species that were both widely distributed and abundantly represented in the available samples, the frequency of such animals was higher in the southern localities where species diversity was greatest. Likewise, the abundance and variety of lizard-eating snakes and birds were greatest in the southern deserts—suggesting that proportionately more of these forms exist in areas of high lizard species diversity.

Pianka's analyses, although unable to establish causal relationships for species diversity, suggest that several rather different mechanisms may operate. Here, four of the five hypotheses discussed showed some possible relationship to the pattern of species diversity in desert lizards. Only one, the productivity hypothesis, was completely rejected. The ecological time hypothesis was felt to apply to only a single genus (represented by one species in any given area). Thus, the major mechanisms appear to be related to stability of environmental conditions, heterogeneity of environmental structure, and intensity of predation.

We should note that examination of diversity patterns in other groups of organisms, or other environments, may show quite different results. The productivity hypothesis, although found to be of little importance in the above situation, should not be considered as unlikely in other situations.

COMMUNITY VARIATION ALONG ENVIRONMENTAL GRADIENTS

The Basic Question

The second major problem posed at the beginning of our discussion of community structure concerned the pattern of change in structure with change in environmental conditions, either in space or time. It is in connection with this pattern that the question of whether or not the community exists as a biological reality has arisen. Much of this argument is semantic and is perpetuated because of failure to define precisely the relationships under consideration. A significant portion of the problem, however, does involve a basic difference of opinion among ecologists about the extent and importance of certain relationships among coexisting species. In this discussion we hope to clarify the semantic confusion and present these differences of opinion as clearly as possible.

We have used the term *community* to refer simply to the group of species populations occurring and interacting within an area of habitat. That such assemblages do exist, and that interactions of major importance do occur among the species involved, is evident. In this sense no argument for the reality of the community exists.

Instead, the real argument centers on the question of whether or not particular "kinds" of communities, or "community types," exist as biological realities. Throughout the period of development of modern ecology, most ecologists assumed, without question, that distinct kinds of communities existed. During this period, one of the major activities of ecologists was that of recognizing, naming, and devising systems of classifying community types. The alternative concept is that community types represent only arbitrarily recognized points or ranges in a continuum of variation of community composition. The question of the reality of the community thus applies to the idea that community types, such as the Beech-Maple forest or any of hundreds of other named communities, exist as other than arbitrarily recognized units.

This question, however, must be formulated even more precisely. If we examine the communities occurring at different locations in an area in which environmental conditions vary, we find differences and similarities between these communities, depending on the differences and similarities of the environments involved. Assemblages of very similar composition and overall structure may be seen in different locations where conditions are similar and where historical factors have provided the same raw materials for community development. Thus, the question of reality of communities does not center on whether or not similar communities may occur at different locations, or times, at which similar environmental conditions reappear.

Rather, this question specifically concerns whether or not there exist particular groups of species significantly coadapted to each other and sharing, to a significant degree, common patterns of adaptation to the physical environment. Coadaptation, in this sense, implies that the species interact so that they are more successful as individual species populations when they occur together than they would if they occurred as members of some other assemblage. In this sense, there is among ecologists a major difference of opinion on the biological reality of an abstract community type such as the Beech-Maple forest.

Hypotheses of Variation with Environment

The concept of coadaptation and coincidence of environmental tolerances allows several alternative hypotheses of the pattern of community variation in space or time to be formulated in specific terms. These may be presented as graphs of the hypothetical behavior of species along an abstract environmental gradient. This gradient may be thought of as representing a scale that combines all of the significant environmental factors in a given area into a single numerical value which, in the graph, changes at a gradual and constant rate from one extreme to the other. For example, when temperature and moisture availability are the major environmental variables, the axis might show continuous, constant change in a temperature-moisture index from hot and dry at one end to cold and wet at the other. For this presentation, it should be noted that the multiplicity of environmental factors important in nature would make the construction of a one-dimensional scale very difficult. Also the heterogeneity of natural environments would suggest that such a gradient would rarely occur in nature.

The behavior of individual species populations along this hypothetical gradient may be taken as corresponding to some specific measure of ecological activity, e.g., productivity. If this model is used, five somewhat different hypotheses (Fig. 8–16) may be presented (Whittaker, 1970b):

1. Dominant species may be somewhat evenly spaced out and may replace each other at critical points along the environmental gradient. Subordinate species may show close correlation to these dominant species in their pattern of distribution, thus forming distinct species assemblages occupying a particular segment of the environmental gradient. Within these segments the composition of the assemblage changes little with change in environmental conditions. At critical points a small change in environmental conditions leads to replacement of one species assemblage by another.

2. Dominant species may be somewhat spaced out but may gradually replace each other along the environmental gradient. Subordinate species may show close correlation to the pattern of success and

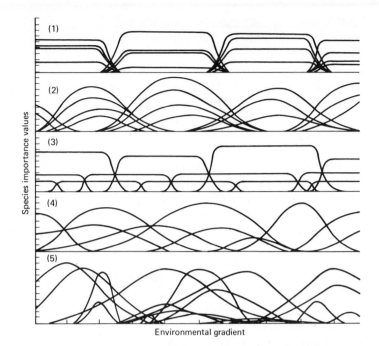

FIGURE 8–16. Five hypotheses of the pattern of change in abundance of species along an environmental continuum (environmental gradient with conditions changing at a gradual, uniform rate from one extreme to the opposite). Panel 1, dominant species evenly spaced and sharply replacing each other with subordinate species distributions strongly correlated to those of dominants. Panel 2, dominant species evenly spaced but gradually replacing each other with subordinate species strongly correlated with dominants. Panel 3, dominant species evenly spaced and sharply replacing; patterns similar for subordinate species but not correlated with dominants. Panel 4, both dominant and subordinate species show bell-shaped abundance curves that tend to be noncoincident for species with similar requirements. Panel 5, both dominant and subordinate species show bell-shaped curves varying randomly in height, extent, and location. See text for further discussion. (Modified from Whittaker, 1970b.)

distribution of these dominants, thus forming recognizable species assemblages occupying a relatively distinct portion of the environmental gradient. Thus, within these segments changes in environmental conditions result in gradual changes in species composition and activity. A more rapid shift in composition from that characteristic of one assemblage to that of another occurs at critical regions of the environmental gradient.

3. Dominant species may be somewhat evenly spaced out and may suddenly replace each other at critical points along the environmental gradient. Groups of subordinate species may show similar patterns not, however, correlated with those of the dominants.

Changes in environmental conditions may thus produce appreciable change in composition and activity of subordinates with little change in dominants, or vice versa.

4. Both dominant and subordinate species may show more or less bell-shaped curves of abundance and activity, the peaks of which tend toward a noncoincidental pattern for groups of similar resource requirements. Changes in environmental conditions consequently produce a gradual and relatively predictable degree of change in community characteristics.

5. Both dominant and subordinate species may show more or less bell-shaped curves randomly located along the gradient. Changes in environmental conditions thus produce a degree of change in community structure varying randomly around some mean value.

Early Community Concepts

These models encompass almost all the major concepts of the nature of community variation. The first presents the general concept formulated by Frederick C. Clements (1916), one of the most influential of American plant ecologists. This concept is often termed *organismic,* since the community was conceived of as representing a supra-organism (see Whittaker, 1957 for a concise summary of the Clementsian system).

This system recognized the *formation* as the major unit. The formation was a climax, regional vegetation type defined on the basis of growth form of the dominant plants (deciduous tree, coniferous tree, etc.). Additional unity was provided by the distribution of species or genera of major importance throughout the range of the formation. The formation thus corresponds to a mature, self-reproducing unit adapted to the climate of a particular region. It was similar to a species in possessing a phylogenetic history. Individual stands showed a pattern of successional development corresponding to the ontogeny of an individual organism. Secondary succession was viewed as a process similar to regeneration of damaged structures by an organism.

Within the formation smaller units known as *associations* were distinguishable. These were recognized on the basis of one or more characteristic dominant species. Some of the associations recognized in the Eastern Deciduous Forest Formation include the Beech-Maple, Oak-Chestnut, and Oak-Hickory Associations. In a mature, climax development each association was considered to be strongly similar throughout in structure and floristic composition. This similarity was considered the result of uniformity of climate and the strong influence of the dominant species in controlling composition of the remainder of the community.

Clements thus envisioned major communities as comprising members having long, common evolutionary histories. Consequently, the co-adjustment among these members was high, and the individuals, species, and community bore a relationship to each other that closely approached that of cells, tissues, and organism. Interspecific relationships, with respect to co-occurrence of species, were in a sense causal rather than simply resultant.

The system of naming and classifying communities (and the community concept underlying it) was almost universally accepted by early ecologists. The system was, indeed, so logical and widely accepted that no name was attached to the basic concept of the community incorporated in it until much later.

Almost simultaneously, however, a strikingly alternative concept was suggested by workers in the United States, Russia, and France (Gleason, 1926; Ramensky, 1926; Lenoble, 1926). The idea, termed the *individualistic* concept, was further developed by Gleason (1939) and Mason (1947) and it constitutes one of the most influential and widely accepted ideas at the present time.

Gleason (1926, 1939) based his interpretation of the nature of the community on several general observations relating to the variability of environment and the patterns of adaptation, reproduction, and dispersal of species. First, he noted that environmental conditions varied in a complex manner, both in time and space. Cyclical diurnal or annual changes, irregular fluctuations, and directional changes over long periods, due to physiographic and climatic processes, combined to make the environment at a given point highly variable through time. He recognized that certain species, the dominants, may modify conditions of the physical environment, but he emphasized that responses of any subordinate species were caused by changed physical conditions—not to the dominants themselves.

Second, he noted that different plant species, even those of a single association, differ in their environmental requirements. Further, each of these species possesses effective mechanisms of reproduction and dispersal. Thus, any given area is continually receiving propagules of species not already established there. These species will grow or not, depending on immediate conditions. Those growing will survive for varying periods of time or will perhaps become permanent members of the community, depending on the extent of environmental variation relative to the tolerances of the species involved.

Gleason (1926) thus concluded that plant communities consisted of species that had been selected only by conditions of the physical environment favoring them. This concept approaches that diagrammed in panel 5 of Fig. 8-16.

Mason (1947) supported this viewpoint with but slight modification. He emphasized the fact that each species is an independent evolutionary unit, except in the cases of symbiosis and parasitism, which he regarded as special cases. Support for this conclusion, and for the individualistic concept of the community, was given by an analysis of the floristic composition of the Redwood forest community through geological time. The fossil Redwood, *Sequoia langsdorfii,* was considered a form directly ancestral to the living species *S. semprivirens,* now restricted to the coastal areas of California and Oregon. The fossil Redwood is represented in many tertiary fossil floras throughout western North America and the arctic. Examination of the other species associated with the Redwood, at present and in these fossil sites, suggested to Mason, however, that the only common feature of this community over geologic time was the presence of the Redwood itself. Thus, the view that the present Redwood forest community is the product of a long, common evolutionary history of the various species now found together was not supported.

Mason recognized that relationships such as symbiosis and parasitism violated a purely individualistic interpretation, but he regarded these as being of minor significance in community structure. He did, however, note that community composition was affected by competition between individuals of different species making parallel demands on the environment. Since this interpretation might be expected to produce some degree of spacing of the performance curves of these species along environmental gradients, Mason's viewpoint is probably represented by panel 4 in Fig. 8–16.

Current Viewpoints

The statement of the individualistic hypothesis was a strong stimulus for study of the pattern of variation in community composition with variation in environment (see Whittaker, 1962, 1967 for historical accounts).

At the University of Wisconsin, beginning in the late 1940's and early 1950's, an extensive series of studies was initiated by J. T. Curtis. Initially, these studies were directed toward an analysis of the forest vegetation of Wisconsin, but since then they have dealt with a variety of plant and animal communities. In examining forest vegetation, stands of mature forest were sampled quantitatively in order to obtain measures of the density, basal area (dominance), and frequency of occurrence of various tree species. These measurements were expressed in relative form, as the percentage which each species constituted of the total for the stand. These relative values—density, dominance, and frequency—could then be summed to give a single measure, termed the *importance value,* for a species in a particular stand.

In order to facilitate comparison of stands, the behavior of each of the tree species was examined, and the species were assigned values, termed *Climax adaptation numbers,* ranging from 1.0 to 10.0. These numbers were based on the frequency of occurrence of species together and, more importantly, on their ability to germinate and grow beneath a forest canopy. Species having the greatest shade tolerance thus were assigned the highest climax adaptation numbers.

The importance values for species in a stand could then be mutiplied by the appropriate climax adaptation numbers, and these products summed to give a single value, the *stand index number,* which reflected the composition of the stand as a whole. This procedure is illustrated in Table 8–9.

TABLE 8–9. PROCEDURE USED BY CURTIS (1957) TO DERIVE STAND INDEX NUMBERS USED IN ORDINATION ANALYSES OF WISCONSIN FOREST VEGETATION.

Species	Sampling data from stand						Importance value (IV)		Climax adaptation number (CAN)		IV x CAN
	Relative density		Relative dominance		Relative frequency						
Acer saccharum	30	+	35	+	25	=	90	x	10.0	=	900
Fagus grandifolia	25	+	35	+	30	=	90	x	9.5	=	855
Fraxinus americana	10	+	15	+	15	=	40	x	6.5	=	260
Ostrya virginiana	20	+	10	+	15	=	45	x	8.5	=	382
Carpinus caroliniana	15	+	5	+	15	=	35	x	8.0	=	280
	100		100		100						
							Stand index number			=	2,677

Species	Climax adaptation numbers for upland forest trees of southern Wisconsin
Acer negundo	1.0
A. rubrum	7.0
A. saccharum	10.0
Carpinus caroliniana	8.0
Carya cordiformis	8.5
C. ovata	4.5
Celtis occidentalis	8.0
Fagus grandifolia	9.5
Fraxinus americana	6.5
Juglans cinerea	7.5
J. nigra	6.5
Ostrya virginiana	8.5
Populus grandidentata	4.5
P. tremuloides	1.0
Prunus serotina	3.5
Quercus alba	3.5
Q. borealis	5.5
Q. ellipsoidalis	1.0
Q. macrocarpa	1.0
Q. muhlenbergii	1.0
Q. velutina	2.5
Tilia americana	7.5
Ulmus americana	7.5
U. rubra	8.0

The stand index number, potentially ranging from 300 to 3,000 in value, provided a scale against which the importance values of a species in individual stands could be examined (Fig. 8–17). These graphs, when combined for all of the major species in an area, suggested a pattern of continuous variation of community composition. Examination of characteristics of the stand environment, including soil moisture capacity, light intensity, and soil nutrient availability showed that a close correlation existed between stand index number and stand environment. This, together with the pattern of continuous variation in composition, was taken as support for the individualistic concept of community organization (Curtis, 1959). Subsequent studies by members of the Wisconsin school have employed more elaborate methods for comparing community composition to environmental variation (e.g., Loucks, 1962; Goff and Zedler, 1968) and have dealt with a variety of both plant and animal communities. In general, these studies have led to similar conclusions.

A similar, but more direct, approach has been taken by Whittaker (1956, 1965, 1967). This approach, termed *gradient analysis,* has involved examination of community composition in relation to directly

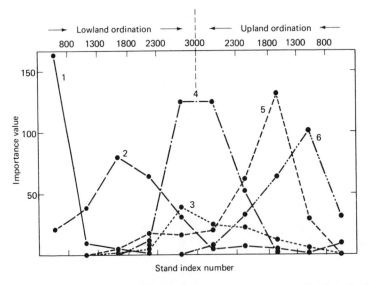

FIGURE 8–17. Behavior of major tree species in a combined ordination of upland and lowland forest stands in southern Wisconsin. In the lowland ordination, increasing stand index numbers are correlated with stands occurring on soils of better drainage. In the upland ordination, increasing stand index numbers reflect improved soil moisture and nutrient conditions. These two ordinations are combined because environmental conditions are similar in stands with maximum stand index numbers in the two ordinations. Species are: 1. black willow, *Salix nigra;* 2. American elm, *Ulmus americana;* 3. slippery elm, *Ulmus rubra;* 4. sugar maple, *Acer saccharum;* 5. red oak, *Quercus borealis;* 6. white oak, *Quercus alba.* (From Curtis, 1959.)

measured variables of the environment. Generally, the technique involves sampling of stands along gradients of moisture, altitude, or other relevant variables, and examination of the behavior of species and the pattern of community change.

The results (Fig. 8–11) obtained by Whittaker show species curves varying in height, spread, and shape, but indicating a pattern of continuous changes in community structure along environmental gradients. Whittaker (1965) notes, however, a strong tendency for species curves to be arranged so that each has an optimum not coinciding with that of a potential competitor, and he suggests that competition has been important in determining the location of adaptation ranges of species along such gradients (Fig. 8–18).

Daubenmire (1966) has supplied some of the most serious criticism of the individualistic concept. He admits that vegetation is a continuous variable. That is, in areas where gradual changes in environmental conditions occur perfectly sharp boundaries between completely different species groups do not occur; instead, the change in composition is more or less gradual. He contends however, that along an environmental gradient there are some segments in which the change in community composition is slow, and there are other segments in which it is high. Points of rela-

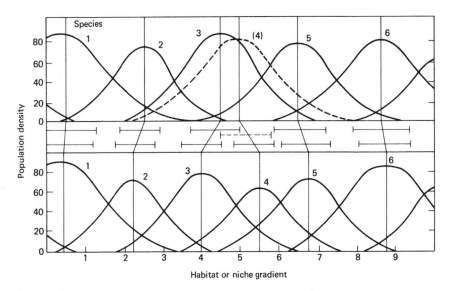

FIGURE 8–18. Establishment of a new species along a community gradient. The new species, number 4, has a potential distribution along the habitat gradient as represented in the dashed line of the upper figure. In competition with species 3 and 5 it fits in between these, as indicated in the lower figure. The bars between the figures represent dispersions, the degrees of deviation or spread of the populations on each side of their mean positions along the gradient. (From Whittaker, 1970b.)

TABLE 8-10. DATA ON PERCENT CANOPY-COVERAGE OF SPECIES IN STANDS SHOWN IN FIG. 8-19.[a]

Median annual precipitation 167 mm → 526 mm
Mean annual temperature 11.2° C → 7.9° C

Species	Artemisia-Agropyron	Artemisia-Festuca	Agropyron-Festuca	Festuca-Symphoricarpos	*	†
Stipa thurberiana	5					
Poa cuspickii	2 9					
Stipa comata	2 + 3 +					
Artemisia tridentata	18 18 13 9 9 14 19 11	13 4	6			
Chrysothamnus viscidiflorus	+	13 8	1 8	+		
Plantago patagonica	11 9	+ 3	3	2 9		
Phlox longifolia	+ 12 14 8 + 7 8 12 5	2 +	7 3	1		
Erigeron filifolia	3	+ 1	+			
Astragalus spaldingii				1		
Poa secunda	40 50 29 61 55 73 36	44 38 5	+ 19 39 16 23 2 13	22 10 6 2 + + 25	1	
Achillea millefolium	+ + 1	+ 1 2	2 3 1 +	57 9 77 10 25	13	
Agropyron spicatum	41 63 46 35 41 55	40 39 26	42 77 33 34 71	79 81 30 41 78 35 40		
Festuca idahoensis		9 23 39	29 7 71 75	+ 2 8 3		
Senecio integerrimus			8	+ 2		
Myosotis stricta				2 5 5		
Haplopappus liatriformis				10 8 + +		
Koeleria cristata				5 3 + 4 4		
Hieraceum albertinum				5 + + 11 1 1		
Lupinus sericeus				6 1 2 3 3		
Festuca scabrella				51 51		
Sidalcea oregana				8		
Castilleja lutescens				+ 2 9 5 +		
Arnica sororia				5		
Solidago missouriensis				29	1	
Balsamorhiza sagittata				41 14 51		
Helianthella uniflora				31 31		
Astragalus arrectus				18 18 18		
Poa ampla				57 53	9	
Rosa nutkana + R. woodsii				1 2 + 1 1	1	
Iris missouriensis				+ 29 2 1	+	
Potentilla gracilis				6 21 45 5 5	2	
Geranium viscosissimum				21 4 +	1	
Galium boreale				+ 11 1 1	1	+
Symphoricarpos albus	4		15	2 11 17 17	42	30

[a]From Daubenmire, 1966.

*Pinus ponderosa-Symphoricarpos albus association. Only those species that occurred in the steppe stands are shown.

†Pseudotsuga menziesii-Physocarpus malvaceus association. Only those species that occurred in the steppe stands are shown here.

Note: Stands are arranged according to longitude, progressing from west to east in a belt 96 × 31 kilometers (153 × 63 miles) that crosses vegetation zones nearly at right angles. Species with coverages never rising to 5% are omitted. Plus (+) sign indicates less than 1%.

tively rapid change in composition represent, in a sense, discontinuities, since they are points at which several species tend to drop out together, and several others appear together. Daubenmire argues that biologically meaningful discontinuities in community composition may occur for the following reasons:

1. The appearance or disappearance of one or more dominant species at points reflecting critical limits of environmental tolerance.
2. A change in population dynamics of a dominant species to or from a condition of competitive superiority and ability to replace itself.
3. The coincident appearance or disappearance of several or many species at the same point on the environmental gradient.

Daubenmire has illustrated these conditions by drawing examples from the steppe and forest vegetation of eastern Washington and Idaho. A series of 21 stands of steppe vegetation were sampled along an environmental gradient of increasing moisture and decreasing temperature (Table 8–10, Fig. 8–19). These stands were selected for analysis because they occurred in undisturbed situations and on deep, nonsaline, upland soils with a slope less than 15%. The data in Table 8–10 are grouped according to the associations recognized by Daubenmire.

Two of the separations between associations are based on the appearance or disappearance of a single species. The *Artemesia-Agropyron* association is separated from the *Artemesia-Festuca* association by the presence of *Festuca* in the latter. This latter association differs from the

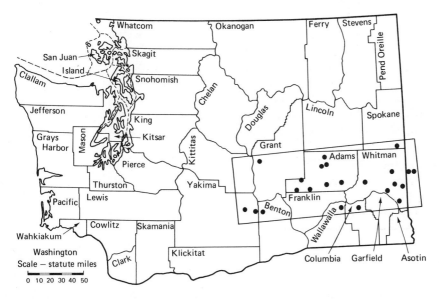

FIGURE 8–19. Locations of stands for which data are presented in Table 8–10. The geographic discontinuity that segregates the four westernmost stands does not correlate with any of the vegetational discontinuities apparent in Table 8–10. (From Daubenmire, 1966.)

Agropyron-Festuca association, in turn, by the loss of *Artemesia*. Daubenmire argues that the presence or absence of these species, *Artemesia tridentata,* a shrubby sagebrush, and *Festuca idahoensis,* a bunch-grass of large size, is significant from the ecosystem standpoint. That is, their presence or absence determines to a great extent what else may be present, and to a degree the pattern of overall community metabolism. For example, when *Artemesia tridentata* drops out, various shrub-dependent birds, epiphytic lichens, and a number of parasites and epiparasites must also disappear.

The third separation, between the *Agropyron-Festuca* and *Festuca-Symphoricarpos* associations, involves the coincident loss of several species and appearance of others. Several of the species appearing in the latter association are perennial dicot herbs and low shrubs, thus producing quite a change in community structure and function.

Daubenmire also examined forest stands from lower to upper timberlines in the Bitterroot Mountains of Idaho. In stands that had not been disturbed by man he noted that, usually, only a single species possessed a population structure indicative of self-replacement. Other species might be present in the canopy, although without reproducing, or they might be present as seedlings that do not survive for long (Table 8–11). From these data he constructed a diagram (Fig. 8–20) suggesting the segments of the elevational gradient in which various species exhibited self-reproducing capabilities. Using this criterion, Daubenmire suggested the recognition of eight forest communities as ecologically meaningful types in this area.

Daubenmire has criticized proponents of the individualistic school on two main grounds. First, he feels that the methods of sampling and analysis are capable of producing apparent continuity of vegetation, even when strong discontinuities occur at critical points on an environmental gradient. These methods include, according to Daubenmire, sampling of disturbed sites, sampling of sites that are internally heterogeneous in environmental characteristics, and analytical techniques that fail to consider the population dynamics of the species involved. Second, he feels that reliance on strictly compositional data, without consideration of the functional significance of the species as members of the overall ecosystem, is inappropriate.

Exponents of the individualistic concept do admit the existence of discontinuities in community composition along environmental gradients (Whittaker, 1967). For the most part, these are regarded as unusual cases involving communities of single-species dominance or environments of extreme or special nature.

Recently, Beals (1969) has compared plant community composition along two natural environmental gradients in Ethiopia. The gradients involved were altitudinal, with the major variable being moisture, which increased with altitude. The two gradients differ in steepness. The steeper

TABLE 8-11. DATA ON POPULATION STRUCTURE OF FOREST STANDS IN THE BITTERROOT MOUNTAINS, IDAHO.[a]

Tree flora	Diameter at breast height (decimeters)									
	0–0.5	0.5–1	1–2	2–3	3–4	4–5	5–6	6–7	7–8	8+
Stand 1 – Near-climax										
Thuja plicata	38	1	3		4	1	4		1	2
Abies grandis	82								1	1
Pseudotsuga menziesii										1
Stand 2 – Climax										
Thuja plicata	1					1	1	1		5
Abies grandis	172*									
Stand 3 – Near-climax										
Abies grandis	37	4	12	7	2	2				
Pseudotsuga menziesii				1	1					
Pinus contorta			2	2	1					
Pinus ponderosa					1					
Stand 4 – Climax										
Abies grandis	50	2	6	5	6	2	1	1	1	
Stand 5 – Near-climax										
Pseudotsuga menziesii	2	11	29	20	1	4	1		1	
Larix occidentalis						4	1			
Pinus ponderosa			1					1		
Stand 6 – Climax										
Pseudotsuga menziesii	107	4	8	2	8	9	2			

[a]From Daubenmire, 1966

*Current-year seedlings, from seeds produced in abundance in contiguous habitat.

Note: Analyses are for trees in areas 15 × 25 meters, each in a different homogeneous stand of forest that has never been logged or evidently grazed by livestock. Data represent the number of trees in each category.

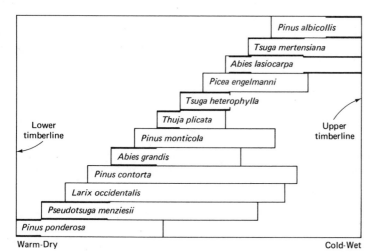

FIGURE 8–20. Coniferous trees in the area centered on eastern Washington and northern Idaho, arranged vertically to show the usual order in which the species are encountered with increasing altitude. The horizontal bars designate upper and lower limits of the species relative to the climatic gradient. That portion of a species' altitudinal range in which it can maintain a self-reproducing population in the face of intense competition is indicated by the heavy lines. (From Daubenmire, 1966.)

one, near the village of Bati, involved a vertical change of 1,250 meters in a horizontal distance of about 20 kilometers. The gentle gradient, near the town of Awash, spanned an altitudinal range of 900 meters in a horizontal distance of about 300 kilometers. The steeper gradient, occurring in an area of rugged topography, was characterized by having soils that were shallower and rockier than those of the gentle gradient.

Beals sampled the plant communities in each gradient at points separated by 10 meter differences in elevation. These were analyzed to provide a composite sample for each 50 meter altitudinal segment. The composition of communities of adjacent segments was then compared by calculation of a similarity coefficient (ranging from 0 to 100). The mean values for these coefficients were nearly the same, 59.1 for the Bati gradient and 62.5 for the Awash. However, the variation in values of this coefficient was much greater for the steeper gradient at Bati (Fig. 8–21). This indicated that over certain segments of the Bati gradient little change occurred, but over other segments of similar altitudinal range change in composition was great. Furthermore, along the Bati gradient the sharp changes in composition were in large part the result of simultaneous appearance or disappearance of several species. In contrast, simultaneous appearance or disappearance of species groups was of lesser frequency on the gentle (Awash) gradient. Beals interpreted the correlated species

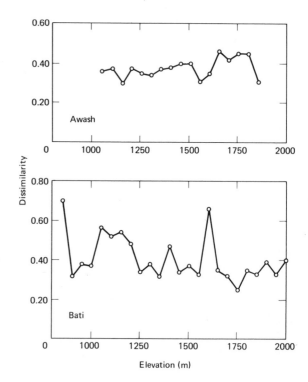

FIGURE 8–21. Degrees of dissimilarity (0 = no dissimilarity, 1.0 = complete dissimilarity) in plant species composition of adjacent segments of gentle (Awash) and steep (Bati) topographic gradients in Ethiopia. High values indicate points of rapid change in vegetational composition with change in altitude; low values indicate points of slow change. (From Beals, 1969.)

limits on the Bati gradient to be the results of direct biotic interrelationships among the species involved or dominance effects of the species in the internal environment of the communities. Thus, the regions of rapid species change represent discontinuities, in the sense defined by Daubenmire.

An Intermediate Viewpoint

Until recently, proponents of the various hypotheses of community variation have failed to state these hypotheses in specific, comparable terms. For example, the conditions under which the conclusion may be made that community types exist as biologically meaningful units have not been clearly stated. Either of the following two conditions, at least, would seem to justify such a conclusion:

1. Coincidence of appearance or disappearance of species groups along an environmental continuum because of their evolutionary coadaptation and correspondence of limits of tolerance to environmental conditions.
2. Coincidence of appearance or disappearance of species groups along an environmental gradient because of the imposition, by certain dominant species, of environmental discontinuities on what otherwise would be an environmental continuum.

If either or both of the above are of general occurrence and major significance in the variation of communities in space and time, it may be concluded that community types do exist. If the above are of very minor importance, the individualistic viewpoint of community variation is supported.

A second major shortcoming of many of the studies presented in support of one community viewpoint or another is their restriction to a particular group of organisms. Many of the individual studies of the Wisconsin school have been restricted, for example, to the canopy trees, the understory vegetation, birds, mosses and lichens, and soil fungi. For each of these groups the results support an individualistic pattern of variation, which, according to Whittaker (1967), parallels that of the forest canopy trees. However, data from these diverse groups, all of which occur together in the same communities, have never been examined together to determine whether or not, along the environmental gradients involved, there are critical points of significant change in overall composition and community function.

Evidence, derived in part from these studies, suggests that examination of such restricted groupings may produce a bias in favor of a highly individualistic community pattern. Whittaker (1965), as noted earlier, has suggested that within such groups interspecific competition may lead to the spacing of optima of the various species so that no two possess identical patterns of response. Such spacing contributes to an individualistic pattern.

Evidence derived from studies of microclimates (e.g., Wolfe, et al., 1949; Cantlon, 1953) indicates that certain species, by virtue of their size and structure, may create environmental discontinuities on what would otherwise be a continuous environmental gradient. Even if the subordinate species are responding to the changed conditions, rather than directly to the species producing modification, such an effect would tend to produce coincidence in the appearance and disappearance of species groups along natural gradients. This would be of greatest significance where major changes in life form of the major species occurred.

Further, it has been demonstrated that many community members, especially animals, are adapted to features of community structure related to the life form of the dominant plants. MacArthur and MacArthur (1961) and a number of subsequent workers have shown that the diversity of the bird species in various communities is related to the number of vegetational strata and to the proportion of foliage in each. Pianka (1967) also found species diversity in lizards to be correlated with the diversity of shrub sizes in desert areas. Although neither worker found a correlation of faunal diversity with plant species diversity, it is apparent that changes in vegetational structure resulting from appearance or disappearance of certain life form groups of plants strongly affect the animal communities occurring in an area.

Recent studies have also suggested that, in addition to simply modifying normal environmental conditions such as temperature and light intensity, certain species create new environmental factors of varying specificity for other potential community members. The phenomena of allelopathy is an example. Whittaker (1970a) has suggested, in fact, that this phenomenon is of importance in most terrestrial communities. These mechanisms clearly represent environmental discontinuities imposed on the environment by organisms themselves, and they may produce coincidence in the appearance and disappearance of groups of species along environmental gradients.

Finally, detailed investigations of interspecific relationships have begun to reveal patterns of coevolution not previously recognized. These patterns range from highly species-specific to specific only for certain species groups, and from a condition of ability of one species to compensate for some detrimental effect of another to a condition of obligatory symbiosis. Because such patterns occur among community members, coincidence of distribution results. Current evidence indicates, however, that these relationships are more than special cases as suggested by Mason (1947).

Therefore, we may tentatively conclude that the pattern of variation of community composition with environment is somewhere between the extreme organismic viewpoint of Clements and the widely held individualistic viewpoint of the 1960's (Whittaker, 1967, 1970b). This intermediate viewpoint recognizes that communities are not composed of species all of which possess highly coincident patterns of distribution along environmental continua. It also denies that distributions of species along environmental gradients, when all community members are considered, are perfectly individualistic in their distribution. Instead, it suggests that because of environmental discontinuities imposed by organisms themselves and interspecific interactions having some degree of specificity, points of significant ecological change may be recognized. These points may involve relatively sharp change in community composition, or more importantly, in basic functional characteristics of the community. This viewpoint recognizes that the frequency and distinctness of such points may vary with many factors, but that in some situations it occurs often enough and to a strong enough degree that existence of biologically meaningful community units must be admitted. It also recognizes that in other cases community composition may vary in a highly individualistic manner over long environmental gradients, warranting recognition only of what may be termed ecoclines (Whittaker, 1967). This intermediate viewpoint is suggested in Figure 8–22.

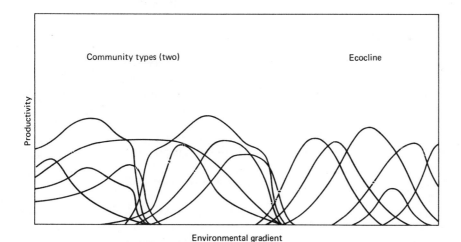

FIGURE 8–22. Possible patterns of productivity of species in relation to position along a hypothetical environmental gradient. In certain cases curves for different co-occurring species may show a degree of congruence that justifies recognition of community types (left half of figure); in other cases gradual and continuous variation in composition of the community and the performance of species warrant only the recognition of an ecocline (right half of figure; see text for further explanation).

REFERENCES

Bartholomew, B. 1970. Bare zone between California scrub and grassland communities: The role of animals. *Science,* **170**:1210–12.

Beals, E. W. 1069. Vegetational change along altitudinal gradients. *Science,* **165**:981–85.

Brower, L. P., and J. Brower. 1964. Birds, butterflies and plant poisons: A study in ecological chemistry. *Zoologica,* **49**: 137–59.

_____ , **and J. M. Corvino.** 1967. Plant poisons in a terrestrial food chain. *Proc. Nat. Acad. Sci. U. S.,* **57**:893–98.

Burkholder, P. R. 1952. Cooperation and conflict among primitive organisms. *Amer. Sci.,* **40**:601–31.

Cantlon, J. E. 1953. Vegetation and microclimates on north and south slopes of Cushetunk Mountain, New Jersey. *Ecol. Monog.,* **23**:241–70.

Clements, F. C. 1916. *Plant Succession, an Analysis of the Development of Vegetation.* Carnegie Inst. Wash. Publ. No. 242. 512 pp.

Cody, M. L. 1968. On the methods of resource division in grassland bird communities. *Amer. Nat.,* **102**:107–47.

Cohen, J. 1968. Alternative derivations of a species-abundance relation. *Amer. Nat.,* **102**:165–72.

Cox, G. W. 1967. *Laboratory Manual of General Ecology.* Dubuque, Iowa: Wm. C. Brown. 165 pp.

_____ . 1971. Species diversity, biotic succession, and the evolution of migration. In preparation.

Curtis, J. T. 1959. *The Vegetation of Wisconsin.* Madison: University of Wisconsin Press. 657 pp.

Daubenmire, R. 1966. Vegetation: Identification of typal communities. *Science,* **151**:291–98.

_____ , **and J. B. Daubenmire.** 1968. *Forest Vegetation of Eastern Washington and Northern Idaho.* Wash. Agr. Exp. Station, Tech. Bull. No. 60. 104 pp.

Ehrlich, P. R., and P. H. Raven. 1964. Butterflies and plants: A study in coevolution. *Evolution,* **18**:586–608.

_____ . 1967. Butterflies and plants. *Sci. Amer.,* **216**:104–13.

Fisher, R. A., A. S. Corbet, and C. B. Williams. 1943. The relation between the number of species and the number of individuals in a random sample of an animal population. *J. Anim. Ecol.,* **12**:42–58.

Gleason, H. A. 1926. The individualistic concept of the plant association. *Bull. Torrey Bot. Club,* **53**:7–26.

_____ . 1939. The individualistic concept of the plant association. *Amer. Midl. Nat.,* **21**:92–110.

Goff, F. G., and P. H. Zedler. 1968. Structural gradient analysis of upland forests in western Great Lakes area. *Ecol. Monogr.,* **38**:65–86.

Greenslade, P. J. M. 1968. Island patterns in the Solomon Islands bird fauna. *Evolution,* **22**:751–61.

Hutchinson, G. E. 1951. Copepodology for the ornithologist. *Ecology,* **32**:571–77.

Janzen, D. H. 1966. Coevolution of mutualism between ants and acacias in Central America. *Evolution,* **20**:249–75.

_____ . 1967. Fire, vegetation structure, and the ant × acacia interaction in Central America. *Ecology,* **48**:26–35.

————. 1969. Allelopathy by myrmecophytes: the ant *Azteca* as an allelopathic agent of *Cecropia*. *Evolution* **50**:147–53.

Kittredge, J. 1948. *Forest Influences*. New York: McGraw-Hill. 394 pp.

Knight, D. H., and O. L. Loucks. 1969. A quantitative analysis of Wisconsin forest vegetation on the basis of plant function and gross morphology. *Ecology*, **50**:219–34.

Lenoble, F. 1926. A propos des associations vegetales. *Bull. Soc. Bot. Fr.*, **73**:873–93.

Loucks, O. L. 1962. Ordinating forest communities by means of environmental scalars and phytosociological indices. *Ecol. Monogr.*, **32**:137–66.

MacArthur, R. 1957. On the relative abundance of bird species. *Proc. Nat. Acad. Sci. U.S.*, **45**:293–95.

————, and J. W. MacArthur. 1961. On bird species diversity. *Ecology*, **42**:594–98.

————, and E. O. Wilson. 1967. *The Theory of Island Biogeography*. Princeton, N.J.: Princeton University Press. 203 pp.

Mason, H. L. 1947. Evolution of certain floristic associations in western North America. *Ecol. Monogr.*, **17**:201–10.

McPherson, J. K., and C. H. Muller. 1969. Allelopathic effects of *Adenostoma fasciculatum*, "Chamise," in the California chaparral. *Ecol. Monogr.*, **39**:177–98.

Muller, C. H., R. B. Hanawalt, and J. K. McPherson. 1968. Allelopathic control of herb growth in the fire cycle of California chaparral. *Bull. Torrey Bot. Club*, **95**:225–31.

Odum, E. P. 1959. *Fundamentals of Ecology*. Philadelphia: W. B. Saunders. 546 pp.

Ovington, J. D. 1965. *Woodlands*. London: English University Press. 154 pp.

Pianka, E. R. 1966. Latitudinal gradients in species diversity: A review of concepts. *Amer. Nat.*, **100**:33–46.

————. 1967. Lizard species diversity. *Ecology*, **48**:333–51.

Pimentel, D. 1968. Population regulation and genetic feedback. *Science*, **159**:1432–437.

Preston, F. W. 1948. The commonness, and rarity, of species. *Ecology*, **29**:254–83.

Ramensky, L. G. 1926. Die Grundmassigkeiten in aufbau der Vegetations decke. *Bot. Centbl., N.S.*, **7**:453–55.

Raunkiaer, C. 1937. *Plant Life Forms*. Oxford: Clarendon Press. 104 pp.

Ricklefs, R. E. 1970. State of taxon cycle and distribution of birds on Jamaica, Greater Antilles. *Evolution*, **24**:475–77.

————, and G. W. Cox. 1972. The taxon cycle in the land bird fauna of the West Indies. *Am. Nat.*, **106**:195–219.

Rowe, J. S. 1951. The level of integration concept and ecology. *Ecology*, **42**:420–27.

Salisbury, E. J. 1942. *The Reproductive Capacity of Plants*. London: G. Bell and Sons. 244 pp.

Salt, G. W. 1953. An ecologic analysis of three California avifaunas. *Condor*, **55**:258–73.

Shultz, A. M. 1967. The ecosystem as a conceptual tool in the management of natural resources. In S. V. Ciriancy-Wantrup and J. J. Parsons (eds.), *Nat-*

ural Resources: Quality and Quantity, pp. 141–61. Berkeley: University California Press.

Simpson, E. H. 1949. Measurement of diversity. *Nature,* **163**:688.

Southwood, T. R. E. 1962. Migration of terrestrial arthropods in relation to habitat. *Biol. Rev.,* **37**:171–214.

Whittaker, R. H. 1956. Vegetation of the Great Smoky Mountains. *Ecol. Monogr.,* **26**:1–80.

―――― . 1957. Recent evolution of ecological concepts in relation to the eastern forests of North America. *Amer. J. Bot.,* **44**:197–206.

―――― . 1962. Classification of natural communities. *Bot. Rev.,* **28**:1–239.

―――― . 1965. Dominance and diversity of land plant communities. *Science,* **147**:250–60.

―――― . 1967. Gradient analysis of vegetation. *Biol. Rev.,* **42**:207–64.

―――― . 1970a. The biochemical ecology of higher plants. In E. Sondheimer and J. B. Simeone (eds.), *Chemical Ecology,* pp. 43–70. New York: Academic Press.

―――― . 1970b. *Communities and Ecosystems.* London: Macmillan. 162 pp.

Wilson, E. O. 1961. The nature of the taxon cycle in the Melanesian ant fauna. *Amer. Nat.,* **95**:169–93.

Wolfe, J. T., R. T. Wareham, and H. T. Scofield. 1949. *Microclimates and Macroclimates of Neotoma, a Small Valley in Central Ohio.* Ohio Biol. Surv., Bull. No. 41. 267 pp.

Part V

THE ECOSYSTEM LEVEL
OF ORGANIZATION

Chapter 9

TROPHIC STRUCTURE AND DYNAMICS OF ECOSYSTEMS

Earlier we discussed processes of energy exchange between individual organisms and their environment, and the adaptations of organisms to their energy environment. This chapter will also deal with energy, but in a different context and on a different scale. Here our principal subjects will be the conversion of radiant energy from the sun into chemical energy by the green plants or *primary producers* in an ecosystem, and the transfer of energy from the primary producers to *primary consumers,* i.e., herbivores, and then to higher levels of consumers.

Qualitative descriptions of the pattern of energy flow through ecosystems are not new. Forbes (1887), for example, discussed the interdependence of organisms in a lake ecosystem in terms of their feeding relationships. Quantitative, analytical studies at the ecosystem level, however, are a much more recent development in ecology. Today, detailed quantitative studies of energy flow have become especially important because of the widespread recognition of man's influence on the productivity of virtually all ecosystems, and because the movement of materials, including pesticides and industrial wastes, as well as necessary nutrients, is closely related to the flow of energy through ecosystems. (This topic will be discussed in the next chapter.)

TROPHIC STRUCTURE

We may abstractly represent the feeding relationships of the producers and consumers of an ecosystem in a diagram showing "who eats whom"

(Fig. 9–1). This abstraction is called a *food web,* and any linear sequence of species in it, such as plants → ptarmigan → arctic fox in Fig. 9–1, is a *food chain.*

Often, the species in a food web can be placed into four or five groups according to how many steps away from the primary producers they are. These groups are called trophic levels; thus, the first trophic level comprises all of the producers in an ecosystem, the second trophic level comprises the herbivores, the third comprises all of the carnivores feeding on herbivores, and so on.

In addition to these groups every ecosystem includes organisms that use the dead remains of other organisms as their source of energy. Many of these organisms are either fungi or bacteria, which are collectively referred to as decomposers, but many animals also obtain the energy they require in this way. Obvious examples include crabs and other marine and aquatic arthropods, many kinds of fly larvae, dung beetles, and birds that feed on carrion. These macroscopic animals have little in common with bacteria and fungi and they are not generally considered as belonging to the same trophic level as the saprophytes. In fact, scavenging animals are often one step below saprophytes in food chains,

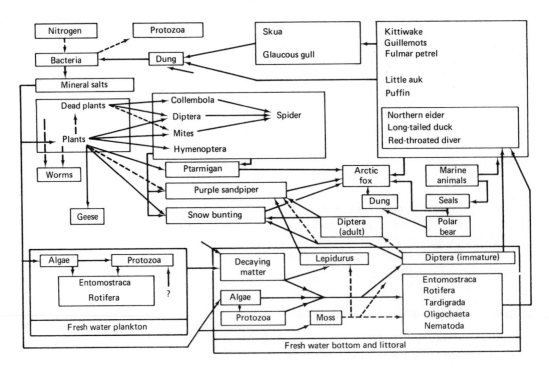

FIGURE 9–1. Simplified food web for Bear Island (near Spitsbergen). (From Summerhayes and Elton, 1923.)

since they may break up the remains of dead plants and animals into organic detritus that is more readily usable by bacteria and fungi.

Although it is often convenient to think of a species as occupying a single trophic level, we must point out that many species cannot be so neatly categorized. Many animals have highly varied diets that include both plant and animal material. Petrusewicz and Macfadyen (1970) mention a wide variety of examples, including species of nematodes, chilopods, collembola, rotifers, and rodents, thus showing that there are no taxonomic boundaries to polyphagia. Still other animals change from one trophic level to another during their life cycle. Tadpoles, for example, are herbivores, but adult frogs are carnivores, and many insects, especially dipterans, show extreme differences in their food habits between the larval and adult stages. Thus, we must recognize that trophic levels are abstractions rather than distinct natural entities; it is especially important to keep this in mind when considering any generalizations that might be made regarding entire trophic levels or relationships among trophic levels. In some instances, it is possible to circumvent these problems by assigning fractions of the energy flowing through a species to different trophic levels in proportion to the amount of food obtained from different sources.

Some Elementary Relationships among Trophic Levels

A cursory examination of most natural communities suggests that the number of individuals within each trophic level is smaller than the number in the preceding trophic level. Assuming this to be true, then a diagram showing the number of individuals in each trophic level as a set of horizontal bars stacked atop one another with primary producers at the bottom will have the appearance of a pyramid. Elton (1927) called this a *pyramid of numbers,* but it is now commonly called an *Eltonian pyramid,* since his early writing stimulated much interest in the area of trophic relationships. In fact, only a few examples of pyramids of numbers have been published, and in these cases the data are weak. Although Elton restricted his original discussion to the animals in a community, other authors have included plants in their pyramids. We suspect that if adequate data were available they would show that herbivores may commonly outnumber primary producers, at least in terrestrial communities, since herbivorous arthropods are often very abundant. In other instances a strong relationship between body size and trophic level has been assumed to exist (e.g., Lindeman, 1942). The resulting graphs are therefore really pyramids of numbers in successive size classes rather than trophic levels.

If the number of individuals in each trophic level is replaced by the biomass, or weight of living material, present in each trophic level, a pyramid-like relationship is more likely to appear. Occasionally, ecosystems are found to have inverted biomass pyramids; that is, the biomass of the primary producers is smaller than the biomass of the consumers. Figure 9–2 shows examples of pyramids of biomass and numbers. The reasons why both numbers and biomass in successive trophic levels would be expected to be pyramidal in form will be discussed below.

PRODUCTIVITY

Information on the number and biomass of organisms in each trophic level gives at best an incomplete picture of an ecosystem, since it shows only the state of the system at a point in time. Additional information is needed to show an ecosystem's dynamic features.

Energy enters an ecosystem at the rate at which the primary producers can convert solar radiation into chemical energy by photosynthesis. This rate is referred to as *gross primary productivity*. Much of this energy may accumulate as new plant tissues are formed, and hence becomes available as food for animals, but some of the energy is used by the plants themselves in their metabolic processes. The rate at which energy becomes available to herbivores is equal to the difference between gross primary production and the plant's respiration rate; this difference is called *net primary productivity*. We shall discuss the details of primary productivity below.

Similarly, we can speak of the rate at which animal biomass at any trophic level is being formed as the *productivity* of that trophic level. (We note, however, that Lindeman (1942) and a few others use productivity as the rate at which energy enters a trophic level, which includes the energy later used in respiration.)

Data for productivity at successive trophic levels can be shown in a

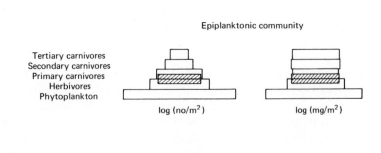

Epiplanktonic community

Tertiary carnivores
Secondary carnivores
Primary carnivores
Herbivores
Phytoplankton

log (no/m²)   log (mg/m²)

Bathyplanktonic community

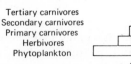
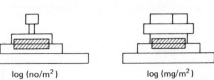

FIGURE 9–2. Pyramids of numbers and biomass in planktonic communities in the Black Sea. Shaded bars are mixed-food consumers. (From Petipa, Pavlova, and Mironov, 1970.)

Tertiary carnivores
Secondary carnivores
Primary carnivores
Herbivores
Phytoplankton

log (no/m²)   log (mg/m²)

pyramid similar to the pyramids of biomass and numbers. Unlike these latter two pyramids, however, a productivity pyramid can never be inverted, because at each trophic level a large amount of energy is being lost as heat due to respiration, and this energy cannot be recaptured by any organism for its use.

Inverted biomass pyramids can now be explained by a consideration of productivities at successive trophic levels. Total productivity at one level is the product of total biomass and productivity per unit biomass. If productivity per unit biomass is very high at one level compared to the next higher level, then total biomass of the first level may be less than the total biomass at the next level although it still maintains higher production than the second level. Usually, the producers in ecosystems having inverted biomass pyramids are plankton whose reproductive rates are high.

Primary Production

Both the plants and animals in an ecosystem depend on energy stored in organic materials in order to maintain themselves, but only the plants are capable of synthesizing these organic materials from inorganic substances. The external energy for the synthesis of these organic compounds usually comes from the sun, but it sometimes comes from the reduction of highly oxidized chemical compounds, such as nitrates or sulfur oxides. Although most ecosystems depend on primary producers within the system for producing these organic materials, a few, such as cave or abysmal depth oceanic communities, have no primary producers and depend on an importation of organic matter from elsewhere for their energy source. Because of the importance of its energy base to the functioning of an ecosystem, much effort has gone into measuring and analyzing the processes of primary production.

The primary production of the world is not evenly distributed over the earth's surface (Fig. 9–3). In terrestrial ecosystems primary production tends to be highest in the tropics and decreases as one moves north and south toward the poles. At any latitude, however, primary production decreases as rainfall decreases (Fig. 9–4). The most productive land areas of the world are fields of irrigated crops or of perennial plants in locations having long growing seasons, and the least productive terrestrial areas are deserts and tundra. Production in tropical rain forests or tropical agricultural land has been estimated as between 13.7 and 20.5 g dry weight organic matter/m²/day,* but in the tundra and deserts production is about 100 g/m²/year, or about 0.3 g/m²/day.

In the ocean productivity is usually limited by a lack of nutrients, and, therefore, the most productive areas are regions of large upwellings, such as off the coasts of Peru and southwestern United States, and the Arctic

*Weights given in the remainder of this chapter will be understood to be for dry organic matter.

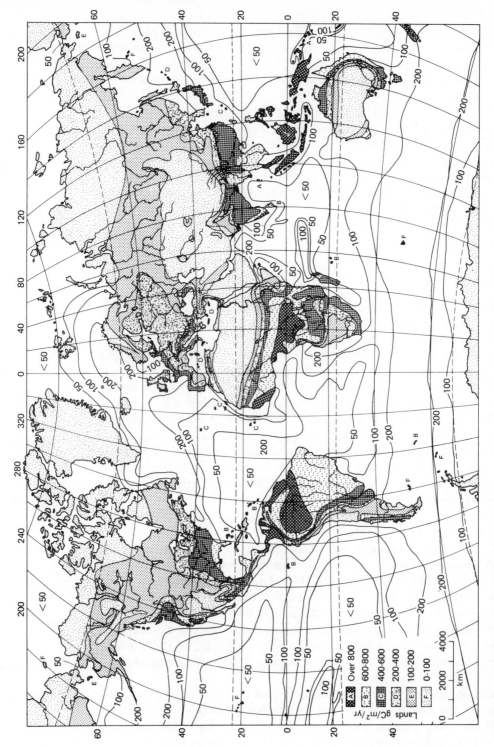

FIGURE 9–3. Worldwide net primary production in grams of carbon per square meter per year. Mapped values are approximate and are in-completely adjusted for losses to consumers, decomposers, and substrate. (Modified from Reichle, 1970.)

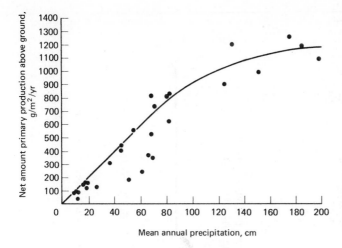

FIGURE 9–4. Relation of net annual primary production to rainfall. Data from various sources; peak productions of unstable communities are excluded. (From Whittaker, 1970.)

and Antarctic Oceans. The least productive oceanic areas, on the other hand, are in the tropics and horse latitudes where no upwellings occur. The Sargasso Sea, for example, is among the least productive regions in the world. Figures reported for annual production of oceans range from over 400 g/m²/year in highly productive areas to about 50 g/m²/year in regions of low productivity (Table 9–1).

Efficiency and Productivity. As we saw in Chapter 3 something between 0 and about 1.2 cal/m² of solar energy strikes the earth's surface, including its vegetation, every minute. If all of this energy could be incorporated into organic material by green plants, the resulting productivity would be prodigious — several orders of magnitude greater than the figures we quoted above. Only a small fraction of incoming solar radiation, however, shows up as primary production, for reasons that will be discussed shortly. At this point, we shall briefly discuss the efficiency of primary production, that is, the ratio of the rate of production of organic matter to the rate of input of solar radiation.

In one of the earliest estimates of the efficiency of primary production, Transeau (1926) estimated the efficiency of gross primary production for a corn crop. According to his calculations 1.6% of the incident solar radiation on a plot of land was converted into organic matter (Table 9–2), and Ovington (1961) estimated that the efficiency of a Scot's pine plantation was 1.3%. Other estimates of the efficiency of primary production under field conditions vary, but they are always low, around 2% or less. Higher efficiencies, sometimes approaching 20%, are attained only in laboratories under very low light intensities. These efficiencies are sometimes quoted in order to support the claim that food can be produced in vast quantities for large human populations by culturing algae in large ponds, but these statements are quite misleading because little energy is

TABLE 9-1. NET PRIMARY PRODUCTION AND PLANT BIOMASS FOR MAJOR ECOSYSTEMS AND FOR THE EARTH'S SURFACE.*

	Area[a] 10^6 km^2	Net primary productivity, per unit area[b] dry g/m^2/yr		World net primary production[c] 10^9 dry tons/yr	Biomass per unit area[d] dry kg/m^2		World biomass[e] 10^9 dry tons
		Normal range	Mean		Normal range	Mean	
Lake and stream	2	100–1,500	500	1.0	0–0.1	0.02	0.04
Swamp and marsh	2	800–4,000	2,000	4.0	3–50	12	24
Tropical forest	20	1,000–5,000	2,000	40.0	6–80	45	900
Temperate forest	18	600–2,500	1,300	23.4	6–200	30	540
Boreal forest	12	400–2,000	800	9.6	6–40	20	240
Woodland and shrubland	7	200–1,200	600	4.2	2–20	6	42
Savanna	15	200–2,000	700	10.5	0.2–15	4	60
Temperate grassland	9	150–1,500	500	4.5	0.2–5	1.5	14
Tundra and alpine	8	10–400	140	1.1	0.1–3	0.6	5
Desert scrub	18	10–250	70	1.3	0.1–4	0.7	13
Extreme desert, rock, and ice	24	0–10	3	0.07	0–0.2	0.02	0.5
Agricultural land	14	100–4,000	650	9.1	0.4–12	1	14
Total land	149		730	109.		12.5	1,852.
Open ocean	332	1–400	125	41.5	0–0.005	0.003	1.0
Continental shelf	27	200–600	350	9.5	0.001–0.04	0.01	0.3
Attached algae and estuaries	2	500–4,000	2,000	4.0	0.04–4	1	2.0
Total ocean	361		155	55.		0.009	3.3
Total for earth	510		320	164.		3.6	1,855.

* From Whittaker, 1970.

Notes:

[a] Square kilometers × 0.3861 = square miles.

[b] Grams per square meter × 0.01 = t/ha, × 0.1 = dz/ha or m cetn/ha (metric centers, 100 kg, per hectare, 10^4 square meters), × 10 = kg/ha, 8.92 = lb/acre.

[c] 8.92 = lb/acre.

[d] Metric tons (10^6 g) × 1.1023 = English short tons.

[e] Kilograms per square meter × 100 = dz/ha, × 10 = t/ha, × 8,922 = lb/acre, × 4.461 = English short tons per acre.

TABLE 9-2. EFFICIENCY OF PHOTOSYNTHESIS OF AN ACRE OF 100-BUSHEL
CORN CROP DURING THE GROWING SEASON.*

Total solar energy available per acre during the growing season	$2,043 \times 10^6$ K cal
Used in photosynthesis	33×10^6
Used in transpiration	910×10^6
Lost by reflection, infrared radiation, and convection	$1,100 \times 10^6$
Total solar energy lost	$2,043 \times 10^6$

*After Transeau, 1926.

available at the low light levels necessary to achieve these high efficiencies. Moreover, a substantial portion of this energy fixed by photosynthesis is used by the plants in respiration.

Under natural conditions, much of the solar energy striking the earth's surface is not available for primary production. As we saw in Chapter 3 some of the solar radiation incident on the earth's surface heats the surface and then the air and soil below. This solar heating is a cost that cannot be avoided and is a cost assigned to the plants rather than the animals that benefit from the warmth generated by the conversion of solar to infrared energy and warm air. The average rate of loss of net infrared energy during the day is between 0.0 and 0.2 cal/cm²/min, since the earth radiates from about 0.4 to 0.7 cal/cm²/min and the sky reradiates about 0.2 to 0.6 cal/cm²/min back to the earth. This net radiation loss is made up from solar energy. Thus, if an area receives 600 cal/cm² of solar radiation each day, about 150 cal/cm²/day is lost as infrared radiation. In addition, in terrestrial ecosystems significant amounts of energy go into convection and evaporation; and finally, photosynthesis utilizes energy only in certain wavelengths. The energy in these wavelengths is about 20% of the total incoming solar energy. In the oceans efficiency is limited by the availability of necessary minerals. Although agricultural crops may have higher production rates than the natural communities they replaced, almost all agricultural communities must be maintained by man, or they will revert to a natural state. Maintenance of agricultural communities generally requires the expenditure of large amounts of energy in addition to what the plants are able to obtain in photosynthesis. These expenditures include manufacturing of fertilizers, running of farm machinery, and pumping irrigation water. The energy for these and a great many other activities necessary for achieving high agricultural production rates comes from fossil fuels, which, of course, are the consequence of primary production in the distant past.

Clements and Shelford (1939) and Lindeman (1942) suggested that production generally increases during succession, although there is disagreement about the specific path that the changes in production would follow. Clements and Shelford believed that production would decrease in the final stages of succession as the community approached the climax. Lindeman postulated that production would increase in lakes and ponds as a body of water progressed from a nutrient-poor (oligotrophic) to a nutrient-rich (eutrophic) condition, but it would then decrease as the lake filled in and became shallower and finally merged into a terrestrial system. Terrestrial production would then increase to above that of the shallow lake, reaching a maximum at the climax condition. However, Westlake (1963) has pointed out that productivity of emergent vegetation, characteristic of the transition from shallow pond to terrestrial conditions, appears to be as great or greater than either completely aerial or submerged communities. The pattern of changing productivity with succession in a terrestrial community may or may not be similar, depending on conditions. Odum (1960) followed net primary production for seven years in an old field succession in Georgia and found that production was highest in the first year after the field was allowed to go fallow. Production decreased rapidly in the first two years as readily available materials in the soil, left from fertilization of the agricultural crop, were used up. Production reached a plateau by the third year. During the several years following, total primary production remained constant, although the species composition changed and species diversity increased. Odum proposed that different vegetation types, such as shrubs, invading the area would be able to exploit mineral resources deeper in the soil, so that production would be increased with each new invading vegetation type and then remain constant until another kind of vegetation became established. The climax community for the area had not been reached by 1960.

Mechanisms Underlying Terrestrial Primary Production. The analysis of the process of primary production has been expedited by developing mathematical models to represent the important underlying processes. The influence of various environmental and vegetational variables on primary production can be ascertained by simulation experiments similar to those presented in Chapter 3 in the discussions of energy exchange with individual organisms. These models have clarified the relative importance of some of the processes and properties of the vegetation. The properties of terrestrial vegetation that have been identified and included in models are: leaf area index, leaf inclination to the horizontal, leaf absorptance, leaf resistance to gas exchange, the rate of photosynthesis of individual leaves at light saturation, the reflection and transmission of light in the canopy, and leaf density. The environmental properties are: the intensity of total solar radiation, the fraction of the total solar radiation which is diffuse, wind, infrared radiation from the sky, air temperature, humidity, and soil temperature. For each simulation of a model, a set of standard conditions for the environment and the canopy was defined and vegetation or environmental properties varied one at a time over a range of values that can be expected in nature.

Models calculating primary production in terrestrial communities can be divided into two categories:

1. Those developing the production of the total canopy from the light-photosynthesis relation of single leaves.
2. Those developing total production from the carbon dioxide profiles within the canopy.

Although these models have produced analyses that should have been interesting to ecologists dealing with the interrelations between the physiognomy of the vegetation, climate, physiology, and production, they have been overlooked for many years.

All the models calculating production of the total canopy from the light-photosynthesis relations of single leaves involve at least four steps:

1. Calculations of light intensity at several levels in the canopy.
2. Calculation of the photosynthetic rate per unit leaf area at the light intensity of each level.
3. Multiplication of this photosynthesis rate by the leaf area in each level.
4. Summation of the total photosynthesis of the levels to obtain a total photosynthesis of the canopy.

Almost all of the models do not include the effects of air temperatures and soil moisture on photosynthesis, and respiratory loss of carbohydrates is usually assumed to occur at a constant rate. The models differ in the specific equations used to describe several important variables such as light profiles in the canopy and light photosynthesis curves and, therefore, can be used to examine the consequences of different assumptions.

A simple model for stand photosynthesis that assumed uniformly overcast skies and a constant coefficent for light penetration into the canopy regardless of the time of day or depth in the canopy was developed by Monsi and Saeki (1953), who used their model to analyze the interrelations between leaf area index (LAI), leaf inclination to the horizontal, incoming light intensity, and primary production. Their analysis showed that under high solar intensities maximum production would occur in canopies having steeply inclined leaves and large LAI's; and under low solar intensities maximum production would occur in canopies having horizontal leaves and small LAI's (Fig. 9–5). The trends predicted by their model were consistent with broad geographic patterns observed in nature, i.e., in areas of low light intensity, such as within forests, species tend to have horizontal leaves and in areas of high light intensity, such as grasslands, leaves tend to be steeply inclined. This trend is apparent both among species of trees in a forest and in leaves on one tree.

The productivity of a canopy would seem to increase as its LAI increases, but this is not entirely true. As leaves are added to a canopy the average amount of light available for photosynthesis per unit leaf area goes down because of the effects of shading. Eventually, a point is

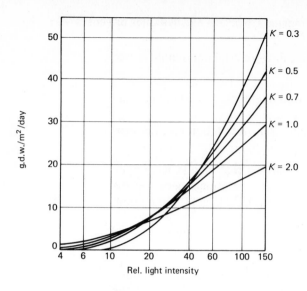

FIGURE 9–5. Interrelation between the extinction coefficient, intensity of solar radiation, and production. The extinction coefficient includes the effects of leaf area index and leaf inclination. (After Saeki, 1960.)

reached when additional leaves do not have enough light to photosynthesize at a rate equal to their respiration rate. Beyond this point an increase in LAI results in a net decrease in productivity.

This suggests that there is an optimum LAI for a particular kind of plant growing under specified conditions. The optimum LAI predicted by Monsi and Saeki's model, however, was usually too small in comparison to values measured for natural canopies or for agricultural crops. The explanation for this discrepancy apparently is that the relationship between respiration rate and temperature is not the same at all levels in a canopy and leaves acclimate rapidly to low light intensities so that net productivity may be positive even at low levels of light (Ludwig et al., 1965: McCree and Troughton, 1966).

Several recently proposed more complex models of terrestrial primary production will form the basis of the following discussion, in which productivity will be expressed in grams of dry organic matter per m² per hour or per day.

Several of the variables affecting production are aspects of canopy architecture that could evolve in response to environmental conditions. These variables are: leaf area index, leaf inclination, canopy density, and leaf width. An interaction between LAI, leaf inclination, and production has been shown in all models. With steeply inclined leaves, production increases as LAI increases; and with more horizontally inclined leaves, production increases to a maximum at a relatively low LAI and then decreases. With high light intensities, horizontal leaves at the top of the canopy are more than saturated with light, and the leaves within the canopy are shaded and thus limited in their productivity, as described above. If the uppermost leaves are inclined, their light absorption is reduced without reducing their photosynthesis and more light is allowed to penetrate into the canopy, thus increasing the production of the lower levels.

Several workers have estimated the rate of increase of productivity per unit increase in LAI for different conditions. DeWitt (1965), for example, assumed a canopy with individual leaves varying in their inclination from 0 to 90 degrees with a mean of 70 degrees. On this basis he estimated that productivity would increase as LAI increased but at a declining rate; at small values of LAI productivity was estimated to increase at the rate of 1.6 g/m²/hr per unit increase in LAI, but at a LAI of above 5, productivity increased at only 0.12 g/m²/hr per unit increase in LAI. Similarly, Miller (1971) calculated that the LAI for maximum production by red mangroves in Florida would decrease as follows:

Leaf inclination (degrees)	85	65	45	15	5
LAI for max production	7	4	3	1.7	1.5

Duncan et al. (1967) showed that production would be increased in a canopy in which leaves at the top were more steeply inclined than were those at the bottom. Maximum production was calculated when the leaf area and the steeply inclined leaves were concentrated at the top of the canopy, and minimum production was calculated with the leaf area and the more horizontally inclined leaves at the top of the canopy. When leaves were concentrated at the top of the canopy, production was more sensitive to the vertical distribution of leaf inclination than when leaves were concentrated at the bottom. Thus, both the maximum and minimum production rates occurred when the leaves were concentrated at the top of the canopy (Table 9–3).

The vertical distribution of leaf area varies for different types of vegeta-

TABLE 9–3. SIMULATED EFFECT OF CANOPY ARCHITECTURE ON PRIMARY PRODUCTION.*

Leaf inclination distribution	Leaf area distribution		
	A	B	C
a	36.2	37.3	38.8
b	33.8	34.0	34.2
c	32.2	31.8	31.6

*From Duncan et al., 1967.

Note: Primary production is in g/m²/day. The canopy used had a total leaf area index of 4.0 changing evenly in ten levels. In leaf area distribution A, the leaf area index was 0.2 in the top stratum and 0.6 in the bottom stratum. In B the leaf area was evenly distributed among levels. In C the leaf area was 0.6 at the top and 0.2 at the bottom. In leaf inclination distribution a, leaves were vertically inclined at the top and horizontal at the bottom. In b, all leaves were at 45°. In c, leaves were horizontal at the top and vertical at the bottom.

tion. Grass-type canopies have their leaf area concentrated at the bottom of the canopy and broad-leaved plants tend to have the leaf area concentrated at the top. In many natural canopies leaves at the top of the canopy tend to be more steeply inclined, even in canopies composed of one species (Miller, 1967).

By considering the energy budget of individual leaves in his model of primary production, Miller (1971) was able to study the effects of leaf shape and size on productivity. Miller's model suggested that as leaf width increases, net productivity decreases, and that this effect is greatest with narrow leaves. Beyond a width of about 10 cm productivity was predicted as becoming relatively insensitive to changes in leaf width. This model also suggests that as leaf width increases, daily transpiration decreases, and the transpiration efficiency; i.e., the ratio of production to transpiration increases to a maximum with 3- to 4-cm wide leaves.

Some of the variables affecting production are physiological, such as the rate of photosynthesis at light saturation, leaf absorptance and reflection, and aging. Of all the structural and physiological characteristics examined, the characteristic having the greatest effect on production appears to be the rate of photosynthesis at light saturation. In general, as the photosynthetic rate at light saturation increases, production increases. According to DeWit's calculations, an increase of 1.0 $g/m^2/hr$ in the light saturated photosynthetic rate should result in an increase in productivity of about 1.6 $g/m^2/hr$ when the LAI is 5.0 (Fig. 9–6). Miller calculated a production increase of about 0.12 $g/m^2/hr$ per 1.0 $g/m^2/hr$ increase in the light saturated photosynthetic rate with a leaf area index of 2.67. Not all the trends of production with variation in one parameter are so consistent in the models. In Miller's model production and evaporation increased as leaf absorptance increased; transpiration efficiency increased as absorptance increased to about 0.55, then levelled off. The increased photosynthesis was due to the increase in absorbed solar radiation, which increased photosynthesis of those leaves that were not light saturated. At the same time, the increase in absorbed solar radiation raised leaf temperatures, which increased the evaporation rate.

As light passes through the canopy, a portion of it is reflected in random directions by the leaves. DeWit examined the effects of this scattering by including a scattering coefficient as an index of reflection from leaves in his model. He found that the loss of light from the canopy increased with an increase of the scattering coefficient from 0 to 0.45 because of the reflection of light upward out of the canopy and downward to the soil. Nevertheless, production increased 1.4 $g/m^2/hr$ because of the more uniform distribution of the absorbed light through the canopy. DeWit pointed out that a high scattering coefficient could be favorable with high light intensities and high leaf area indices, but unfavorable with low light intensities and low leaf area indices.

The effect of aging of leaves was also included in DeWit's model by reducing the light saturated photosynthetic rate of older leaves. The rate

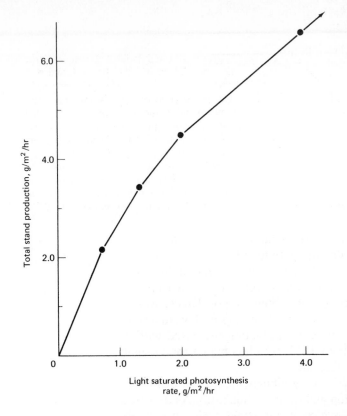

Total stand production, g/m²/hr

Light saturated photosynthesis rate, g/m²/hr

FIGURE 9–6. Effect of the maximum rate of photosynthesis with light saturation (AMAX) on the rate of production of the stand. (After DeWit, 1965.)

was decreased linearly from 2.0 g/m²/hr for leaves at the top of the canopy to zero for those below a leaf area index of 10. This modification had little effect on production when the LAI was small, i.e., less than one, and reduced production about 20% when the LAI was between 5 and 10 (Fig. 9–7).

In the above discussion the question asked was, "Given an environment, what is the optimum canopy structure or physiology for maximum production?" Now the question will be asked, "Given a particular canopy structure and physiology, what will be the effects of changing environmental variables?" The first question has fundamental interest in the evolutionary and competitive relationships of vegetation types in a climatic region, and the second question has practical importance in predicting possible consequences of man's alterations of the environment. Man is decreasing incoming solar radiation with atmospheric dust and pollution, increasing infrared radiation by increasing dust and carbon dioxide in the air, increasing air temperatures and decreasing precipitation and vapor density in cities, and increasing precipitation on the lee of cities. Man is developing the technology of weather modification to the extent that attempts at increasing the precipitation in an area by cloud seeding are becoming common. Some insight into possible ecological consequences of these environmental changes can be given with the models of primary production that we have been discussing.

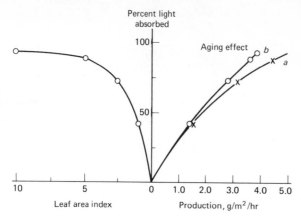

Percent light
absorbed

Aging effect

Leaf area index

Production, g/m²/hr

FIGURE 9–7. Interrelation between leaf area index, percentage of light absorbed, and production for a canopy in which the light saturated photosynthesis rate is constant with height (a) and one in which the light saturated decreases with depth in the canopy (b) representing and effect of aging of older leaves. The graph can be read beginning with the leaf area index, going up to the curve and across to the percentage of light absorbed, then across to the curves with or without aging, and down to the production estimate. (From DeWit, 1965.)

Miller's model suggests that changes in air temperature and vapor density have the greatest effect on productivity of any of the environmental variables. In his simulations of mangroves production decreased by about 8 g/m²/day when air temperatures were increased from 5°C below to 10°C above the usual conditions. When vapor density was increased by 20 g/m³, production increased by about 4 g/m²/day. In a simulation of production in the arctic tundra, production increased with increasing air temperature up to about 15°C, which is higher than usually occurs in area studied (Miller and Tieszen, 1971).

Air temperature affects productivity by influencing leaf temperatures, which in turn affect both respiration and transpiration rates. As transpiration increases, the plants may develop water stress which causes the stomates to close, thus limiting the exchange of CO_2 needed for photosynthesis. An increase in leaf temperature also increases respiration rates, which reduces net production. An increase in vapor density will also increase leaf temperatures and therefore raise respiration rates, but this is counterbalanced by a reduction of water stress which permits gas exchange and hence photosynthesis to proceed at a higher rate.

If the intensity of solar radiation remains constant and the proportion of diffuse solar radiation increases, primary production increases (Fig. 9–8). Diffuse light penetrates into the canopy more than does direct light, particularly when the sun is low in the sky, because the diffuse beam comes from all parts of the sky, instead of a point source. The greater penetration of light increases the illumination of the lower leaves and increases their rates of photosynthesis. Calculations made by DeWit (1965) suggest that the increase in primary production per unit increase in solar radiation is almost twice as great with overcast skies as with clear skies (Fig. 9–9). Similarly, Miller calculated an increase of about 13 g/m²/day per percent increase in diffuse radiation through the day as this increased from 5 to 35%.

Variations in infrared radiation can also affect production and the magnitude of the effect depends on leaf size (Miller, 1971; Miller and Tieszen, 1971). Increasing infrared radiation may have no influence on production by canopies of narrow leaves, but it may reduce production by canopies

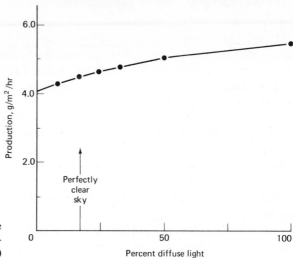

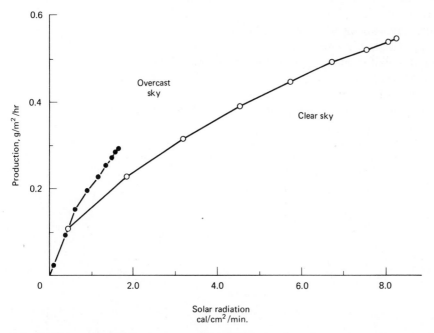

FIGURE 9–8. Response of production to the percentages of diffuse solar radiation in the incoming solar beam. (From DeWit, 1965.)

FIGURE 9–9. The effect of increasing solar radiation on production under clear and overcast skies. (From DeWit, 1965.)

of broad leaves. Narrow leaves can dissipate heat more rapidly by convection than broad leaves, and thus they are less affected by changes in radiation. Miller showed that production may decrease by 0.08 g/m²/day for a change of 0.01 cal/m²/min of infrared variation through the day. The significance of this effect can be appreciated by noting that a cloud passing overhead on an otherwise clear night may reduce infrared radiation by 0.05 cal/cm²/min, and variations twice this large may occur at one location through the year because of changing sky conditions

(Miller, 1971). Even larger differences may exist between geographic areas; for example, a three-fold difference in infrared radiation has been shown to exist between the arctic and south Florida during the growing season. Geographic variations of this size are sufficient to cause a difference of 3.2 g/m²/day in productivity.

Changes in air resistance above the canopy can also affect production in a rapidly photosynthesizing canopy but not in a canopy photosynthesizing at a low rate. DeWit (1965) studied this effect by considering the effect of air resistance above the canopy on the flux of carbon dioxide into the canopy. He did this by assuming in his model that wind speed above the canopy had a logarithmic profile that permitted air resistance to be calculated (as was discussed in Chap. 3). He found that production dropped from 5.1 to 3.3 g/m²/hr when air resistance was increased from 0.25 to 2.0 sec/cm. With the same change in air resistance, production dropped by less than 0.1 g/m²/hr when its initial value was only 1.5 g/m²/hr (Fig. 9–10).

The models discussed above were all based on processes with a single leaf. A different approach has been taken by Lemon (1960, 1967), who has developed methods of predicting production from micrometeorological assumptions. Lemon (1960) first developed a theory of the turbulent transfer of heat and water and an aerodynamic method of measuring evapotranspiration in a canopy. If the logarithmic expression for the wind

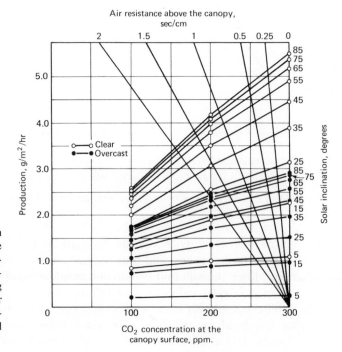

FIGURE 9–10. Interrelations between solar inclinations, air resistance above the canopy, carbon dioxide concentration at the canopy surface, and production. The graph can be read by locating the intersection between the lines of air resistance and solar inclination and reading carbon dioxide concentration and production. (After DeWit, 1965.)

profile is valid (Chap. 3), the flux of carbon dioxide to or from a vegetation canopy can be calculated from:

$$P_{net} = \frac{k^2 (u_2 - u_1) [(CO_2)_1 - (CO_2)_2]}{[ln (z_2 - d) / (z_1 - d)]^2}$$

where k is a proportionality constant, u_1 and u_2 are wind speeds at distances z_1 and z_2 from the ground, $(CO_2)_1$ and $(CO_2)_2$ are the carbon dioxide concentrations of the air at the two levels, and d is the zero plane displacement (p. 114). Lemon applied this approach to growing corn and was able to calculate a daily course of production. The method is promising because it is nondestructive and alters the natural configuration of the stand relatively little during the measurements. But in canopies with low productivity, with high winds, or with a large roughness parameter, such as most natural communities, the carbon dioxide concentrations must be measured precisely because the carbon dioxide gradients are small. In addition, the flux calculated by aerodynamic methods will equal productivity only if carbon dioxide production from the soil is zero. Furthermore, the turbulent transfer method measures total production by a stand and does not allow this production to be partitioned into the contributions of different strata in the canopy or according to species.

However, Lemon (1967) has further developed the aerodynamic method to allow net production to be partitioned into levels in the canopy. This newer approach requires measurements of the carbon dioxide profiles within the canopy. This method has been applied to fields of corn, but it has not yet been used in natural stands of vegetation.

Primary Production in Aquatic Ecosystems

Although there are basic similarities between aquatic and terrestrial primary production, there are also several major differences. Aquatic plants are in thermal equilibrium with the water they inhabit, and, of course, they do not experience water stress as do terrestrial plants. The complex interactions among leaf temperature, stomatal opening, water stress, etc., that are so important in the productivity of terrestrial plants do not occur in aquatic plants. A second major difference between primary production in terrestrial and aquatic ecosystems is that in the latter, the most important producers are often unicellular algae that are moved about by water currents and may sink below the depth of light penetration.

As in their terrestrial counterparts, photosynthetic rates of aquatic plants are dependent on light intensity. The rate of attenuation of light by water is far greater than by air, however, so that even in clear seawater there is insufficient light below a depth of about 120 m for photosynthesis to occur at a greater rate than respiration. The depth at which light is sufficient for photosynthesis to equal respiration is called the *compensation point,* and no net growth is possible. The water above the compensation point is called the *euphotic zone.*

The rate of attenuation of light in a body of water, and hence the depth of the euphotic zone, depends not only on the physical properties of water, but also on the amount of particulate matter suspended in the water. Usually the amount of light at depth z can be expressed by the equation

$$I_z = I_o e^{-kz}$$

where I_0 is the amount of light incident on the water's surface and k is the *extinction coefficient* (Hutchinson, 1957). The extinction coefficient for clear ocean water is about 0.04–0.05 when depth is in meters. Typical values for temperate oceans range from about 0.10 to 0.20 and may be as great as 1.0 in coastal regions (Ryther, 1963). Unusually turbid lakes may have extinction coefficients as high as 3 or 4.0.

A high extinction coefficient may be the result of silt and other suspended inorganic matter, but it can also be due to a high density of phytoplankton (Fig. 9–11). Here productivity may be high in spite of a shallow euphotic zone. In fact, very clear water is generally indicative of low productivity (Fig. 9–12). This suggests that the availability of light, although important, is not the main factor controlling productivity in many aquatic systems.

In contrast to many terrestrial systems, the availability of nutrients appears to be exceedingly important in determining the productivity of aquatic systems. Concentrations of important nutrients such as phosphates and nitrates are generally low in unpolluted aquatic systems.

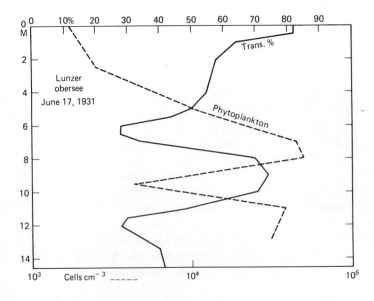

FIGURE 9–11. Relative horizontal transmission and number of plankton organisms at various depths, Lunzer Obersee (From Hutchinson, 1957.)

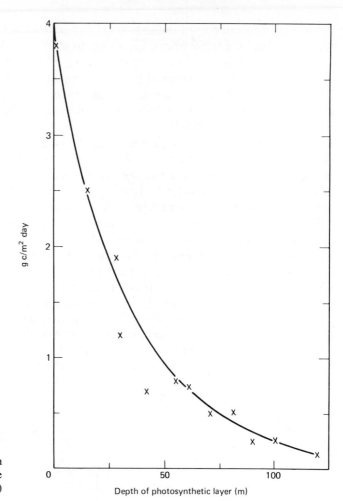

FIGURE 9–12. The relation between productivity and the depth of the euphotic zone. (From Ryther, 1963.)

Ryther (1963) has suggested that nitrate concentrations in seawater are typically only 1/10,000 as great as nitrate levels in fertile soils. The importance of nutrients as limiting factors can be demonstrated experimentally by adding known amounts of nutrients to lake or seawater samples and comparing photosythetic rates with controls. The high productivity of bodies of water that are polluted by organic wastes also demonstrates the importance of nutrient availability, as does the productivity of oceanic waters in regions of upwellings that carry nutrients back into the euphotic zone.

Nutrient availability is so important to primary production in lakes that limnologists use it to characterize lakes of different types. The two categories most often referred to are *oligotrophic* lakes, which have very low concentrations of nutrients, and *eutrophic* lakes, which have an abundance of nutrients. In the geological and ecological history of a lake there tends to be a trend from oligotrophy to eutrophy. As a lake ages it usually accumulates sediments from erosion of the surrounding area, and, depending on the nature of the parent material, nutrients may also be accumulated. When sedimentation has proceeded far enough, rooted aquatic plants may become established, and their productivity is added to that of the phytoplankton. The rate at which this gradual change from an oligotrophic to a eutrophic state takes place is highly dependent on geological conditions. In addition, in many areas the transition of lakes to a eutrophic state has been enormously accelerated by man's proclivity to use the nearest body of water as a dump for sewage and agricultural runoff, both of which are rich in nutrients.

Figure 9–13 shows profiles of photosynthesis for several lakes representing various trophic states. Lake Vanda, located in Antarctica, is one of the least productive lakes in the world. Although it remains frozen throughout the year, enough light penetrates the ice during the summer for a small amount of photosynthesis to occur. Lake Tahoe, a very deep alpine lake, is known for its clarity. This lake is located in a granitic basin in the Sierra Nevada Mountains, and it receives a minimum of nutrients from runoff, although human activities have unfortunately changed this situation in recent years. Castle Lake lies in the Klamath Mountains of northern California. As the graphs show, it is intermediate

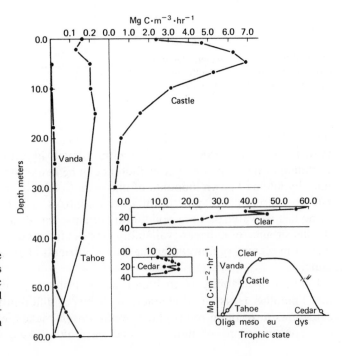

FIGURE 9–13. Profiles of the change in photosynthesis with depth in five lakes during summer. The general trophic state of these lakes is also indicated in relation to their relative carbon assimilation per unit of surface area. (From Goldman, 1968.)

in productivity. Clear Lake, also in California, is a highly eutrophic, shallow lake that is often anything but clear. Cedar Lake in Siskiyou County, California, is well beyond the eutrophic state. That is, it has become very shallow, is being invaded by rooted vascular plants, and is not far removed from becoming a bog. The adjective *dystrophic* is usually applied to this type of lake.

Of course, lakes cannot be neatly pigeon-holed into discrete categories. These categories should be thought of as loose terms of convenience that communicate some of the salient features of a lake. Figure 9–13 shows a continuum of productivity with these particular examples as points along the continuum.

A final major difference between primary production in aquatic and terrestrial systems is that rates of change of species composition and population densities are generally much higher in aquatic systems (Figs. 9–14 and 9–15). The small organisms that make up the phytoplankton are capable of very high growth and reproduction rates, but they are also subject to rapid population declines when conditions become unfavorable. The net result is that most aquatic systems are characterized by seasonal changes in both productivity and species composition. These seasonal changes are the result of complex interactions among seasonal fluctuations in physical conditions, such as water temperature, water circulation, light, and biological processes. Among the latter depletion of available nutrients by rapidly growing plankton and grazing of phytoplankton by zooplankton are probably the most important.

Some appreciation of the major features of these interactions can be gotten from a model of phytoplankton population dynamics proposed by Riley. We shall present only the simplest form of this model which appeared in 1946, but later versions incorporating various modifications have appeared more recently (Riley, 1963, 1965). Riley's model begins with an equation for the rate of change of phytoplankton population, P:

$$\frac{dP}{dt} = P(P_h - R - G)$$

where P_h is the photosynthetic rate per unit phytoplankton population, R is the respiration rate of the phytoplankton per unit of P, and G is the rate of grazing on the phytoplankton by zooplankton per unit P. Photosynthetic rate is assumed to be a linear function of light when nutrients are not limiting; i.e., $P_h = pI_z$, where I_z is the amount of light at depth z and p is an empirically determined constant. Substituting the equation given above for light as a function of depth gives

$$P_{hz} = pI_o e^{-kz}$$

This equation is then integrated over the depth of water from the surface to the bottom of the euphotic zone, depth z_1. Dividing this integral by

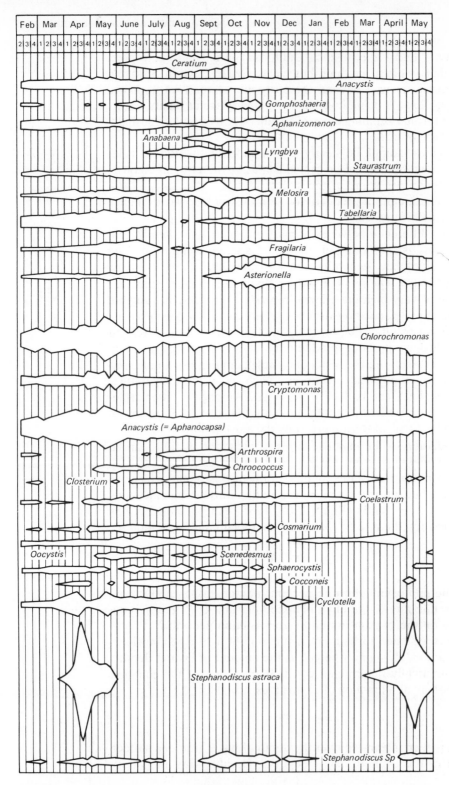

FIGURE 9–14. Seasonal variation of phytoplankton in Lake Mendota 1916–1917, relative abundances plotted in proportion to the cube roots of concentration. (From Hutchinson, 1967.)

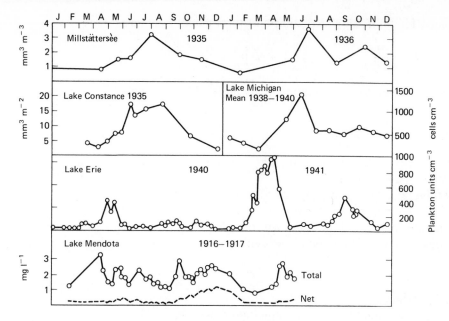

FIGURE 9–15. Seasonal variation of total phytoplankton in several lakes. (From Hutchinson, 1967; data from various sources.)

z_1 then gives the average photosynthetic rate per unit depth in the euphotic zone:

$$\bar{P}_h = \frac{pI_0}{kz_1}\left(1 - e^{-kz}\right)$$

The effect of depletion of only phosphate is included in this model. Although it has recently been shown that nitrates rather than phosphates may be more commonly limiting in marine environments (Ryther and Dunstan, 1971), availability of the two nutrients is often correlated and more data are available for phosphates. The effects of phosphate depletion are included by multiplying by the expression $1 - N$ when phosphorus ≤ 0.55 μg atoms where

$$N = \frac{0.55 - \mu\text{g atoms phosphorus}/l}{0.55}$$

This is a rough estimate based on laboratory experiments; expressed in this form, $1 - N$ decreases linearly from 1 to 0 as phosphorus concentration decreases from 0.55 to 0.0 μg $-$ at/l.

Vertical turbulence, which carries phytoplankton beneath the euphotic zone, is included by multiplying the photosynthetic rate by the ratio z_1/z_2 when $z_1 \leq z_2$, where z_2 is the depth of the mixed layer of water. Including the expressions for phosphate depletion and turbulence then gives

$$\bar{P}_h = \frac{pI_0}{kz_1} \left(1 - e^{-kz} \right) \left(1 - N \right) \frac{z_1}{z_2}$$

Respiration rate is assumed to be an exponentially increasing function of temperature:

$$R_T = R_0 e^{rT}$$

where R_0 and R_T are respiration rates at $0°$ and $T°$, respectively, and r is an empirically determined constant.

The zooplanktonic grazers are mostly filter-feeders that, according to Riley, tend to filter water at an approximately constant rate, regardless of the density of food present. Therefore, the quantity of plankton removed per unit time is a linear function of the density of zooplankton, Z:

$$G = gZ$$

where g is a fitted constant.

Combining these expressions into a single equation gives

$$\frac{dP}{dt} = P \left[\frac{pI_0}{kz_1} \left(1 - e^{-kz_1} \right) \left(1 - N \right) \frac{z_1}{z_2} - R_0 e^{rT} - gZ \right]$$

This equation has been applied to several sets of data from different locations with reasonable success; i.e., seasonal trends are approximately predicted (Fig. 9–16), suggesting that the major factors determining plankton productivity have been included in a fairly realistic manner.

We must note, however, that zooplankton density, phosphate concentration, and light are included in the model as empirically measured variables. That is, there is no provision in the model for phytoplankton density, P, to influence these variables, although the phytoplankton actually present in the locations studied did influence the data obtained for phosphate concentration, k, and z. These and other considerations are explored to some extent in the later papers by Riley mentioned above.

Productivity of Nonplanktonic Aquatic Plants. Phytoplankton form the base of the majority of aquatic food chains, and, consequently, a great many studies of the productivity of these plants have been made. In contrast, much less is known about the productivity of nonplanktonic aquatic plants, although they are undoubtedly important in shallow marine and freshwater communities.

A wide variety of taxonomic groups is represented in this group of producers, ranging from several phyla of algae to flowering plants. Included in the flowering aquatic plants are the so-called *emergents*, that is, vascular plants rooted in the water but having most of their leaves

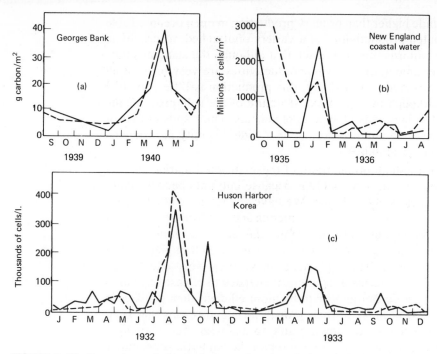

FIGURE 9–16. Comparison of observed seasonal cycles of phytoplankton (solid lines) with theoretical cycles (dotted lines) computed according to Riley's equations. (From Riley, 1963.)

above water. Cattails (*Typha*) and bulrushes (*Scirpus*) are examples of emergent plants. Primary production by these plants has much more in common with terrestrial plants than with other aquatic plants, and we shall not discuss them further here.

Nonplanktonic algae are found widely in both marine and fresh waters, but only in shallow water such as ponds, the margins of lakes, streams, estuaries, and coastal areas of the seas. Familiar examples include (1) the filamentous algae that sometimes form dense mats on the surface of ponds and estuaries, particularly eutrophic waters; (2) algae growing on pilings, rocks, boat hulls, etc.; and (3) the large kelps found along the coasts of North and South America.

Although much is yet to be learned about these nonplanktonic algae, one point seems clear—some of them are extremely productive (Blinks, 1955; Westlake, 1965). Ryther (1963) suggests that world-wide productivity of benthic algae may equal 10% that of marine phytoplankton, even though benthic algae are found in perhaps 0.1% of the area of the ocean. The dense kelp beds along the coast of California are certainly among the most productive areas in the world, with production rates estimated to be as great as 33 g carbon/m²/day (Blinks, 1955). This is roughly two orders of magnitude higher than primary production in open oceans. Table 9-4 gives estimated productivities of a variety of attached, macrophytic algae from the California coast. We see that although the standing crops tend to be higher among the browns, productivities are very high for all three phyla represented. Surprisingly, this production is little exploited by herbivores, although the forest-like structure of the kelp provides the habitat essential to hundreds of species of animals. Thousands of tons of kelp are also harvested annually for use in a wide diversity of products, from agar to ice cream stabilizers.

The reason for the high productivity of these attached marine algae is not precisely known, but it probably is because nutrients become available to them at a relatively high rate. We have already noted that the regions near the coasts of western South America and California are areas of upwellings that bring nutrients back into the euphotic zone. Blinks (1955) also notes that the attached marine algae are rich in pigments in addition to chlorophylls. These pigments make them very effective at intercepting light, but in terms of light quanta needed to release one molecule of oxygen, they do not appear to be more efficient than planktonic algae. Nevertheless, because of their growth form and organs that keep them buoyed up in the water they are able to maintain their photosynthetic apparatus in a dense layer near the top of the euphotic zone, rather than being dispersed throughout the euphotic zone, as phytoplankton tend to be.

Production at Secondary and Higher Levels

In many ecosystems the relatively large part of the chemical energy fixed by the primary producers passes on to the decomposers and a rela-

TABLE 9-4. PRODUCTIVITIES OF CALIFORNIA ALGAE (DRY WEIGHT) AS DETERMINED BY CARPELAN.*

Type		Standing crop g/m^2	Production $g/m^2/day$	Time for crop days
Green	Ulva	70	3–7.2	10–23
Red	Porphyra	300	11–21	15–26
Red	Gigartina	750	54	15
Red	Iridophycus	760	19	40
Brown	Egregia	800	25	32
Brown	Alaria	1,800	14	130
Brown	Fucus	2,130	19–42	50–112
Brown	Pelvetia	4,240	35	120
Brown	Laminaria (blades)	4,400	66	66

*From Blinks, 1955.

tively small fraction is consumed by animals, but this fraction is of great importance to us, since we ourselves are part of the consumer food web, as are the animals that supply us with useful commodities, compete with us for our harvest, or simply provide us with the intangible enjoyment of their presence.

How much energy an animal is able to sequester for its use and how it uses this energy are certainly of major importance in determining the animal's ecological success. Moreover, these considerations are closely interrelated, since to obtain energy (food) requires an expenditure of energy, as do adaptations to avoid being eaten. Thus, examination of how animals budget the energy available to them can add to our understanding of their role in the trophic dynamics of an ecosystem.

The various pathways for energy flow are shown diagramatically in Fig. 9–17. The biomass of an individual animal or population can be

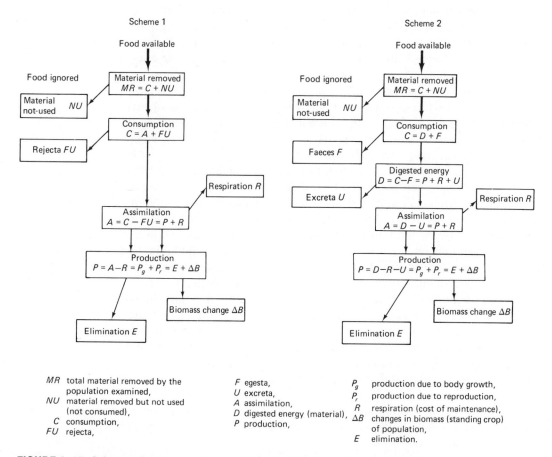

FIGURE 9–17. Scheme of matter (and energy) flow through an ecological unit. (From Petrusewicz and Macfadyen, 1970.)

thought of as a reservoir of chemical energy into which energy is flowing in the form of food and from which energy flows by several pathways. Energy flow through all of the animal populations in an ecosystem can then be diagramatically represented by many modules, such as that shown in Fig. 9–17, linked together in a network.

From the law of conservation of energy we see that if energy is not being accumulated or lost from the biomass reservoir, the rate of energy entering the reservoir must equal the sum of the rates of energy flowing from the reservoir. This principle allows us to construct a balanced *energy budget* showing rates of energy flow, together with the rate of energy accumulation or loss from the reservoir. The construction of an energy budget is not, however, an end in itself, but a tool that may be helpful in exploring ecologically interesting questions about the structure of food webs and the dynamics of energy flow through them.

The analysis of energy budgets of individual animals belongs to the study of environmental physiology, but since the energy budget of a population (and, ultimately, an entire community) is a composite of energy budgets of individual organisms, it is necessary to bring out two general points. First, although animals of one species will have similar energy budgets, individual differences in attributes such as size, age, and sex may result in significant differences among the energy budgets of members of a single population. Since the composition of a population changes through time with respect to these attributes, the energy budget of the entire population as a whole may likewise change. Second, the energy budget of a free-living animal depends on its environment and its activities within this environment. Laboratory data, depending on the conditions under which they were obtained, may give a highly inaccurate picture of what an animal's energy budget would be under natural conditions.

These points may be illustrated by considering energy lost in respiration, which is by far the largest expenditure of energy for almost all animals. Respiration rates are most commonly measured in the laboratory under standard conditions, that is, when the animal is quiet, in a post-absorptive state, and not under physiological stress, such as might be caused by a humidity level unusual to the animal. For homeotherms, ambient temperature is held within the animal's thermoneutral zone, that is, the temperature range within which respiration rate is not elevated because of either heat or cold stress. Since poikilotherms have no thermoneutral zone in this sense, their respiration rates are usually measured at several temperatures spanning the range occurring in the usual habitat of the species.

Studies of many species have shown that respiration rate, under the conditions described above, is often approximately related to body size by the equation

$$y = ax^b \qquad (9-1)$$

where x is the animal's weight and y is its respiration rate and a and b are empirically determined constants. The magnitude of b is usually around 0.7. [Much has been written in the physiological literature concerning the significance of this figure. See Kleiber (1961) or Gordon et al. (1968) for further discussion.] The size of a depends on the units chosen for x and y.

Figure 9–18 shows respiration rates of successive instars of the flour beetle *Tribolium castaneum*. These data illustrate both the appropriateness of Eq. (9–1) for instars that are similar except for size, i.e., larvae II–VI, and also the fact that considerable differences may exist among different life cycle stages within a species. Thus, we see that eggs, first instar larvae, prepupae, pupae, and adults of this species have markedly lower rates than would be expected from the equation fitted to data for larvae II–VI.

The second general point stated above can also be illustrated with respect to energy expenditure in metabolism. An animal moving about in search of food, constructing a nest, or engaging in any other routine activity will expend energy at a rate greater than would be predicted by Eq. (9–1), both because it is performing work (in the physical sense) and because it will often be operating under environmental conditions that are not optimal with respect to its heat balance.

Because of the technical difficulties involved, few estimates of energy expenditures by active animals have been made. Those that are available indicate that an energy budget that does not make adequate allowances for increases in respiration rate due to normal activity may be in great error. Tucker (1968) has shown, for example, that budgerigars consume oxygen about 12.8 times as fast while in level flight at 34 km/hr than when at rest, and that even more oxygen is used in level flight at higher or lower speeds (Fig. 9–19). Tucker was further able to show that metabolic rate of budgerigars changes with ascending or descending flight, even when these deviations are only five degrees.

Even when good estimates of respiration rates of active animals can be made, data showing how much time is spent in various activities are rarely available. Moreover, we usually lack accurate information on important physical parameters such as solar radiation, wind speed, and ambient temperature actually impinging on the animals we are interested in. As was shown in Chapter 3, these parameters may take on significantly different values at locations only a few centimeters apart, and, therefore, an energy budget that makes use of environmental data from the general area the species is found in may be quite misleading. In a few instances good progress has been made toward overcoming these problems. For example, Pearson (1960) placed a treadle that triggered a camera in the runways of the meadow mouse *Microtus californicus*. Humidity and temperature recorders were placed adjacent to the runways so that they were photographed along with the mouse as it passed over the treadle,

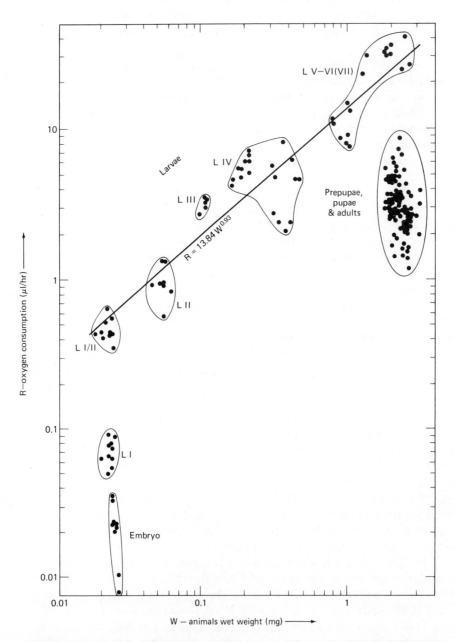

FIGURE 9–18. Size dependence of respiration rate of different developmental stages of *Tribolium castaneum*. (Modified from Klekowski, Prus, and Zyromska-Rudzka, 1967.)

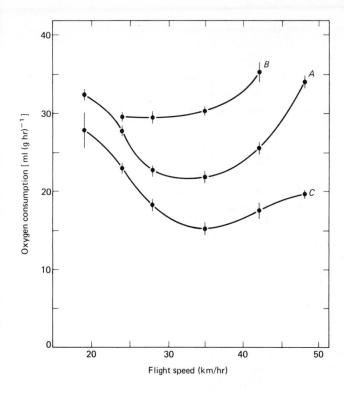

FIGURE 9–19. Mean oxygen consumption of two budgerigars during level (*A*), ascending (*B*) and descending (*C*) flight at different speeds. The angle of ascent and descent is 5°. The vertical bars represent two standard errors on either side of the mean. (From Tucker, 1968.)

thus simultaneously providing information on both the animal's activity patterns and its microenvironment. Using Pearson's data on temperatures, McNab (1963) was able to construct an approximate energy budget for *Peromyscus maniculatus,* which commonly uses the same runways. McNab concluded that a 19 g *Peromyscus* would require roughly 14 kcal/day to fulfill its energy demands, and that about 80% of this energy is used for thermoregulation and basal metabolism, leaving only about 2.8 kcal for activity. Other workers in environmental physiology have obtained data from sensors placed in birds' nests, rodent burrows, and similar places, and some behavioral and physiological data are being obtained by telemetry, but this information has not yet been used in studies of energy flow at the population or ecosystem level such as we are concerned with in this chapter.

A second major source of difficulty in constructing energy budgets lies in determining how much energy a free-living animal actually obtains from its food. Most data on what free-living animals eat are qualitative; that is, we often know what is included in an animal's diet, but we don't know how much of each food item is eaten, and we don't usually know how the availability of various food alters the animal's selectivity. This lack of information would be of no great consequence (at least to energy flow determination) if all food eaten by an animal had the same caloric value and digestibility, but, unfortunately, this is far from the case. A great many studies, particularly of domestic grazers such as sheep and cattle have demonstrated this. Among recent investigations of wild species, that of Drożdż (1967) is a good example. Drożdż investigated food preferences and digestive ability for the bank vole *(Clethrionomys glareolus)* and the yellow-necked field mouse *(Apodemus flavicollis)* living in a beech forest in Poland. Table 9–5 shows the results of feeding experiments on captive animals in this study and Figure 9–20 shows the results of stomach content analysis of animals trapped in the woods. (The mixed diet referred to consisted of ground hazelnuts and plants from the herb layer of plants, mixed in approximately the same ratio as yearly averages of stomach contents indicated each species had in its natural diet). Based on these data and gross consumption figures (which were not given) Drożdż calculated that each bank vole consumed from about 13.01 to 15.03 kcal/day, depending on its diet, and of this, from 11.58 to 12.77 kcal/day were assimilated. For yellow-necked field mice, these figures ranged from 12.14 to 16.96 kcal/day for intake and from 9.42 to 15.49 kcal/day for assimilation.

The difficulties in constructing accurate energy budgets for individual animals that we have briefly discussed (and there are many more that were not mentioned) may give the impression that obtaining reasonable estimates of energy flow through populations or trophic levels may be hopelessly complex. It is probably true, however, that changes in energy flow that are the consequences of changes in population sizes are of greater importance to the dynamics of energy flow through ecosystems than are changes associated with differences among individual energy budgets. At any rate, this question warrants further investigation.

The study of Wiegert (1964) of the energetics of meadow spittlebug populations is an excellent example of the energy budget approach applied to populations. Wiegert constructed energy budgets for populations of spittlebugs living in two different habitats, and as we shall see, this comparative approach revealed interesting aspects of their energetics that would not have been disclosed if only one population had been studied. One spittlebug population lived in an alfalfa field, the other lived in an "old field" that had not been cultivated for nearly 40 years.

Table 9–6 shows energy budgets expressed in calories per day for individual spittlebugs. The figures given in "Ingestion" are based on laboratory experiments in which spittlebugs fed on the xylem sap of tomato plants (the natural populations, however, fed on other species of plants). Alternative estimates of ingestion rates were obtained for field populations by summing the rates of respiration, growth, and egestion. This sum is the total output, which should be equal to the input, and as the table shows, this indirect estimate of egestion agrees well with the rate of ingestion obtained for nymphs on tomatoes. The lack of agreement for adults

TABLE 9-5. DIGESTIBILITY, ENERGY CONTENT, AND ABUNDANCE OF FOODS OF THE BANK VOLE (*CLETHRIONOMYS GLAREOLUS*) AND THE YELLOW-NECKED FIELD MOUSE (*APODEMUS FLAVICOLLIS*).

Digestibility and assimilation of natural diets in voles and mice (in percent of energy intake).

Diets	Feces	Digested energy	Urine	Assimilated energy
Clethrionomys glareolus				
Beechmast	7.04	92.96	3.98	88.98
Mixed	12.31	87.69	2.90	84.79
Oatmeal	11.39	88.61	3.30	85.28
Apodemus flavicollis				
Hazelnuts	6.96	93.04	1.69	91.35
Mixed	8.75	91.25	1.99	89.26
Acorns	18.65	81.35	2.75	77.60

Caloric values of used diets, feces, and urine. Diets and feces in Kcal/g ash free weight. Urine in cal/g of liquid.

Kind of diets	Food	Feces	Urine
Clethrionomys glareolus			
Beechmast	7.212	5.200	170
Mixed	6.998	5.026	164
Oatmeal	4.619	5.318	65
Apodemus flavicollis			
Hazelnuts	8.032	5.340	200
Mixed	7.431	4.863	330
Acorns	4.159	4.479	132

Net primary productivity of the beech forest and food available to rodents. All values in kcal/ha/year $\times 10^3$.

Kind of food	Net productivity of beech forest	Food of voles	Food of mice
1. Herb layer vegetation	1,083	918	512
2. Tree leaves	13,427	671	324
3. Tree twigs (trunks and branches)	3,584 (25,557)	107	?
4. Tree seeds	225 (28.5–456)	202 (22–360)	202 (22–360)
5. Fungi and invertebrates	74	51	47
6. Total food supply (1–5)	44,000 (= 10.3 tons)	1,949 (1,769–2,007)	1,085 (905–1,234)

*From Drożdż, 1967.

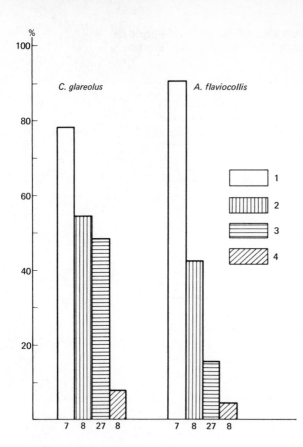

%

FIGURE 9–20. The food preference of rodents. Bars represent mean degree of consumption in percent. Figures indicate the number of species tested in given group: 1—seeds of trees, 2—fresh seeds and fruits, 3—plants and shrubs of herb layer, 4—twigs and buds. (From Drożdż, 1967.)

TABLE 9-6. INDIVIDUAL ENERGY BUDGETS FOR THE NYMPHS AND ADULTS OF *PHILAENUS SPUMARIUS.* *

Temp. °C	Insect	Ingestion	Respiration	Growth	Egestion	Σ Resp. + growth + egestion	% Assim- ilated
23	3rd instar (0.39 mg)	0.8–2.7	0.34	0.25	0.43–1.35	1.02–1.94	30–58
25	4th instar (1.16 mg)	2.1–6.9	0.76	0.74	1.12–3.48	2.62–4.98	30–57
21	Adult (3.87 mg) excreting uric acid	1.8–5.9	1.00	—	0.41	1.41	71
21	Adult (3.87 mg) excreting urea	1.8–5.9	1.00	—	0.25	1.25	80

*From Wiegert, 1964.

Note: Values in cal/day. Respiration growth, and egestion are independent estimates. Ingestion based on feeding rate and known range of caloric values for tomato xylem sap.

was probably due to differences in the caloric values of alfalfa and grass xylem sap as compared to tomato xylem sap.

Rates of oxygen utilization were measured in a manometer-type respirometer and then converted to rates of energy utilization by using a respiratory quotient of 0.82. This figure is appropriate if amino acids are being catabolized, as appeared to be the case. Growth was measured gravimetrically, with increases in weight being converted into caloric values. To make this conversion, Wiegert multiplied weight gains by the caloric value of a unit weight of spittlebug tissue, as determined by combusting spittlebugs in a bomb calorimeter.

Much of the energy lost by nymphs through egestion is in the form of organic material in a foamy, viscous liquid that covers the body of each nymph. (The appearance of this material gives these insects their common name. Presumably, this substance provides some protection from predators.) Since the amount of organic material in the spittle varies and the energy released by burning small samples in a calorimeter is difficult to measure, there is a large range in the rates of energy loss shown for egestion in Table 9-6.

Table 9-7 shows energy budgets for populations of spittlebugs living in the old field and a small alfalfa field. These energy budgets were constructed by combining estimates of population densities, body growth rates, mortality rates, and other field data with energy budget data obtained for individual spittlebugs.

The much higher rates of energy flow in the alfalfa field as compared to the old field is immediately obvious. In addition, there is a subtle but important difference between the two habitats. First, notice that within

TABLE 9-7. POPULATION ENERGY BUDGETS FOR MEADOW SPITTLEBUGS (*PHILAENUS SPUMARIUS*) ON THE OLD FIELD 1959-1960 AND THE SMALL ALFALFA FIELD 1960.*

Energy category	Old field – 1959			Old field – 1960			Small alfalfa field – 1960		
	Nymphs	Adults	Total	Nymphs	Adults	Total	Nymphs	Adults	Total
Eggs previous fall	4			8			399		
Growth	46			93			15,125		
Loss to molt	2			3			838		
Production	48			96			15,963		
Respiration	38	496	534	79	992	1,071	12,216	10,386	22,602
Assimilation	86	496	582	175	992	1,167	28,179	10,386	38,565
Excretion	173	248	421	350	496	846	56,358	5,193	61,551
Ingestion	259	743	1,002	525	1,488	2,013	84,537	15,579	100,116
Increment to Adults	6			11			10,120		
Mortality	41	475	516	82	1,056	1,138	5,005		

*From Wiegert, 1964.

Note: All values in calories/m^2/year.

each column the energy budget balances. For example, in the first column the total energy ingested for the year (259 cal/m²) equals the sum of the production, respiration, and excretion. Now compare figures in different columns. Column 1 shows 6 cal/m² as increment to adults. This figure represents what the total biomass per m² of the adult population would be if no adults died or emigrated in that year. If no immigration of other adults into the old field took place, then the 8 cal/m² of eggs shown at the top of column 4 would have to have been produced by the 6 cal/m² of adults, or even fewer adults, since some adult mortality undoubtedly occurred. A female spittlebug weighs about 3.9 mg and lays about 22 eggs, each weighing about 0.030 mg, and only about half of the adults are females. Simple arithmetic using rough estimates of about 6 cal/mg for adult spittlebugs and their eggs shows that 6 cal/m² of adults cannot possibly produce 8 cal/m² of eggs.

Another apparent discrepancy appears when we note that the 6 cal/m² of new adults in column 1 seem to have respired 496 cal/m² during their life (column 2), whereas the 10,120 cal/m² of new adults in the alfalfa field (column 7) respired only 10,386 cal/m². The ratios for these pairs of numbers are 82.7 and 1.0 for the old field and alfalfa field, respectively.

These apparent discrepancies suggest that there was a large flux of adult spittlebugs into the old field and that these immigrants add their eggs and respiration to the residents of the old field. These immigrants probably originated in the alfalfa fields from which they were forced to disperse when the alfalfa was mowed in the late spring and late summer. Figure 9–21 shows changes in population density, energy flow, and biomass for the two fields studied by Wiegert (1964). The curves for the alfalfa field do not show drops produced by mowing, since this particular field was not mowed, but the sharp increases in the curves in June for the old field coincide with the time of mowing of other alfalfa fields in the area.

At this point the reader might wonder what is the impact of a population of spittlebugs on their host. No data are available for answering this question in the case of the old field population, but Wiegert (1964) estimates that the alfalfa field population decreased the gross photosynthesis of the alfalfa by about five times the amount ingested by the spittlebugs. This amounts to a decrease of about 420 kcal/m² or about 100 grams of alfalfa per m².

Few studies have been made of energy flow through an entire ecosystem, for a study of any except the simplest ecosystems requires a prodigious amount of work extending over several years. Even the work done on very simple ecosystems has not progressed much beyond a descriptive phase. Cold springs were selected for several of these studies because physical conditions in them change much less through time than do conditions in terrestrial ecosystems, and, equally important, the number of species is usually quite small.

The work of Tilly (1968) on Cone Spring, Iowa is a good example of a

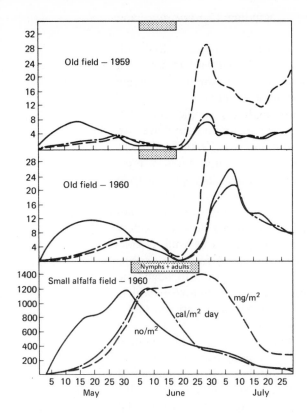

FIGURE 9–21. Fluctuations in energy flow, numbers, and biomass of populations of *Philaenus spumarius* on the old field 1959–1960 and the small alfalfa field 1960. The stippled bars indicate periods when both nymphs and adults were present. Diel energy flow in cal/m^2 (–––––––); number of individuals/m^2 (————); biomass in mg/m^2 (–––––––). (From Wiegert, 1964.)

study of energy flow in a simple ecosystem. The total area of this spring was 141 m^2 and its depths ranged from 0.5 to 16.5 cm. Flow remained essentially the same throughout the year at 1 liter/sec. Water temperature ranged from 5 to 18.5°C, dissolved CO_2 ranged from 1.5 to 11.2 mg/liter, which was the widest range of any of the physical properties reported, and dissolved oxygen ranged from 5.4 to 11.3 mg/liter. Other physical conditions showed less variation than these.

The primary producers comprised three flowering plants: *Bacopa rotundifolia,* a perennial similar to watercress in growth form; *Impatiens capensis,* a rapidly growing annual that reached a height of 2 meters; and *Lemna minor* (duckweed) that was found throughout the year on the water's surface in the spaces between the *Bacopa* leaves and in the open areas of low flow where *Bacopa* was absent. No algae of importance were found.

Somewhat surprisingly, no herbivores were found to feed directly on any of these primary producers. Rather, the major pathway of energy flow from the autotrophs to the heterotrophs was through several detritus feeders. These were the amphipod *Gammarus pseudolimneus,* the trichopteran *Frenesia missa,* the pulmonate snail *Physa integra,* several species of Chironomid (Tendipedid) flies (including *Cardiocladus* and *Pentaneura*).

Feeding on the detritus feeders were found the planarian *Phagocata*, the Megalopteran *Chauliodes*, an unidentified larval Heleid fly, and, interestingly, the fly *Pentaneura*. Thus, we have in this simple ecosystem a good example of a species occupying two trophic levels.

Measuring energy flow through even this relatively simple food web was not an easy task. Tilly used a variety of techniques that we can only briefly mention. Net primary production was estimated from sampling the biomass of each important species of plant at five-day intervals, or, in the case of *Impatiens*, by measuring changes in height and then converting this to biomass.

Estimating energy flow through the animal population was more complicated. As we pointed out in our discussion of energy flow in meadow spittlebug populations (Wiegert, 1964), mortality rates as well as consumption rates must be estimated. Mortality rates, *d*, for each species with relatively large populations changing unabruptly in size through time were estimated by first assuming that the interval of time over which population size increased the most represented a period of zero mortality. The growth rate of population, *b*, in the absence of mortality, i.e., with $d = 0$, was then assumed to follow the equation $N_{t+1} = N_t e^b$ This equation was solved for *b* (see p. 149) with N_t and N_{t+1} having been estimated by sampling. During other intervals of time, mortality rate was estimated by replacing *b* with $b - d$ in the same equation and solving for *d*. This approach has obvious shortcomings. If, for example, reproduction did not occur during a time interval, mortality would be significantly overestimated. In addition, animals produced and dying between samplings would not be accounted for. In the absence of additional data, however, the approach used by Tilly was not unreasonable. Net production by less abundant species was estimated by other indirect means that we shall not discuss.

Estimates of instantaneous mortality rates were then multiplied by population sizes for each interval of time and the resulting number of animals dying summed for one year. This total, together with the net change in population size, was used to estimate annual production of biomass by these species.

Energy loss from the system by animal respiration was estimated for some species by placing individuals in closed containers of water and then measuring the dissolved oxygen by microtitration at the start of the experiment and then again after several hours. Respiration rates were assigned to other species on the basis of published values for closely related species.

Respiration rates for the entire community were estimated by placing entire sampling units, including bottom material, into glass jars that were darkened, and then measuring the amount of dissolved oxygen at the beginning and end of one hour. From these measurements respiration rates of microorganisms were estimated by subtracting oxygen consumption by microorganisms.

As we have already mentioned, detritus was very important in Cone Spring. The amount present in the system was estimated by first separating detritus from inorganic material and live organisms, principally by sieving and hand sorting, and then combusting the remaining material in a muffle furnace.

The results of Tilly's study are summarized in Fig. 9–22 and Fig. 9–23 and Table 9–8. These data constitute an energy budget for this relatively simple ecosystem. This energy budget, like all others, must balance, and, as depicted in Fig. 9–23, it does. The data in Table 9–5 show some discrepancies, however, since they show that the total respiration of consumers is greater than the sum of net primary production import of organic material and net change in the standing crop of the consumers. Tilly con-

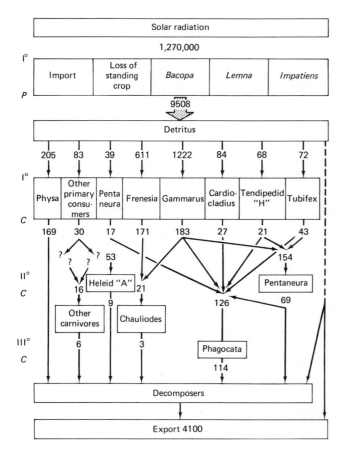

FIGURE 9–22. Quantitative food web indicating relations of major species in Cone Spring simplified and quantified in terms of income and yield estimated as described in the text. Boxes represent component units recognized; numbers above boxes are income estimate, numbers below are yield estimates. Trophic levels are symbolized as follows: I°P = primary producers, I°C = primary consumers, II°C = secondary consumers, and III°C = tertiary consumers. Only the main directions of energy flow in kcal/m²/yr have been indicated by arrows; minor species have been lumped. (From Tilly, 1968.)

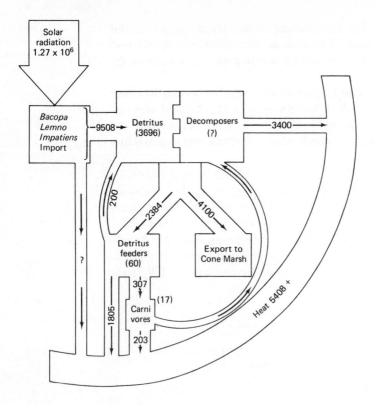

FIGURE 9–23. Energy flow diagram summarizing the major pathways through the Cone Spring ecosystem. Numbers in parentheses represent mean monthly standing crops. All values are kcal/m². Flow rates per year are indicated by arrows. (From Tilly, 1968.)

TABLE 9–8. ECOSYSTEM ENERGY BUDGET ITEMS
kcal/m²/yr.

Energy budget items	Kcal/m²/yr
Macrofauna respiration	2,008
Microorganism respiration	3,400
Total consumer respiration	5,408
Exported Organic matter	4,100
Change in standing crop	
Bacopa	−259
Lemna	+5
Consumers	+7
Total (net)	−247
Net primary production	
Bacopa	204
Lemna	134
Impatiens	730
Total	1,068
Import	626

*From Tilly, 1968.

sidered it highly likely that the combination of net primary production and import was underestimated, and that if his estimates of consumer respiration were in error, they were too low. He, therefore, used the figure of 9,508 kcal/m²/yr as the rate of energy flow from producers to consumers. Tilly's interpretation seems reasonable to us, if we recall that his estimates of primary production were based on measurements of standing crop made at five-day intervals. This is not a short interval of time relative to the rates at which rapidly developing animals can expend energy, and it is likely that a significant amount of plant material was produced or imported and consumed in the interval between samplings.

In view of these considerations, we shall assume that Fig. 9–23 is a sufficiently reasonable representation of energy flow through Cone Spring to illustrate our discussion.

An important question regarding Cone Spring (or any other ecosystem being studied) is whether its yearly energy budget represents a steady state; that is, Are these significant changes from one year to the next in the amount of biomass in the various components of the ecosystem? Tilly's data show a net decrease for the year of 259 kcal/m² in standing crop by *Bacopa* and some smaller changes in some of the other populations, but Tilly considered these to be only temporary deviations from an essentially steady state. Large fluctuations in standing crops, however, did occur within the year in almost all of the species studied. We shall say more about this later.

A steady-state condition, on a year to year basis, is probably common among ecosystems that have not been disturbed for a long time, although sufficient data do not exist to be certain of this. It is possible that even in the absence of disturbance some communities may not achieve a steady state. At any rate, we can be certain that in an ecosystem undergoing succession a steady state for energy flow does not prevail. As we saw in Chapter 7, the latter part of a successional sequence is characterized by, among other things, an accumulation of biomass at a gradually decreasing rate. This means that for some period of time the overall rate of energy input into the organisms that make up the ecosystem is greater than their output in metabolic heat, and that this difference gradually decreases.

Figures 9–22 and 9–23 show that virtually all of the organic matter produced by plants in Cone Spring becomes detritus before being consumed by animals. This lack of true herbivores may seem surprising, but it is not so very different from other ecosystems as might be supposed. In aquatic ecosystems having large populations of herbivores the primary producers are usually algae having high turnover rates, but Cone Spring, like many springs, contained only negligible numbers of algae. In a study of energy flow in a saltmarsh ecosystem in which the main primary producer was Spartina, Teal (1962) found that less than 5% of primary production was used directly by herbivores, and W. Odum (personal communication) found that little herbivory occurs in mangrove swamps.

Few other studies of energy flow through either aquatic or terrestrial ecosystems have been carried out. We shall briefly summarize two additional studies of springs, each different in its own way from Cone Spring, so that comparisons can be made among them. The techniques used in studying these other springs were similar to those already mentioned and, therefore, will not de described.

Perhaps the best-known investigation of energy flow at the ecosystem level is H. T. Odum's (1957) work on Silver Springs, Florida. This system is far larger than either Cone Spring or Root Spring, which will be discussed below. Water enters Silver Springs in several boils, and the total flow rate is about 80,000 m³/hr, whereas the flow rate of Cone Spring is only about 3.6 m³/hr and Root Spring is even smaller. The chemical and physical properties of the water entering Silver Springs remain nearly constant throughout the year. The bulk of primary production is performed by rooted flowering plants, mainly *Sagittaria lorata* (eelgrass) and algae growing in a mat (referred to as *aufwuchs*) on blades of *Sagittaria*.

In marked contrast to Cone Spring, there are significant populations of herbivores in Silver Springs, including several species of fish, snails, turtles, shrimp, and insects. Carnivores include fish (including one species that is also an important herbivore), insects, and coelenterates. Decomposers are principally bacteria, as in Cone Spring and Root Spring, but Odum also classified crayfish as important decomposers because they feed mainly on the remains of dead animals. Although bacteria make up a small part of the total standing crop of organisms in Silver Springs, they are second only to the green plants in rate of energy utilization.

Even though the chemical and physical properties of the water entering Silver Springs are nearly uniform throughout the year, the input of solar radiation is not. Because of seasonal variations in solar radiation primary production shows a marked seasonal pattern (Fig. 9–24). Unlike Cone Spring, however, numbers and biomass of consumers do not change very much, although seasonal variations in energy available to consumers may be related to seasonal variations in reproduction in some of the animal populations (Figure 9–25). Many of the consumers in Silver Springs are relatively large and long-lived, and it seems likely that these characteristics may dampen seasonal fluctuations in population sizes and biomasses.

Energy flow through Silver Springs is shown in Fig. 9–26. The pattern appears to be similar to the pattern of energy flow through Cone Spring (Fig. 9–23); that is, the number of trophic levels, orders of magnitude of standing crop and of flow rates are essentially the same in both springs. Decomposers are seen to be more important in both systems than would be indicated by just their biomasses.

The rate of downstream export from Silver Springs is notably less than in Cone Spring, which may be surprising in view of the vast difference in water flow rate between them. The export rates, however, are expressed

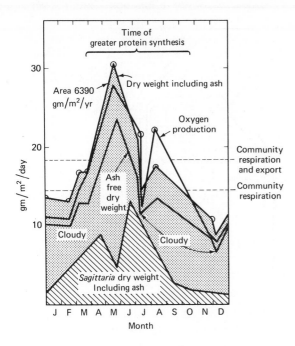

FIGURE 9–24. Annual sequence of primary production. The lowermost curve is the net production of *Sagittaria* from growth measurements in cages. Times of protein synthesis are indicated on the basis of photosynthetic quotients. Community compensation points are indicated by horizontal dashed lines. The annual production is obtained from the shaded area under the curve of dry weight production. (From Odum, 1957.)

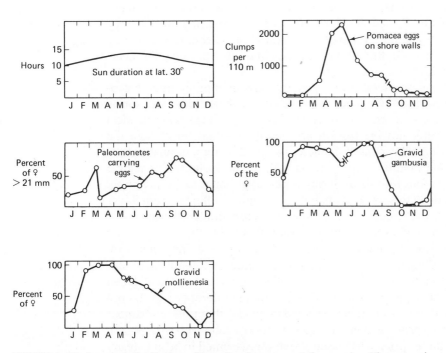

FIGURE 9–25. Seasonal distribution of breeding in representative species. The approximate day period is given in the upper graph. (From Odum, 1957.)

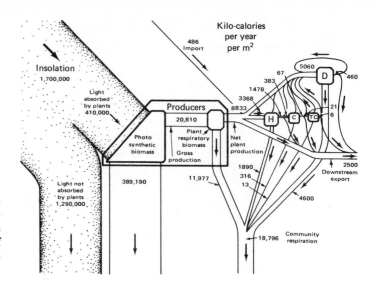

FIGURE 9–26. Energy flow diagram summarizing the major pathways through the Silver Springs ecosystem. Estimates of rates of flow in kcal/m²/yr are shown on arrows. (From Odum, 1954.)

in kcal/unit area, not in terms of total export from the system, and, therefore, these rates are dependent on velocity of water movement, not on volume per unit time. Solar radiation input is seen to be 25% greater in Silver Springs than in Cone Spring, but the rate of energy flow to consumers is about the same. This does not necessarily mean that the primary producers in Cone Spring are more efficient in converting solar energy to chemical energy; more probably, it is because of a higher rate of energy entering Cone Spring in the form of leaves, twigs, etc., produced by nearby terrestrial plants.

Teal (1957) has provided data on energy flow in another aquatic ecosystem—Root Spring in Concord, Massachusetts. This cold-water spring is similar to others described above in that the physical and chemical properties of water flowing from it are nearly constant throughout the year. Root Spring, whose basin diameter is only 2 meters, is smaller than Cone Spring and minute in comparison with Silver Springs. The regime of solar energy input into Root Spring is perhaps similar to Cone Spring, although no data on this were given, but it is certainly different from Silver Springs because of the considerable differences in latitude between them. More important, however, is the fact that about three-quarters of the energy input to Root Spring was debris (leaves, twigs, etc.) from a nearby apple orchard falling into the water and only one-quarter was from photosynthesis by aquatic plants (principally filamentous and colonial green algae and diatoms) in the spring. Despite these differences in energy input and size, there was nothing especially unusual about the assortment of animals in the spring or their arrangement into a food web (Fig. 9–27).

Each of the three springs discussed above was assumed to be in a steady state, at least on a year-to-year basis; that is, a tally sheet showing the

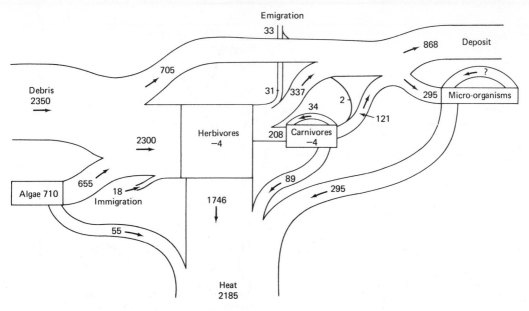

FIGURE 9–27. Energy flow diagram for Root Spring, Concord, Mass., in 1953–1954. Figures in kcal/m²/yr; numbers inside boxes indicate changes in standing crops; arrows indicate direction of flow. (From Teal, 1957.)

standing crops of the various species would not show much change from one year to the next. If we examine what is happening within the span of a year, however, we will see that there are fairly large changes taking place in each of these ecosystems. Figures 9–28 and 9–29 demonstrate this for the more important populations (in terms of energy flow) in Cone Spring. The fact that large fluctuations occur and yet these ecosystems return to much the same state at yearly intervals strongly suggests the existence of homeostatic interactions among the components of each ecosystem.

Other ecosystems are not in a steady state, however, but are undergoing succession, which shows that the interactions among their constituent parts are different in some important respects from steady state ecosystems. Assuming that the interactions of the species in an ecosystem are reflected in the pattern of energy flow through the system, it is reasonable to try to quantify these patterns in some way that allows comparisons to be made.

The first approach used toward this end was to lump the species of an ecosystem into trophic levels and to compare energy flow through these greatly simplified, abstract systems. Most studies of this kind have also made the simplification of dealing only with yearly energy budgets. These restrictions should be kept in mind when interpreting such comparisons.

In comparing a diversity of ecosystems it is difficult to utilize raw data

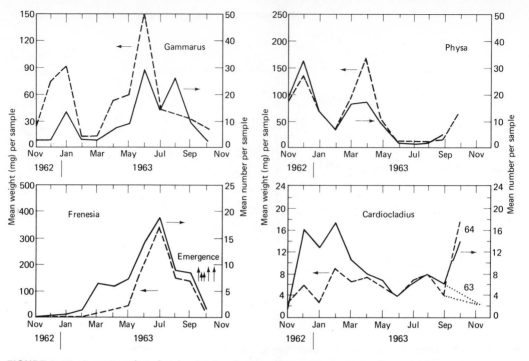

FIGURE 9–28. Annual cycles of major detritus feeders. Changes in mean standing crops of numbers (solid lines) and biomass (dashed lines) blotted wet weight per 0.025 m² sample. (From Tilly, 1968.)

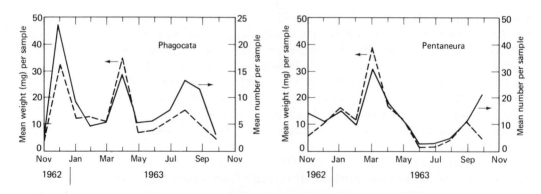

FIGURE 9–29. Annual cycles of major carnivores. Changes in mean standing crops of numbers (solid lines) and biomass (dashed lines) blotted wet weights per 0.025 m² sample. (From Tilly, 1968.)

directly. What we are interested in, however, are usually not absolute rates of flow of energy but relative flow rates; for example, how much energy is being consumed by primary carnivores relative to how much energy is being consumed by secondary carnivores per unit time? Thus, in comparing ecosystems, it makes more sense to compare the ratios of two energy flow rates rather than to use raw data directly.

A ratio of two energy flow rates is generally called *efficiency*. For example, the ratios of energy ingested by secondary consumers to energy ingested by primary consumers per unit time for each of the aquatic ecosystems we have discussed are 0.129 for Cone Spring, 0.09 for Root Spring, and 0.113 for Silver Springs. Thus, in these three ecosystems roughly 90% of the energy obtained by the primary consumers from plants never reaches the next trophic level. This great loss of energy in each step upward from one trophic level to the next is generally the case and accounts for the fact that top carnivores are only a few steps away from the primary production—within a few steps there simply is not enough energy available to support another population of predators.

Beginning with Lindeman (1942) ecologists have attempted to discover patterns or trends relating trophic level with efficiency. For example, Lindeman (1942) examined relationships between trophic level and the ratio in several lake ecosystems.

$$\frac{\text{Energy assimilated by trophic level } n}{\text{Energy assimilated by trophic level } n - 1}$$

He called this ratio *progressive efficiency,* but it is now commonly called *Lindeman's efficiency.* Lindeman suggested that progressive efficiency increases with each successively higher trophic level. Only a few studies since Lindeman's have been made that can be used to test this hypothesis. Almost all of them have been reviewed by Kozlovsky (1968), who has also provided a convenient summary of the confusing array of overlapping terms that have been applied to various ratios (Table 9–8). Figure 9–30 shows some of the graphs from Kozlovsky's paper.

Figure 9–30(a) shows progressive efficiency plotted against pairs of trophic levels. Lindeman's hypothesis would predict a definite upward trend in the points from the left to right, but this does not appear to be the case. Thus, we must agree with Kozlovsky's conclusion that Lindeman's hypothesis is not supported by the evidence now available.

Figure 9–30(b) shows an upward trend in the ratio of respiration rate to ingestion rate at successively higher trophic levels. The reason for this trend is not definitely known, but we can suggest one likely possibility. If biomass in successive trophic levels forms a pyramid, as we have suggested, then the average density of food decreases at successive trophic levels. Therefore, animals at higher trophic levels may have to expend more energy (in respiration) to acquire each unit of food they ingest. Other factors certainly would affect the ratios plotted in Fig. 9–30(b), however. For example, if the average size per animal is greater at higher trophic levels, then we would expect respiration rate per unit biomass to decrease at higher trophic levels.

Figure 9–30(c) shows a downward trend in the ratio of net production to ingestion from the second trophic level (herbivores) to the fourth. This line is not independent of that shown in Fig. 9–30(b), since, of the energy taken in, the more that is lost in respiration, the less that is left for net production.

Several comments are in order regarding the interpretation of the data in Fig. 9–30(a–f) and other similar data that may become available. First, note that some graphs are dependent on one another. For example, graph 9–30(e) is almost the same as graph 9–30(f) turned upside down. This is because net production, NP, is equal to assimilation, A, minus respiration, R. Therefore, in 9–30(e) it is R_n/A_n that is being plotted while in 9–30(f) it is $(A_n - R_n)/A_n$ that is being plotted.

Second, an efficiency may be defined in such a way that it becomes zero by definition for the highest trophic level. In Fig. 9–30(d), for example, the numerator, P_n, is the rate of energy flow from level n to level $n + 1$. When n represents the highest trophic level, F_n is zero and, therefore, F_n/I_n is also zero. Other cases where this occurs can be seen by examining the definitions in Table 9–9. Thus, the evidence for an apparent downward trend, such as Fig. 9–30(d) shows going from level 2 to 4, may be tenuous at best.

Third, the same definition of an efficiency that is used for consumers cannot be directly applied to producers, since the biological processes involved are quite different. The choice of what is used as the efficiency of the primary producers will affect the appearance of the graph resulting from plotting this efficiency along with efficiencies for consumers. In Fig. 9–30 (b), for example, ingestion for consumers is replaced by incident solar radiation. This figure would be considerably greater than the amount of light penetrating to the plant community which excludes albedo and absorption by water, and it might well be argued that this lesser figure is a more reasonable one to use than incident radiation. Clearly, the interpretation of the graph produced becomes equivocal.

Fourth, the number of ecosystems examined is extremely small; in effect, we are trying to make inferences about ecosystems from a sample of only about six ecosystems at best and often only two or three. In this vein, we point out the almost total lack of data on energy flow through entire terrestrial ecosystems. Even the work of Golly (1960), which is one of the most comprehensive examinations of the trophic dynamics of a terrestrial ecosystem, is not very comprehensive. This lack of information on terrestrial ecosystems is especially important since homeotherms usually play a much more important part in them than in aquatic ecosystems, and homeotherms differ in ways that may significantly affect energy flow through the ecosystems in which they live. For example Turner (1970) convincingly argues that the efficiency of

$$\frac{\text{Energy passed on to trophic level } n + 1}{\text{Energy ingested by trophic level } n} \times 100$$

(which is called *ecological efficiency* by several authors) cannot exceed 2–3% for homeotherms, in contrast to the figures of 8–12% that had been

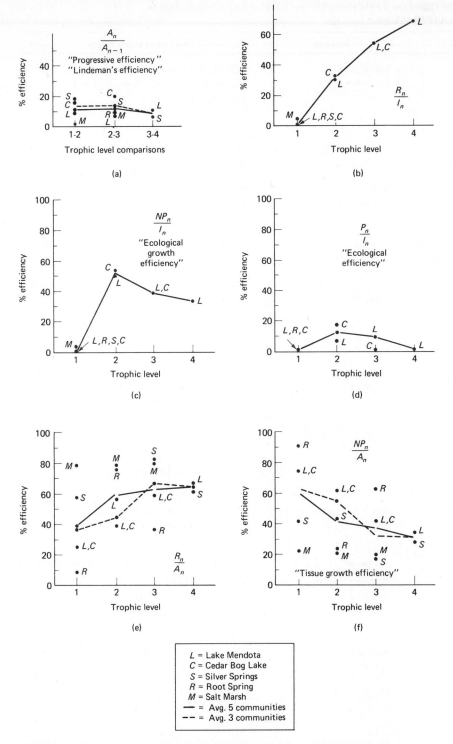

FIGURE 9–30. Various efficiency ratios, named with the expression(s) most commonly found in the literature. (From Kozlovsky, 1968.)

TABLE 9-9. DEFINITIONS OF EFFICIENCIES THAT HAVE BEEN PROPOSED IN THE ECOLOGICAL LITERATURE.*

Abbreviation	Meaning	Equivalent terminology
$\dfrac{A_n}{I_n} = \dfrac{GP_n}{I_n} = \dfrac{S_5}{S_0}$	$\dfrac{\text{assimilation at } n}{\text{ingestion at } n}$	Assimilation efficiency Trophic level energy intake efficiency Efficiency of digestion
$\dfrac{S_5}{S_2}$	$\dfrac{\text{assimilation at } n}{\text{ingestion at } n}$	Efficiency of primary production
$\dfrac{S_5}{S_4}$	$\dfrac{\text{assimilation at } n}{\text{ingestion at } n}$	Assimilation efficiency Trophic level energy intake efficiency
$\dfrac{R_n}{I_n} = \dfrac{A_n - NP_n}{I_n} = \dfrac{S_5 - NP_n}{S_0}$	$\dfrac{\text{respiration at } n}{\text{ingestion at } n}$	Energy degradation or respiration efficiency
$\dfrac{NP_n}{I_n} = \dfrac{S_5 - R_1}{S_0}$	$\dfrac{\text{net productivity at } n}{\text{ingestion at } n}$	Energy coefficient of growth of the first order Ecological growth efficiency Efficiency of assimilation Gross efficiency of growth
$\dfrac{S_5 - R_1}{S_4}$	$\dfrac{\text{net productivity at } n}{\text{ingestion at } n}$	Efficiency of photosynthesis in relation to visible radiation available to photosynthetic pigments
$\dfrac{P_n}{I_n} = \dfrac{P_1}{S_0}$	$\dfrac{\text{energy passed to } n+1}{\text{ingestion at } n}$	Efficiency of transfer to next level in terms of ingested energy Ecological efficiency Gross efficiency of yield to ingestion
$\dfrac{TA_n}{I_n} = \dfrac{TA_1}{S_0}$	$\dfrac{\text{tissue accumulation at } n}{\text{ingestion at } n}$	Efficiency of conversion to standing crop in terms of ingested energy
$\dfrac{NP_n}{A_n}$	$\dfrac{\text{net productivity at } n}{\text{assimilation at } n}$	Energy coefficient of growth of the second order Growth efficiency Tissue growth efficiency Net efficiency of growth
$\dfrac{R_n}{NP_n}$	$\dfrac{\text{respiration at } n}{\text{net productivity at } n}$	Respiratory coefficient
$\dfrac{P_n}{NP_n}$	$\dfrac{\text{energy passed to } n+1}{\text{net productivity at } n}$	Efficiency of transfer to next level in terms of assimilated energy
$\dfrac{TA_n}{NP_n}$	$\dfrac{\text{tissue accumulation at } n}{\text{net productivity at } n}$	Increase efficiency Efficiency of conversion to standing crop in terms of assimilated energy

422

TABLE 9-9. (continued)

Abbreviation	Meaning	Equivalent terminology
$\dfrac{I_n}{I_{n-1}} = \dfrac{I_n}{S_0}$	ingestion at n over ingestion at $n-1$ (this is equivalent to EE 5–0, in the absence of input from outside the system)	Lindemann's efficiency, ratio of intakes of trophic levels
		Ecological efficiency
		Efficiency of transfer to next higher level in terms of ingested energy
		Gross efficiency of yield to ingestion
$\dfrac{A_n}{A_{n-1}}$	assimilation at n over assimilation at $n-1$	Progressive efficiency
		Biological efficiency, progressive relative efficiency at a given level in terms of relative productivities
		Lindemann's efficiency, ratio of intakes of trophic levels
		Trophic level energy intake efficiency, trophic level assimilation efficiency
$\dfrac{R_n}{A_n}$	respiration at n over assimilation at n	
$\dfrac{P_n}{A_n}$	energy passed to $n+1$ over assimilation at n	
$\dfrac{R_n}{R_{n-1}}$	respiration at n over respiration at $n-1$	
$\dfrac{I_n}{NP_{n-1}}$	ingestion at n over net productivity at $n-1$	Consumption efficiency
		Utilization efficiency
		Ecotrophic coefficient (approximately)
$\dfrac{A_n}{NP_{n-1}}$	assimilation at n over net productivity at $n-1$	Efficiency of energy conversion
		Trophic level production ratios
		Trophic level production efficiency

*From Kozlovsky, 1968. References to authors using each term may be found in Kozlovsky.

Parameters used:

S_0 = incident radiation
S_1 = visible radiation
S_2 = light penetrating to the community (excludes albedo and absorption by water)
S_3 = light absorbed by photosynthetic tissues
S_4 = light available to photosynthetic pigments
S_5 = energy fixed in photosynthesis (= gross primary productivity, GPP, = assimilation by plants)
I = ingestion
A = assimilation
Df = defecation or nonassimilation ($I = A + Df$)
R = respiration
NP = net productivity ($NP = A - R$)
P = production; energy passed to next higher trophic level
Dc = decomposition
TA = tissue accumulation; increase in biomass
L = loss from system

proposed as generalization for consumers (Slobodkin, 1968) on the basis of data from poikilotherms. And finally, we must never lose sight of the quality of the original data. Even a quick reading of the original reports will show that the technical difficulties involved have forced the investigators to use many rough approximations, to draw conclusions from small samples, and, at times, to assign values to some rates in order to make a balanced energy budget. These are not meant to be criticisms of those who did the studies, but simply precautions against placing more confidence in the numbers they provide than is warranted.

Because of these problems we must conclude that studies of efficiencies have not yet been very helpful in answering ecologically interesting questions. It seems to us that a much tighter comparative approach will have to be used if studies of efficiencies are to yield useful information.

Within the last few years, several ecologists have taken a very different approach to examining the homeostatic interactions among parts of an ecosystem. In this approach an ecosystem is analyzed as a dynamic system, and many of the same techniques that a control systems engineer employs are used. That is, sets of equations, generally first-order differential equations, are used to express the rates of change of energy in the various parts of the ecosystem. Ideally, these equations reflect the causal mechanisms for these changes. The use of computer simulation techniques then allows a wide variety of specific questions to be asked, such as: What is the theoretical equilibrium state of the ecosystem? Does this agree with our observations? Does the system tend to oscillate? If so, do these oscillations have a pattern similar to fluctuations observed in the real ecosystem? What are the effects of perturbations on the system; for example, if incoming solar radiation decreases by a certain percent, as it might due to an increase in atmospheric turbidity, how will this affect the ecosystem?

It is clear from the small amount of work that has been done along these lines that models including only energy flow are not sufficient to answer these questions satisfactorily. Many other additional factors must be included in the models, such as nutrient exchange. We will return to a discussion of this approach to ecosystem analysis in Chapter 11.

REFERENCES

Birkebak, R., and R. Birkebak. 1964. Solar radiation characteristics of tree leaves. *Ecology,* **45**:646–49.

Blinks, L. R. 1955. Photosynthesis and productivity of littoral marine algae. *J. Marine Res.,* **14**:363–73.

Clements, F. E., and V. E. Shelford. 1939. *Bioecology.* New York: Wiley. 425 pp.

Davidson, J. L., and J. R. Philip. 1958. Light and pasture growth. In UNESCO, *Climatology and Microclimatology,* pp. 181–87. Paris: UNESCO.

DeWit, C. T. 1965. *Photosynthesis of Leaf Canopies*. Agr. Res. Rept. No. 663, Inst. Biol. Chem. Res. on Field Crops and Herbage, Waginengen.

———— . 1966. Photosynthesis of crop surfaces. *Advanc. Sci.*, **23**:159–62.

Drożdż, A. 1967. Food preference, food digestibility and the natural food supply of small rodents. In K. Petrusewicz (ed.), *Secondary Productivity of Terrestrial Ecosystems (Principles and Methods)*, Vol. I. Warszawa: Państwowe Wydawnictwo Naukowe. 379 pp.

Duncan, W. G., R. S. Loomis, W. A. Williams. and R. Honau. 1967. A model for simulating photosynthesis in plant communities. *Hilgardia*, **38**:181–205.

Elton, C. 1927. *Animal Ecology*. London: Sidgwick and Jackson, Ltd. 207 pp.

Federer, C. A., and C. B. Tanner. 1966. Spatial distribution of light in the forest. *Ecology*, **47**:555–60.

Forbes, S. A. 1887. The lake as a microcosm. Republished 1925. *Illinois Nat. Hist. Survey Bull.*, **15**:537–50.

Goldman, C. R. 1968. Aquatic primary production. *American Zool.*, **8**:31–42.

Golley, F. B. 1960. Energy dynamics of a food chain of an old-field community. *Ecol. Monogr.*, **30**:187–206.

Gordon, M. S., G. A. Bartholomew, A. D. Grinnell, C. B. Jorgensen, and F. N. White. 1968. *Animal Function: Principles and Adaptations*. New York: Macmillan. 560 pp.

Hutchinson, G. E. 1957. *A Treatise on Limnology*, Vol. I. New York: Wiley. 1015 pp.

———— . 1967. *A Treatise on Limnology*, Vol. II. New York: Wiley. 1115 pp.

Kleiber, M. 1961. *The Fire of Life: An introduction to Animal Energetics*. New York: Wiley. 454 pp.

Klekowski, R. Z., T. Prus, and H. Żyromska-Rudzka. 1967. Elements of energy budget of *Tribolium castaneum* (Hbst) in its developmental cycle. In K. Petrusewicz (ed.), *Secondary Productivity of Terrestrial Ecosystems (Principles and Methods)*, Vol. II. Warszawa: Państwowe Wydawnictwo Naukowe. 379 pp.

Kozlovsky, D. C. 1968. A critical evaluation of the trophic level concept. 1. Ecological efficiencies. *Ecology*, **49**:48–60.

Lemon, E. R. 1960. Photosynthesis under field conditions. II. An aerodynamic method for determining the turbulent carbon dioxide exchange between the atmosphere and a corn field. *Agron. J.*, **52**:697–703.

Lindeman, R. L. 1942. The trophic-dynamic aspect of ecology. *Ecology*, **23**:399–418.

Ludwig, L. J., T. Saebi, and L. T. Evans. 1965. Photosynthesis in artificial communities of cotton plants in relation to leaf area. I. Experiments with progressive defoliation of mature plants. *Anat. J. Biol. Sci.*, **18**:1103–18.

McCree, K. J., and J. H. Troughton. 1966. Prediction of growth rate at different light levels from measured photosynthesis and respiration rates. *Plant Physiol.*, **41**:559–66.

McNab, B. K. 1963. A model of the energy budget of a wild mouse. *Ecology*, **44**:521–32.

Miller, P. C. 1967. Leaf transpiration, leaf orientation and energy exchange in quaking aspen (*Populus tremuloides*) and Gambell's Oak (*Quercus Gambellii*) in central Colorado. *Oecol. Plant.*, **2**:241–70.

———— . 1971. Bioclimate, leaf temperature, and primary production in red mangrove canopies in South Florida. *Ecology*, **53**:22–45.

———— , and L. Tieszen. 1971. A preliminary model of processes affecting primary production in the arctic tundra. *Arctic Alpine Res.* **4**:1–18.

Monsi, M., and T. Saeki. 1953. Lichtfalstar in den Pflanzengesellchaften und seiner Bedentung fur die Stoffproduktion. *Japan J. Botany*, **14**:22–52.

Monteith, J. L. 1965. Light distribution and photosynthesis in field crops. *Ann. Bot. N.S.*, **19**:17–37.

Odum, E. P. 1960. Organic production and turnover in old-field succession. *Ecology*, **41**:34–49.

Odum, H. T. 1957. Trophic structure and productivity of Silver Springs. *Ecol. Monogr.*, **27**:55–112.

Ovington, J. D. 1961. Some aspects of energy flow in plantations of *Pinus sylvistris* L. Ann. Bot. Lond. N.S., **25**:12–20.

Pearson, O. P. 1960. Habits of *Microtus californicus* revealed by automatic photographic records. *Ecol. Monogr.*, **30**:231–49.

Petipa, T. S., E. V. Pavlova, and G. N. Mironov. 1970. The food web structure, utilization, and transport of energy by trophic levels in the planktonic communities. In J. H. Steel (ed.), *Marine Food Chains*. Berkeley: University of California Press.

Petrusewicz, K., and A. Macfadyen. 1970. *Productivity of Terrestrial Animals. Principles and Methods.* IBP Handbook No. 13. Oxford: Blackwell. 190 pp.

Reichle, D. E. 1970. *Analysis of temperate forest ecosystems.* New York: Springer-Verlag. 304 pp.

Riley, G. A. 1946. Factors controlling phytoplankton populations on Georges Bank. *J. Marine Res.*, **6**:54–73.

———— . 1963. Theory of Food-chain Relations in the Ocean. In M. N. Hill (ed.), *The Sea*, Vol. 2. New York: Wiley Interscience. 554 pp.

———— . 1965. A mathematical model of regional variations in plankton. *Limn. Ocean.*, **10** (Suppl):R202–R215.

Ryther, J. H. 1963. Geographic variations in productivity. In M. N. Hill (ed.), *The Sea*, Vol. 2. New York: Wiley Interscience. 554 pp.

———— , and W. M. Dunstan. 1971. Nitrogen, phosphorus, and eutrophication. *Science*, **171**:1008–13.

Saeki, T. 1960. Interrelationships between leaf amount, light distribution, and total photosynthesis in a plant community. *Botan Mag.*, **73**:55–63.

Slobodkin, L. B. 1968. How to be a predator. *Amer. Zool.*, **8**:43–51.

Stanhill, G. 1962. The effect of environmental factors on the growth of alfalfa in the field. *Netherlands J. Agric. Sci.*, **10**:247.

Summerhayes, V. S., and C. S. Elton. 1923. Contributions to the ecology of Spitsbergen and Bear Island. *J. Ecol.*, **11**:214–86.

Teal, J. M. 1957. Community metabolism in a temperate cold spring. *Ecol. Monogr.*, **27**:283–302.

———— . 1962. Energy flow in the salt marsh ecosystem of Georgia. *Ecology*, **43**:614–24.

Tilly, L. J. 1968. The structure and dynamics of Cone Spring. *Ecol. Monogr.*, **38**:169–97.

Transeau, E. N. 1926. The accumulation of energy by plants. *Ohio Jour. Sci.*, **26**:1–10.

Tucker, V. A. 1968. Respiratory exchange and evaporative water loss in the flying budgerigar. *J. Exp. Biol.,* **48**:67–87.

Turner, F. 1970. The ecological efficiency of consumer populations. *Ecology,* **51**:741–42.

Weaver, D. F. 1969. *Radiation Regime Over Arctic Tundra, 1965.* Department of Atmospheric Science, University of Washington, Seattle. (n.p.)

Westlake, D. F. 1963. Comparisons of plant productivity. *Biological Review,* **38**:385–425.

Whittaker, R. H. 1970. *Communities and Ecosystems.* New York: Macmillan. 162 pp.

Wiegert, R. G. 1964. Population energetics of meadow spittlebugs (*Philaenus spumarius* L.) as affected by migration and habitat. *Ecol. Monogr.,* **34**:225–41.

Chapter 10
BIOGEOCHEMICAL CYCLES

In the previous chapter we examined the flow of energy through the ecosystem. This represents only one major aspect of ecosystem function. The processes of nutrient cycling constitute a second. These processes are termed *biogeochemical cycles* to emphasize the fact that interchanges of the chemical substances involved occur among both the biotic and abiotic components of the ecosystem. We shall use the term biogeochemical cycle to refer to patterns of nutrient cycling at both the level of local ecosystem units and the level of the biosphere as a whole.

Biogeochemical cycles may be recognized for any of the chemical elements that occur in living systems and for certain other substances of more complex chemical structure. Biogeochemical cycles may be arbitrarily divided into two groups: those concerned with macronutrients and those with micronutrients. Macronutrient cycles involve the nutrient substances of major importance in the structure and metabolic activity of all organisms. Included here are cycles of carbon (C), hydrogen (H), oxygen (O), nitrogen (N), calcium (Ca), silica (Si), phosphorus (P), potassium (K), sodium (Na), sulfur (S), and magnesium (Mg). Since hydrogen usually enters and leaves living systems in the form of water, this element is usually analyzed in the context of the hydrologic, or water, cycle. Micronutrient cycles include cycles of the remaining chemical elements, such as molybdenum (Mo), iron (Fe), and many others, which are important for the life activities of all or many organisms but are required in much smaller quantities. Our discussions will concentrate on macronutrient cycles, for which the most data are available. The principles of biogeochemical function, however, apply to both macronutrients and micronutrients.

The activities of man have resulted in introduction to the biosphere of a variety of novel chemical substances, many of which establish patterns of biogeochemical cycling. These include radioisotopes of naturally occurring elements, as well as complex organic compounds such as pesticides and polychlorinated biphenyls. Furthermore, human activities have greatly modified cycles of biologically active elements such as mercury, lead, and cadmium. Many of these substances have important ecological effects and show patterns of movement and concentration in components of the ecosystem that are best considered in the context of biogeochemical cycling.

BASIC CHARACTERISTICS OF BIOGEOCHEMICAL CYCLES

Biogeochemical cycles may be described in terms of *pools* and *flux rates*. A biogeochemical pool consists of the quantity of the particular chemical substance in some ecosystem component, either biotic or abiotic. For example, in a lake ecosystem the quantity of phosphorus in the water mass may be considered to be one pool and that in the phytoplankton a second pool. These pools are interrelated by processes that transfer the substance involved from one pool to another. These transfers may be described in terms of flux rates, defined as the quantity of material passing from one pool to another per unit time and per unit area or volume of the system. These relationships are illustrated for a simple aquatic system in Fig. 10–1.

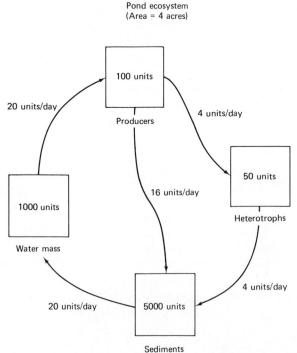

FIGURE 10–1. Biogeochemical pools and flux rates for a hypothetical nutrient cycle.

Fluxes of nutrients among ecosystem pools and across the boundaries of the ecosystem may be described in several ways. For purposes of measurement and modelling these fluxes are usually described in absolute terms as the quantity of a nutrient passing between two components per unit time and per unit area or volume of the system. For example, in Fig. 10–1 the transfer of the nutrient from the water mass of the pond to the producer component would be equal to 5 units/acre/day. Other such fluxes for this hypothetical system are shown in Table 10–1. For purposes of describing the relative importance of a particular flux process in relation to the pools involved, however, the use of *turnover rates* and *turnover times* is valuable. The turnover rate is calculated as the flux rate, into or out of a pool, divided by the quantity of nutrient in the pool. Thus, the transfer rate describes the importance of the flux process in relation to pool size. Turnover rates for the cycle shown in Fig. 10–1 are given in Table 10–1. These rates show the effect of various fluxes on the pools from which they leave and those which they enter. In this system the higest turnover rates are for the flux from the water mass to the producers (0.20) and from the producers to the sediments (0.16). From this observation it can be inferred that the pool most subject to short-term disturbance will be that of the nutrient in the producer component of the system.

An alternative manner of looking at the same relationship is by calculation of the turnover time. This value is calculated as the quantity in the pool divided by the flux rate. The turnover time thus describes the time required for movement of a quantity of nutrient equal to that in the pool. In the simple system analyzed in Table 10–1, the turnover time is shortest for the input to the producer pool from the water mass (4 days) and the output from the producer pool to the sediments (6.25 days). These relationships directly correspond to the high turnover rates for these same fluxes.

The term *cycle* used in connection with these processes implies a pattern distinct from that seen for energy flow through ecosystems. Energy flow was seen to be a one-way process – energy entering the eco-

TABLE 10–1. ABSOLUTE FLUX RATES, TURNOVER RATES, AND TURNOVER TIMES FOR THE HYPOTHETICAL BIOGEOCHEMICAL CYCLE SHOWN IN FIG. 10–1.

Transfer route		Absolute flux rate (units/acre/day)	Turnover rate		Turnover time (days)	
From	To		Departure pool	Recipient pool	Departure pool	Recipient pool
Water mass – Producers		5	0.02	0.20	50.	4.
Producers – Sediments		4	0.16	0.0032	6.25	312.5
Producers – Heterotrophs		1	0.04	0.08	25.	12.5
Heterotrophs – Sediments		1	0.08	0.0008	12.5	1250.
Sediments – Water mass		5	0.004	0.02	250.	50.

system as solar radiation and ultimately leaving as heat energy unavailable for use by living systems. In contrast, movement of nutrient substances is, in many cases, truly cyclical. The same units of matter may repeatedly pass from one pool to another and back again. Thus, the flow of matter may be considered to be the vehicle of energy flow. Like a bus following a definite route through consumption of fuel, the bus repeatedly covers the route while the fuel follows a one-way path—in as gasoline and out as the waste products of combustion.

Further, we suggest that major biogeochemical cycles, prior to the interventions of man, must have been in a steady-state condition. This condition means that inflows to major pools must be balanced almost exactly by outflows, as suggested in Fig. 10–1. Of course, such a balance cannot always be expected to exist over short periods or for systems of limited size. From our discussion of the process of ecological succession it is apparent that during this process balanced flux rates are the exception. However, in climax ecosystem units, and for major geographic regions and the biosphere as a whole, a very close balance of inputs and outputs from pools must exist.

This conclusion is supported by several observations. For example, in the carbon cycle one of the major routes of flux is the removal of CO_2 from the atmosphere and its fixation in organic compounds by photosynthetic organisms. This process alone removes a quantity of CO_2 equal to the total amount present in the atmosphere in slightly over one year (Cole, 1958). Clearly, this action is closely balanced by processes returning CO_2 to the atmosphere. In a similar manner, the process of fixation of molecular nitrogen (conversion from N_2 to ammonia) by certain groups of bacteria and blue-green algae converts a quantity equal to the total present in the atmosphere in less than 1,000,000 years. In terms of the history of life on the earth, this is a short period, and it is clear that nitrogen fixation is balanced, on the geological time scale, by processes that return molecular nitrogen to the atmosphere.

Since almost all of the major substances contributing to the solute load of the oceans are participants in biogeochemical cycles, this generalization suggests that the solute composition of ocean water is an equilibrium situation. At first glance this seems unlikely because of the obviously large quantities of material entering the oceans through runoff from the continents. These processes are balanced, however, by sedimentation of materials and by other processes of lesser magnitude. Goldberg (1963) has given estimates of the turnover times of major components of seawater, based on estimated sedimentation rates (Table 10–2). These turnover times are calculated by dividing the amount of the material in solution in the world's oceans by the annual rate of deposition of the material in marine sediments. These values range from 8,000 years for silicon to 206 million years for sodium, but they are far below the known minimum age of the oceans. Thus, it is apparent that the quantities of solutes present must have been replaced several to many times during the

TABLE 10-2. RESIDENCE TIMES OF ELEMENTS IN SEA-WATER AS CALCULATED BY RATES OF SEDIMENTATION ON THE OCEAN FLOOR.*

Element	Quantity in ocean water ($\times 10^{20}$ grams)	Residence time (millions of years)
Na	147.8	260
Mg	17.8	45
K	5.3	11
Ca	5.6	8
Si	0.052	0.008

* From Goldberg, 1963.

life of the oceans. It is likely that the composition of the oceans was established early in the geological history of the earth and that it has been nearly constant throughout recent geological time.

It may be noted that the greatest rate of sedimentation occurs on continental shelf areas where the sediments produced may later be uplifted. However, much sedimentation also occurs over the floor of deep ocean basins from which return to the land is unlikely. This removal is, in effect, permanent and represents an exception to the cyclical pattern for behavior of nutrient substances described above. These losses to deep-water sediments, however, appear to be balanced by gains due to weathering of primary igneous rocks and release of materials by volcanic action.

PERFECT AND IMPERFECT BIOGEOCHEMICAL CYCLES

On a functional basis, different biogeochemical cycles may be grouped as either perfect or imperfect, or, better, arranged on a spectrum between these two extremes. This classification refers to the degree of regulation of processes in the cycle or to the extent to which the pools or flux rates in the cycle can reestablish a normal condition following some disturbance. Perfect biogeochemical cycles show a strong degree of regulation.

Perfect biogeochemical cycles are characterized by two features:

1. A large available abiotic pool of the substance;
2. The existence of many negative feedback controls.

An available abiotic pool is a quantity of the particular nutrient to which the living ecosystem components have free access. Biogeochemical cycles of C, N, and O possess such pools, which correspond to the form of the nutrient existing in the atmosphere. These pools are available in the sense that only the metabolic ability to convert the substance from abiotic to biotic form is necessary for the withdrawal of quantities of the nutrient from this pool. Except in extremely arid regions, the hydrologic cycle may be considered to possess such a pool, corresponding to the water present in the soil of terrestrial systems or constituting the medium of aquatic systems. Certain of the mineral nutrients, such as Si and Ca, may also possess large available pools in the soil, where soils are derived by weathering of substrates rich in these elements.

The second characteristic of perfect cycles is the occurrence of regulatory control. These controls act in a negative feedback manner. If the size of a pool, or the flux rate between pools, is disturbed by either an increase or a decrease, the control operates to restore the original condition. These controls are analogous to those operating within the organism to maintain normal body temperature levels or to regulate the composition of body fluids. In order to exert a regulatory function and to reestablish an original condition, controls must be capable of varying their intensity of operation much as density-dependent population controls vary their intensity of operation. That is, the control must be capable of response (increase or decrease in intensity) proportionately greater than the change produced by the disturbance. As a result, negative feedback controls on biogeochemical processes, like most density-dependent population controls, are frequently biotic in nature. They usually occur in situations in which a flux from one pool to another is mediated by some group of organisms. The growth or decline of populations of these organisms is one of the major factors in their ability to show a variable intensity of operation.

The nitrogen cycle is an excellent example of a cycle presumed to possess such controls. Figure 10–2 indicates the major pools and transfer processes for nitrogen in the biosphere. Nitrogen exists in chemical forms varying from highly oxidized (+5 valence; NO_3^-) to highly reduced (−3 valence; NH_3). In organic compounds of living tissues it occurs in the reduced state (−3 valence). Valence states intermediate between these extremes are shown by molecular nitrogen (0 valence) in the air and water, and by nitrite (+3 valence). Conversion of nitrogen from one chemical form to another is mediated, in almost all cases, by metabolic activities or organisms. In many cases these are specific groups of microorganisms. The functions served by some of these conversions are energy yielding, with the energy being used to drive chemosynthetic processes. Others involve the assimilation of nitrogen in the synthesis of organic compounds such as proteins, nucleotides, and vitamins. In other cases nitrogen-containing compounds participate in chemical reactions permitting other heterotrophic or chemosynthetic processes to function. In effect, each form of nitrogen represents a resource that may be advantageously utilized for the life activities of different organisms (Table 10–3). Disturbance of some portion of the nitrogen cycle thus means the modification of resource availability for some group of organisms which utilizes nitrogen. The organisms involved are capable of changing their rate of nitrogen use, either sharply reducing or sharply increasing it, so

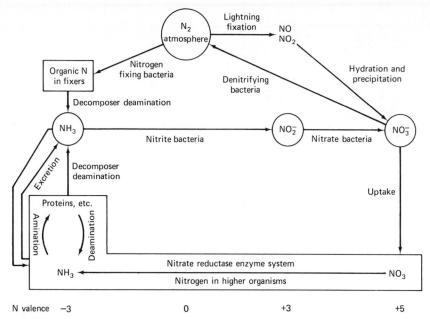

FIGURE 10-2. Major processes of the nitrogen cycle within the biosphere. Circular figures indicate abiotic environmental pools, rectangular figures biotic pools.

that the normal amounts of nitrogen in the pools involved are reestablished.

Other biogeochemical cycles, including those of C, O, H_2O, and P, show the complex involvement of organisms in the chemical conversion of transfer of materials from pool to pool. The extent to which these two characteristics, the presence of a large available pool and the occurrence of negative feedback control, can accommodate the massive interventions being produced by man, however, is poorly known. Also, as pointed out by Hutchinson (1948), perfect and imperfect biogeochemical cycles differ in the extent to which the effects of disturbance are predictable. For many of the nutrients with sedimentary cycles lacking strong negative feedback control, the effects of human disturbance can be predicted accurately. The consequences of mining of certain mineral resources can be predicted within the limits of our knowledge of the extent of these resources. The consequences of the disturbance of the carbon cycle, however, are very difficult to predict. It appears that regulatory processes in this cycle can compensate to a degree for man-produced disturbance. If the limits of operation of existing negative feedback controls are exceeded, however, the possible consequences are difficult to determine. In a sense, the very self-regulatory processes of the cycle may accentuate the magnitude of disturbance if their capacities for regulation are exceeded. We will examine the operation of processes of biogeochemical regulation more fully in Chapter 11.

TABLE 10-3. CONVERSION PROCESSES, REACTION PATTERNS, ENERGY YIELD, AND FUNCTION OF NITROGEN CONVERSION PROCESS FOR VARIOUS ORGANISMS INVOLVED IN THE NITROGEN CYCLE.

Group of organisms	Reaction type	Typical reactions (generalized)	Energy yield (kcal/mole of N)	Value to organism
Nitrogen-fixing bacteria and blue-green algae	Nitrogen fixation (reduction)	$2N + 3H_2 \rightarrow 2NH_3$	-147	N at -3 valence for building organic compounds (including those of host in case of symbiotic bacteria)
Nitrite bacteria (chemosynthetic)	Oxidation	$NH_3 + 1\frac{1}{2}O_2 \rightarrow HNO_2 + H_2O$	66	Energy yielded for chemosynthesis (carbon fixation)
Nitrate bacteria (chemosynthetic)	Oxidation	$KNO_2 + \frac{1}{2}O_2 \rightarrow KNO_3$	17.5	Energy yielded for chemosynthesis (carbon fixation)
Denitrifying bacteria (obligate anaerobes)	Reduction	$5C_6H_{12}O_6 + 24KNO_3 \rightarrow$ $30CO_2 + 18H_2O + 24KOH + 12N_2$	$570^\dagger$	In absence of O_2 nitrates supply hydrogen acceptor for oxidation of carbohydrates. In some reaction sequences, N_2O produced instead of N_2, with lower energy yield
Denitrifying bacteria (chemosynthetic)	Reduction	$5S + 6KNO_3 + 2CaCO_3 \rightarrow$ $3K_2SO_4 + 2CaSO_4 + 2CO_2 + 3N_2$	132^*	Nitrate reduction permits sulfur oxidation with a net energy yield used for chemosynthesis (carbon fixation)
Decomposer	Ammonification	$CH_2NH_2COOH + 1\frac{1}{2}O_2 \rightarrow 2CO_2 + H_2O + NH_3$ (Glycine)	176	Ammonia released as waste by-product in oxidation breakdown of carbon skeleton of amino acids (and other N-containing compounds)
All nonnitrogen-fixing organisms	Nitrate reduction and/or amination	Complex	Variable cost	NO_3 or NH_3 used at -3 valence in building organic compounds

† per mole of glucose
* per mole of S

ANALYSIS OF BIOGEOCHEMICAL CYCLING

Investigations of biogeochemical processes are currently being pursued at two major levels. The first major approach involves intensive studies of the processes occurring within local ecosystem units, such as forest stands or lakes. Some of these have involved the selection of particular species and the study of their role in cycling of specific nutrients. More recently, however, attempts have been made to examine processes of cycling of a number of macronutrients through whole ecosystems. In these studies emphasis has been placed on analysis of inputs and outputs from the system as a whole and on the exchanges occurring among major biotic and abiotic components within the system. The latter studies, by necessity, group populations of species having a similar general function into single components such as producers, decomposers, herbivores, etc.

The second major approach involves analysis of cycles at the level of the biosphere. Studies of this kind are being attempted for cycles such as those of water, C, O, N, and P. These have been stimulated by observations of the impact of man on portions of them, combined with their importance to life in general.

We will examine a series of studies at both of these levels. Through these discussions we will examine the patterns and causal factors of biogeochemical cycles in general and of man's influence upon them in particular.

NUTRIENT CYCLING THROUGH INDIVIDUAL SPECIES POPULATIONS

One approach to the study of biogeochemical processes has been the analysis of the role of selected species in the cycling of particular nutrients. For animal populations, one of the few comprehensive studies is that of Kuenzler (1961a, 1961b) on the mussel, *Modiolus demissus,* in tidal marshes along the coast of Georgia.

In this study Kuenzler examined the patterns of both energy flow and phosphorus cycling through the mussel population. This species occurs attached to hard substrates in muddy intertidal flats and feeds by filtering particulate matter from the water during periods of high tide. For the analysis of energy flow, data were accumulated on the rates of recruitment, growth, and mortality of animals in the field and on the size-frequency structure of populations throughout the year.

To gain additional data necessary for an examination of phosphorus cycling, analyses were made of the phosphorus content of body tissues of animals differing in body size. Three components were chemically analyzed: shell, body tissues, and the liquor. The liquor consisted partly of internal body fluids and partly of seawater located within the gill-filtration system. For body tissues, analyses were made throughout the year, since tissues differing in phosphorus content (e.g., gonadal tissues) varied in degree of development seasonally (Fig. 10–3). The phosphorus

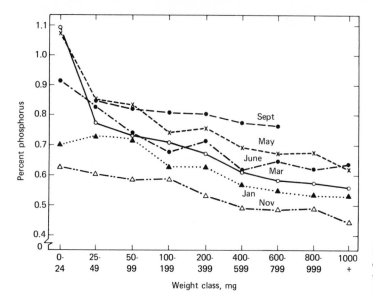

FIGURE 10–3. Mean percentage of phosphorus in the body fraction of *Modiolus*. (From Kuenzler, 1961b.)

content of body tissues also varied to some extent with size (Fig. 10–3).

The available phosphorus in the water of the marsh existed in three forms: particulate matter containing phosphorus, inorganic phosphorus (phosphate), and dissolved organic substances which contained phosphorus. Water samples, taken during both flood and ebb tides to give the mean condition during the period of feeding activity of the mussels, were analyzed chemically to gain estimates of each of the three phosphorus components in the marsh water.

In the laboratory a series of experiments, using the radioisotope ^{32}P, were conducted to determine the rate at which filtering of particles and absorption of phosphate from water occurred. To determine filtration rates, ^{32}P-labeled diatoms were suspended in clean ocean water and mussels placed in the tank. From the rate of reduction of the activity of the water the volume of water cleared of diatoms per unit time could be calculated. In similar experiments the uptake of ^{32}PO$_4$ from solution in the water was determined.

The rates of elimination of phosphate, dissolved organic phosphorus, and feces were also determined in the laboratory by placing freshly-collected animals in clean seawater which had been analyzed to determine the content of PO$_4$ and dissolved organic phosphate, and which contained little or no particulate phosphate. The increase in these three components in the water (stirred to place feces in suspension as particulate matter) gave estimates of elimination rates.

With data from these analyses, the average characteristics of phosphorus pools in the water and in the mussel population could be determined, and the major flux rates between these two pools could be estimated (Fig. 10–4). From measurements of the average depth of water above

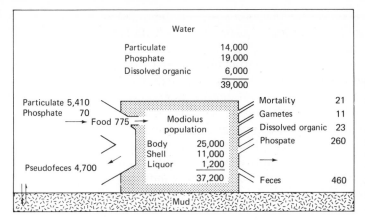

FIGURE 10–4. Diagram of phosphorus flow through the mussel population. Values for the water and the mussel population are μg P/m²; rates are μg P/m² day. The flux rates of phosphorus in food and pseudofeces are calculated values necessary to balance the other, measured flux rates. (From Kuenzler, 1961b.)

mussel beds, and those of the amounts of particulate phosphorus, phosphate, and organic phosphorus in the water, the total phosphorus pool in the water was calculated to be 39 mg/m². Likewise, from measurements of the phosphorus concentration in mussel tissues, combined with measurements of the density of mussels, the phosphorus pool in the mussel population was calculated as 37.2 mg/m².

Flux rates from the mussel population to the water occurred through mortality, shedding of gametes, and the elimination of feces, phosphate, and dissolved organic materials. Data on size-specific mortality rates, combined with phosphorus-content measurements of tissues of such animals, gave an estimate of 21 μg/m²/day as the loss of phosphorus from the mussel pool by mortality. An estimate of the loss through shedding of gametes was determined indirectly. Measurements of the phosphorus content of tissues of mussels at different times of year were combined with data on the size-frequency structure of the population in order to give a seasonal curve of total phosphorus content in mussel tissues. During July and August, the season at which gametes are shed, this curve indicated a net loss of phosphorus by the population. This decrease was taken as a quantitative estimate of the phosphorus lost in gametes. Spread over the entire year, this equalled 11 μg/m²/day.

Losses of phosphorus in feces, as phosphate, and as dissolved organic phosphorus were determined by multiplying experimentally determined losses per gram per day by the average standing crop biomass of mussels. These estimates equalled 460 μg/m²/day through feces, 260 μg/m²/day as phosphate, and 23 μg/m²/day as dissolved organic phosphorus. Altogether, phosphorus losses from the mussel population totalled 775 μg/m²/day.

The phosphorus intake as food was taken as equal to the total of the above losses, on the assumption that the mussel population was nearly stable from year to year. Mussels, however, filter out a much larger quantity of material than is actually ingested. Much of the particulate matter trapped by the mucous secretions of the filtration system is deposited as *pseudofeces* which have not passed through the digestive tract. In large part this represents particulate matter too large to enter the mouth. The total amount of phosphorus in material filtered from the water can be estimated as the product of the concentration of particulate matter in the water, the experimentally determined filtration rate, the biomass of the population, and the time during which the mussel beds are covered by water. The calculated amount of particulate phosphorus removed from the water by filtration was 5,410 $\mu g/m^2/day$. In addition, phosphate was absorbed directly from the water. This amount, calculated as the product of average biomass and the experimentally determined phosphate uptake rate, equalled 70 $\mu g/m^2/day$. Thus, a total of 5,480 $\mu g/m^2/day$ of phosphorus were removed from the water. Of these 775 were actually taken in as food, leaving 4,705 $\mu g/m^2/day$ to be deposited as pseudofeces.

An indication of the degree of influence of the mussels on phosphorus content of the water is given by calculation of turnover times for phosphorus in the water as a function of various processes of the mussel population. These values are termed *participatory turnover times* because they indicate the contributory effect of the mussel population alone in phosphorus exchange with the water. Turnover times were calculated for all three phosphorus components of the water (Table 10–4).

Participatory turnover times were shortest for particulate phosphorus, intermediate for phosphate, and longest for dissolved organic phosphorus. Thus, the major effect of the mussel population was in removal of particulate phosphorus from the water. This is borne out by the observation that the concentration of particulate phosphorus is greatest in the seaward portions of the marsh and decreases toward the shore.

Earlier, Kuenzler (1961a) had estimated the total energy flow of the mussel population to be 56 $kcal/m^2/year$. This value was close to that of several other animal populations in the marsh. However, no other filter feeder approaches the abundance of the mussel, and, thus, the rates of particulate phosphorus filtration and deposition of pseudofeces by mussels are probably far above those of any other organism (Kuenzler,

TABLE 10–4. PARTICIPATORY TURNOVER TIMES OF PHOSPHORUS IN SEAWATER DUE TO THE MUSSEL POPULATION IN A GEORGIA SALT MARSH.*

Phosphorus fraction	Route	Participatory turnover time (days)
Particulate	Filtration	2.6
Phosphate	Elimination	73
Dissolved organic	Elimination	260

*From Kuenzler, 1961b.

1961b). This species appears to be considerably more important as a biogeochemical agent than as an energy transformer in the salt marsh ecosystem.

An analysis of the role of a particular plant species in nutrient cycling in a terrestrial ecosystem is given by Thomas (1969). This study, dealing with the role of the Flowering Dogwood, *Cornus florida,* in the cycling of calcium in forests of the southeastern United States, emphasizes the variety of measurements required to clarify such a relationship, as well as demonstrating the unique importance of this species. The Flowering Dogwood is a widely distributed, deciduous tree of the forest understory throughout most of eastern North America. This species has been acknowledged as valuable because of the high calcium content and rapid decomposition rate of its litter. The small size at maturity—generally less than 10 m in height—further facilitates the conduct of experimental studies of all stages of the life cycle of the species.

In this study Thomas combined analyses of stable Ca in different components of the tree-soil system with an experimental approach in which the radioisotope ^{45}Ca was introduced into trees at the beginning of the growing season and its subsequent movements followed. Chemical analyses were made to obtain the stable Ca content of various tree components, including leaves, the bark and wood fractions of twigs, branches, and trunks, and roots of various size classes. The quantities of stable Ca in the soil and in precipitation above the forest canopy were determined. Collection devices were used to sample the precipitation passing through the canopy carrying Ca leached from the foliage (leachate) and to collect the water running down branches and trunks to the ground (stemflow). Losses from the leaves by grazing of herbivorous insects were also estimated from measurements of the fractions of leaves eaten.

Measured quantities of ^{45}Ca were introduced into the base of the trunks of a number of trees in early May. The concentration of ^{45}Ca was periodically sampled in the various major components of the tree-soil system throughout the remainder of the growing season.

The difficulties in acquiring data adequate to define the role of the Dogwood, as it exists as a natural component of a forest ecosystem, are enormous. For example, the concentration of Ca in various tree tissues varies considerably, and the fractions of tissues falling into different categories (leaves, wood, bark, roots, etc.) depend on tree size. Furthermore, in a deciduous species such as the Dogwood, several types of materials contribute to the litter (leaves, twigs, bracts, fruits), each of which is characterized by distinct Ca concentrations and decomposition rates. Decomposition rates, and their quantitative importance to the forest ecosystem, further depend on the type and quantity of litter from other species in the forest system. Consequently, Thomas was able only to obtain evidence suggestive of the general importance of the Dogwood in Ca cycling.

Analyses of the transfer of ^{45}Ca away from the innoculation sites in the trunks of experimental trees (Fig. 10–5) showed that almost all (89%) was translocated to other parts of the tree during the growing season. Almost three-quarters (73%) was transferred to the leaves. An estimated 3% of this was lost to insect grazing, and 6% to leaching of Ca from leaves by rainfall. Contrary to the typical pattern of withdrawal of certain nutrients, such as sodium and potassium, from the foliage prior to shedding of leaves, the remaining Ca was transferred by leaf fall to the litter layer. Stemflow was found to be a very minor pathway of Ca movements (0.01%). During the growing season, very little of the ^{45}Ca reached the mineral soil (<0.01%), since the bulk of the ^{45}Ca in the leaves did not reach the litter layer until leaf fall in the autumn.

Estimates of the total annual loss of Ca from Dogwood trees showed that these varied from 114% (for seedlings) to 71% (for mature trees) of the total Ca accumulated during the life of the tree in nonfoliar tissues. Direct measurements of the decomposition rates of litter materials derived from Dogwood indicated that approximately 79% of Ca in these materials was released during the first year of decomposition.

Further evidence of the relative importance of Dogwood in the cycling of Ca in the forest ecosystem was provided by an analysis of the quantities of Ca in aboveground components of Dogwood and Pine in one of the stands studied (Table 10–5). This analysis gave estimates of the dry weight and calcium content of plant tissues for the Dogwood and Pine populations (both leaves and woody parts), and for the leaf-litter layer. Although living Dogwood tissues comprised only 0.2% of the total dry weight for these three components, 1.8% of the Ca was localized in the Dogwood component.

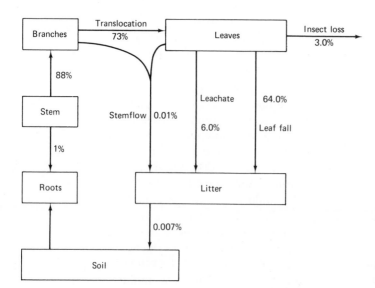

FIGURE 10–5. Percentages of ^{45}Ca transferred, during one growing season, from innoculation site in stems of Flowering Dogwood trees to various locations in the Dogwood-soil system. (From Thomas, 1969.)

TABLE 10-5. ABOVE-GROUND WEIGHT AND CALCIUM CONTENT OF COMPONENTS OF AN 18-YEAR-OLD LOBLOLLY PINE STAND IN TENNESSEE.*

Component	Oven-dry weight (kg/ha)	Calcium (kg/ha)	Percentage of total	
			Oven-dry weight	Calcium
Dogwood trees	1,030	12.5	0.2	1.8
Pine trees	382,940	525.0	89.3	74.1
Litter	45,020	170.6	10.5	24.1
Totals	428,990	708.1	100.0	100.0

*From Thomas, 1969.

These data indicate that the Dogwood, although constituting a relatively minor component of the ecosystem from the standpoint of biomass and, probably, energy flow, performs a more important role in Ca cycling. Thomas concluded that the Dogwood essentially functions as a "pump" that promotes an active cycling of Ca through the active upper soil layers. This "pump" acts to transport Ca from deeper layers of the mineral soil to the surface and acts to deposit it there in an organic form from which it is rapidly released by composition for use by other organisms.

The two studies just examined (Kuenzler, 1961; Thomas, 1969) have in common the conclusion that regularly occurring species may carry a function in nutrient cycling disproportionate to that in energy dynamics. Since, as previously shown, evolutionary processes tend to lead to complex patterns of ecological isolation and resource partitioning among ecosystem members, it is likely that such cases are quite frequent. In the case of producers, which are exploiting resources defined in terms of energy and chemical nutrients, the probability that each species carries a uniquely important role in nutrient cycling is great. This possibility must obviously be kept in mind in the design of comprehensive studies of ecosystem function.

BIOGEOCHEMICAL CYCLES AT THE ECOSYSTEM LEVEL

Study of biogeochemical cycles within an ecosystem first requires an examination of processes affecting the quantity of particular nutrients in active circulation or in available pools within the system. The total quantity of nutrients in biotic pools and in available abiotic pools is termed the *nutrient capital* of an ecosystem. The nutrient capital of an ecosystem may be relatively stable, or it may be changing in quantity, depending on the net gain or loss of nutrients by various input and output processes (Table 10–6).

TABLE 10-6. MAJOR ROUTES OF INPUT AND OUTPUT OF NUTRIENTS FROM ECOSYSTEMS.

	Input	Output
Natural routes	Fixation from atmosphere	Release to atmosphere
	Weathering of substrate	Loss by leaching or sedimentation to deep strata
	Precipitation or inflow	Runoff or outflow
	Particulate fallout from atmosphere	Particulate loss by wind
	Biotic immigration	Biotic emigration
Routes caused by man	Fertilization	Harvest
	Increase in volume of active ecosystem	Reduction in volume of active ecosystem

Nutrient input and output routes vary in nature with the type of ecosystem (e.g., terrestrial, running water, lake, etc.). Likewise, the chemical form in which a nutrient enters or leaves an ecosystem may vary. Some inputs and outputs may occur in gaseous or dissolved abiotic form, some as inorganic or organic particulates, and some in tissues of living or dead organisms. Major input routes in natural ecosystems involve chemical fixation of nutrients from the atmosphere, release of nutrients into active pools by weathering processes, precipitation or inflow of water (carrying nutrients), fallout of particulate matter from the atmosphere, and biotic immigration. Outputs occur by conversion of nutrients into volatile form and their loss to the atmosphere, leaching of nutrients to deep strata or their deposition in deep sediments, runoff or outflow of water (carrying nutrients), blowoff of particulate matter by wind, or biotic emigration.

In addition to modifying the nature and importance of many of these routes in particular ecosystems, man has added two routes basically different from the above: inputs by fertilization and outputs as harvested material. In certain situations man has also modified ecosystem nutrient capital by increasing or decreasing the active volume of the ecosystem within which nutrients may be circulated. Deep plowing of agricultural land may destroy impervious soil layers that inhibit nutrient exchanges between surface soils and deeper layers. By processes of dredging or bottom-sealing of lakes the quantity of nutrients in active sediment pools may be reduced and the nutrient capital of the system depleted. One of the major objectives of study of biogeochemical cycling is to gain an understanding of the manner in which these input and output routes function and of the effects of human management techniques on the nutrient capital of both natural and exploited ecosystems.

Analysis of biogeochemical cycling, however, is not restricted to routes of input and output. Understanding the processes of transfer of nutrients among components within the system, as well as those across system boundaries, is a major objective of current studies.

A variety of techniques must be used to obtain measurements of biogeochemical flux rates. In general, these must be grouped into three categories:

1. Direct measurement.
2. Measurement by difference.
3. Measurement by use of radioactive tracers.

Direct measurement is possible when independent estimates of major physical or biotic processes are available. For example, when inputs such as precipitation or inflow of water, or outputs such as runoff and outflow of water can be measured, these may be combined with measurements of nutrient concentrations in the water to give estimates of nutrient flux rates. Likewise, estimates of overall biological processes such as primary production, or, for populations of individual species, mortality and excretion, may be combined with measurements of nutrient concentration in the tissues or products involved to give estimates of total nutrient flux.

Estimates for the transfer rate by a given route may also be calculated as the difference between other measured rates. Obviously, this is only possible when the rates of all but the process in question are known. For example, when inputs and outputs across the boundaries of a terrestrial system are known, together with the rate of change in total quantity of the nutrient in active pools, the rate of addition of the nutrient by weathering processes in the soil may be calculated. Similar techniques may be used for nutrient flow into or out of individual biotic or abiotic components within an ecosystem. Techniques of this kind have the major fault of lacking an independent check on the value obtained by difference. This value may strongly reflect the measurement errors associated with the other processes used in making the calculations.

Transfer rates may also be determined by using radioactive tracers. This technique may be used only when radioisotopes of the nutrient are available or when an available radioisotope, such as ^{137}Cs, shows strong similarity to a particular nutrient, in this case, K, in its activity. The technique here involves the introduction of a quantity of the radioisotope into a particular nutrient pool. The rate of disappearance from this pool and appearance in other pools is then followed. Data from such experiments may be used to calculate turnover rates and turnover times for the nutrient pools involved. Combined with estimates of the total size of the nutrient pools involved, estimates of absolute flux rates may also be made.

In the following section examples of studies utilizing these techniques will be discussed. This examination will concentrate on studies of terrestrial ecosystems and especially of temperate deciduous forests in which some of the more intensive work is currently being pursued.

Nutrient Cycles in Small Watershed Ecosystems

The most comprehensive study of nutrient inputs and outputs from terrestrial ecosystems has been conducted by Bormann and Likens (1967) on a series of small, forested watersheds at the Hubbard Brook Experimental Forest in central New Hampshire. This forest covers an area of over 3,000 hectares in the White Mountains. The portion of the forest chosen for these studies consists of dense second-growth northern hardwood forest consisting primarily of sugar maple, *Acer saccharum,* beech, *Fagus grandifolia,* and yellow birch, *Betula allegheniensis,* with some red spruce, *Picea rubens* and balsam fir, *Abies balsamea.* The region possesses a humid continental climate with short, cool summers and long, cold winters. Annual precipitation averages 123 cm, of which one-third to one-fourth occurs as snow. Despite the cold winters, the insulating effects of snow pack and litter layers of the forest floor combine to keep the soils unfrozen throughout the winter. The underlying bedrock is a metamorphic gneiss, which in most areas is overlain by glacial till several meters in depth.

The geological features of the area are important in providing a watertight basement layer that prevents major losses of water and nutrients by deep seepage. They thus create a situation in which nearly the entire geological loss of water and nutrients occurs through streamflow.

At the outset of these experiments six small watersheds having sharply defined topographic boundaries were selected for study (Fig. 10–6). These watersheds ranged from 12 to 43 hectares in area and were similar in both altitude (500–800 m) and vegetational characteristics. At the base of each watershed a concrete weir, anchored in bedrock, was constructed to allow the outflow of water to be measured and its solute and particulate loads sampled. Precipitation gauges were placed throughout the watersheds, one for each 12.9 hectares, to monitor the amount and nutrient characteristics of incoming precipitation.

This design permitted measurement of the major inputs and outputs of water and nutrients by the routes of precipitation, streamflow, and weathering of parent materials in the lower soil layers. The availability of several similar watersheds permitted examination of the degree of variability in the behavior of normal watersheds and the comparison of watersheds receiving experimental treatments with similar, untreated watersheds.

The initial series of studies was designed to measure the processes occurring under undisturbed conditions. These studies have revealed several interesting facts about the hydrologic and nutrient cycles of ma-

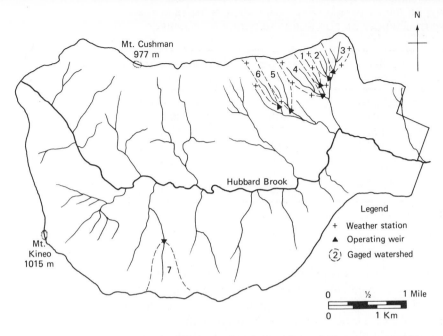

FIGURE 10–6. Outline map of the Hubbard Brook Experimental Forest in central New Hampshire. Watersheds used in nutrient balance studies are indicated by dashed outlines. Watershed number 2 was used in the clear-cutting experiment. (From Bormann and Likens, 1971.)

ture forest ecosystems. Data have now been obtained on the movement of water through the six watersheds over a period of five years (Table 10–7). Over this period, annual precipitation varied from a low of 94.9 cm to a high of 141.8 cm, averaging close to the long-term average of 123 cm. Stream outflow varied in a similar manner. On the assumption that losses by deep seepage were negligible, made reasonable by the watertight geological structure of underlying bedrock, the combined losses by evaporation and transpiration (= evapotranspiration) could be calculated. These varied only between 46.1 and 51.9 cm per year (Table 10–7). Thus, the forest itself shows a high degree of homeostasis in water loss by evaporation. Annual fluctuations in total precipitation consequently are reflected almost entirely in variations in total streamflow. The fraction of the water input lost through evapotranspiration averaged 41.25% over the five-year period. This value is much lower than estimates previously obtained for temperate forest areas, suggesting that, in these other cases, an adequate accounting of losses by deep seepage and by other watershed and ground water movements has not been obtained.

Analysis of nutrient balance required periodic measurements of the concentration of the substances involved both in precipitation and in

TABLE 10-7. INPUT OF WATER IN PRECIPITATION AND OUTPUTS IN STREAM-FLOW AND EVAPOTRANSPIRATION FOR HUBBARD BROOK WATERSHEDS FROM 1963-1968.*

Water year (June 1–May 31)	Precipitation (cm)	Stream outflow (cm)	Evapotranspiration (cm)
1963–64	117.1	67.7	49.4
1964–65	94.9	48.8	46.1
1965–66	124.5	72.7	51.8
1966–67	132.5	80.6	51.9
1967–68	141.8	89.4	52.4
Average	122.2	71.8	50.4

*From Likens et al., 1970.

the stream outflow. Input via precipitation was appreciable for most of the major nutrients. In order to obtain some idea of the relative importance of nutrients entering the system in actual precipitation as opposed to those entering by fallout of fine particulate matter two kinds of precipitation gauges were used. One kind was continuously open and could, therefore, accumulate particulate matter, such as dust and aerosols; the other opened only during actual precipitation periods. Comparisons of the nutrient concentrations from these two collection devices suggested that virtually all of the inputs in the Hubbard Forest area occurred in actual precipitation.

Some variation in concentrations of nutrients in precipitation at different times was noted. This was especially pronounced for sodium and magnesium ions. Since these are ions abundant in seawater, it appears likely that variation in their concentration is related to variation in the extent to which precipitation is derived from marine air masses from time to time.

Concentrations of nutrients in stream outflows were sampled at weekly intervals. Measurements were obtained for both dissolved nutrients and nutrients in particulate matter, both organic and inorganic, carried in suspension in the outflow. Because of the great variation in volume of streamflow, both within and between years, analyses of nutrient concentrations in relation to the volume of streamflow at the time of sampling were performed. For dissolved nutrients these analyses showed concentrations to be relatively independent of flow volume (Fig. 10–7). For calcium and magnesium there was no significant relationship with streamflow. For sodium the relationship was inverse, i.e., showing somewhat lower concentrations at higher flow volumes. The reverse appeared to be true for potassium; higher flow volumes tended to be accompanied by higher concentrations of this element. Nevertheless, a surprising degree of constancy was seen in nutrient concentrations at different stream-

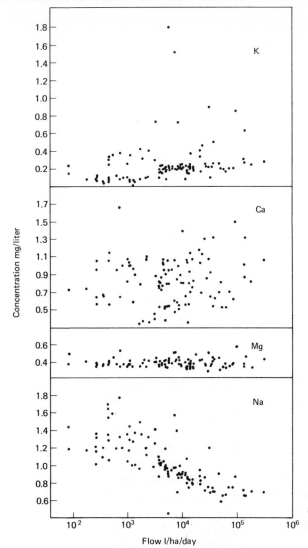

FIGURE 10–7. Relationship between cation concentration and volume of flow for stream water in watershed number 4 from 1963 to 1965. (From Likens et al., 1967.)

flow levels. In these watersheds, even during periods of intense precipitation, direct runoff of water over the surface is uncommon. Almost all water reaches stream channels by percolation through the soil. As this occurs the water is exposed to chemical conditions in the soil that buffer the concentrations of various ions, especially cations of the alkaline earth group, in soil water.

For particulate materials, however, a positive relationship was found for streamflow volume and nutrient concentration (Fig. 10–8). For the entire year, despite appreciable losses during periods of high water flow,

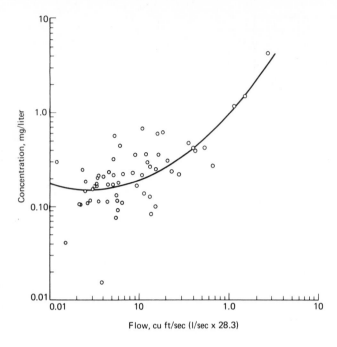

FIGURE 10–8. Curve showing the relationship between concentration of filtered material and flow rate. Open circles show data collected with single filters; solid circles show data collected with double filters. (From Bormann, Likens, and Eaton, 1969.)

particulate losses were of minor importance in nutrient outflow (Table 10–8). For the major cations of calcium, magnesium, nitrogen, potassium, sodium, and sulfur, losses in particulate matter were generally less than 10% (18.0% for potassium).

Balance sheets of these processes could be constructed for the major nutrients from the combined data on precipitation inputs and stream outputs (Table 10–9). These balance sheets indicate that precipitation inputs are appreciable in relation to streamflow losses of many ions. In the cases of nitrogen and chlorine, precipitation inputs exceeded stream losses. If we assume that the nutrient capital of the forested watersheds is nearly stable, the differences in precipitation input and output processes indicate the magnitude of processes operating within the forest ecosystem to increase or decrease nutrient capital. For the mineral ions of calcium, magnesium, potassium, sodium, sulfur, and silicon, these presumably are additions through the weathering of primary minerals in the soil. For nitrogen, carbon, and, perhaps, chlorine, the processes may be more complicated and may involve additional patterns of biological fixation from, and release into, the atmosphere.

The nature of the cycle of calcium within these ecosystems may be examined more completely by combining the observations on nutrient input and loss with data on the calcium capital of forest ecosystems and transfer rates between various intrasystem components. Data on the latter relationships were taken primarily from Ovington (1962) and used

TABLE 10-8. LOSSES OF NUTRIENTS IN DISSOLVED AND PARTICULATE FORM FROM WATERSHED NUMBER 6 AT HUBBARD FOREST IN 1965–1966 AND 1966–1967.*

| | Percent of total losses | | |
| | Particulate matter | | |
Element	Organic	Inorganic	Dissolved
Calcium	0.7	1.8	97.5
Magnesium	2.0	3.7	94.3
Nitrogen	5.9	a	94.1
Potassium	0.5	17.5	82.0
Sodium	0.0	2.8	97.2
Silicon	b	18.8	81.2
Sulfur	0.1	0.1	99.8

*From Bormann et al., 1969.

Notes:

[a]Not measured, but very small.

[b]Not measured.

TABLE 10-9. INPUT BY PRECIPITATION AND OUTPUT IN STREAMFLOW FOR VARIOUS NUTRIENTS IN AN UNDISTURBED WATERSHED (NUMBER 6) AT HUBBARD BROOK FOR 1966–1967 AND 1967–1968.*

| Nutrient | 1966–67 | | | 1967–68 | | |
	Precipitation input	Stream outflow	Net loss or gain	Precipitation input	Stream outflow	Net loss or gain
Water	133	85	—	171.8	89.4	—
Ca	2.4	10.7	−8.3	3.0	12.2	−9.2
Mg	0.4	2.9	−2.5	0.8	3.4	−2.6
K	0.6	1.7	−1.1	0.8	2.4	−1.6
Na	1.3	6.8	−5.5	1.8	8.8	−7.0
N (NH_4)	1.9	0.3	+1.6	2.6	0.2	+2.4
N (NO_3)	4.6	1.3	+3.3	5.2	2.8	+2.4
S (SO_4)	14.4	17.1	−2.7	16.0	19.3	−3.3
Cl	6.9	4.6	+2.3	5.2	5.3	−0.1
C (HCO_3)	a	0.4	−0.4	a	0.5	−0.5
Si (S_1O_2)	a	17.2	−17.2	a	17.0	−17.0

*From Likens et al., 1970.

[a]Not measured, but very low.

by Bormann and Likens (1970) to derive the analysis presented in Fig. 10–9. This diagram shows a balance of calcium inputs and losses. Inputs of calcium occur from the atmosphere, through precipitation and dust fallout, and amount to 2.6 kg/hectare year. Weathering of primary and secondary minerals releases nutrients equal to 9.4 kg/hectare year, 9.1 kg of which are added to the pool of available soil calcium and 0.3 kg of which is stored in an increasing biomass pool of the forest. These inputs

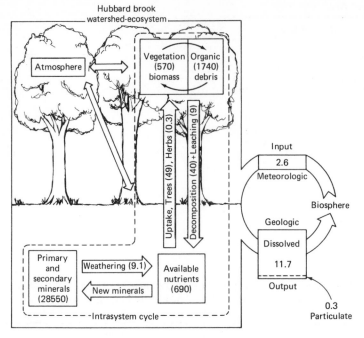

FIGURE 10-9. Calcium cycle is depicted for an undisturbed ecosystem. Numerals represent the average number of kg per hectare per year. Thus the meteorological input to the ecosystem in precipitation and dust is 2.6 kg per hectare annually. A substantial amount of calcium is in soil and rocks; 9.1 kg is released annually by weathering. Vegetation takes up 49.3 kg; 49 kg is returned to the soil by decomposition and leaching. Gross loss in stream drainage is 12 kg, so the net loss is 9.4 kg. (From Bormann and Likens, 1971.)

total 12.0 kg/hectare year, which is balanced by a loss of 12.0 kg/hectare year in stream outflows.

The rate of weathering of primary and secondary minerals could also be estimated from data on inputs and outputs, using the equation:

$$W = \frac{D_i}{C_i - S_i}$$

where

W = Weight of primary and secondary minerals undergoing weathering, per unit time
D_i = Amount of element i removed in solution per unit time
C_i = Concentration of element i in unweathered material
S_i = Concentration of element i in weathered products

For calculations of weathering rate, D_i was determined as the loss of the element i in streamflow (dissolved form). The concentration of the element in unweathered material was determined by chemical analysis of the underlying bedrock and unweathered glacial till (including its rock components). The concentration of the element in weathered products was determined by analyzing the chemical composition of mineral soil from the A_2-soil horizon, which is the layer, located beneath the humus-rich A_1 horizon, subject to most intense leaching of soluble materials.

This layer contains the minimum concentrations of nutrients and, therefore, represents the final result of weathering processes. In essence, this formula gives the weight of materials that must be weathered in order to release nutrients to replace those leached from the soil (mainly from the A_2-horizon) and carried out of the system in streamflow.

Calculations were made for W, using data for several nutrient elements, but primary importance was placed on the estimates derived from data for calcium and sodium. This was done because of the lack of involvement of ions of these elements with inorganic reactions involved in clay formation in deeper soil layers. In other words, ions of these elements disappearing from the A_2 soil horizon could really be regarded as having been removed from the active nutrient pool. Calculations for these two elements (Table 10–10) suggested that each year about 770–800 kg/hectare of rock and unweathered till undergo weathering.

Following the elucidation of basic nutrient budgets for the undisturbed condition, one of the watersheds in this study was selected for a major disturbance study. In the winter of 1965–1966, all woody vegetation was cut and left in place. Growth of higher plants was inhibited by herbicide treatments during the following summer. The nutrient inputs and losses from this watershed were then monitored and compared with those of undisturbed watersheds and with the pattern shown prior to cutting.

As expected, runoff from this watershed increased, reflecting the reduction of water loss due to transpiration. Increased runoff amounted to about 40% and contributed to an increase in outflow of dissolved and particulate material. However, more important changes occurred in the concentrations of dissolved nutrients in the outflow (Fig. 10–10). Concentrations of almost all cations were greatly increased; those of some anions such as sulfate were decreased. The total export of dissolved inorganic substances from the watershed was 6–8 times that of an undisturbed system (Likens et al., 1970). The relationships permitting this loss were related closely to major modifications of the nitrogen cycle and to the interaction of this cycle with cycles of other nutrients.

In the undisturbed forest ecosystem the processes of nitrification

TABLE 10-10. CALCULATED CHEMICAL WEATHERING RATES FOR THE HUBBARD BROOK STUDY AREA DURING 1963-1967.*

Element	D (4-year mean) (kg/ha/yr)	C[a]	S[a]	W (4-year mean) (kg/ha/yr)
Ca	8.0 ± 0.7	0.014	0.004	800 ± 70
Na	4.6 ± 0.4	0.016	0.010	770 ± 70

*From Johnson et al., 1968.
[a]Decimal fraction of element in unweathered or weathered material.

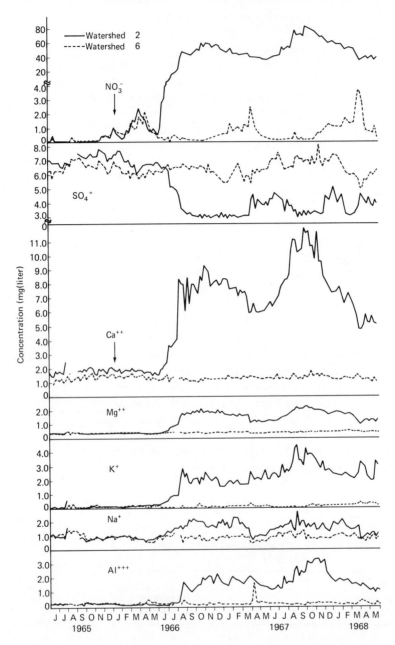

FIGURE 10–10. Measured stream water concentrations of various ions in watersheds number 2 (deforested at time indicated by arrow) and number 6 (normal). Gaps in graphed lines indicate periods of negligible water discharge. (From Likens et al., 1970.)

(Fig. 10–11) are of relatively minor importance. Decomposition of organic matter results in the gradual production of ammonium ions (NH_4^+) which are tightly held in the soil and can be directly absorbed and used by higher plants. Almost all of the nitrate input to the system is, in fact, accountable for as precipitation input. Thus, very little nitrate is

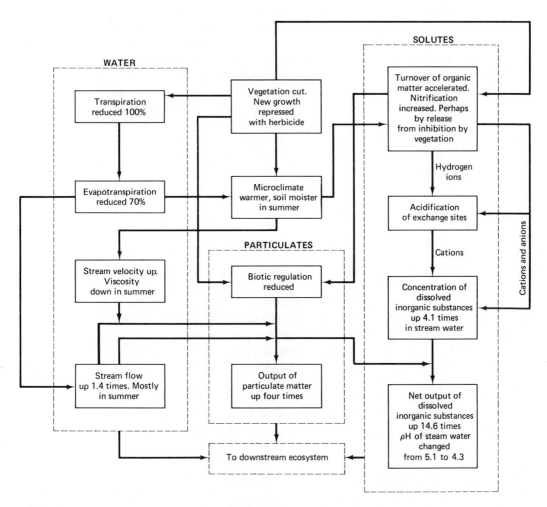

FIGURE 10–11. Effects of deforestation on ecosystem processes in watershed number 2 in the Hubbard Brook Experimental Forest. Effects are shown on the output of water, dissolved nutrients (solutes), and particulate material carried in streamflow. Normally cations (positively charged nutrient ions) are held tightly adsorbed to soil colloids. Accelerated loss of these cations appears to occur largely as a result of increased nitrification. Nitrification results in production of hydrogen ions that displace other cations and permit them to be leached away. (Modified from Bormann and Likens, 1970.)

actually produced within the soil. This is important because nitrate is a highly soluble form of nitrogen and, thus, is susceptible to being removed by leaching.

In contrast, in the cleared watershed decomposition suddenly produces large quantities of ammonium ions that are not taken up by plants. Ammonium is then converted in large quantities to nitrate by nitrite and nitrate bacteria in the soil. The nitrate so produced is subject to intense leaching. In addition, the conversion of ammonium to nitrate releases two hydrogen ions. These hydrogen ions replace other cations, such as Ca^+, Mg^+, Na^+, and K^+, held on the surface of colloidal soil particles. The replaced ions enter the ground water, and, as a consequence, are subjected to rapid leaching. Thus, it is clear that the capacity of a forest ecosystem to retain nutrients is closely related to the maintenance of particular patterns of nutrient cycling. Here it is clear that the disturbance of normal nitrogen cycling processes, rather than the mechanics of erosion, was the most important factor in causing depletion of ecosystem nutrient capital following forest cutting. The sequence of events involved in this depletion is summarized in Fig. 10–11.

Nutrient Cycling within Temperate Deciduous Forests

At the present time, a number of major studies of nutrient cycling within major ecosystems are being conducted in connection with the Biome Studies of the International Biological Program. These studies are designed to provide information on the quantities of nutrients in various biotic and available abiotic pools and on the rates of transfer among these pools. Duvigneaud and Denaeyer-De Smet (1969), concentrating on two major study areas in Belgium, have recently provided a detailed summary of studies of these relationships in European deciduous forests. Data for these two areas will serve to give an idea of some of the objectives and conclusions of these studies.

The two forests studied were a mature oak-ash forest (*Quercus robur, Fraxinus excelsior*) with canopy trees varying in age from 115 to 160 years and a mixed oak-beech-hornbeam forest (*Q. robur, Fagus sylvatica, Carpinus betulus*) with trees ranging in age from 30 to 75 years. The oak-ash forest occurred on deep, rich soils derived from shale and had favorable moisture conditions. In contrast, the mixed forest was developed on an area of thin, calcareous soils. In pursuing studies of these areas a variety of techniques was employed to measure the biomass and productivity of the forest vegetation and to determine the nutrient content of various plant materials, litter, and soil. Many of the specific techniques utilized in these studies were similar to those previously discussed in connection with analysis of energy flow and nutrient cycling in terrestrial ecosystems. It should be emphasized that a very extensive series of measurements was necessary to obtain the summary values discussed below.

Data for these two forests are compared in Table 10–11 and the annual nutrient fluxes shown diagramatically in Figs. 10–12 and 10–13. The more mature oak-ash forest showed a much greater total biomass of plant

TABLE 10–11. BIOMASS, NET PRIMARY PRODUCTION, SIZE OF NUTRIENT POOLS, AND ANNUAL NUTRIENT FLUX RATES FOR TWO DECIDUOUS FORESTS IN BELGIUM.*

		Oak-ash forest (age 115–160 years)	Mixed oak-beech-hornbeam forest (age 30–75 years)	Ratio
Total soil weight		6,318 t	1,360 t	4.65:1
Soil pools	K	185 t	26.8 t	6.90:1
	Ca	35 t	33 t	1:4.03
	Mg	50 t	6.5 t	7.70:1
	N	14 t	4.5 t	3.11:1
	P	2.2 t	0.9 t	2.44:1
Total biomass		380 t	156 t	2.44:1
Vegetational pools	K	624 kg	342 kg	
	Ca	1,648 kg	1,248 kg	
	Mg	156 kg	102 kg	
	N	1,260 kg	533 kg	
	P	95 kg	44 kg	
Net primary production		14.3 t	14.4 t	
Annual uptake	K	99 kg	67 kg	1.43:1
	Ca	129 kg	201 kg	1:1.56
	Mg	24 kg	19 kg	1.26:1
	N	123 kg	92 kg	1.34:1
	P	9.4 kg	6.9 kg	1.36:1
Annual return to forest floor	K	78 kg	53 kg	
	Ca	87 kg	127 kg	
	Mg	19 kg	13 kg	
	N	79 kg	62 kg	
	P	5.4 kg	4.7 kg	
Annual retention in increased biomass	K	21 kg	16 kg	
	Ca	42 kg	74 kg	
	Mg	5 kg	6 kg	
	N	44 kg	30 kg	
	P	4.0 kg	2.2 kg	

*From Duvigneaud and Denaeyer–DeSmet, 1969.

Note: Units are metric tons (t) or kilograms (kg) of nutrients or dry weight of plant materials.

tissues—about 2.44 times that of the younger mixed forest. The deeper soils of the oak-ash forest were also reflected in the greater total soil weight per hectare for this system (4.65 times as great as that of the younger forest). This greater soil weight was correlated with larger soil pools of all nutrients except calcium. The younger mixed forest, it may be remembered, occurred on soils derived from calcareous rock, and it thus contained much greater amounts of calcium.

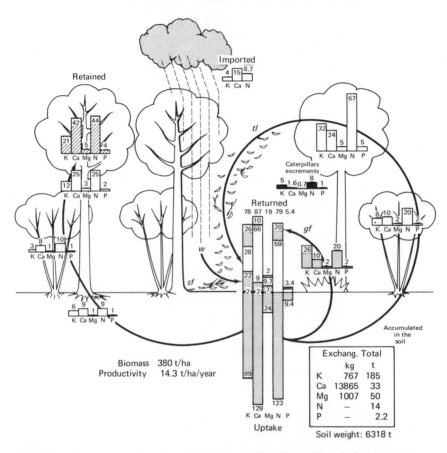

FIGURE 10–12. Annual mineral cycling of K, Ca, Mg, N, and P (in kg/ha) in a *Quercus robur-Fraxinus excelsior* forest with coppice of *Corylus avellana* and *Carpinus betulus* at Wavreille-Wève Belgium. Retained: in the annual wood and bark increment of roots, 1-year-old twigs, and the above-ground wood and bark increment. Returned: by tree litter (*tl*), ground flora (*gf*), washing and leaching of canopy (*w*), and stem flow (*sf*). Imported: by incident rainfall (not included). Absorbed (uptake): the sum of quantities retained and returned. Macronutrients contained in the crown leaves when fully grown (July) are shown on the right-hand side of the figure in italics; these amounts are higher (except for Ca) than those returned by leaf litter, due to reabsorption by trees and leaf-leaching. Exchangeable and total element content in the soil are expressed on air-dry soil weights of particles < 2mm. (From Duvigneaud and Denaeyer–De Smet, 1969.)

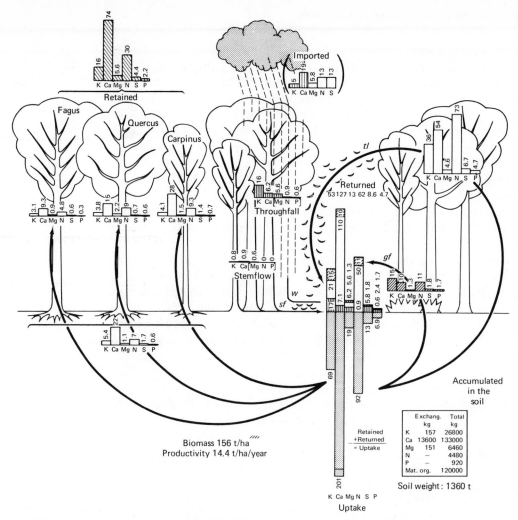

FIGURE 10–13. Annual mineral cycling in kg/ha of macronutrients in the "mixed oak-wood" ecosystem at Virelles, Belgium. Retained: in the annual wood and bark increment of roots and aerial parts of each species (total is hatched). Returned: by tree litter (*tl*), ground flora (*gf*), washing and leaching of the canopy (*w*), and stem flow (*sf*). Imported: by incident rainfall (not included). Macronutrients contained in the crown leaves when fully grown (July) are shown on the right-hand side of the figure in italics; these amounts are higher (except for Ca) than those returned by leaf litter. Exchangeable and total element content in the soil are expressed on air-dry soil weights of particles < 2mm. (From Duvigneaud and Denaeyer–De Smet, 1969.)

Net primary production was nearly the same for the two forests, amounting to somewhat over 14 tons per hectare per year. It might be expected that similarity in level of net primary production would be reflected in similarity of total uptake of specific nutrients. However, this was not the case. The younger mixed forest showed a total uptake of calcium 1.56 times that of the more mature oak-ash forest. Likewise, the older oak-ash forest showed uptakes of K, Mg, N, and P at much higher levels than for the younger forest. Thus it is apparent that uptake is related in part to the quantity of nutrients available in soil pools for these two forests. When certain nutrients are available in abundance (for example, calcium), they may be taken up and actively cycled in quantities greater than those actually required, thus demonstrating "luxury" use of the nutrient in question. It may be noted, however, that the ratios in quantities of nutrients taken up for the two forest areas vary less than do the ratios of quantities of nutrients in soil pools (Table 10–11). Whether these different uptake patterns reflect differences between nutrient requirements of different species, influences of availability and uptake patterns for one nutrient upon those of other nutrients, or "luxury" use is not known.

The nutrient uptake of forest vegetation is partitioned into two categories, separable according to the fate of the net primary production of the vegetation. One portion is returned to the forest floor, largely through leaf and dead branch fall and the death of herbaceous vegetation. In addition, appreciable amounts are also leached from leaves and branches by water and carried to the forest floor as indicated in Figs. 10–12 and 10–13. The nutrients not returned to the forest floor are retained by the vegetation and correspond to the nutrient content of the living biomass added to the forest during the year. Estimates of annual return to the forest floor and annual retention in increased biomass of the forest ecosystem are given in Table 10–11. The younger mixed forest shows a somewhat higher retention (about 33%) of the nutrients taken up than does the older oak-ash forest (about 30%). Presumably, this reflects the fact that the older forest is approaching somewhat more closely a steady-state condition in which uptake would be balanced by processes returning nutrients to the forest floor.

Patterns of Biogeochemical Cycling in Terrestrial Ecosystems

A number of studies, similar in approach and objectives to that just discussed, have been carried out for various ecosystem types. The accuracy and completeness of these studies vary greatly, and major gaps exist for both geographical areas and ecosystem types. Rodin and Bazilevich (1968) have attempted to piece together the available data, however, in order to provide a preliminary picture (for terrestrial ecosystems) of world patterns of ecosystem energetics and nutrient cycling.

Rodin and Bazilevich first compiled available data on the biomass of living tissues, quantity of litter materials, rate of primary production, the size of various biotic and environmental nutrient pools, and the magnitude of major nutrient fluxes. From an examination of these data they selected five relationships as a basis for characterizing major patterns of ecosystem function. These were:

1. Predominant mineral elements in litter fall.
2. Standing crop biomass of the vegetation.
3. Quantity of annual litter production.
4. Litter decay rate.
5. Ash content of litter fall.

We will first examine the general significance of each of these.

The specific element of greatest abundance in litter produced in a particular ecosystem differs for various ecosystem types. The four elements, or ions, which in different ecosystem types may show the greatest abundance are nitrogen, calcium, silica, and chloride. Predominance of different elements in different ecosystems presumably reflects both changed patterns of environmental availability of nutrients and changed structual and physiological requirements for plant growth. The abundance of these elements in litter materials was chosen on the assumption that the nutrient composition of this material, which is easily collected and measured, generally reflects the pattern of uptake and fixation of nutrients in primary production.

Standing crop biomass of the living vegetation was regarded as indicative both of the complexity of vegetational structure and of the size of pools of nutrients and energy in the living portion of the system.

The quantity of litter produced annually is of importance as an index to the level of net primary production, especially in systems approaching a climax state.

Litter decay rate represents the rate at which nutrients are regenerated from dead tissues and made available again for use by living plants. The index of litter decay rate used by Rodin and Bazilevich was the ratio of *standing crop of litter* to *annual litter fall*. A high value for this ratio indicates that a long period passes before the litter produced in a given year becomes completely decomposed, with the result that an accumulation equal to many years' production develops. Likewise, a value for this ratio of <1 indicates that, on the average, decomposition of litter requires less than one year.

Finally, the quantity of mineral ash, expressed as a percentage of dry weight of freshly fallen litter, gives a measure of the concentration of nutrients occurring in a given unit of plant tissues.

Other important characteristics can certainly be proposed; however, these five constitute a reasonable basis for a preliminary classification. Furthermore, they represent characteristics for which data are frequently available from past studies.

For characteristics 2–5, Rodin and Bazilevich devised a numerical importance scale ranging from 1 to 10 (Table 10–12). On this scale low values reflect low quantities or rates of function and vice versa. The ranges of actual values corresponding to scale numbers were selected subjectively to cover the full range seen in the available data.

TABLE 10-12. TEN-POINT SCALE FOR CLASSIFICATION OF BIOMASS, ANNUAL LITTER FALL, LITTER DECAY RATE, AND LITTER ASH CONTENT FOR VARIOUS ECOSYSTEMS.*

Scale number	B (Biomass) (100 kg/ha)	L (Annual litter fall) (100 kg/ha)	D (Decay rate) Standing crop / Annual fall	A (Ash content of litter fall) Percent
1.	<25	<10	>50	<1.5
2.	26–50	11–25	21–50	1.6–2.0
3.	51–125	26–35	16–20	2.1–2.5
4.	126–250	36–45	11–15	2.6–3.5
5.	251–500	46–75	6–10	3.6–5.0
6.	501–1,500	76–100	1.6–5	5.1–6.5
7.	1,501–3,000	101–125	0.8–1.5	6.6–8.0
8.	3,001–4,000	126–225	0.3–0.7	8.1–9.5
9.	4,001–5,000	226–400	0.1–0.2	9.6–12.0
10.	>5,000	>400	<0.1	>12.0

*From Rodin and Bazilevich, 1968.

Using the information on predominant nutrient components of litter fall and the scale values for the remaining four relationships, Rodin and Bazilevich were able to distinguish a number of ecosystem types of distinctive functional characteristics (Table 10–13). For example, a number of ecosystem types showed nitrogen as the principal nutrient component of litter fall. The most extreme case involved arctic tundra in which nitrogen was by far the most prominent nutrient. In other systems, described as calcic-nitric types, nitrogen was of greatest importance, followed closely by calcium. These systems were generally found in arctic and boreal regions and in deserts in which the primary vegetation was of semi-woody shrubs. A subtropical nitric group was recognized, although this group may actually constitute an artifact associated with an intermediate pattern of calcium and silica use (see below).

In temperate deciduous forests calcium is the primary constituent of litter. In a variety of ecosystem types, including grasslands, deserts having prominent representation of annuals, savannas, and tropical evergreen forests show that silica is the most abundant nutrient.

For all of the above communities nitrogen is one of the two most abundant elements. This is probably the result of its physiological importance, since it occurs primarily as a constituent of proteins. Calcium and silica, on the other hand, are utilized by plants for primarily structural purposes. Thus, the major groupings of ecosystems according to predominant nutrients in a sense reflect the quantitative requirement for structural materials by the dominant plants, together with a major climatic gradient

TABLE 10-13. PRELIMINARY CLASSIFICATION OF MAJOR ECOSYSTEM TYPES ACCORDING TO PREDOMINANT NUTRIENT ELEMENTS IN LITTER FALL, STANDING CROP BIOMASS, QUANTITY OF ANNUAL LITTER FALL, DECAY RATE OF LITTER, AND ASH CONTENT OF LITTER.*

Type class	Type group	Vegetation type	B (Biomass)	L (Litter fall)	D (Decay rate)	A (Ash content)
Boreal nitric	Nitric	Tundra	2–5	1–2	1–4	2–3
	Calcic-nitric	Boreal coniferous forests	6–8	3–5	3–9	1–2
	Calcic-nitric	Boreal deciduous forests	7	5	5	4
Desert nitric	Calcic-nitric	Semi-shrub desert	2	2	10	4
Subtropical nitric	Calcic-nitric	Subtropical deciduous forests	9	8	8	4
Subboreal calcic	Nitric-calcic	Temperate deciduous forests	8	5–6	6	4
Steppe silicic	Nitric-silicic	Steppe (grassland)	3–4	4–7	7	3–4
Desert silicic	Nitric-silicic	Semishrub desert with annuals	3	6	10	5
Tropical silicic	Nitric-silicic	Savanna	5–6	5–7	9	4
	Nitric-silicic	Tropical rain forests	10	9	9	4
Chloridic	Sodic-chloridic	Saline deserts	1	1	10	10

*Modified from Rodin and Bazilevich, 1968.

Note: The numbers in the last four columns correspond to scale numbers from Table 10–12.

in the specific nutrient (calcium or silica) utilized. Why a climatic gradient in relative importance of calcium and silica should occur is, at present, uncertain. The predominance of nitrogen over other nutrients in sub-tropical deciduous forests may result from this system's occupying an environment close to the point where changeover from calcium to silica occurs. In these forests the sum of the quantities of calcium and silica in the fresh litter greatly exceeds the quantity of nitrogen, and predominance of nitrogen may reflect simply a nearly even split in structural use between calcium and silica.

In desert areas characterized by saline soils the primary minerals in plant tissues are chloride and sodium. These materials are accumulated in large quantities, presumably because of their abundance in the soil and, perhaps, their physiological function in increasing the osmotic concentration of plant fluids, thus facilitating the uptake of water from soils having similarly high solute concentrations. Otherwise, the nutrient relationships correspond to a calcic-nitric pattern seen for desert shrub vegetation on nonsaline soils.

Thus, in arctic, subarctic, temperate forest, and desert shrub communities calcium is the most abundant structural mineral. Differences among these ecosystems relate to the quantitative relationships between calcium and nitrogen, which are physiologically important, and to conditions of salinity of the soil. By contrast, in grassland, desert grassland, savanna, and tropical forest communities silica is the primary structural component, usually in quantities exceeding those of nitrogen, which, however, remains an important constituent. Subtropical deciduous forests show an intermediate calcium-silica relationship that probably corresponds to environmental conditions in which the changeover between these two nutrients occurs.

Biogeochemicals and Environmental Pollutants

It has recently become apparent that man has introduced many novel chemical substances into the biosphere and that these substances have often become involved in major patterns of biogeochemical movement. These include radioisotopes such as ^{137}cesium and ^{90}strontium, produced by atomic testing, and a variety of complex organic chemicals such as the chlorinated hydrocarbon pesticides. Man has also greatly increased the quantities of naturally occurring substances in active circulation in the biosphere, e.g., lead and mercury. The behavior of these materials is closely tied to processes of biogeochemical cycling.

The patterns of circulation of radioisotopes in the biosphere are related to the source of the radioisotopes and to the similarity or identity of the isotope to elements utilized as nutrients by organisms. The radioisotope ^{32}phosphorus, which has entered the environment primarily as a waste from atomic reactors, is taken up and utilized by organisms exactly as normal phosphorous. The radioisotope ^{131}iodine is concentrated by higher animals which utilize normal iodine as a constituent of the thyroid hormone. Fortunately, in this case, the half-life of ^{131}iodine is only 8 days, and thus it constitutes only a temporary danger as an environmental pollutant.

The isotopes ^{137}cesium and ^{90}strontium, however, possess half-lives of 30 and 28 years, respectively. These isotopes, together with ^{131}iodine, are fission products and are released into the atmosphere primarily by atomic testing (fission products are basically left-over fragments of uranium atoms that have undergone atomic fission). They are thus injected into atmospheric circulation systems through which they may be distributed world-wide. Fallout of such materials, however, is greatest in latitudinal belts close to that of its atmospheric introduction. For example, fallout from Russian testing in arctic areas is concentrated in arctic areas around the world. The biological uptake of these isotopes is related to their chemical similarity to other elements. Cesium behaves chemically much like potassium and is taken up and concentrated in tissues, such as muscle, in which potassium is an important functional element. Strontium, similar in its chemistry to calcium, is taken up and concentrated in tissues, such as bone and milk, in which calcium is present in large quantities.

Hanson (1967) has summarized the pattern of movement of ^{137}cesium in the tundra ecosystem of Alaska. The ^{137}cesium here was derived mainly from the Russian atomic testing at high latitudes. The isotope ^{137}cesium was found to be accumulated first by ground lichens of various types. These plants, by virtue of their longevity, persistences of aerial parts, and dependence on nutrients from precipitation, are very efficient in concentration of fallout materials. The biological half-life (taking into account both atomic decay and physiological exchanges of cesium with the environment) of ^{137}cesium in these lichens was estimated to be about 13 years.

Lichens constitute the primary winter food of caribou and reindeer in arctic tundra areas. Thus, ^{137}cesium is concentrated in tissues of these animals more than in those of other arctic tundra herbivores. Maximum concentrations occur in Alaskan caribou in April and May when the animals are leaving their winter range. On the summer range, where other foods are utilized, the caribou show rapid decline in ^{137}cesium levels. The biological half-life of ^{137}cesium in this species is about 3–5 weeks.

Many caribou, however, are harvested by Eskimo hunters during their movements from winter to summer ranges. These animals provide a major source of meat for certain Eskimo villages for much of the summer. Consequently, Eskimos show maximum ^{137}cesium body burdens during the summer. Like caribou, humans show fairly rapid exchange of ^{137}cesium, with the biological half-life being about 65 days.

Not only does ^{137}cesium pass along the food chain in this manner, it is also concentrated in the process. Hanson estimated that for each link in the food chain, a concentration by $2X$ occurs. This is clearly seen in analysis of the flesh of wolves, which are almost entirely dependent on caribou. Eskimos, deriving about half of their meat intake from caribou, show about the same body burdens, per unit weight, as do caribou.

This process of biological concentration or magnification of materials is a general property of food chains and may occur for a variety of types of chemical substances. At the root of it is the fact that the chemical substances composing a unit weight of tissue at one trophic level are derived from several times that total amount of food materials supplied by the trophic level below.

The chlorinated hydrocarbon pesticides, including DDT, Dieldrin, Aldrin, Heptachlor, and many others, illustrate the fact that synthetic organic materials released into the environment by man may become involved in biosphere-wide patterns of circulation. For DDT the side effects resulting from the pattern of movement and accumulation in various parts of the world ecosystem are now beginning to become apparent (Wurster, 1969).

The special problem presented by the chlorinated hydrocarbons results from a combination of four features:

1. Chemical stability
2. Solubility characteristics
3. Dispersal mechanisms
4. Broad toxicity and biological activity

Chlorinated hydrocarbons are known as persistent pesticides because of their chemical stability. The chemical half-life of these compounds in the environment is difficult to measure, but for DDT it is estimated to be 10–15 years (Wurster, 1969). Chemical breakdown involves first the conversion of DDT to DDE and DDD, which are related chemical substances that still retain considerable toxicity and biological activity. Thus, disappearance of DDT itself does not necessarily mean the disappearance of all chlorinated hydrocarbon compounds.

DDT has a very low solubility in water—about 1.2 ppb (parts per billion)—but is highly soluble in fats and lipids. Consequently, as it moves through the environment, DDT tends to accumulate in materials composed of fats or lipids. Since concentrated deposits of these materials occur only in tissues of living organisms, this solubility characteristic leads to a net "flow" of DDT from the abiotic to the biotic components of an ecosystem.

Dispersal of chlorinated hydrocarbons through the biosphere occurs along a number of routes. Although water solubility and vapor pressure of pure DDT are both very low, in time, large amounts can be transported in dissolved or vapor form by water or air movements. Furthermore, since DDT is usually delivered to the environment as finely divided dusts or sprays, the transport of large quantities in suspension in air or water is facilitated. DDT also becomes adsorbed to particulate matter, further enhancing its potential for dispersal in suspension in air or water. In water samples taken in the field, concentrations up to 10^6 times those of maximum solubility have been recorded, indicating the importance of transport in suspended form.

Recently, it has also been found that DDT codistills with water. Codistillation involves an increased rate of vaporization of DDT from an aqueous solution or suspension over that expected on the basis of the vapor pressure of pure DDT. In an aqueous suspension of a few parts per billion of DDT, exposed to the air, over one-half of the DDT present may be lost within a 24-hour period (Wurster, 1969).

Through the operation of these various routes of dispersal DDT and other commonly used chlorinated hydrocarbons have now become worldwide in distribution. Their solubility characteristics have resulted in a pattern of concentration along food chains similar to that shown by the radioisotope ^{137}cesium discussed earlier. For DDT, however, the concentration factor for each link in the food chain, due largely to the extreme difference between solubility in water and fats, may range from as low as 2–3 times to as high as 1,000 times.

The side effects of DDT are both direct and indirect. Direct side effects are those resulting from the exposure of nontarget organisms to applications of DDT intended to kill some pest organism. For example, the use of DDT to control outbreaks of the spruce-budworm in Canadian coniferous forests has resulted in nearly complete mortality of aquatic insect populations and severe mortality of young salmon in streams within the treated areas (Wurster, 1969).

Indirect side effects involve the action of DDT after it has been translocated to distant parts of the biosphere or concentrated by food chain processes. Use of DDT in programs to control the Dutch Elm Disease by spraying trees in an attempt to reduce populations of an elm bark beetle which carries the disease agent, a fungus, from one tree to another has led to heavy mortality of certain songbirds. The American Robin, *Turdus migratorius,* has been one of the most severely affected species. Mortality of this species has been shown to result from their feeding on earthworms which concentrate DDT carried to the soil by rain and leaf fall.

Although DDT is used almost entirely in terrestrial situations, its varied means of dispersal have led to indirect side effects in quite distant locations. These indirect effects have been demonstrated for organisms of all trophic levels. Wurster (1969) has shown that concentrations of a few ppb may reduce the photosynthetic rates of a variety of species of marine phytoplankton. Other aquatic organisms, such as brine shrimp, are sensitive to DDT in concentrations as low as one part per trillion.

General environmental contamination, together with food chain concentration, has threatened populations of a number of species of carnivorous birds. In the case of some species, such as the Herring Gull, the concentrations of DDT built up in body tissues are enough to cause direct mortality, especially if stress conditions require individuals to metabolize fat reserves containing stored DDT (Hickey et al., 1966). Direct mortality of this sort apparently results from toxic effects of DDT on the central nervous system (Wurster, 1969). At sublethal levels, however, DDT has effects that are probably of greater danger to these species than this direct toxicity. DDT and other chlorinated hydrocarbons induce the production of certain hepatic enzymes which, among other effects, lead to the breakdown of steroid hormones (Wurster, 1969). Since steroid hormones control various aspects of the reproductive process, including the uptake and storage of calcium for egg shell production, this results in major disturbances of reproductive function. DDT, furthermore, has been shown to be a strong inhibitor of the enzyme carbonic anhydrase, which mediates the deposition of calcium by the shell gland region of the avian oviduct. Consequently, one of the general effects of DDT on carnivorous birds has been the production of thin-shelled eggs that are easily broken in the nest. For several species, such as the Brown Pelican, *Pelecanus occidentalis,* this effect has reduced reproduction to the extent that extinction of major regional populations is now occurring.

Lest it be assumed that the chlorinated hydrocarbons represent an exceptional case, we may note the recent recognition of other widespread, long-life pollutants. In the course of chemical analyses for chlorinated hydrocarbons, the presence of another hydrocarbon derivative as a major environmental pollutant was accidentally discovered (Risebrough et al., 1968; Peakall and Lincer, 1970). This group of substances, known as polychlorinated biphenyls (PCBs) is an ingredient of plastics, resins, varnishes, waxes, and many other materials. They are presumably released into the environment largely by the burning of wastes and trash. Like chlorinated hydrocarbon pesticides PCBs are concentrated along food chains and are strong inducers of liver enzymes which decompose steroid hormones. PCBs are widespread in the environment, and although their long-distance dispersal ability apparently does not equal that of DDT, in areas close to major population centers the concentration of PCBs in animal tissues may exceed that of DDT.

BIOGEOCHEMICAL CYCLES AT THE BIOSPHERE LEVEL

It is now apparent that man's intrusions into biogeochemical processes rival in scale, or even exceed, naturally occurring processes at the level of the biosphere. Man has roughly doubled the rate of input of mercury into the world oceans through processes of waste disposal and burning of fossil fuels (Klein and Goldberg, 1970). The rate of input of lead to the oceans is thought to be 40 times the naturally occurring rate, largely as a result of the use of tetraethyl lead in gasoline (Patterson, 1965).

Perhaps it is not surprising that human activities rival natural processes for elements such as mercury and lead, which are of minor importance in living systems but are of major importance in modern industrial society. Human influences, however, extend to biogeochemical cycles of the principal constituents of living systems, such as C, O, N, P, and H_2O. Since even slight disturbances of these cycles may have profound effects upon man, we will examine their quantitative features and evaluate the extent and possible consequences of human intervention.

Carbon Cycle

The short-term cycling of carbon within the biosphere is closely tied to the pattern of energy flow, since nearly 50% of organic matter, by dry weight, consists of carbon. In reduced chemical form in organic compounds carbon constitutes the major vehicle for energy transfer through living systems. When the carbon cycle is viewed over longer periods of geological time, however, a number of other processes become important. During the history of the earth, quantities of carbon greatly exceeding that in active circulation at the present time have been stored in sedimentary rocks in the form of coal, oil, and carbonate minerals. Man is now releasing portions of this stored carbon into the active carbon cycle.

The major quantitative features of the carbon cycle are shown in Fig. 10–14. The greatest fraction of carbon in the crust of the earth is that located in sedimentary rocks, primarily in the form of calcium and magnesium carbonates of limestones and dolomites. These rocks were deposited in large part during the early Paleozoic Era. It is likely that their deposition was correlated with the introduction of molecular oxygen into the atmosphere in quantity, and with the simultaneous depletion of the CO_2 content of the atmosphere. Deposition of limestones and dolomites may also have been favored by increase in the pH of ocean waters accompanying the increase of the O_2 content of the atmosphere and hydrosphere (Cloud and Gibor, 1970).

Carbon is also stored in sedimentary rocks in the reduced, organic form. Part of this fossil carbon is in the form of commercially exploitable deposits of coal and oil. A much greater portion, however, is dispersed in very low concentrations through sedimentary rocks such as shales. Certain of these rocks, such as the oil shales of Colorado and Wyoming, may eventually be used as sources of oil.

Within the actively circulating portion of the biosphere the greatest quantity of carbon is located in seawater. Some of this carbon exists as CO_2 in solution, but almost all of it is in the form of carbonate and bicarbonate ions. The pool of carbon in the oceans is approximately 50 times that of the atmosphere. The atmospheric and oceanic pools are in equilibrium with each other, however, and continuous interchange occurs.

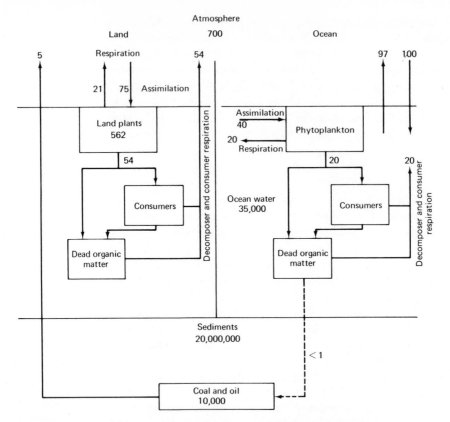

FIGURE 10-14. Quantitative relationships of the world carbon cycle. Values are in billions of metric tons (quantities in pools or annual flux rates). (Modified from Bolin, 1970, using data from Olson, 1969, and Ryther, 1969.)

Calculations based on the disappearance rate of ^{14}C from the atmosphere following atmospheric nuclear tests suggest that the annual exchange between the atmosphere and the oceans amounts to about 100 billion tons of carbon, or an amount equal to about $\frac{1}{7}$ the content of the atmospheric pool (Bolin, 1970).

Carbon exists in the atmosphere almost entirely in the form of CO_2. Atmospheric CO_2 is continually being removed by photosynthetic uptake by terrestrial plants and returned by respiratory processes of plants, consumers, and decomposers. The pool of reduced carbon in the living tissues of terrestrial organisms and in undecomposed organic matter on land is roughly equal to (or somewhat greater than) that in the atmospheric pool.

The processes of biological uptake and release of CO_2 influence the atmospheric CO_2 concentration both diurnally and seasonally. Diurnal CO_2 changes, in fact, have been used to measure the metabolic activity of whole ecosystems (Woodwell, 1970). The mean concentration of CO_2 in the atmosphere is about 320 ppm. At night, concentrations near the surface of the ground may rise to levels exceeding 500 ppm, however. Figure 10–15 shows measurements of the CO_2 concentration at various heights above the ground in an oak-pine forest on Long Island, New York (Woodwell, 1970). These data were obtained during periods when temperature inversions and low wind intensities during the night permitted the accumulation of CO_2 produced by respiratory processes in the forest ecosystem. From the rate of increase in CO_2 concentration during the night the overall respiration of the forest ecosystem could be estimated.

A seasonal pattern may also be seen in temperate zone areas of the earth, especially those of the northern hemisphere. During the summer, fixation of CO_2 exceeds the return to the atmosphere by respiration and decomposition; during the winter, the reverse is true. These differences lead to a seasonal variation of as much as 20 ppm in CO_2 concentration of air at the surface of the earth (Fig. 10–16).

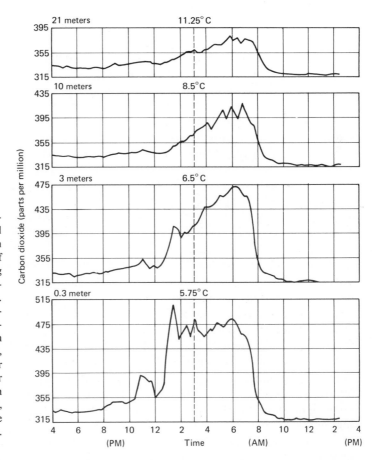

FIGURE 10–15. Rate of respiration of the forest was determined by measuring the rate at which carbon dioxide, a product of respiration, accumulated during nights when the air was still because of a temperature inversion. The curves give the carbon dioxide concentration at four elevations in the course of one such night. (Note that the temperature, recorded at 3:00 A.M. was lower near the ground than at greater heights.) The hourly increase in carbon dioxide concentration, which was calculated from these curves, yielded rate of respiration. (From Woodwell, 1970.)

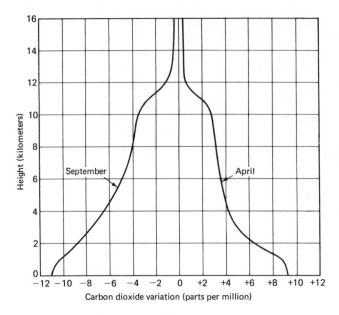

FIGURE 10–16. Seasonal variations in the carbon dioxide content of the atmosphere reach a maximum in September and April for the region north of 30 degrees north latitude. The departure from a mean value of about 320 ppm varies with altitude as shown by these two curves. (From Bolin, 1970.)

Beginning in the seventeenth century, man introduced a new pathway into the carbon cycle. This pathway is the release into the atmosphere of CO_2 by the burning of fossil fuels. The annual rate of release of carbon from fossil fuels is now about 5 billion tons, or an amount equal to about $\frac{1}{10}$ that of carbon fixation by photosynthesis of terrestrial plants. This addition has not been fully compensated for by an increased rate of removal of CO_2 from the atmosphere, with the result that a significant increase in CO_2 concentration in the atmosphere has occurred. Over the past century an estimated 200 billion tons of fossil carbon have been added to the atmosphere by fossil fuel combustion. Current measurements suggest that about $\frac{2}{3}$ of this have been removed from the atmosphere. It is likely that almost all of this removal has occurred by movement of CO_2 into the oceans, but some may have been removed by an increased photosynthetic rate of terrestrial plants. The pool of atmospheric CO_2 has thus increased by an amount equal to about $\frac{1}{3}$ of the 200 billion tons of released carbon. This suggests that the present atmospheric concentration of CO_2 of 320 ppm is an increase of about 30 ppm over the level prior to the use of fossil fuels.

At present, the annual rate of increase in average CO_2 concentrations of the atmosphere is about 0.7 ppm. The rate of use of fossil fuels is, however, increasing, and almost $\frac{1}{5}$ of the total increase of CO_2 level of the atmosphere occured during the 1960's. It is estimated that by the end of the century the atmospheric CO_2 level will have risen to between 375 and 400 ppm (Bolin, 1970).

The actual and potential effects of increased CO_2 content of the atmosphere have been the subject of considerable speculation. Carbon dioxide absorbs infrared radiation leaving the surface of the earth and reradiates this energy. Part is radiated back toward the surface of the earth, thus conserving heat much as does the glass in a greenhouse roof (Sargent, 1967). Based on this effect an increased CO_2 level would be expected to produce an increase in mean world temperature. It has been estimated that from the late 1800's to about 1940 an increase of as much as $1.6°C$ may have taken place (Sargent, 1967). During this period an appreciable increase in CO_2 level in the atmosphere also occurred. Since 1940, however, mean world temperature has decreased about $0.4°C$ — an effect attributed to increase in particulate matter levels in the atmosphere which reflect a portion of incoming solar radiation. It is suggested that the latter effect has, in essence, overridden the warming influence of CO_2 increase (Sargent, 1967; Wendland and Bryson, 1970).

Abelson (1967), projecting possible future dangers, has suggested that increased CO_2 levels in the atmosphere might increase world temperatures by as much as 4.0°C by 2000 A.D. More recent estimates (Landsberg, 1970) suggest, however, that a doubling of the CO_2 content of the atmosphere, from 300 to 600 ppm, may be required to produce a $2.0°C$ increase in mean world temperatures. Major difficulties exist in predicting these effects because of the involvement of other variables such as humidity, cloudiness, and dustiness of the atmosphere.

Oxygen Cycle

The oxygen cycle is closely tied to that of carbon, and thus to the pattern of energy flow through the biosphere. Like the carbon cycle, however, processes of a primarily geological and chemical nature assume importance when the cycle is considered through geological time (Fig. 10–17).

The total quantity of molecular oxygen in the atmosphere is about 0.8×10^{15} tons (Johnson, 1970), the percentage composition averaging 20.946 per cent for dry air. An additional 0.2×10^{15} tons of molecular oxygen exists dissolved in the oceans (Johnson, 1970). Thus, there are about 10^{15} tons of molecular oxygen extant on the earth's surface at the present time. Despite the present abundance of molecular oxygen in the atmosphere and hydrosphere, it is probable that the primordial atmosphere of the earth lacked free oxygen. The principal evidence for this is the absence of oxidized minerals, such as iron in the ferric state (Fe_2O_3) in the earliest sedimentary rocks (Cloud, 1968; Cloud and Gibor, 1970). Furthermore, since the gases released into the atmosphere from volcanos are, in general, in a reduced chemical form, it is unlikely that free oxygen entered the atmosphere from this source.

The introduction of molecular oxygen in the atmosphere thus occurred through processes operating on the surface of the earth. These processes are photosynthesis by green plants and photodissociation of water in the upper atmosphere (Johnson, 1970; Olson, 1970).

Estimates of the rate of production of O_2 through photodissociation of

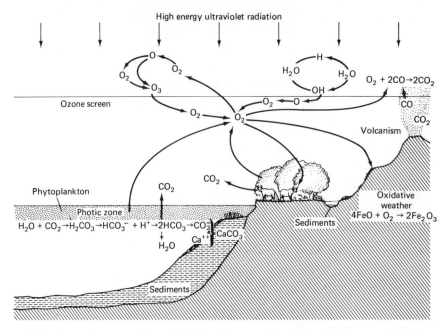

FIGURE 10–17. Oxygen cycle is complicated because oxygen appears in so many chemical forms and combinations, primarily as molecular oxygen (O_2), in water and in organic compounds. Some global pathways of oxygen are shown here in simplified form. (From Cloud and Gibor, 1970.)

H_2O suggest that only about 10^{15} tons of molecular oxygen may have been produced during the history of the earth, assuming that a rate comparable to that at present has prevailed through geological time (Johnson, 1970). It is possible that rates were higher in the geological past. However, estimates of the total quantity of reduced carbon in living and dead organic tissues, in total reserves of high-grade coal and oil, and in low concentrations in sedimentary rocks suggest that an excess production of organic carbon corresponding to 13×10^{15} tons of oxygen has occurred. Thus, approximately 14 times as much O_2 has apparently been produced — primarily by photosynthesis — during the history of the earth than exists at present. Apparently, about 13×10^{15} tons of this has been utilized in the oxidation of minerals released by weathering of igneous rocks or of gases released into the atmosphere by volcanic action. Considerable uncertainty exists about the latter use of oxygen. For example, the exact proportion in which CO and CO_2 have been released into the atmosphere affects in a major way the amount of O_2 expended in the oxidation of CO. At the present time, all that can be said is that the amount of materials occurring in oxidized form in sediments, the hydrosphere, and the atmosphere reasonably account for about thirteen-fourteenths of the molecular oxygen produced during the history of the earth (Fig. 10–18).

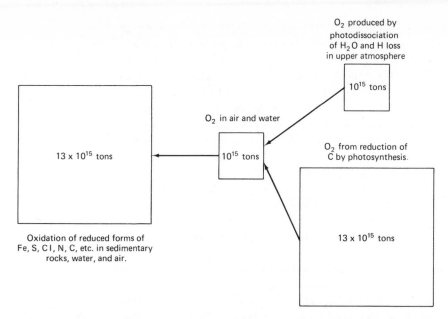

O$_2$ produced by
photodissociation
of H$_2$O and H loss
in upper atmosphere

10^{15} tons

O$_2$ in air and water

O$_2$ from reduction of
C by photosynthesis.

13 x 10^{15} tons

10^{15} tons

Oxidation of reduced forms of
Fe, S, C l, N, C, etc. in sedimentary
rocks, water, and air.

13 x 10^{15} tons

FIGURE 10–18. Estimates of the quantity of molecular oxygen released into the atmosphere during the history of the earth and the quantity subsequently removed by processes of oxidation. (From Johnson, 1970.)

It appears that the oxygen cycle is very nearly in a steady-state condition, in spite of the atmospheric disturbances being induced by man. It is estimated that 8×10^{10} tons of oxygen are released yearly by photosynthesis. At this rate of photosynthetic production, storage of only one part in 10^4 of the carbon fixed would account for the fossil carbon stores that have accumulated over geological time. At present, of course, a much greater amount than this is being released from the fossil carbon store by burning (about 2,000 times the average annual fossil carbon storage rate). This greatly increased rate of removal of O$_2$ from the atmosphere, however, has had no measurable effect on the O$_2$ concentration over the period from about 1910 to 1970 (Machta and Hughes, 1970). Calculations of the effect of combustion of fossil fuels over this period suggest that, at most, a decrease of 0.005% by volume should have occurred.

Redfield (1958) has suggested that sulfate-reducing bacteria occurring in anaerobic environments may operate as a long-term negative feedback control for the oxygen cycle. These bacteria are able to use the sulfate ion as an oxygen source for oxidation of organic matter, according to the general formula:

$$SO_4^= + 2C \rightarrow 2CO_2 + S^=$$

Although this reaction does not release free oxygen, it does allow the decomposition of organic matter with consequent release of nutrients such as P, N, and CO_2. The CO_2 may later be transported to a site where it may be utilized in photosynthesis, thus leading to the release of molecular oxygen. Presumably, changes in the oxygen content of the atmosphere would be reflected in the increase or decrease of extent of anaerobic habitats and would be inversely related to the extent of O_2 release by sulfate reduction. For example, decreased atmospheric O_2 would lead to an increase in extent of anaerobic conditions and increased sulfate reduction, which would release increased amounts of O_2 from sulfate. Unfortunately, the rate of oxygen return to the active pool of molecular oxygen by this route is unknown at present. Furthermore, since this process occurs largely in anaerobic situations that are heavily insulated from the atmosphere by organic deposits (as in swamps and bogs), its actual regulatory role may be very small.

Redfield (1958) has also suggested that the atmospheric concentration of O_2 corresponds to a particular overall relationship of the nutrients P, N, and O in seawater. The atmospheric O_2 concentration corresponds to an equilibrium level of dissolved oxygen nearly equal to that required to allow aerobic decomposition of the quantity of organic matter which could be synthesized on the basis of the quantities of P and N present in the water. Redfield suggests that this level of atmospheric oxygen has been determined, during the history of life on the earth, by the nutrient relationships of synthesis and decomposition in the world oceans.

Nitrogen Cycle

Earlier we examined the nitrogen cycle in terms of the interchange of nitrogen among several chemical forms within the biosphere caused by the metabolic activities of various groups of organisms. We now wish to examine the quantitative relationships of this cycle at the biosphere level.

Quantitative estimates of pool size and flux rates, formulated by Delwiche (1970), are shown in Fig. 10–19. In addition to the flux routes discussed earlier, nitrogen enters and leaves the biosphere in appreciable quantities. Nitrogen gas, termed *juvenile nitrogen*, enters the biosphere through volcanic action. Removal of nitrogen from the biosphere occurs through the incorporation of nitrogen-containing materials in deep-water sediments. The rates for these two processes are estimated to be about the same—approximately 2×10^5 tons per year.

Nitrogen fixation, the conversion of molecular nitrogen to nitrogen at the -3 valence level (organic nitrogen), is still an imperfectly understood process. Quantitative estimates of this flux rate are quite uncertain, especially for the oceans. The most reasonable current estimate suggests that it amounts to about 54 million tons annually. Atmospheric fixation, leading to production of nitrates which may reach the continents or oceans

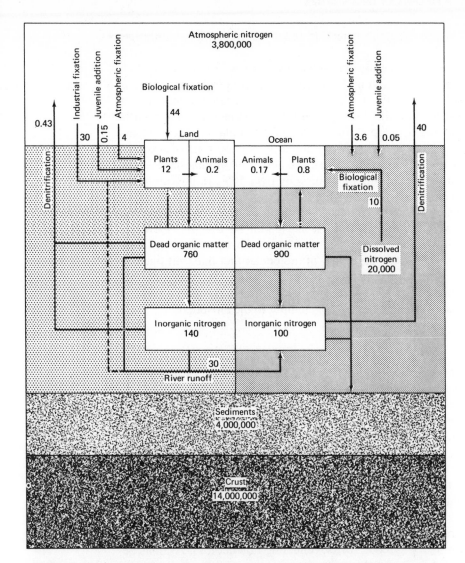

FIGURE 10–19. Distribution of nitrogen in the biosphere and annual transfer rates can be estimated only within broad limits. The two quantities known with high confidence are the amount of nitrogen in the atmosphere and the rate of industrial fixation. The apparent precision in the other figures shown here reflects chiefly an effort to preserve indicated or probable ratios among different inventories. Thus the figures for atmospheric fixation and biological fixation in the oceans could well be off by a factor of 10. The figures for inventories are given in billions of metric tons; the figures for transfer rates are given in millions of metric tons. Because of the extensive use of industrially fixed nitrogen, the amount of nitrogen available to land plants may significantly exceed the nitrogen returned to the atmosphere by denitrifying bacteria in the soil. A portion of this excess fixed nitrogen is ultimately washed into the sea, but it is not included in the figure shown for river runoff. Similarly, the value for oceanic denitrification is no more than a rough estimate that is based on the assumption that the nitrogen cycle was in overall balance before man's intervention. (From Delwiche, 1970.)

in precipitation, is estimated to convert about 7.6 million tons of nitrogen annually.

It is interesting to examine the activities of man in relation to these processes of nitrogen fixation. Prior to 1914, man was dependent on nitrate deposits, principally in Chile, for fixed nitrogen required for fertilizers and explosives. In 1914, however, an industrial process (the Haber Process) was developed for fixation of molecular nitrogen. Annual industrial fixation of nitrogen is now about 30 million tons of nitrogen, principally for use in fertilizers. Thus, the activities of man have apparently increased the global rate of fixation by over 50%. In the future this rate is likely to increase to much higher levels.

The nitrogen cycle has been described as an example of a perfect biogeochemical cycle. If this is true, it may well be that the increased fixation by man will be counterbalanced by increased activity by denitrifying organisms, so that the pools of molecular and fixed nitrogen remain the same. It is also apparent, however, that the increased fixation by man has produced at least local conditions of imbalance as a result of greatly accelerated rates of fixed nitrogen input to the biosphere.

Phosphorus Cycle

Of the elements generally required by living organisms, and which do not possess atmospheric pools of major importance, phosphorus appears to be most often limiting to ecosystem function. The phosphorus cycle (Fig. 10–20) involves release of phosphorus from igneous and sedimentary rocks by weathering, leaching and transport by water from the continents to the ocean basins, deposition in marine sediments, and the ultimate return by uplift of a portion of those marine sediments deposited on continental platforms. Some fraction of the phosphorus deposited in deep ocean sediments, however, is effectively lost to the biosphere since it is unlikely that such sediments will ever be returned to the continents. Thus, the average concentration of phosphorus in sedimentary rocks is lower than in primary igneous rocks, reflecting this net loss. As soils become increasingly derived from sedimentary deposits through geological time, phosphorus thus becomes an increasingly scarcer element within the biosphere.

This process alone is probably unimportant to man. However, man's disturbance of the phosphorus cycle has become important. Hutchinson (1948) has estimated the annual flow of phosphorus into the oceans in river water to be 20 million tons. Other workers estimate this loss more conservatively at about 3.5 million tons (Cole, 1958). The value of the flux rate for phosphorus from the continents to the ocean prior to the acceleration of erosion processes by man is uncertain. It is known, though, that man has accelerated the loss of phosphorus through his agricultural practices, and that this acceleration alone probably amounts

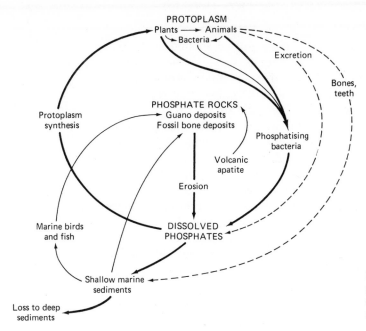

PROTOPLASM

FIGURE 10–20. Major pools and pathways of the biogeochemical cycle of phosphorus. (From Odum, 1959.)

to a large fraction of the rate prior to disturbance. Man has also mined sources of phosphorus to make phosphate fertilizers. This use now amounts to between 5 and 6 million tons of phosphorus annually (Brown, 1970) and is rapidly increasing. It is likely that much of the fertilizer applied by man ultimately finds its way into the ocean. Phosphorus in other forms, such as constituents of detergent chemicals, has also been introduced into streams, and ultimately the ocean, in large quantities. Thus it is likely that man has increased the flux rate of phosphorus from the continents to the ocean by a factor of perhaps 2–3.

At the same time, return of phosphorus to the continents is essentially unchanged. Hutchinson (1948) has estimated that about 10,000 tons of phosphorus are returned as guano deposited on land by sea birds, and that 60,000 tons are returned to the continents in fish caught by man. In the latter case it is probable that much of the phosphorus is rapidly returned to the ocean in sewage wastes, rather than being added to the activity circulating pool of terrestrial nutrients. Disturbance of the phosphorus cycle has already led to important environmental problems. Phosphorus has been considered one of the key nutrients in the process of eutrophication of freshwater systems because of its increased quantities in runoff and wastes disposed by man. The supplies of phosphate-containing rock, although not yet critical, are limited, and it is possible that in the future this element will become limiting to productivity of terrestrial agriculture.

The Hydrologic Cycle

Despite the fact that water is the most abundant compound in the bio-sphere, man has attempted to modify its abundance and movement ever since the invention of agriculture. At first these attempts at modification of the hydrologic cycle were local, but now it is clear that regional or world-wide changes in characteristics of the cycle are possible. It is quite likely that man's activities have already produced basic changes in the hydrologic cycle, although proof of such changes is lacking. A large por-tion of the difficulty of pin-pointing the effects of man on the hydrologic cycle is the fact that the events of the water cycle show great variation both in space and time. Month-to-month and year-to-year variability in precipitation and streamflow is great. In addition, the hydrologic cycle reflects long-term trends in climate. For example, change in mean tem-perature may influence evaporation rates and melting of glacial ice. Identifying the effects of man obviously requires that such effects ex-ceed the magnitude of variation produced by the above factors.

Estimates (Kalinin and Bykov, 1969) of the quantities of water in different portions of the biosphere, and the flux rates between these por-tions, are given in Fig. 10–21. These data suggest that the total amount of water occurring as vapor in the atmosphere is a relatively small portion of the total water of the earth. Dividing the total annual precipitation (over both land and ocean) by the quantity of water in the atmosphere at any one time suggests that replacement of the moisture of the atmosphere occurs about 37 times a year, or once every 9–10 days (Penman, 1970).

Figure 10–21 also shows that evaporation exceeds precipitation over the oceans and the opposite over the continents. The difference represents

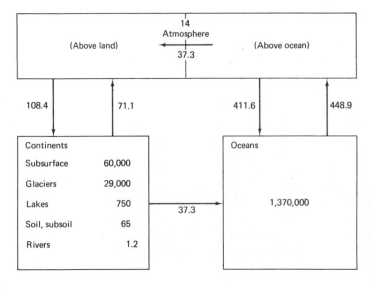

FIGURE 10–21. Quantities of water in various biogeochemical pools and annual flux rates for the world hydrologic cycle. Numbers represent thousands of cubic kilometers.

the net transport of water in the air flow from oceans to continents and, likewise, the net discharge of rivers from continents into the oceans. The actual flow of moisture from the oceans to the continents by air movements is estimated to be about 10 times the net flow (Penman, 1970). Of this, nine-tenths passes across the continents and is reincorporated in air masses above the oceans.

Water is stored in various forms on the continents. The largest quantity is that stored beneath the surface. Much of this lies so deep that very little actual interchange occurs with actively circulating water. The second largest component is that tied up in snow and ice, principally in glaciers. This amount, 29 million cubic kilometers, would, if melted, correspond to a water layer 50 meters in depth over the entire surface of the earth. Rapid melting of this ice would clearly bring about major changes in sea level.

Human activities are now capable of major impacts on the hydrologic cycle. Revelle (1963) has estimated that in the United States the total withdrawal of water from streams for human use amounts to about 25% of the total volume of flow (Fig. 10–22). About two-sevenths of this withdrawal returns to the atmosphere by evaporation. Revelle (1963) estimates that by the year 2000 withdrawal from streams will increase to an amount equal to 75% of the total flow. It is likely that these withdrawals will be subjected to uses that will increase the proportion of water lost by evaporation. Considerable quantities are likely to be used for irrigation in the arid southwest, for cooling of power plants, and other processes involving considerable evaporation. Thus, one major modification of the hydrologic cycle in the future will likely be an increased rate of evaporation in continental areas.

Kalinin and Bykov (1969) have prepared comparable estimates for

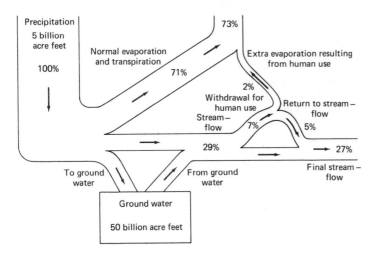

FIGURE 10–22. Annual human water use in the United States (exclusive of Alaska and Hawaii) in relation to the hydrologic cycle. (Data from Revelle, 1963.)

world use of water by 2000 A.D. (Table 10–14). These estimates suggest that withdrawal from streamflow for human use will equal about one-half the total volume of streamflow, and that over one-fourth of this will be lost by evaporation.

The effect of this increased evaporation will likely be a slight increase in average precipitation for the continental areas involved. This results from the fact that a portion of the precipitation in a continental area represents water that entered the atmosphere by evaporation from the continent itself. The relative total quantity of precipitation for a continental area resulting from an initial quantity of precipitation (due to the return to the atmosphere by evaporation) is termed the *coefficient of the hydrologic cycle* (Table 10–15). For North America this value is 1.35. This value means that about one-quarter (0.35/1.35) of the total continental precipitation consists of water evaporated from the continent itself.

It is clear that in the future water will become important to man in ways little appreciated at present. For example, water is now used as a coolant for certain industrial and power-generating plants. If we compare the solar radiation input to the earth's surface with the quantity of energy released by industry (energy from fossil fuels and radioactive sources), the latter is only one part in 2,450 at present. However, if this release of energy were to increase by 10% annually, it would become equal to the solar energy input to the entire earth in about 100 years. Thus, water evaporation may be required in greatly increased amounts in order to dissipate the added heat.

The consequences of modification of the energy balance, hydrologic cycle, and atmospheric composition of the earth by such factors are difficult to predict. One can appreciate that changes in cloudiness (and thus the albedo of the earth), the area and current patterns of the oceans, atmospheric circulation patterns, and atmospheric chemistry may interact to produce changes totally unexpected on the basis of current knowledge.

TABLE 10–14. TOTAL ANNUAL WORLD WATER
REQUIREMENTS BY 2000 A.D.*

Use	Water required (cu km)	
	Total	Lost by evaporation
Irrigation	7,000	4,800
Domestic	600	100
Industrial	1,700	170
Dilution of wastes	9,000	
Other	400	400
Totals	18,700	5,570

*From Kalinin and Bykov, 1969.

EVOLUTION OF BIOGEOCHEMICAL CYCLES

In our development of the concept of the ecosystem we have emphasized the interrelationships existing between the biotic and abiotic components of ecological systems. We have shown that these relationships are not one-way, but involve influences of living organisms on the physical environment as well as their dependence on conditions of the physical environment. In fact, recognition of the role of organisms in modifying conditions of the physical environment can be considered to date the origin of the ecosystem concept (Macfadyen, 1962).

The investigations of paleoecologists and geochemists are now demonstrating that the basic geochemistry of the modern world, and the complexity of certain biogeochemical cycles, are products of gradual evolutionary processes. In other words, organic evolution has been accompanied by biogeochemical evolution. The original chemical composition of the atmosphere, oceans, and continents has been profoundly determined by the activities of living organisms.

The age of the earth is now estimated at between 3.5–3.6 billion years, the age of the oldest dated rocks, and 4.6 billion years, the inferred age of the solar system (Cloud, 1968). The chemical nature of the earliest sedimentary rocks, deposited over 3 billion years ago, suggests that at the time of their deposition atmospheric weathering processes and a well-developed hydrosphere existed (Cloud, 1968). The mineral composition of sedimentary rocks deposited between 1.8 and 3.0 billion years ago, however, suggests that the composition of the atmosphere was different from what it is now. For example, the absence of ferric iron and other oxides in these sediments suggests that free oxygen was almost completely absent. Although several different hypotheses about the chemistry of the primordial atmosphere have been proposed, the most probable composition is one consisting of juvenile gases known to be released by volcanic and hot-spring sources. The constituents of such an atmosphere would be N_2, CO, CO_2, SO_2, H_2O, and other trace gases (Cloud, 1968).

TABLE 10–15. CALCULATED HYDROLOGIC CYCLE COEFFICIENTS FOR CONTINENTAL AREAS.*

Continental area	Area (millions of square kilometers)	Hydrologic coefficient
Australia	7.96	1.15
South America	17.98	1.30
North America	20.44	1.35
Africa	21.81	1.45
Eurasia	51.95	1.65

*From Kalinin and Bykov, 1969.

Under conditions of an atmosphere devoid of oxygen and lacking a high-altitude ozone layer, high-energy ultraviolet radiation would have reached the surface of the earth in intensities adequate to prevent life. This same energy source, however, may have promoted the abiotic synthesis of a variety of organic compounds—and constituted the driving agent for a period of chemical evolution in the oceans of the early earth. Presumably, this period of chemical evolution extended from the period of formation of the oceans, estimated to lie between 3.0 and 3.5 billion years ago (Hutchinson, 1970), through the period when the first living organisms appeared. Some of the earliest sedimentary rocks, slightly over 3 billion years in age, contain fossils of bacteria-like structures (Hutchinson, 1970).

These earliest living organisms were almost certainly restricted to aquatic situations in which they were protected from intense ultraviolet radiation. They were very likely heterotrophic in their metabolism. Their food source presumably consisted of the supply of abiotically formed organic molecules in their aquatic environment. Initially their metabolism must have involved the simplest kind of fermentation processes because of the absence of free oxygen in the environment. The efficiency of these first metabolic systems must have been low, and the substrates used as food must have been the most abundant and energy-rich. The chemical reactions of metabolism must have been those furnishing the greatest release of usable energy.

Considerable evolution of these first metabolic systems must have occurred in order to perfect even the pattern of fermentation as we now know it. In this process organisms having metabolic systems operating with greater efficiency in extracting energy, or capable of utilizing a greater variety of organic materials, must have been favored. As the first organisms increased in abundance in the world oceans, their original abiotically produced food molecules must have become depleted, thus favoring forms possessing the ability to utilize other kinds of organic compounds as energy sources.

Autotrophic metabolic systems arose gradually by extension of the processes just described. It is possible that certain of these heterotrophs acquired the ability to utilize light energy to drive metabolic processes through which preformed organic molecules in the environment were assimilated by the organism (Olson, 1970). This system could have been a simple, cyclic electron transport system (Fig. 10–23), which released no oxygen to the environment and simply gave the organism an increased efficiency in assimilation of external organic molecules. The ability of a variety of modern algae and autotrophic bacteria to carry out such activities has been demonstrated (Olson, 1970).

As the environmental supply of complex organic molecules was further depleted, an evolutionary advantage must have accrued to organisms able to utilize less reduced organic compounds. Such a change would have required an external electron donor, permitting the reduction of

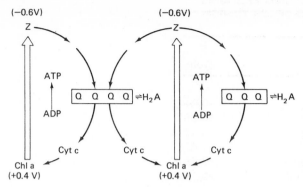

FIGURE 10-23. Electron transport scheme (cyclic) for the hypothetical ancestral photobacterium. Abbreviations used are Q, quinone; Chl, chlorophyll; Cyt, cytochrome; Z, unknown electron acceptor; H_2A, electron donor. (From Olson, 1970.)

carbon compounds to be completed. Initially, the source of electrons in this process was probably one or more inorganic compounds having low oxidation potentials for chemical conversions releasing electrons. Since the oxidation potential for decomposition of water molecules with the release of molecular oxygen is +0.82 volts, this source was probably not utilized at first. The earliest sources may have been certain nitrogen compounds, such as hydrazine or hydroxylamine, which have redox potentials of +0.25 and −0.04 volts, respectively, for conversion to nitric oxide (Olson, 1970). As these compounds were utilized in the earliest noncyclic photosynthetic processes, their supply in the environment must have been depleted. Consequently, selection probably favored the increased ability of photosynthetic systems to acquire electrons through oxidations with higher potentials and from substances of greater overall abundance in the environment. Eventually, this sequence enabled the development of a photosynthetic process in which the external electron source was water, possibly through a series of intermediates such as those indicated in Table 10–16.

These evolutionary processes must have produced a corresponding pattern of change in the biogeochemistry of the hydrosphere, particularly in the cycle of nitrogen. Ultimately, with water as the external electron source for photosynthesis, free molecular oxygen was released into the environment for the first time.

It is thought that the first photosynthetic organisms still were restricted to deeper regions of aquatic habitats by intense ultraviolet radiation. Thus oxygen, when first released, entered the surrounding water. This period of the evolution of life may be represented by a series of sedimentary rocks, the banded iron formations, deposited between 3.0 and about 1.8 billion years ago. Banded iron formations consist of alternating strata of iron-rich and iron-poor silicates. The iron is in the ferric state, and it is likely that oxidation to this state was permitted by the first release of oxygen by photosynthesis.

As oxygen-releasing photosynthesis increased in importance, release of molecular oxygen into the atmosphere occurred. This is evidenced,

TABLE 10-16. POSSIBLE SEQUENCE OF ELECTRON-RELEASING REACTIONS PROCEEDING FROM REACTIONS OF LOW TO HIGH REDOX POTENTIAL.*

	Reaction	E_7 (volt)
1.	$NH_2OH \rightarrow NO + 3H^+ + 3e^-$	
	$NH_3OH^+ \rightarrow NO + 4H^+ + 3e^-$	-0.04
2.	$N_2H_5^+ + 2H_2O \rightarrow 2NO + 9H^+ + 8e^-$	+0.25
3.	$NO + H_2O \rightarrow NO_2^- + 2H^+ + e^-$	+0.37
4.	$NO_2^- + H_2O \rightarrow NO_3^- + 2H^+ + 2e^-$	+0.42
5.	$2NO_2^- + H_2O \rightarrow NO_3^- + NO_2 + 2H^+ + 3e^-$	+0.58
6.	$NO_2 + 2H_2O \rightarrow NO_2^- + O_2 + 4H^+ + 3e^-$	+0.79
5 + 6.	$NO_2^- + 3H_2O \rightarrow NO_3^- + O_2 + 6H^+ + 6e^-$	+0.69
7.	$2H_2O \rightarrow O_2 + 4H^+ + 4e^-$	+0.82

*From Olson, 1970.

Note: Such a sequence may have been followed in the evolution of oxygen-producing photosynthesis by green plants.

geologically, by the appearance of *red beds,* which are sedimentary rocks rich in ferric iron, suggesting the oxidation of iron, early in the cycle of erosion, by oxygen in the atmosphere. These deposits occur in the geological record from about 1.8 billion years to the present (Cloud and Gibor, 1970).

The appearance of molecular oxygen must have also permitted certain evolutionary improvements in systems of heterotrophic metabolism. Prior to this time respiration must have been largely fermentative. The appearance of oxygen permitted evolution of the complex system of oxidative metabolism, in which oxygen acts as the final acceptor in the electron transport system.

The appearance of oxygen in the atmosphere also led to the development of the ozone layer in the upper atmosphere. Cloud and Gibor (1970) have suggested that when the oxygen concentration of the atmosphere reached 1% of its present level, an ozone screen effective enough to permit organisms to occupy surface waters existed. This probably occurred about 1 billion years ago. Continued increase in atmospheric oxygen presumably led to the increase in effectiveness of this ozone screen, thus permitting the appearance of land organisms, for the first time, about 400 million years ago.

Although the details of the above sequence are hypothetical, what is certain is the fact that organic evolution was accompanied by biogeochemical evolution. Further, it seems likely that the evolution of complex metabolic systems, such as photosynthesis and oxidative respiration, has evolved by gradual step-wise processes. Evolutionary acquisition of metabolic ability to utilize one substance has led to the depletion of supply of the substance, favoring organisms having metabolic systems able to use precursors of the substance. These processes have very likely been accompanied by patterns of changing abundance of available nutrients in the environment. Thus, the complex, interrelated, and balanced biogeochemical cycles of the modern biosphere are not the result of an accidental fitness of the environment, but rather the product of a complex evolutionary process coincident with the evolution of life. It is this complexly evolved system that is receiving the massive impacts of the human technological and population explosions.

REFERENCES

Abelson, P. H. 1967. Global weather. *Science,* **155**:153.

Bolin, Bert. 1970. The carbon cycle. *Scientific American,* **223**(3):124–32.

Bormann, F. H., and G. E. Likens. 1967. Nutrient cycling. *Science,* **155**:424–29.

———. 1970. The nutrient cycles of an ecosystem. *Scientific American,* **223**(4): 92–101.

———. 1971. The ecosystem concept and the rational management of natural resources. *Yale Scientific,* **45**:2–8.

———, and J. S. Eaton. 1969. Biotic regulation of particulate and solution losses from a forest ecosystem. *BioScience,* **19**:600–610.

Brown, L. R. 1970. Human food production as a process in the biosphere. *Scientific American,* **223**(3):161–70.

Cloud, P. E., Jr. 1968. Atmospheric and hydrospheric evolution on the primitive earth. *Science,* **160**:729–36.

———, and A. Gibor. 1970. The oxygen cycle. *Scientific American,* **223**(3): 110–23.

Cole, L. C. 1958. The ecosphere. *Scientific American,* **198**(4):83–92.

Delwiche, C. C. 1970. The nitrogen cycle. *Scientific American,* **223**(3):136–46.

Duvigneaud, P., and S. Denaeyer-De Smet. 1969. Biological cycling of minerals in temperate deciduous forests. In D. E. Reichle (ed.), *Analysis of Temperate Forest Ecosystems,* pp. 199–225. New York: Springer-Verlag.

Goldberg, E. D. 1963. The oceans as a chemical system. In M. N. Hill (ed.), *The Sea,* pp. 3–25. New York: Wiley.

Hanson, W. C. 1967. Cesium-137 in Alaskan lichens, caribou, and eskimos. *Health Physics,* **13**:383–89.

Hickey, J. J., J. A. Keith, and F. B. Coon. 1966. An exploration of pesticides in a Lake Michigan ecosystem. *J. Appl. Ecol.,* **3**(Suppl.):141–54.

Hutchinson, G. E. 1948. On living in the biosphere. *Sci. Monthly,* **67**:393–97.

———. 1970. The biosphere. *Scientific American,* **223**(3):44–53.

Johnson, F. S. 1970. The balance of atmospheric oxygen and carbon dioxide. *Biol. Cons.,* **2**:83–89.

Johnson, N. M., G. E. Likens, F. H. Bormann, and R. S. Pierce. 1968. Rate of chemical weathering of silicate minerals in New Hampshire. *Geochim. et Cosmochim. Acta,* **32**:531–45.

Kalinin, G. P., and V. D. Bykov. 1969. The world's water resources, present and future. *Impact,* **19**:135–50.

Klein, D. H., and E. D. Goldberg. 1970. Mercury in the marine environment. *Env. Sci. Tech.,* **4**:765–68.

Kuenzler, E. J. 1961a. Structure and energy flow of a mussel population in a Georgia salt marsh. *Limnol. Oceanogr.,* 6:191–204.

———. 1961b. Phosphorus budget of a mussel population. *Limnol. Oceanogr.,* 6:400–15.

Landsberg, H. E. 1970. Man-made climatic changes. *Science,* 170:1265–274.

Likens, G. E., F. H. Bormann, N. M. Johnson, D. W. Fisher, and R. S. Pierce. 1970. Effects of forest cutting and herbicide treatment on nutrient budgets in the Hubbard Brook watershed-ecosystem. *Ecol. Monog.,* 40:23–47.

Likens, G. E., F. H. Bormann, N. M. Johnson, and R. S. Pierce. 1967. The calcium, magnesium, potassium and sodium budgets for a small forested ecosystem. *Ecology,* 48:772–85.

Macfadyen, A. 1962. *Animal Ecology: Aims and Methods,* 2nd ed. London: Pitman. 344 pp.

Machta, L., and E. Hughes. 1970. Atmospheric oxygen in 1967 to 1970. *Science,* 168:1582–584.

Odum, E. P. 1959. *Fundamentals of Ecology,* 2nd ed. Philadelphia: W. B. Saunders. 546 pp.

Olson, J. M. 1970. The evolution of photosynthesis. *Science,* 168:438–46.

Olson, J. S. 1970. Carbon cycles and temperate woodlands. In D. E. Reichle (ed.), *Analysis of Temperate Forest Ecosystems,* pp. 199–225. New York: Springer–Verlag.

Ovington, J. D. 1962. Quantitative ecology and the woodland ecosystem concept. *Adv. Ecol. Res.,* 1:103–92.

Patterson, C. C. 1965. Contaminated and natural lead environments of man. *Arch. Env. Health,* 8:270–77.

Peakall, D. B., and J. L. Lincer. 1970. Polychlorinated biphenyls: Another long-life widespread chemical in the environment. *BioScience,* 20:958–64.

Penman, H. L. 1970. The water cycle. *Scientific American,* 223(3):98–108.

Redfield, A. C. 1958. The biological control of chemical factors in the environment. *Amer. Sci.,* 46:205–21.

Revelle, R. 1963. Water-resources research in the federal government. *Science,* 142:1027–33.

Risebrough, R. W., P. Reiche, S. G. Herman, D. B. Peakall, and M. N. Kirven. 1968. Polychlorinated biphenyls in the global ecosystem. *Nature,* 220:1098–1102.

Rodin, L. E., and M. I. Bazilevich. 1968. *Production and Mineral Cycling in Terrestrial Vegetation.* London: Oliver and Boyd. 288 pp.

Ryther, J. H. 1969. Photosynthesis and fish production in the sea. *Science,* 166:72–76.

Sargent, F. 1967. A dangerous game: Taming the weather. *Scientist and Citizen,* 9:81–88.

Thomas, W. A. 1969. Accumulation and cycling of calcium by dogwood trees. *Ecol. Monog.,* 39:101–20.

Wendland, W. M., and R. A. Bryson. 1970. Atmospheric dustiness, man, and climatic change. *Biol. Cons.,* 2:125–28.

Woodwell, G. M. 1970. The energy cycle of the biosphere. *Scientific American,* 223(3):64–74.

Wurster, C. F., Jr. 1969. Chlorinated hydrocarbon insecticides and the world ecosystem. *Biol. Cons.,* 1:123–29.

Chapter 11

THE INTEGRATION OF
ECOSYSTEM STRUCTURE
AND FUNCTION

INTRODUCTION

Our preceding discussions have dealt with a variety of relationships at
the ecosystem level of integration. These have concentrated on individual
aspects of ecosystem function, such as energy flow and nutrient cycling,
or on specific components of ecosystems, such as the producers. These
discussions have provided the necessary background for an integrated
consideration of ecosystem structure and dynamics. It is essential that
we make such an examination, since only in this way can we fully understand the extent and seriousness of man's effects on the biosphere. It may
often appear that human activities affect only particular processes or
portions of an ecosystem. Usually these are the processes or portions of
the system most closely related in space or time to the activities involved.
Contrary to this belief, many of man's actions have led to major reorganization of entire ecosystems over wide areas. In this chapter we shall concern ourselves with these overall ecosystem responses.

We must first recognize that man's population growth and technological development have affected all major ecosystems to some extent. These
effects range from virtually complete control of biotic structure and nutrient cycling processes in agricultural systems to the inconspicuous, but
potentially serious, consequences of chemical pollution of the major air
and water circulation systems of the earth (Wurster, 1969).

In order to survive man must learn to manage the world environment so
that major biotic and abiotic processes necessary for his survival are protected. In this context the term *manage* refers broadly to whatever

degree of intervention or nonintervention proves necessary to achieve the goal of preserving these processes. Examples of the processes involved include photosynthesis and energy release by respiration and decomposition, nutrient release into the soil by weathering and loss by erosion, release of gases into the atmosphere and their removal, and many others. To preserve these functions, we must learn how different processes are integrated into complete ecosystems.

Two aspects of this integration that have attracted the recent attention of ecologists are *stability* of structure and function and *predictability of future change* of these characteristics. The concept of stability has proved to be one exceedingly difficult to formulate in precise terms (Margalef, 1969). In one sense, it may be thought of as simple constancy of structure and function regardless of whether or not the system is tested by varying conditions or disturbance. Such a condition may be exemplified by undisturbed ecosystems occupying environments essentially constant in their conditions. In a more dynamic sense, it may be thought of as the ability of a system to maintain an overall constancy of function through internal changes in the kinds, abundances, and activity of member species in the face of changing environmental conditions or human disturbance (Margalef, 1969). In general, we will use the concept in this latter sense.

With this in mind, we may pose several questions toward which our examination of integrated function of ecosystems will be directed:

1. What factors contribute to stability at the ecosystem level, and how do these factors operate in order to produce different degrees of stability in ecosystems of different structure and dynamics?
2. What are the effects of man's activities on ecosystem stability and on the patterns of stability of different ecosystem types?
3. How do important ecosystem characteristics change through time, what are the causes of such changes, and how do the rate and predictability of such changes vary for different ecosystem types?
4. What are the effects of man's activities on the patterns of change in ecosystem characteristics through time?
5. What procedures are required to gain information and develop methods for the ecologically sound management of major ecosystems?

ECOSYSTEM HOMEOSTASIS, ECOSYSTEM MATURITY, AND SYSTEMS ECOLOGY

The questions outlined above direct us to an examination of two major concepts relating to ecosystem function: *homeostasis* and *maturation*. Some systems have the tendency to remain in a steady state if undisturbed and to return to that state, through the modification of internal system processes, if disturbed. This tendency is termed homeostasis. Applied to the ecosystem, it refers to the capacity for maintaining or reestablishing a particular pattern of structure and function, by biological mechanisms, in the face of some disturbing influence. Ecosystem types vary greatly in the degree of homeostasis exhibited. For systems having a high degree of homeostasis, disruption of the normal pattern of structure and function is difficult, and internal disturbances that are induced are rapidly corrected. In other words, the response of the system is more dependent on the characteristics of the system itself than on the nature of the disturbance (Margalef, 1963). The opposite is true for systems having a low degree of homeostasis. Changes in the physical or biotic environment produce a stronger and more lasting effect on ecosystem structure and function, and they are thus of greater importance in determining the response of the system. The mechanisms of ecosystem homeostasis, and their capacity for dealing with both natural and man-caused environmental disturbances are the topics to which questions (1) and (2) above are directed.

In the absence of major disturbance ecosystems tend to show a consistent pattern of change in structure and function through time. This process of change—the result both of ecological and evolutionary processes—may be termed maturation (Margalef, 1963). Studies of different ecosystem types have suggested that several major patterns of change in characteristics tend to occur, regardless of the type of ecosystem involved. The most important of these are thought to be:

1. Increase in the complexity of biotic structure.
2. Increase in the efficiency of utilization of nutrients and energy.
3. Increase in the degree of homeostatic regulation.

Complexity of biotic structure refers both to the variety of different species present and to the complexity of physical structure resulting from the living tissues of organisms and their nonliving products. For example, the structure of a coral reef ecosystem is very complex, both in terms of the variety of organisms present and the physical size and complexity of the reef structures produced by coral animals and calcareous algae. Complexity would also be great in a forest ecosystem having a large plant biomass of living trees, shrubs, and ground plants, as well as appreciable quantities of dead organic matter in the form of leaf litter and humus. In contrast, the organic structure of an open water, planktonic ecosystem would be relatively simple.

Efficiency of utilization of energy includes the efficiency of storage of incoming solar energy by producers as well as the cost, in respiratory energy, for the maintenance of a unit of biomass. This latter characteristic may be expressed as the ratio of *Biomass/Gross Primary Production* or *Biomass Accumulation Ratio* (BAR). Increased efficiency is reflected in an increase in this ratio, indicating that a given amount of primary production is supporting a greater total biomass within the system. Efficiency in the utilization of nutrients refers to the degree to which nutrients are concentrated in biotic pools of the ecosystem and prevented from being carried out of the system. These characteristics, together with increase in ecosystem homeostasis, relate to questions (3) and (4) above.

Analysis of homeostatic mechanisms and maturation processes at the ecosystem level will require highly organized research programs whose objectives are clearly defined. The techniques employed must allow the simultaneous examination of many environmental variables. The development and testing of methods of ecosystem management must also be accompanied by research programs of comparable scope. The need for these programs has stimulated growth of the field of systems ecology. In studying ecosystem processes the approach of the systems ecologist is to first recognize major functional components of the system. The relationships among these components are then evaluated as functions of the initial state of the system, the inputs and outputs of matter and energy for the system, the exchanges occurring within the system, and time (Van Dyne, 1966).

In the following pages we will examine current knowledge and theory relating to the processes of ecosystem homeostasis and maturation. Following this discussion we will examine in more detail the approach of systems ecology to the study of ecosystem dynamics.

ECOSYSTEM HOMEOSTASIS

There are two major mechanisms operating at the ecosystem level to produce homeostasis: *biodemographic* and *biogeochemical* (Hutchinson, 1948). Biodemographic regulation results in the control of the numbers and biomasses of organisms belonging to various trophic groups and occurs through the operation of density dependent factors on populations of the organisms involved (Fig. 11–1). Biogeochemical regulation occurs through the action of negative feedback controls which operate to reestablish the normal concentrations of chemical nutrients in available ecosystem pools following a disturbance (Fig. 11–2). Both types of controls are usually biological in nature and result from the abilities of the organisms to exert regulatory action by varying their behavior and physiological activity and by increasing or decreasing in abundance.

Biodemographic and biogeochemical controls differ somewhat, however, in their manner of operation. In order to restore a condition existing prior to a disturbance, a biodemographic control must be capable of a response proportionately stronger than the change induced by the disturbing factor. Thus, for a carnivore to exert density dependent control on a herbivore population, the total predation rate by the carnivore population must more than double when the herbivore population doubles (see Chap. 5). Biogeochemical regulation, on the other hand, involves primarily the control over fluxes into and out of the active pools of an ecosystem. Almost all disturbance of nutrient cycles involve depletion or augmentation of the total nutrient capital (see Chap. 10) of the system. Biogeochemical regulation thus involves the capability of the system either to divert excess nutrient into unavailable pools (Fig. 11–2) or to reduce the losses to unavailable pools so that a depleted nutrient capital may be increased.

The effectiveness of biogeochemical and biodemographic controls in maintaining stability of structure and function in an ecosystem depends on the strength and rapidity of these responses. Poor regulation may be seen in the failure of systems to compensate for disturbance and for the changes resulting from such disturbances to produce violent fluctuations in system structure and function. For example, Cole (1954) has demonstrated mathematically that delayed feedback of increased population density on reproduction and mortality of a population can cause cyclical fluctuations in population density. Time lags in the operation of negative feedback and density dependent controls are precisely of this nature.

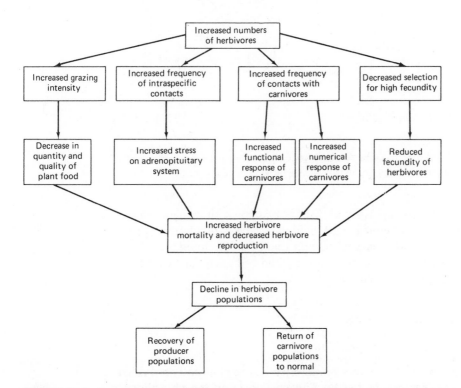

FIGURE 11–1. Potential routes of biodemographic regulation of herbivore component of an ecosystem through density dependent population controls and changing selective pressures.

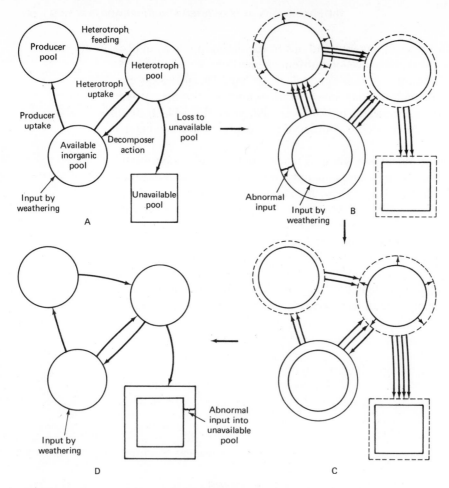

FIGURE 11–2. Operation of negative feedback controls to maintain a constant nutrient capital in the available pools of an ecosystem biogeochemical cycle.

In the following sections we will discuss examples showing contrasting degrees of biogeochemical and biodemographic regulation. From these examples the weaknesses of current theory on the integrated functioning of ecosystems can be clearly seen and the need for systematic and intensive study made apparent.

BIOGEOCHEMICAL REGULATION

Phosphorus Cycle in a Eutrophic Lake

An excellent example of effective biogeochemical regulation is provided by a series of studies conducted by Einsele (1941) on the phosphorus cycle of the Schleinsee, a small lake in Germany. Einsele investigated the fate and biological effects of phosphorus added to the surface waters of the lake during the summer. The Schleinsee has a surface area of 14.9

hectares and a maximum depth of about 11 meters. During the summer, it develops a sharp temperature-density stratification. This stratification is similar to that shown by almost all temperate-zone lakes and involves the development of a surface zone of warmer, lighter water, termed the *epilimnion*, and a bottom region of colder, denser water, known as the *hypolimnion*. The transition region between these two zones is known as the *thermocline*. Here the temperature in summer may decrease from 15–20°C to about 4°C (the temperature of maximum water density) in 1–2 meters. Much of the importance of this stratification results from the fact that circulation of water between these two layers is almost completely prevented. Consequently, little or no exchange of nutrients and dissolved gases may occur between epilimnion and hypolimnion during the summer stratification period. In the Schleinsee the thermocline occurred at a depth of about 6 meters.

The biological consequences of this stratification in large part depend on the level of productivity of the lake. Lakes may be classified according to level of nutrient content and productivity as *oligotrophic* or *eutrophic*. Oligotrophic lakes are those which have a small nutrient capital, show low levels of primary production, and possess clear water because of the absence of an abundant plankton community. Eutrophic lakes are nutrient rich, show high levels of primary production, and possess low transparency because of the abundance of organisms in the water. In actuality, of course, these two lake types represent the ends of a continuum of conditions relating to productivity.

One of the major features of eutrophic lakes, resulting from the high productivity levels and summer stratification, is depletion of oxygen in the hypolimnion during the summer. Organic matter is produced in such quantity in the surface water and transported in such quantity to the hypolimnion that the oxygen supply of the bottom waters may be greatly reduced or exhausted by processes of decomposition and respiration. In this respect the Schleinsee constituted a good example of a eutrophic lake. During the summer stratification period the hypolimnion suffers severe oxygen depletion but contains relatively high concentrations of nutrients, including phosphorus. At the same time the surface waters show almost complete depletion of inorganic phosphate and very low concentrations of phosphorus in dissolved organic form.

The importance of phosphorus in basic metabolic processes in organisms led Einsele to conduct a series of experiments involving the addition of phosphorus to surface waters of the Schleinsee during the summer. These experiments were designed to investigate the role of this nutrient in the overall productivity of the Schleinsee.

During the summer of 1937, a preliminary experiment was conducted in which 14 kg of inorganic phosphorus, in the form of superphosphate, were added to the surface waters of the lake on July 3 (Table 11–1). This amounted to slightly under 1 kg/hectare of surface. Just prior to this introduction inorganic phosphate was undetectable in the epilimnion, and

TABLE 11-1. PHOSPHORUS CONTENT, IN mg/m², OF THE EPILIMNION (SURFACE TO 6 m DEPTH) OF THE SCHLEINSEE BEFORE AND AFTER ADDITION OF 14 kg OF INORGANIC PHOSPHORUS ON JULY 3, 1937.

Phosphorus fraction	Before addition	Date of addition (calculated)	After addition		
	June 29	July 3	July 7	July 14	July 21
Inorganic	0	100	30	13	5
Organic	87	87[a]	147	163	120
Total	87	187	177	176	125

[a]Assumed equal to the June 29 value.

the total organic phosphorus content, including that in living as well as nonliving material, was 86 mg/m² (water column beneath one square meter to a depth of 6 meters). The introduced phosphorus was equivalent to an additional 100 mg/m². In the period between July 3 and July 14, this inorganic phosphorus was rapidly taken up by organisms, and possibly by detrital material, so that the decrease in concentration of inorganic phosphorus was nearly balanced by the increase in organic phosphorus. After July 14, the total quantity of phosphorus in the epilimnion declined rapidly due to sedimentation of planktonic material into the hypolimnion. This decline continued until, in August, the total quantity of phosphorus in the epilimnion was equal to that normally occurring in late summer. Furthermore, the added phosphorus, although more than doubling the existing phosphorus pool of the epilimnion, had little detectable effect in increasing productivity.

The phosphorus removed from the epilimnion through sedimentation was, in addition, apparently diverted into unavailable pools. Even following the fall and spring overturns, when temperature and density become uniform over the entire depth profile and currents transport nutrients from the bottom waters into the surface waters, the phosphorus content of the epilimnion was not increased beyond its normal level.

In the summer of 1938, Einsele conducted a second experiment in which 94 kg of phosphorus (compared to 14 kg in 1937) were added. This equalled about 6.3 kg/hectare of lake surface. The addition was made on May 30 and June 30 in lots of 47 kg each. Each of these additions was equivalent to an increase in the concentration of phosphorus in the epilimnion of 380 mg/m². The combined addition thus represented a greater than six-fold augmentation of the phosphorus concentration of the surface waters.

Following these additions (Fig. 11-3), uptake of inorganic phosphorus and removal from the epilimnion through sedimentation occurred in a

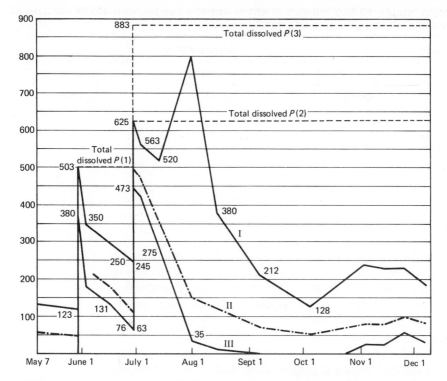

FIGURE 11–3. Changes in the phosphorus concentration of the epilimnion (mg/m² to a depth of 6 m) of the Schleinsee following addition of 47 kg of phosphorus on May 30, 1938 and the same amount on June 30, 1938. Curve (I) gives total organic and inorganic phosphorus, curve (II) inorganic + dissolved organic, and curve (III) dissolved inorganic phosphorus. (From Einsele, 1941.)

pattern similar to that seen in 1937. The rapid uptake of inorganic phosphorus was evidenced by the change in the amount of phosphorus and the ratio of nitrogen to phosphorus in samples of net plankton (Table 11–2). Prior to phosphorus additions the nitrogen: phosphorus ratio varied from 10:1 to 19:1. On July 14, after both additions, it had dropped to 4:1. This change suggests that planktonic organisms are capable of rapidly taking up and storing phosphorus in quantities greatly exceeding their immediate needs. The uptake of planktonic organisms resulted in a rapid decline in the inorganic phosphorus concentration in the epilimnion following each addition (Fig. 11–3), so that, by the beginning of September, no detectable amount remained. The total inorganic and organic phosphorus content of the epilimnion also declined rapidly after the first addition and for a period of about two weeks after the second addition. In the period following July 14, however, the total phosphorus pool of the epilimnion increased. This increase was attributed to the release of phosphorus from decomposition of organic material sedimented into shal-

TABLE 11-2. RATIOS OF NITROGEN TO PHOSPHORUS IN NET PLANKTON BEFORE AND AFTER ADDITIONS OF PHOSPHORUS TO THE EPILIMNION OF THE SCHLEINSEE IN 1938.

	Before additions			After additions								
	May 27			July 14			August 1			September 7		
	mg/m^3			mg/m^3			mg/m^3			mg/m^3		
Depth (m)	N	P	Ratio	N	P	Ratio	N	P	Ratio	N	P	Ratio
0	19	1	19:1	10	2.6	4:1	38	5.7	6:1	126	7	18:1
1 + 2	46	3.8	12:1	62	15.0	4:1	96	14.5	6.5:1	170	12	14:1
3 + 4	73	5.6	13:1	100	24.0	4:1	117	26.0	4.5:1	185	15	13:1
5 + 6	71	7.0	10:1	110	27.0	4:1	160	40.7	4:1	290	19	15:1

low water areas within the epilimnetic depth range. The chemical nature of this released phosphorus was apparently stimulating to the growth of planktonic organisms and led to an increased rate of primary production and a major phytoplankton bloom in the epilimnion. This bloom is evidenced by the greatly increased amounts of organic nitrogen recorded in the epilimnion during August and September, 1938 (Table 11–2).

Toward the end of this bloom, however, sedimentation of organic matter into deep water led to a rapid decline in the total phosphorus content of the epilimnion and an increase in the ratio of nitrogen to phosphorus. Although the fall overturn in November caused a minor increase in inorganic and organic phosphorus in the surface water, almost all of the phosphorus added during the summer of 1938 had apparently been permanently removed from the active biogeochemical cycle of the lake. At the start of the summer stratification period in 1939, the phosphorus content of the entire lake was only about 30% above that prior to fertilization in 1938. The total phosphorous content and productivity of the surface waters were significantly higher than normal during the summer of 1939. By September, 1939, however, the total phosphorus content of the lake had returned to the level found for several years prior to 1937 and 1938.

These results demonstrate that the phosphorus cycle of the Schleinsee is highly regulated as a result of complex biological and chemical processes occurring in the surface waters, bottom waters, and sediments. These processes, in combination with normal inputs and losses of phosphorus from the active pools of the cycle, function to restrict the recycling rate of phosphorus within rather narrow limits. Individual disturbances, such as the additions carried out by Einsele, although causing temporary disturbances in biotic structure in specific parts of the system, are inadequate to produce basic reorganization of the entire cycle and result in the diversion of the added phosphorus to biologically inactive pools.

Thus, it appears that to effect a permanent change in the active phosphorus pool, continued increase in the input of phosphorus would be necessary (Einsele, 1941). Thus, it may be that in the case of lakes suffering artificial eutrophication caused by man's activities reestablishment of original conditions of nutrient input may initiate a return toward conditions of oligotrophy (Margalef, 1968). We will consider this problem in detail later in this chapter.

It should be noted that the experiments of Einsele, while suggesting that strong homeostatic regulation of the phosphorus cycle occurs in the Schleinsee, do not tell us much about the specific physiological and chemical processes producing this regulation. For example, it is apparent that the decomposition processes occurring in shallow water and deep water sediments function in quite different manners in the return of phosphorus to active parts of the cycle. Likewise, these experiments fail to clearly define the limits of the regulatory processes involved and the relationships between the phosphorus cycle and cycles of other important nutrients, such as nitrogen. It is apparent that more broadly oriented and detailed studies will be required in order to gain an understanding of biogeochemical cycling and homeostatic regulation adequate to allow development of ecologically sound management techniques.

Population Cycles in Arctic Areas

Terrestrial ecosystems of arctic and subarctic regions are characterized by a high degree of biotic instability. In these areas cyclical fluctuations in population density of various birds and mammals are strikingly developed, and severe outbreaks of forest insect pests are frequent. In general, these fluctuations do not appear to be correlated with fluctuations of the climate. Instead, certain of them appear to result largely from weak homeostatic regulation of the ecosystems involved.

Cyclical fluctuations of population density in various species of arctic lemmings (species of the genera *Lemmus* and *Dicrostonyx*) have attracted the attention of ecologists for many years. In these cycles, for which the period between successive peak populations averages 3–4 years, the lemming population densities may change by a factor of 50–100. Many different explanations for these cycles have been proposed, and many have been discarded. Early workers suggested that these cycles reflected cyclical variation in extrinsic factors such as weather conditions or solar activity. Current theories, however, regard these fluctuations as being the result of factors intrinsic to arctic ecosystems or to the lemming populations themselves. Several divergent explanations of this type have been suggested. These relate to predator-prey interaction (Pearson, 1966), disruption of endocrine function of lemmings as a consequence of behavioral stress factors appearing at high population densities (Christian and Davis, 1964), changing genetic constitution of lemming populations due to changing selective pressures at different population densities (Chitty, 1960), and the nutrient cycling patterns inherent in arctic tundra ecosystems (Schultz, 1964). Although these different ideas cannot all be compared and evaluated here, the last, known formally as the *nutrient recovery hypothesis* of arctic lemming cycles, is supported by the most broadly based series of studies on the dynamics of arctic tundra ecosystems (Pitelka, 1964; Schultz, 1964). Nevertheless, the reader should bear in mind the tentative, and even speculative, nature of this hypothesis.

The nutrient recovery hypothesis is based on studies initiated by Pitelka and a number of coworkers at Point Barrow, Alaska in 1950. These studies have followed the changes occurring over several complete cycles of lemming abundance and have considered many important characteristics of the tundra ecosystem. Specific factors examined have been densities of lemming populations, densities and feeding activity of important predator species, the quantity, composition, and nutrient quality of the vegetation, and the depth, nutrient characteristics, and decomposer flora of the zone of soil (active layer) above the permafrost layer, from which the plants draw nutrients during summer growth.

Results of these studies suggest that the primary interaction leading to lemming cycles involves the lemmings, the quantity and nutrient quality of their plant food, and the availability of soil nutrients (Schultz, 1964, 1969). The arctic tundra ecosystem lacks large available pools of most nutrients and has only small annual inputs for most nutrients. The consequences of these relationships can best be described by examining the changes occurring in basic ecosystem characteristics during a complete cycle of lemming abundance (Fig. 11–4). At the low point of the lemming

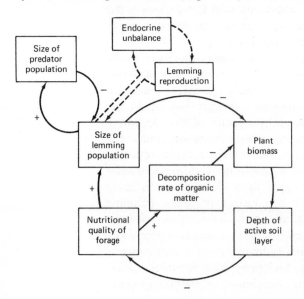

FIGURE 11–4. Diagrammatic representation of the relationships postulated as causal factors of arctic lemming cycles under the nutrient threshold hypothesis. The + and − symbols indicate that the state of a given compartment either amplifies (+) or counteracts (−) the change in state of the next. (From Schultz, 1969.)

cycle, following a population crash, the vegetation is sparse and of low nutrient content. Much of the nutrient capital of the ecosystem is tied up in dead animal tissues, animal waste products, and other detritus. The soil decomposer flora adapted to utilizing these materials is expanding, but because of the cold, wet soil conditions and the short summer, the release of tied-up nutrients is slow. As these nutrients are regenerated, however, the vegetation begins to recover. At this time the grazing intensity by lemmings is low, and the new growth, utilizing the released nutrients, is of high nutrient quality. As vegetation growth continues, the depth of the active layer becomes less because of the insulating effect of the vegetation cover. Furthermore, as a dense plant cover develops, the decomposer flora of the soil shifts to a dominance of forms adapted to utilizing dead plant material. Following recovery of the vegetation, and its increased nutrient quality, the lemming population begins to increase.

The growing lemming population, the members of which feed and reproduce year-round, soon outstrips the available food supply. The nutrient-rich plant species are rapidly depleted. Nutrients again become tied up in animal tissues and wastes. Behavioral interference among feeding lemmings results in the wastage, by prostration, of much plant food which is cut but not eaten. This plant material constitutes a major contribution to the accumulation of organic detritus. In some locations the accumulation of cut plant material may insulate the soil surface to the extent that the depth of the thawed, active layer is reduced. The soil nutrient pool is thus depleted and the new plant growth is of low nutrient quality. Reduction of the quantity and quality of vegetation continues until widespread malnutrition and starvation occur among the lemmings. These individuals become subject to all major causes of mortality, including starvation, exposure, predation, and disease. The soil decomposer flora gradually shifts to a dominance of forms adapted to animal tissues and wastes, but too late to re-supply nutrients for plant growth to support the high lemming population. Because of heavy grazing and the resultant thinning of the vegetation cover, the active soil layer increases in depth. This, however, has little beneficial effect on plant growth because lower soil layers contain low nutrient concentrations. As a result of these changes the lemming population enters the "crash" phase of the cycle.

Thus, if we assume that the above observations are accurate, the primary cause of lemming population cycles appears to be the delayed feedback of nutrients by decomposers to the producers in the tundra ecosystem. This delay is presumably caused, in large part, by the cold, wet soils and short summer, which greatly restrict the activity of soil organisms. The importance of this delay is amplified by the presence of a homoiothermic herbivore which is active year-round and is capable of consuming large quantities of plant material. Evidence from preliminary experiments (Schultz, 1969) suggests that the lemming cycle can be obliterated by artificial fertilization of the tundra ecosystem, so that abundant nutrients are available for plant growth even under conditions of high lemming populations.

It should be emphasized that this hypothesis is still largely untested and that detailed studies of the rates of nutrient regeneration by specific decomposer groups, as well as studies of the physiological nutrient requirements of lemmings, will be necessary in order to evaluate it.

Similar situations, in which delayed feedback of nutrients leads to major fluctuations in ecosystem characteristics, may be seen in other ecosystem types. One such example is the plankton bloom occurring in deep lakes following periods of lake overturn and the reintroduction of nutrients into the epilimnion from the bottom waters. Likewise, in many terrestrial ecosystems of semiarid regions, such as the California chaparral, nutrient cycling is greatly slowed by the accumulation of nutrients in living and dead tissues of woody plants. This accumulation results from the inefficiency of utilization of woody tissues by herbivores and the slow rates of decomposition because of the lack of moisture. In many of these systems the periodic occurrence of fire results in the rapid release of certain nutrients and greatly increases productivity among species adapted to fire recovery.

BIODEMOGRAPHIC REGULATION

Theories of Regulation at the Community Level

Biodemographic regulation results from the operation of density dependent controls on individual species populations (Hutchinson, 1948). If we extend this principle to the community level, we can reason that the community having the greatest proportion of individual species populations which are acted upon by density dependent factors capable of strong and rapid response will show the highest stability. Likewise, since density independent factors, such as severe weather conditions, tend to counteract the effects of density dependent factors by producing erratic patterns of population fluctuation, it may be reasoned that stability will be greatest in a system having the lowest frequency of such events. Since the biotic structure of a system may buffer the effects of fluctuations in the overall physical environment, the effectiveness of this buffering effect will also be significant for community stability.

The question of biodemographic regulation thus resolves itself into one of the specific characteristics of composition and structure of communities that produce effective density dependent control and environmental buffering. Many ecologists have suggested that diversity of species composition is directly related to community stability (Hutchinson, 1959; Pimentel, 1961). This hypothesis is based on the idea that a community of greater species diversity will possess a more complex food web than will a community of low species diversity. Since most density dependent controls involve food chain relationships such as predation, parasitism, and competition for food resources, a complex food web will thus incorporate many relationships of a potentially density dependent nature. Furthermore, in a system having a complex food web the food chain importance of individual species populations is minimized. The failure of any single species through accidental reasons may be compensated for by minor shifts in the feeding activity of the remaining community members.

This hypothesis has been formalized by MacArthur (1955) in an attempt to obtain an index of community stability. MacArthur suggested that stability is related to the number of *trophic pathways* in an ecosystem and to the proportion of total energy flow passing along each pathway. Trophic pathways are defined as potential routes of energy flow from the producer to the highest consumer trophic levels in the system.

Figure 11–5 illustrates the concept of trophic pathways and shows the effects of various factors on the number of trophic pathways and the equitability of energy flow along different pathways. Panel *A* depicts the manner in which an increase in species diversity may affect the number of trophic pathways. Addition of species belonging to trophic levels already represented in the system increases the number of trophic pathways. The addition, however, of a species belonging to a new trophic level, such as a top carnivore, to the system shown in the left half of panel *A* would not increase the number of trophic pathways.

Changes in the feeding characteristics of community members may also affect the number of trophic pathways present in the system. A change from restricted to omnivorous feeding habits of the species present increases the number of trophic pathways. Panel *B* depicts the increase in number of trophic pathways resulting from the utilization, by members of a particular trophic level, of a wider range of food species at the trophic level below. Panel *C* illustrates the change that would occur if consumer species change food habits so that they are able to utilize forms of any lower trophic level for food. This latter change, together with increasing the number of possible routes along which food energy may flow, results in a blurring of the trophic level occupied by such consumers. This pattern is shown by many large mammals, including man, which draw food from herbivore, carnivore, and producer levels.

Panel *D* illustrates the concept of equitability of energy flow along different trophic pathways. The system shown in the left half of this panel possesses only one functionally important pathway, along which 97% of the total food energy passes. Obviously, in this system disturbance of any one of the species in this pathway would result in serious repercussions for the entire system. The system shown in the right half of panel *D* possesses four trophic pathways of equal importance. In this system disruption of one species population could be more easily compensated for by small shifts in the energy flow along other pathways.

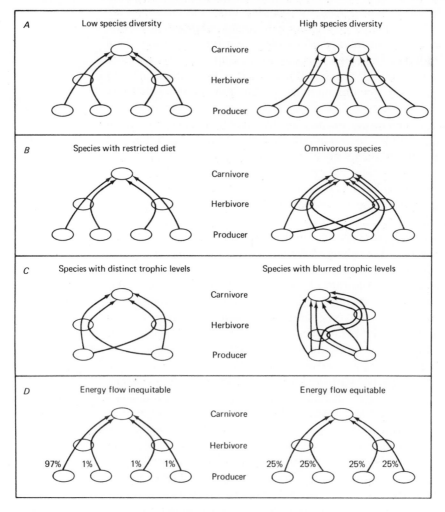

FIGURE 11–5. Trophic pathways and equitability of energy flow in food webs of varying structure (see text for explanation).

Using this model as a basis, MacArthur (1955) has suggested that a quantitative measure of community stability may be obtained by using the Shannon-Wiener Function derived from information theory. This measure is given by the formula:

$$\text{Stability index} = -\Sigma\, pe_i \log_2 pe_i$$

where

$$pe_i = \text{Decimal fraction of total energy flow}$$
$$\text{passing along trophic pathway } i$$

The summation term in this formula covers forms of the expression $pe_i \log_2 pe_i$ computed for each trophic pathway represented in the system. This formula operates in such a manner that increase in either the number of trophic pathways or in the equitability of energy flow along existing pathways will produce an increased stability index.

Since direct measurements of energy flow along trophic pathways are virtually impossible to obtain, this formulation can rarely be used. However, on the assumption that the variety and relative abundance of species are closely related to energy flow along trophic pathways in a system, the diversity index calculated by the above formula may be a useful measure of community stability. This calculation follows the form:

$$\text{Diversity index} = -\Sigma \, pn_i \log_2 pn_i$$

where

$$pn_i = \text{Decimal fraction of total individuals}$$
$$\text{belonging to the } i^{th} \text{ species}$$

A practical method for making this calculation is given by Cox (1967). It should be remembered that MacArthur's formulation represents a hypothesis that stability and community structure are related in this manner and not a proven relationship.

More recently, Watt (1965) has presented a series of hypotheses regarding community stability which are, in part, contradictory to those of MacArthur (1955). Development of these hypotheses was stimulated by the observation that, contrary to what would be expected from the MacArthur model, many pest species and species having unstable population densities have a large number of parasites and predators. Examples of such forms include grasshopper species of semiarid grasslands, the spruce budworm in coniferous forest areas, and microtine rodents of arctic and subarctic regions. These observations led Watt to analyze the number of species of host and enemies associated with various Canadian forest insect pests, a group for which many data are available.

From this analysis Watt concluded that three relationships were important in determining the degree of stability shown by a pest species. The most important of these involved the proportion of the total environment containing food for the pest species. Instability was greatest in species for which the greatest proportion of the environment provided food. This occurred either when a pest was restricted to one or a few tree species, but these species occurred in nearly pure stands, or when the pest was capable of feeding on many different tree species. It may be noted that for these latter species the hypotheses of MacArthur and Watt are in conflict. The MacArthur model predicts that addition of more tree species to a stand containing a pest species that can feed on them would increase the number of trophic pathways, and thus increase system stability. Watt's hypothesis predicts that the effect of adding more tree species would depend on whether or not the species involved can be utilized by the pest.

Secondly, Watt concluded that stability of a pest species was increased by the presence of competitor species (other pests utilizing the same host). For example, tree species having a high diversity of pest species experienced serious damage by the pests only infrequently. Apparently, interactions among individuals of different pest species in some manner stabilized the overall pest populations. Since increased numbers of species would likely lead to increased numbers of trophic pathways, this conclusion is consistent with the MacArthur model.

Finally, Watt concluded that herbivorous insect pests that have few enemies are more stable than similar pests that have many enemies. In a manner similar to that for interactions among pest species utilizing a single host, interactions among the many enemies of a single pest stabilize populations of these enemies, thus reducing their capacity for exerting density dependent control upon the pest population. This conclusion is in direct conflict with the MacArthur model. Furthermore, this third relationship is critical for the strategy of biological control. If the conclusion by Watt is correct, it may be deleterious to introduce several rather than one or a few biological control agents for a single pest.

The ideas of MacArthur (1955) and Watt (1965) point up the weakness of theory relating to regulation at the community or ecosystem level. The practical importance of an understanding of these relationships is also clear, since pest species outbreaks represent failures of operation of such regulation. It is apparent, however, that stability is related to features of the biotic structure and function of ecosystems, and that diversity of species composition is important to at least some degree. These observations are the reason for the ecologists' current interest in the controlling factors and biological significance of species diversity. An understanding of the important relationships may allow more effective structuring of agricultural, grazing, and forest ecosystems for maximum stability and productivity.

Studies of Regulation at the Community Level

Despite the importance of understanding the determinants of ecosystem stability, almost all available information is of an anecdotal or subjective nature. Few workers have attempted to describe and measure stability in natural situations, and even fewer have attempted experimental studies of this problem. Much of the older information on this topic is summarized by Elton (1958) and Pimentel (1961).

Recently, several workers have attempted to measure the degree of stability in natural systems and to correlate the observed values with other system characteristics. One such study, conducted by Brewer (1963), is an analysis of year-to-year fluctuations in breeding populations of songbirds in five forest areas for which long census sequences were available. These areas (Table 11–3), all located in the Eastern United States, ranged from a dry, upland oak forest of low tree species diversity (located in Minnesota) to a rich flood plain forest in Maryland. Brewer found the year-to-year variability in the total songbird population to be greatest in the dry, oak forest and least in the flood plain forest. This relationship was inversely related to bird species diversity in these forests, which was greatest in the flood plain and lowest in the oak forests. Thus, a direct relationship between stability and species diversity is suggested by this study. However, as noted by Brewer, the five forest areas lie along a gradient of climatic stability, with the oak forest occurring at the location of greatest instability, and the flood plain forest near the point of greatest stability. This analysis illustrates the difficulty involved in studying such problems by observational techniques. Here a good correlation exists between stability and several ecosystem characteristics, the causal significance of which cannot easily be sorted out.

Several initial attempts have been made to study the stability-diversity relationship experimentally. Pimentel (1961) conducted a study of the role of species diversity in preventing insect outbreaks on collards, one of the cultivated forms of *Brassica oleracea*. This crop plant was selected because it was subject to a variety of insect pests and because it has broad, relatively flat leaves on which insect densities and damage could be measured readily.

The design of this study involved comparison of the frequency of pest species outbreaks on collards in three experimental plots: a mixed-species plot and two single-species plots differing in the spacing of plants. The mixed-species plot was located in an abandoned field that had grown up in weedy species over a period of 10 years. This plot, 500 ft. by 600 ft. in size, contained an estimated 300 species of vascular plants and 3,000 species of heterotrophs. Within this plot individual collard plants were

TABLE 11–3. SPECIES DIVERSITY AND STABILITY OF TOTAL SONGBIRD BREEDING POPULATION IN FIVE DECIDUOUS FOREST AREAS OF THE EASTERN UNITED STATES.*

	Upland oak	Maple-elm-oak	Beech-maple	Lowland beech-maple	Flood plain
Location	Minnesota	Illinois	Ohio	Ohio	Maryland
Area (acres)	23.4	55	65	55.3	18.75
Years censused	7	10	18	8	10
Mean number of breeding species per year	15.6	23.3	24.7	25.8	27.6
Coefficient of variation for total population[a]	18.0	15.3	14.2	8.1	8.2

*Adapted from Brewer, 1963.

[a]Calculated as 100 x (standard deviation of total population size/ mean total population size).

planted at 9 ft. intervals in a grid arrangement. The planting sites were fertilized and the ground surface immediately surrounding the plants covered with a plastic shield in order to prevent intense competition between collards and surrounding native species.

The single-species plantings were made in bare agricultural fields. A sparse planting was located in a plot 300 ft. by 900 ft. in size, with individual collard plants spaced in a 9 ft. grid, as in the mixed-species planting. A dense single-species planting was also made in a plot 75 ft. by 100 ft. in size, with close spacing of the individual plants. Insect populations on collard plants were sampled over a period of two growing seasons.

The results of this experiment showed that the collard plants in the mixed-species planting, despite the presence of a much greater number of herbivores in the surrounding vegetation, were not subject to a greater number of herbivore pests (Table 11–4). More importantly, in the single-species plantings outbreaks of aphids and several species of flea beetles occurred (Fig. 11–6), while no such outbreaks were noted in the mixed-species planting. Differences in spacing of plants in the single-species plantings had little influence on the occurrence of these pest species outbreaks. As a result of the herbivore insect outbreaks in the single-species plantings, a greater variety of parasite and predator taxa were apparently attracted to these areas (Table 11–4). However, the ratios of population densities of parasite and predator forms to those of herbivores were generally greater in the mixed-species than in the single-species plantings (Table 11–4), except for the parasite herbivore relationship in 1957. These higher ratios suggest that parasites and predators usually exerted more effective control over herbivores in the mixed-species planting.

The results of this experiment suggest the importance of diversity in

TABLE 11-4. NUMBERS OF INVERTEBRATE TAXA AND RATIOS OF PREDATOR AND PARASITE TO HERBIVORE POPULATION DENSITIES ON COLLARDS PLANTS IN MIXED-SPECIES AND DENSE SINGLE-SPECIES PLANTINGS IN 1957 AND 1958.*

	1957		1958	
	Mixed	Single	Mixed	Single
Taxa				
Herbivores	20	21	25	27
Parasites	4	14	6	11
Predators	3	15	8	12
Total	27	50	39	50
Population ratios				
Parasites:herbivores	8.1:100	18.9:100	12.0:100	4.6:100
Predators:herbivores	9.5:100	1.1:100	102.5:100	4.5:100

*See text for discussion.

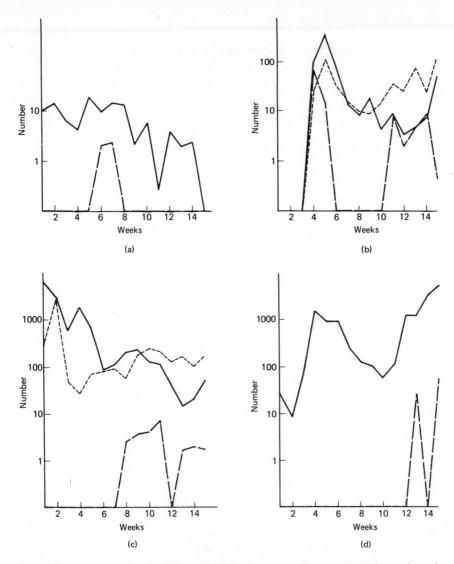

FIGURE 11–6. Numbers of aphids and flea beetles per unit plant area in single-species and mixed-species plots during 1957 and 1958. (a) The log number of flea beetles per unit plant area in the mixed (— — —), single (————), and sparse-single species (– – – –) stands during 1957. (b) The log number aphids per unit plant area in the mixed- (— – — —), single- (————), and sparse-single species (– – – – –) stands during the 1958 season. (c) The log number of flea beetles per unit plant area in the mixed (— — —), single- (————), and sparse-single species (– – – –) stands during 1958. (d) The log number of aphids per unit plant area in the mixed (— — —) and single species (————) stands during the 1957 season. (From Pimentel, 1961.)

favoring stability in complex biotic assemblages. The experimental design, however, does not permit evaluation of the contribution to stability by diversity within specific trophic groups. In fact, it does not tell us whether the important factor was the greater overall complexity of the food web in the mixed-species planting or simply the presence of specific organisms that were effective biological control agents of the particular pest species that reached outbreak levels in the single-species areas. Thus, this experiment does not distinguish between the hypotheses of MacArthur (1955) and Watt (1965). Nevertheless, it does provide a starting point for more carefully designed field studies of ecosystem stability.

A somewhat different approach to the experimental study of the diversity-stability relationship has been taken by Hairston et al. (1968). This approach has involved the analysis of stability in simple laboratory cultures of bacteria and protozoa. The advantage of such systems is the control that can be maintained over species composition. In the initial experiments by these workers three species of bacteria, three *Paramecium* species which feed primarily on bacteria, and the ciliate protozoans *Didinium* and *Woodruffia*, which are predators of *Paramecium*, were used.

In one set of experiments cultures consisting of two trophic levels were established. Different cultures were designed to contain one, two, or three species of bacteria and either two or three species of *Paramecium*, in all possible species combinations. These experimental cultures were placed in deep-depression slides with a volume of 1.0 ml. From two to six replicates of each species combination were observed. Supplementary quantities of the bacteria cultures were added every other day. At regular intervals over a 20-day period, the cultures were sampled and the presence and abundance of *Paramecium* species noted.

One of the best indicators of stability of these systems was found to be the simple persistence of the *Paramecium* species. In all cases it was observed that, by the end of 20 days, one *Paramecium* species had reached a high density and the remaining species either disappeared or persisted at much lower densities. The frequency of persistence of these less abundant species was thus used to compare stability of the different systems.

In cultures differing in the number of bacterial food species the persistence of *Paramecium* species was greatest when the most bacteria species were present. With one species of bacterium, the less abundant *Paramecium* species persisted in only 32.1% of the possible cases. With two species of bacteria, persistence increased to 61.3%, and with three species to 70.0%. Increased diversity at the lower trophic level thus appeared to favor stability at the higher level. This result suggests that increased diversity of food species favors stability by permitting feeding specialization and thus reducing competition for food among members of the higher trophic level.

Analysis of these same data according to the number of *Paramecium* species present indicated that increase in number of species at this level did not increase stability but may have decreased it. In systems having two *Paramecium* species the less abundant species persisted in 57.3% of the possible cases, and in three-species systems they persisted in only 41.1% of possible cases. This result suggests that an increase in diversity at one level, with no increase at the level below, may lead to reduced stability by increasing the probability of loss of species by competitive displacement.

In a second set of experiments the effect of addition of one or both species of predators to such cultures was explored. In these systems the most frequent result was the complete disappearance of all species of *Paramecium,* together with their predators, within a short time. No increase in stability was noted, and no differences were apparent between systems having one or both of the predator species.

The drawback of these experiments is the simplicity and restricted size of the experimental environment. For the particular species used this environment may be so abnormal that the results obtained have little relevance, even to very simple natural systems. The experimental design further emphasizes the peculiar biological characteristics of the few species chosen, which may not be representative of those of species generally occurring together in nature. It is worth noting, however, that it is possible that stability patterns in natural systems reflect primarily the peculiar biological characteristics of the species involved, rather than the general characteristics of species diversity, relative species abundance, and trophic level which form the basis for the stability models proposed earlier.

It is apparent, though, that the problem of biodemographic regulation at the ecosystem level is open to experimental investigation, and that it will represent one of the most important areas of future research.

ECOSYSTEM MATURATION

As emphasized earlier in our discussion of community dynamics, few, if any, ecological systems have reached a perfect steady-state condition in which the structure and function of the system remain constant through time. Long-term fluctuations and trends in regional climates, together with the geological processes of erosion and deposition, slowly but continually modify conditions of the physical environment in any given area. Sudden disturbances of environmental conditions or ecosystem structure result from the occurrence of fire, flooding, and landslides, or from the activities of grazing animals and insect pests. Ever since the appearance of man disturbances resulting from his exploitation and pollution of the environment have affected nearly all world ecosystems.

Ecosystems show two major patterns of response to the changes produced by these agents: successional and evolutionary. These responses are similar in that both appear to lead toward the development of systems showing steady-state conditions, maximum efficiency in energy and nutrient utilization, and maximum stability. As stated earlier, the degree of development of these characteristics may serve as an index of the level of maturity of the ecosystem (Margalef, 1963).

An additional effect of disturbance, when localized, is to introduce a major element of spatial variation in the characteristics of ecosystems occupying a given geographical area. This variation is magnified by the fact that successional and evolutionary changes in ecosystem maturity proceed at quite different rates in different environments. Thus, ecosystems of widely differing degrees of maturity may exist adjacent to each other, or even intermingled. This fact has been of evolutionary significance to the systems involved, and important interactions, related to the differences in maturity involved, occur among them.

In the following discussions we shall consider the patterns and causes of successional and evolutionary changes in ecosystem maturity, and we shall examine the relationships existing among ecosystems of differing maturity.

Successional Patterns of Change in Maturity

Our earlier examination of the detailed patterns and causes of ecological succession (Chap. 7) demonstrated that during this process major changes occur in overall system characteristics. Many ecologists feel that a number of these changes represent basic successional trends which tend to occur, in some form, regardless of the particular characteristics of the system undergoing succession. That is, regardless of whether one is following succession in a laboratory culture, a natural aquatic system, or a bare land area, certain basic patterns of change are almost always shown. It is, in fact, this idea that gave rise to the concept of ecosystem maturity.

Recently, a number of workers have attempted to define these trends (Margalef, 1963, 1968; Odum, 1969; Woodwell, 1967). We will examine these suggested patterns. In doing so the reader should bear in mind that many of them represent tentative generalizations, which may be violated in the case of certain ecosystem types, or in certain cases, hypothetical or inferred patterns, which have not yet been adequately tested.

These trends can best be considered by comparing structural and functional characteristics of systems of early, intermediate, and late successional status (Table 11–5). Ecosystems of early successional status occupy areas of newly created habitat or areas recently subjected to some severe disturbance such as fire. The physical environment of these areas is characterized by extremes of environmental conditions and sharp fluctuations in resource availability (Woodwell, 1967). In terrestrial environments factors such as temperature, humidity, and wind are not buffered by a dense cover of vegetation. Soil moisture levels fluctuate widely as a result of the direct exposure of the soil to rainfall and to the drying action of wind and solar radiation. In aquatic environments water currents are unbuffered by aquatic vegetation, and nutrient concentrations vary widely with the inflow of water from surrounding drainage areas. The total nutrient capital in these systems may vary

TABLE 11-5. A TABULAR MODEL OF ECOLOGICAL SUCCESSION: TRENDS TO BE EXPECTED IN THE DEVELOPMENT OF ECOSYSTEMS.*

Ecosystem attributes	Developmental stages	Mature stages
Community energetics		
1. Gross production/community respiration (P/R ratio)	Greater or less than 1	Approaches 1
2. Gross production/standing crop biomass (P/B ratio)	High	Low
3. Biomass supported/unit energy flow (B/E ratio)	Low	High
4. Net community production (yield)	High	Low
5. Food chains	Linear, predominantly grazing	Weblike, predominantly detritus
Community structure		
6. Total organic matter	Small	Large
7. Inorganic nutrients	Extrabiotic	Intrabiotic
8. Species diversity — variety component	Low	High
9. Species diversity — equitability component	Low	High
10. Biochemical diversity	Low	High
11. Stratification and spatial heterogeneity (pattern diversity)	Poorly organized	Well-organized
Life history		
12. Niche specialization	Broad	Narrow
13. Size of organism	Small	Large
14. Life cycles	Short, simple	Long, complex
Nutrient cycling		
15. Mineral cycles	Open	Closed
16. Nutrient exchange rate, between organisms and environment	Rapid	Slow
17. Role of detritus in nutrient regeneration	Unimportant	Important
Selection pressure		
18. Growth form	For rapid growth ("r-selection")	For feedback control ("K-selection")
19. Production	Quantity	Quality
Overall homeostasis		
20. Internal symbiosis	Undeveloped	Developed
21. Nutrient conservation	Poor	Good
22. Stability (resistance to external perturbations)	Poor	Good
23. Entropy	High	Low
24. Information	Low	High

*From Odum, 1969.

widely, depending on the nature of the environment. In areas of habitat never before occupied by organisms, however, the nutrient capital is typically very low. Nutrients may be relatively abundant in areas formerly occupied by ecosystems of relatively high maturity levels. In either case the bulk of the nutrients in the system is in an inorganic form (Wood-well, 1967).

The community of organisms occupying these environments is characterized by low species diversity and, frequently, high abundances of those species that are present (Margalef, 1968). The total biomass of living organisms is low, as is the biomass of dead organic matter derived from them. The member species are highly adapted for dispersal into and colonization of unoccupied and extreme habitats. For example, species which are good island colonists tend to possess high reproductive potentials, efficient dispersal mechanisms, and short life cycles (MacArthur and Wilson, 1967). Physiologically, they are tolerant of wide fluctuations in conditions of the physical environment and are relatively nonspecialized in their patterns of utilization of environmental resources, including energy and nutrients (Williams, 1969). Furthermore, patterns of utilization of resources by different species tend to be broadly overlapping, and ecological isolating mechanisms are poorly developed (Margalef, 1968). Such characteristics are indicative of a low degree of interspecific competition for resources and suggest that these species are adapted to show a high level of productivity at the price of low efficiency of resource utilization (Margalef, 1968).

Correlated with low species diversity, especially evident among heterotrophs, and the low total biomass of organisms, the members of early successional systems show a low frequency of close interspecific relationships. As a result of weak development of competition and interspecific interactions of various types, the dispersion patterns of species populations tend to be random or aggregated because of the species' patterns of reproduction or heterogeneity of the environment. These species thus conform to the characteristics of fugitive species (Hutchinson, 1959) or opportunistic species (MacArthur, 1960).

These characteristics are clearly shown in communities of planktonic organisms colonizing newly prepared nutrient cultures (Margalef, 1968). Such cultures are colonized by a few species that rapidly increase in population density to produce a bloom. The producer members of these systems are characterized by a predominance of chlorophyll a, which is capable of a high rate of energy conversion, but exhibits low efficiency in utilization of the total incident light energy.

This combination of species characteristics results in a unique pattern of ecosystem function. The rate of primary production per unit biomass of producer organisms is high because of the small biomass, high productive capacity, and predominance of photosynthetic over nonphotosynthetic tissues of the producers present. Fluctuations of species populations are great, owing to the unbuffered effects of varying conditions of the physical environment. The poor representation of heterotrophs, resulting from the instability of physical and biotic conditions, is correlated with an incomplete utilization of the available net primary production. Thus, gross primary production exceeds total community respiration, permitting an increase of total biomass to occur within the system. The weak development of interspecific relationships such as predation, parasitism, and mutualism, which may function as density dependent controls on the species populations involved, contributes further to the biotic instability of the system.

These pioneer systems are thus less predictable in their behavior and more highly influenced by inputs and outputs resulting from the action of the physical environment and agents outside the ecosystem. Spatial variation in the structure and function of such systems is great, even in very similar environmental situations, and is closely related to the characteristics of surrounding ecosystems from which the pioneer species are derived (Margalef, 1968).

As succession proceeds a number of important trends occur in ecosystem characteristics. Diversity of species increases, with that for the producer level often beginning earlier than that for the dependent heterotrophs. Maximum species diversity among producers may actually be reached at an intermediate stage in succession (Odum, 1969). Increase in diversity is accompanied by the addition of new trophic levels to the system and by increases in the total organic biomass and level of primary production (Margalef, 1968). In terrestrial systems increase in biomass of producers is correlated with increase in the size of individual organisms and by increase in the ratio of nonphotosynthetic to photosynthetic tissues (Woodwell, 1967). As these changes occur, the ratio of gross primary production to total community respiration declines and the ratio of biomass to gross primary production (Biomass Accumulation Ratio) increases. However, at an intermediate point in succession the absolute difference between gross primary production and community respiration reaches a maximum, allowing the greatest rate of accumulation of organic matter in the system (Odum, 1969).

In theory these trends lead toward the development of systems of maximum organization, stability, and predictability (Margalef, 1963, 1968). As the total biomass of living organisms increases, and maximum utilization of environmental resources is approached, the intensity of interspecific competition for resources increases. A premium is thus placed on efficiency of utilization of resources and on the effectiveness of mechanisms of ecological isolation among species. Consequently, species diversity may decline as maximum maturity is approached because of the competitive displacement of poorly adapted species. Despite this, organization, in the sense of maximum efficiency and specialization in resource utilization, reaches a maximum in mature systems. Correlated with these resource relationships, dispersion patterns related to biotic interactions become more prominent. Intraspecific patterns, such as uniform dispersion of individuals due to competition, and interspecific patterns, such as positive and negative associations among species, reach highest frequencies.

In mature ecosystems the physical structure associated with the biomass of living organisms reaches maximum size and complexity. In terrestrial ecosystems this is seen in characteristics of height, density, and stratification of the vegetation. This development is accompanied by an increase in the ratio of nonphotosynthetic and support tissues to productive tissues. This, together with the increased representation of heterotrophs in mature systems, eventually results in an increase in community respiration in order to balance gross primary production. A metabolic balance between production and consumption is thus approached in mature systems (Woodwell, 1967).

These characteristics of mature systems are reflected in patterns of adaptation of the member species. The complex physical structure of the biotic portion of the system serves to buffer fluctuating conditions of the physical environment (Connell and Orias, 1964). The intricate biotic organization of the system leads to more effective biogeochemical and biodemographic regulation of biotic components of the system. These conditions permit the existence of species having narrow tolerances to extremes of physical factors and specialized patterns of resource utilization. Increased importance of interspecific relationships favors species having effective adaptations for competition, predator avoidance, and efficiency in energy utilization (Margalef, 1968; Cody, 1966; Salt, 1957). As a result, member species of mature ecosystems tend to show lower reproductive potentials, larger body size and lower metabolism per unit weight, longer life cycles, and greater survivorship. The increased buffering and regulation within mature systems also results in an extension of the period of productivity and activity during the annual environmental cycle (Dunbar, 1968). This leads to an increased representation of species that are permanent rather than migratory residents or that are active year-round (Cox, in preparation).

These changes can be seen even in simple successional sequences, such as those occurring in the nutrient cultures mentioned earlier (Margalef, 1968; Odum, 1969). With time, the species diversity in these systems increases, and the presentation of phytoplankton species possessing chlorophylls b and c, which have a lower potential production rate but operate more efficiently in combination with various accessory pigments, increases.

Thus, mature systems exhibit a higher degree of homeostatic regulation than do those of early successional status. Their behavior is more predictable and is less a function of inputs and outputs from the physical environment and surrounding ecosystems than of the characteristics of the system itself.

The contrasting structural and functional characteristics of ecosystems of different levels of maturity lead to a predictable pattern of interchange of energy and nutrients where such ecosystems occur adjacent to each other. In general, this pattern involves a net flow from the system of lower maturity to the system of higher maturity (Margalef, 1963). The net flow simply reflects the fact that there is a net export of organic food materials, with their energy and nutrients, from the less to the more mature system. This is made possible by the lower intensity of competition and the excess of production over consumption that characterize systems of low maturity. Conversely, the intense competition for resources, and the metabolic balance of production and community respiration, of mature systems favor species able to exploit the excess production of adjacent systems of low maturity. This may be accomplished in a variety of ways. Attached benthic forms in aquatic ecosystems frequently exploit the open water planktonic systems of lower maturity through mechanisms of filter-feeding. Likewise, many benthic species produce free-swimming larvae that exploit planktonic systems during their period of development. In terrestrial ecosystems similar patterns are seen in many vertebrates which carry out reproductive activities in more mature, stable ecosystems but forage in habitats of early successional status. To some degree this exploitation of less mature by more mature ecosystems must act to limit the rate of development shown by these systems of lower maturity (Margalef, 1963, 1968).

In describing the patterns of successional development and increase in system maturity little has been said about factors limiting the rate of succession or the maximum level of maturity achieved in particular situations. The increase in maturity of an ecosystem may be limited either by total available resources or by intense action of factors of the physical environment (Woodwell, 1967). Ultimately, the energy input to the system, in the forms of solar radiation and imported organic matter, sets a limit to development. Exhaustion of nutrients may, however, set lower limits to the level of maturity that can be achieved. The action of the physical environment may further restrict the potential level of maturity by limiting the extent to which the organisms present can modify and buffer fluctuations of the physical environment (Odum, 1962, 1969; Woodwell, 1967). For example, in open water ecosystems the absence of substrates for attachment and the instability of the medium prohibit the growth of plants of large size. This, in turn, prevents the biotic community from modifying currents within the system to a significant degree. Thus, the modifying effects of organisms on their local chemical environment tend to be cancelled out rapidly. As a result the successional sequence is of short duration and leads to a system of relatively low maturity (Odum, 1962). In moist, temperate terrestrial environments, on the other hand, the development of a dense forest vegetation makes possible continued modification of substrate and microclimate over long periods of time. These systems may thus reach a high level of maturity and exhibit significant successional change for upward of 1,000 years (Odum, 1962; Olson, 1958).

Human Influences on Ecosystem Maturity

The operation of these principles of ecosystem development is clearly seen in the effects of exploitation and pollution of world environments by man. Through agriculture, grazing of his animals, and lumbering he has depleted the nutrient capital of many systems. Through pollution and the use of fertilizers he has increased nutrient capital in other systems. Deliberate use of chemicals, such as pesticides, and the discharge of industrial and urban wastes have created new environmental factors that influence the structure and function of ecosystems over wide areas. To illustrate the impact of these activities on ecosystem processes, we will examine three patterns of ecosystem disruption:

1. The effects of radiation on open field and forest ecosystems.
2. The effects of shifting agriculture on the nutrient capital of tropical forest ecosystems.
3. The eutrophication of aquatic ecosystems.

Many other similar examples can be cited, but these three illustrate the diversity of ways in which man is capable of modifying the structure and function of complex ecosystems.

Effects of Radiation on Ecosystems. Woodwell (1967) has summarized an extensive series of studies dealing with the effects of gamma radiation on ecosystems of early and late successional status. These studies, conducted at the Brookhaven National Laboratory on Long Island, New York, are part of a series of similar experiments being conducted by various workers in order to determine the effects of atomic radiation on major ecosystem types.

The studies by Woodwell contrasted the effects of radiation on pioneer and late successional ecosystems in the field-to-forest successional sequence on Long Island (Fig. 11 -7). The pioneer system consisted of an abandoned agricultural field colonized by species of weedy plants during the first year after abandonment. The late successional system was an oak-pine forest typical of that developing on such an area 100–150 years after abandonment. Radiation sources were located in the center of experimental plots in each area. These sources were designed to deliver a radiation dosage varying from several thousand roentgens per day near the source to near-background levels at distances of about 300 meters from the source. Radiation effects were observed over a six month period of exposure.

The two systems showed striking differences in radiation sensitivity. Changes in species diversity along the radiation gradient within each sys-

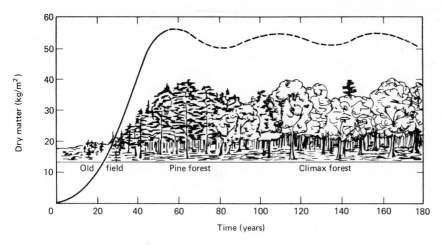

FIGURE 11–7. Field-to-forest succession in the eastern United States. The oscillations of the climax are assumed. (From Woodwell, 1967.)

tem proved to be one of the most consistent indices of radiation effect. These changes are similar to the patterns of retrogression, discussed in Chapter 7, in conjunction with biotic succession. After six months, a 50% reduction in species diversity occurred in the abandoned field at distances at which the radiation dosage was 1,000 roentgens per day. Some species, however, tolerated dosages over 3,000 roentgens per day. In contrast, a 50% reduction in species diversity in the forest ecosystem occurred at less than 160 roentgens per day, and no species survived at dosage levels exceeding 350 roentgens per day (Fig. 11–8). Thus, as evidenced by species diversity, sensitivity of the forest was about ten times that of the abandoned field.

The abandoned field ecosystem showed even greater resistance in terms of total productivity. At a dosage level of 1,000 roentgens per day, the standing crop of plant tissue at the end of the growing season was 800 grams per square meter, approximately double that shown by an unirradiated control plot. This occurred in spite of the 50% reduction in species diversity in the irradiated area. Thus, exclusion of certain radiation-sensitive species was more than compensated for by increased activity of those tolerant of the radiation dosage. In the forest plot standing crop declined at a rate parallel to that of species diversity, since the longer-lived, slow growing, woody species were unable to compensate as rapidly for the lost productivity of the radiation-sensitive species that were eliminated.

Radiation sensitivity of plant species in the two communities was found to be correlated with the overall life form of the plant and with the interphase chromosome volume of the cells. Species having prostrate

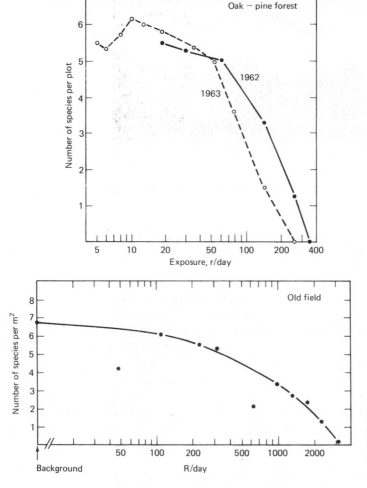

FIGURE 11–8. Species diversity along radiation gradients in the abandoned field and oak-pine forest ecosystems after six months exposure. (From Woodwell, 1967.)

growth form showed greater resistance, perhaps as a result of the lower exposure profile presented to the radiation source. Species having smaller interphase chromosome volumes were also more resistant (Fig. 11–9). These forms predominated in the abandoned field system. These observations suggested that sensitivity to radiation damage is closely related to sensitivity to other environmental factors that may act as mutagens or that may disturb basic cellular processes. Since environments colonized by pioneer species are characterized by extremes of physical factors, it is possible that these pioneer species have evolved characteristics specifically to reduce their sensitivity to such factors. Consequently, they may thus show low sensitivity to effects of gamma radiation.

Of major significance, however, was the high sensitivity shown by the

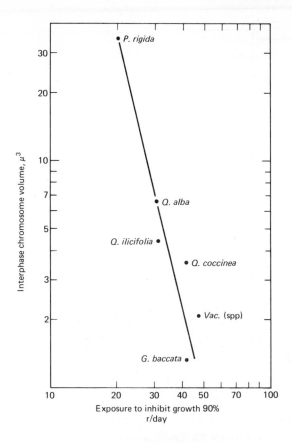

FIGURE 11-9. Relation between radiosensitivity, measured as inhibition of growth (90%), and interphase chromosome volume in the irradiated oak-pine forest. (From Woodwell, 1967.)

forest ecosystem. The complex biotic structure of the forest ecosystem, which contains populations of many species of heterotrophs and shows a high ratio of nonphotosynthetic to photosynthetic plant tissues (Fig. 11–10), in effect constitutes a mortgage on which annual photosynthetic payment must be made (Woodwell, 1967). Any unusual stress applied to the system is likely to reduce the capacity for production by the photosynthetic tissues of producers, while affecting heterotrophic components less. If this disturbance continues, the system will be returned to a state of lower maturity in which a lower photosynthetic mortgage is required.

Disturbance of mechanisms of biotic stability was evident in the forest ecosystem. Outbreaks of a number of herbivorous insects occurred, even in areas receiving light radiation dosages. In one case the abundance of aphids on oak leaves increased to about 200 times normal in areas receiving only 9.5 roentgens per day. These outbreaks suggest damage to biological control systems normally functioning to keep such organisms in check.

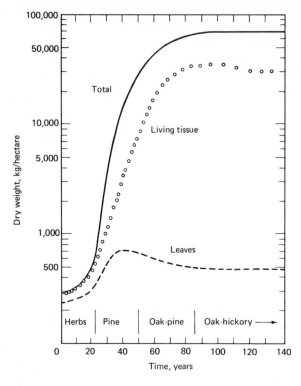

FIGURE 11–10. Approximate relations between total above-ground standing crop of plants, weight of living tissue, and weight of leaves in a normal field-forest succession of eastern North America. (From Woodwell, 1967.)

These studies illustrate the extent to which new environmental factors resulting from man's technology may cause the overall reorganization of major ecological systems. They further suggest that the systems most sensitive to disturbance may be those of greatest maturity and highest productivity.

Tropical Ecosystems, Fire, and Shifting Agriculture. Modification of the structure and function of major ecosystem types by man is not restricted to regions of advanced technological development. The impact of primitive agricultural practices on the vegetation of humid tropical areas clearly illustrates this fact. In the tropics, certain major vegetation types, although appearing natural, are actually the direct result of agricultural practices that have favored depletion of soil nutrient capital and have increased frequency of fire. These vegetation types, including especially savannas and pine forests, are maintained by recurrent fires.

Tropical forest ecosystems exhibit the highest degree of complexity of any major world ecosystem type. The soils on which these forests develop are nutrient-poor, largely because of the rapid rate of decomposition and intense leaching of nutrients from the surface soil which occur under the warm, moist climatic conditions. The major pool of nutrients in the system is thus the pool occuring in the tissues of living organisms.

Shifting tropical agriculture, or *slash and burn agriculture,* is a primitive agricultural system geared to these conditions of nutrient availability. This system has been evolved in a parallel manner by primitive peoples in different parts of the tropics. The slash and burn practice involves the cutting of trees on a small forest plot and their removal or, more frequently, their burning. The plot is then planted in crops for a period of two to three years. In most cases the stumps of cut trees are left in place, and the soil is only lightly tilled with hand implements. Burning of the felled timber serves as a technique of fertilization by making mineral nutrients available in the ash. The short period of use of these plots results from the rapid loss of nutrients by leaching, which causes a rapid drop in productivity. After the two to three years of use the plots are abandoned and undergo secondary succession.

Kowal (1966) has studied the impact of this agricultural system, together with the effects of more recent practices, on the vegetation of the Cordillera Central of Luzon in the Philippine Islands. This region, lying mostly between elevations of 1,000 and 2,400 meters, has a monsoon climate, with well-defined wet and dry seasons. Shifting agriculture, or *kaingin* as it is termed in the Philippine Islands, has been practiced since before the Spanish explorers first visited the islands.

The original vegetation of the Cordillera Central consisted of tropical lowland forest below elevations of about 1,000 meters, grading into montane forest and, at the highest elevations, mossy forest (Fig. 11–11). These three types of forest vegetation consist of broad-leafed tree species. The mossy forest is composed of very dense, dwarfed trees thickly covered with epiphytes. Patches of pine forest, which is successional to broad-leafed forest in most tropical montane regions where both occur, were probably restricted to local areas of steep slope and unstable soil.

Kowal concluded that the effect of kaingin agriculture on the native vegetation of level or gently sloping mountain land was probably not great. Since, in the clearing of agricultural plots, the stumps of cut trees are not removed and the dense network of roots not greatly disturbed by tilling, erosion tends to be slight. Resprouting of the cut stumps is rapid, and regrowth of the original forest occurs quickly after the plot is abandoned.

On steeper slopes, however, kaingin promotes erosion and fire. Heavy monsoon rains cause erosion of the surface soil and the exposure of the underlying parent material. The resulting surfaces, exposed and dry, then constitute a habitat that can only be invaded by grasses and pines tolerant of dry conditions. Succession on such areas usually leads to pine forest. Maintenance and expansion of these pine forests are favored by frequent ground fires of accidental and deliberate origin. These fires spread rapidly up steep slopes, killing most tree species of the broad-leafed forest but not seriously damaging the resistant pines. Kaingin

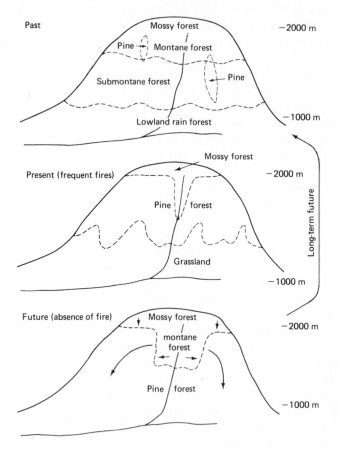

FIGURE 11–11. Probable vegetation of the Cordillera Central, Luzon, Philippine Islands; past, present, future. The diagrammed future of the vegetation is based on the assumption of absence of fire. (From Kowal, 1966.)

agriculture has thus resulted in the replacement of submontane and montane broad-leafed forest by pine forest over much of the Cordilla Central (Fig. 11–11). At the lowest elevations, the greater frequency of fires and other human disturbances has led to the elimination of all forest vegetation and to the creation of extensive grassland areas known as *kogonales.*

Since the turn of the century there has been an increasing trend toward the development of permanent gardens in the Cordillera Central because of the increasing demand for vegetable crops to supply expanding urban areas. Establishment of permanent gardens involves complete clearing of stumps, intensive tilling of the soil, and the use of artificial fertilizers. Farming of this type, in the absence of adequate means for the prevention of erosion, seems likely to intensify the processes described above and to lead to the complete replacement of broad-leafed montane forest by pine forest and grassland.

These vegetational changes again represent the replacement of systems of greater maturity by systems of lower maturity — a result of disturbance by man. Here the less mature systems, the pine forest and grassland communities, show a much greater degree of instability than does the broad-leafed forest. The variation of productivity from wet to dry season is greater in the pine communities, and they are subject to sudden changes in structure and productivity resulting from the occurrence of fire. The ultimate cause of these changes appears to be the depletion of nutrient capital and destruction of mature soil profiles by erosion. Reversal of these trends will require extensive changes in agricultural practices in the Cordillera Central, and even then recovery of soil nutrient levels will be very slow.

Eutrophication of Aquatic Ecosystems. Concurrent with the depletion of nutrient capital in certain terrestrial ecosystems, many aquatic ecosystems, into which these nutrients ultimately find their way, are undergoing enrichment. This process of enrichment, termed eutrophication, is a normal occurrence in the geological history of most lakes. When newly formed, almost all lakes possess a small nutrient capital and have a low productivity. Thus, they are oligotrophic. With time nutrients are accumulated from the watershed, sediments are built up in the lake bottom, and the nutrient capital of the lake increases. As this occurs, productivity of the lake also increases. These patterns constitute natural eutrophication, and they involve very slow changes in lake characteristics.

Eutrophication can be induced artificially by man through heavy inputs of nutrients in sewage wastes, fertilizers carried by runoff from agricultural land, and increased erosion. When continued inputs from these sources exceed the regulatory capacity of the biogeochemical cycles involved, major changes occur in ecosystem characteristics. Although these changes are related to increased levels of lake productivity, many are undesirable from the standpoint of multiple uses of aquatic ecosystems by man.

Some of the undesirable effects of eutrophication can be indicated by a comparison of physical and biotic conditions of oligotrophic and eutrophic temperate zone lakes. Because of their small nutrient capital oligotrophic lakes support a small biomass of planktonic organisms in the surface waters. Since much of the absorption of light, as it passes through water, is by these organisms, oligotrophic lakes are characterized by deep light penetration, and by a clear, blue appearance. The small biomass and low rate of production of plankton result in a low rate of sedimentation of dead organic matter into the bottom waters of the lake where decomposition occurs. Thus, the rate of decomposition in the bottom waters is low and does not lead to significant depletion of the oxygen content of the bottom water, even during the summer stagnation period when temperature-density stratification of the lake prevents transfer of oxygen from the surface to the bottom waters of the lake. Lakes of this type constitute the optimal habitat for a number of game fish, including various species of trout (*Salmo* spp.) and whitefish (*Coregonus* spp.). Consequently, these lakes possess many recreational attractions and, in addition, furnish water of high quality for various uses, including municipal, by man.

In contrast, in highly eutrophic lakes the biomass and productivity of planktonic organisms in the surface waters is high. Consequently, light penetration is low. Certain species of Blue-green Algae, especially *Oscillatoria rubescens,* together with various species of Diatoms, predominate in the phytoplankton. These species are apparently favored by the high nutrient concentrations and are good indicators of eutrophication (Sawyer, 1966). Species diversity among planktonic organisms, however, tends to be low. Moreover, the efficiency of utilization of much of the net production of phytoplankton by herbivorous zooplankton and fish is low. As a result, mats of algae may form on the surface and be blown onto the shore where decomposition occurs. The odor from this decomposing material may constitute a severe nuisance. Large quantities of dead algal material, as well as tissues and wastes from consumer trophic groups, may be sedimented to the lake bottom. Decomposition of this material in the bottom water may result in the depletion or complete removal of oxygen from this part of the lake during the period of summer stratification. Mortality of fish and bottom invertebrates may occur as a consequence.

These conditions favor a fauna of coarse fish, such as perch, catfish, and carp. The productivity of these fish populations is usually greater than that in oligotrophic lakes. Often, however, there is incomplete utilization of the net primary production by herbivores, and masses of algae accumulate and drift onto shore. In general, the changes induced by artificial eutrophication, although increasing the level of primary production, involve reduction of biotic diversity, reduced efficiency of resource utilization, and reduced stability of biotic and physical conditions. Thus, artificial eutrophication may be viewed as a reduction of maturity in the aquatic ecosystems involved.

The changes occurring with eutrophication frequently reduce the desirability of the lake for various recreational uses. In addition, some of the predominant species of algae release organic substances into the water that give municipal drinking water a bad taste. These effects, which are now occurring in many aquatic ecosystems, have led to the recognition of eutrophication as a serious environmental problem.

The first detailed documentation of the effects of artificial eutrophication was obtained for Lake Zurich in Switzerland (Hasler, 1947). This lake consists of two basins, the Obersee and Zürichsee, connected by a narrow channel (Fig. 11–12). Prior to 1900, both basins were typical oligotrophic lakes having clear waters and a complex fauna of salmonid fishes. Although the Obersee has changed little since then, the Zürichsee

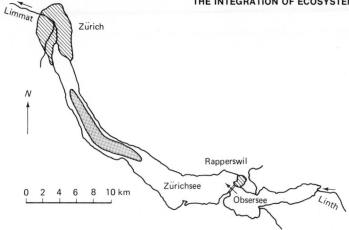

FIGURE 11–12. Lake Zurich, Switzerland, showing the positions of the Obersee and Zürichsee. (From Hasler, 1947.)

has undergone eutrophication because it has received sewage wastes from several communities located along its shore. Eutrophication of this basin has been accompanied by the occurrence of blooms of nuisance algae, reduction of light penetration, depletion of oxygen in the bottom waters, and replacement of almost all of the species of salmonids by coarse fish (Hasler, 1947). Similar patterns of eutrophication have subsequently been noted in many small lakes and reservoirs (Hasler, 1947; Sawyer, 1966).

Artificial eutrophication has now begun to affect very large bodies of water, e.g., the St. Lawrence Great Lakes (Beeton, 1965). At the present time, three of these lakes, Superior, Huron, and Michigan, are still clearly oligotrophic in character. One, Lake Erie, is eutrophic. Lake Ontario is intermediate in character. Because Lake Ontario is deeper than Erie, oxygen depletion of the bottom waters is not so easily effected.

In all lakes except Superior, however, increases in the total quantities of dissolved solids, some of which constitute important nutrients, have occurred in recent years (Fig. 11–13). Historical data accurate enough to show changes in the species composition of the plankton are not available for Lakes Huron, Superior, and Ontario. Several important changes in the occurrence and abundance of planktonic forms have been noted for Lake Michigan, however, and very extensive changes have occurred in Lake Erie (Beeton, 1965).

Lake Erie has been the most severely affected of the Great Lakes. Depletion of oxygen in the bottom waters during the summer is now severe. Formerly, such depletion did not occur. Now an estimated 70% of the central basin of the lake develops concentrations of 3 ppm or less. This level is inadequate to support certain fish and invertebrate species, which have consequently declined markedly in abundance. Nymphs of the mayfly, *Hexagenia*, have been largely replaced in the western part of the lake by midge fly larvae and tubificid worms adapted to water of low oxygen content (Fig. 11–14). The decline of these mayflies, which formerly emerged in immense quantities and constituted an important summer food and nutrient supply in terrestrial systems along the lake shore, may also have affected these ecosystems (W. D. Stull, personal communication).

The fish faunas of all of the Great Lakes have undergone major change in recent years. In several of the lakes this has been correlated with invasion of the sea lamprey. In Lake Erie, however, the sea lamprey has not been an important factor because of the small number of tributary streams offering suitable breeding areas to this species. Major changes in the fish fauna of this lake appear to relate to eutrophication processes and commercial exploitation of fisheries species. The composition of the commercial fish harvest in Lake Erie has shown an almost complete change since 1900 (Table 11–6). Several species that formerly contributed more than one million pounds annually to the commercial harvest have declined to positions of negligible importance, largely because of the failure of these species to reproduce (Beeton, 1965). Decline of these species has been parallelled by increases in freshwater drum, carp, yellow perch, and smelt, so that the present harvest is about 50 million pounds annually.

Thus, increases in nutrient capital, as well as decreases, may cause severe disturbance of ecosystem processes. This example clearly illustrates the system-wide nature of the responses of ecosystems to exploitation and pollution by man. It also indicates the broad scale on which major changes in ecosystem characteristics are occurring as a result of

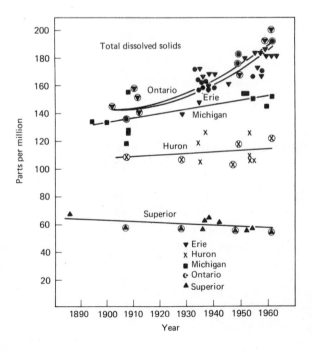

FIGURE 11–13. Concentrations of total dissolved solids in the Great Lakes from the late 1800's to 1961. (From Beeton, 1965.)

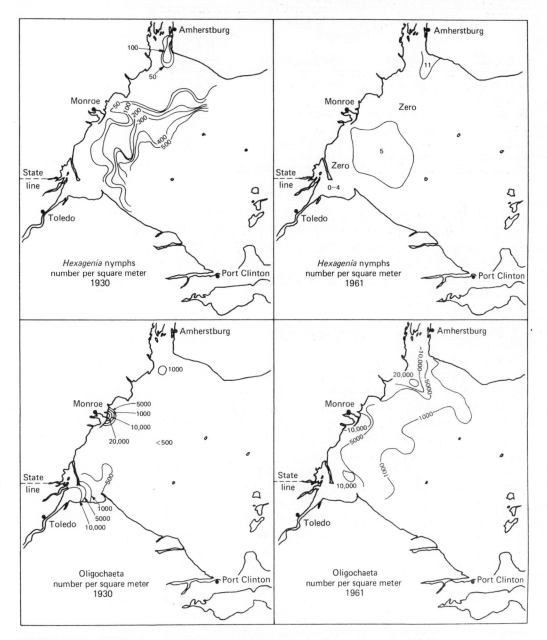

FIGURE 11–14. Distribution and abundance of *Hexagenia* mayfly nymphs and tubificid oligochaetes in western Lake Erie, 1930 and 1961. (From Beeton, 1965.)

TABLE 11-6. DECLINE IN COMMERCIAL HARVEST OF LAKE ERIE FISH SPECIES AS A RESULT OF LAKE EUTROPHICATION.

Species	Peak contribution to Lake Erie fisheries		Commercial catch in 1962 (pounds)
	Period	Catch (pounds)	
Lake herring (Cisco)	Prior to 1925	20–48.8 million	7,000
Whitefish	Prior to 1948	2 million	13,000
Sauger	Prior to 1946	1 million	1–4,000
Walleye	1956	15.4 million	1 million
Blue pike	Prior to 1958	15 million	1,000

very basic patterns of human activity. The immediate effects of these activities on the patterns of maturity shown by major ecosystem types are obviously great. Ecosystem maturity, however, is also affected by processes of evolutionary change, and the effects of human activities must be examined in this context as well. This problem will be dealt with in the following pages.

Evolutionary Patterns of Change in Maturity

The process of natural selection acts within the context of the ecosystem. As a result, evolutionary changes in characteristics of individual species populations are related not to single factors of the physical or biotic environment but to the entire complex of physical and biotic conditions within the ecosystem. Likewise, evolutionary change in individual species populations affects the entire ecosystem. It is in this sense that evolutionary change may be said to occur at the ecosystem level.

Since many of the important functional relationships within ecosystems, which constitute agents of natural selection, are related to conditions of ecosystem maturity, many of the resulting evolutionary processes are likewise related to this pattern. Selection operates to improve patterns of adaptation of species to the unique sets of conditions existing in ecosystems of different states of maturity. In addition to the action of natural selection in modifying characteristics of species already present in an ecosystem, the processes of geographic dispersal and speciation are continually making new species available for inclusion in the system.

The results of these processes are the varying patterns of ecosystem maturity and successional change seen in different environments and different geographical areas. A model of the integrated operation of these processes over geological time has been devised by Connell and Orias (1964). This model incorporates the action of factors both favoring and limiting increase in organization and maturity of ecosystems, and thus attempts to account for existing differences in these characteristics in different environments. Basically, these authors suggest that the maturity of an ecosystem is determined by the rate of energy flow through its food web, which is in turn related to the stability of environmental conditions and to the action of factors directly limiting energy input to the system. In examining the details of Connell and Orias' model, the reader should bear in mind that it is highly speculative and represents, in fact, a hypothesis of the manner in which ecosystem characteristics are influenced by evolutionary processes.

The basic relationships incorporated in this model are shown diagrammatically in Fig. 11–15. In describing this model we may start by as-

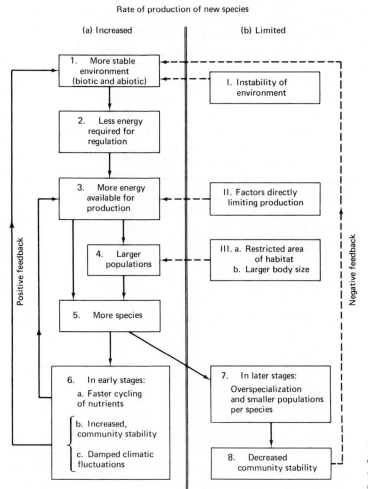

FIGURE 11–15. A model of the evolutionary development of maturity in ecological systems. (From Connell and Orias, 1964.)

suming the existence of two ecosystems differing in stability of physical and biotic conditions. In the more stable system it can be seen that less energy will be required by individual organisms for the regulation of their internal environment in the face of fluctuating external conditions. That is, a smaller fraction of their energy intake will be expended in the operation of mechanisms for countering the variable challenges, both physical and biotic, of the environment. Thus, a greater fraction of the food energy assimilated by these individuals, and by the species population to which they belong, will instead appear as new growth and newly reproduced offspring (Fig. 11–16). Thus, in the more stable environment, a given species will be able to have a higher net production, maintain greater population densities, grow to larger body size, or realize some combination of the above.

When this difference results in increased population densities, conditions favoring an increased rate of speciation may result. An increase in population density results in a larger gene pool, which can produce and hold a greater amount of genetic variability. Increased population densities also lead to increased importance of interspecific interactions such as predation, parasitism, and mutualism. These interactions become either more profitable or more likely as the population densities of the species involved increase. Finally, the greater efficiency of utilization of food by individuals means that minimum food requirements can be satisfied within a smaller area of habitat. Reduced mobility of individuals is thus favored. This in turn results in enhancement of the effectiveness of geographical barriers to dispersal. As mechanisms of geographical isolation become more effective, increased genetic divergence of the isolated populations occurs. Ultimately this change results in an increased rate of speciation in the more stable environment.

With increase in population densities or total biomass of species populations, and the addition of new species to the ecosystem, several posi-

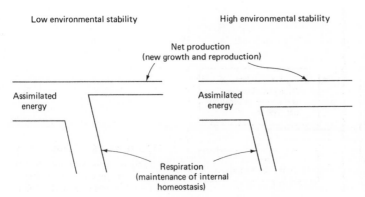

FIGURE 11–16. Partitioning of assimilated energy into net production and respiration by populations of a hypothetical species in environments of low and high stability.

tive feedback relationships assume importance. Persistence of forms produced by speciation is possible only if the forms demonstrate distinct and efficient patterns of resource utilization. Their persistence is thus indicative of a more rapid and complete utilization of food materials within the system and, thus, a more rapid return of nutrients to the primary producers of the system. This, in turn, may lead to an increased rate of production at the producer level. The addition of new species may also increase biotic stability by relationships involving food web complexity. Furthermore, increases in biomass and population densities will produce a more complex biotic structure that will further buffer fluctuations of the physical environment within the system. All of these changes reinforce the stable conditions originally present and favor further changes in the direction of increased maturity.

Obviously, limits exist to the degree of maturity that can be achieved in a given environment and to the rate at which evolutionary increase in maturity can occur. Instability of the physical environment may be. so great, as in the case of the desert, that little buffering of conditions can be achieved. The total energy input to the ecosystem may be low, as in the case of cave or deep water systems. There may exist severe restrictions on the size of the area that can be occupied by an ecosystem type, as in the case of island areas, thus limiting the sizes of gene pools of the member species. Moreover, an increase in the fraction of assimilated energy, which may appear as net production, may, in some species, appear as increased body size of individuals rather than increased numbers of individuals. As a result no increase in size of the species' gene pool may occur.

More importantly, limits to the increase in number of species may be set by the specialization in resource utilization accompanying their integration into the ecosystem. At some point increased specialization may balance increase in species number, so that no net increase in complexity of the food web occurs and no increase in biotic stability is produced.

Finally, it should be noted that the degree of maturity shown by a particular ecosystem type reflects to some degree the length of geological time over which the above process has been operating (Dunbar, 1968). Conditions of low maturity in environments such as the Arctic may, in part, reflect the recency of serious disturbances such as Pleistocene glaciation. Recently, observations such as these have given rise to the *stability-time hypothesis* of species diversity (Sanders, 1968, 1969; Whittaker, 1969). This hypothesis suggests, first, that in environments that possess favorable and constant conditions, organisms will be forced to deal primarily with biological challenges. Through evolutionary time, co-adjustment of species to each other occurs and new species, through evolutionary acquisition of specializations, are able to enter the system. Diversification in this manner is self-augmenting; the addition of species makes possible still other ways of life for additional new invaders.

Sanders (1968) has illustrated the results of this process by examples from marine and freshwater faunas. Lake Baikal, USSR, is the deepest, and, with an age of between 1 and 30 million years, one of the oldest freshwater lakes in the world. The deeper waters of this lake are nearly constant in physical and chemical conditions. Lake Baikal possesses a fauna of 240 species of Gammarid Amphipods, almost all of which are endemic to the lake. In all of North America, where lakes are all shallow and recent, only 28 species of Gammarid Amphipods occur. It appears that this accumulation of related species in Lake Baikal can only be explained by the combination of stable conditions and long evolutionary history.

Evolutionary, as well as successional, trends toward increased maturity thus may be seen in ecosystems. These trends result from processes of natural selection which favor species characteristics leading to increased efficiency of resource utilization. In addition to producing disturbances of ecological systems existing at the present time, the exploitation and pollution of the environment by man have modified conditions under which these evolutionary processes are occurring. In maintaining agricultural communities of low species diversity and high productivity, he has placed an evolutionary premium on precisely those characteristics important in pest species: effective dispersal mechanisms, high reproductive potential, broad feeding habits, rapid growth, and resistance to artificial chemical controls employed by man (Geier and Clark, 1960). He has thus created conditions favoring evolutionary adaptation of species to systems of low maturity by increasing the extent and permanence of such systems. At the same time, ecosystems of high maturity are being severely damaged or destroyed. These evolutionary implications of man's use of the environment are likely to provide the most severe test of his ability to establish a harmonious working relationship with the natural world.

SYSTEMS ECOLOGY AND THE ANALYSIS OF ECOSYSTEM FUNCTION

The magnitude and complexity of the relationships operating at the ecosystem level of integration require investigational tools of a new order of capability. Systems ecology is the developing branch of ecology concerned with the dynamics of complex ecological systems. Although difficult to define because of the different characteristics of the systems studied, systems ecology has as its goal the development of predictive mathematical models of ecological systems. This approach utilizes many of the techniques developed in business and industry for the analysis of complex operations. The systems approach is currently being applied to problems ranging from predator-prey interactions (Holling, 1959, 1964) to the nutrient and energy flow processes of major ecosystem types (Van Dyne, 1966).

To illustrate the application of the systems approach to the study of ecosystem processes, we will examine the design of current studies, developed in connection with the International Biological Program, aimed at functional analysis of major ecosystem types. The objective of these studies is to develop a model of, for example, the grassland ecosystem, which will permit prediction of the effects of various kinds of human manipulation on the quantities of nutrients and energy existing in different components of the ecosystem and on the rates of transfer of nutrients and energy among these components. Carbon, nitrogen, phosphorus, and water represent some of the basic nutrients for which such predictive models will be attempted.

The first step in such a study is to define the components of the system to be studied. These components, whether biotic or abiotic, must be parts of the system within which energy or nutrients share some common feature of origin or chemical form and between which significant transfers occur. The number of components recognized depends on the degree of detail desired, or possible, in the analysis of function. Almost all ecosystem studies will begin with an examination of a few major components, defined in broad terms. For example, the major biotic components may be defined as the various trophic groups of producers, herbivores, carnivores, and decomposers. Later, as a general understanding of ecosystem function is achieved, resolution of the system into components corresponding to individual species may be attempted.

The second step in this analysis is to determine the state conditions for the various ecosystem components. These state conditions correspond to the measurable characteristics of ecosystem components in a given situation. These include the standing crop of energy or nutrient material in the ecosystem component at a given point in time. In an analysis of ecosystem energetics, for example, state conditions would refer to the caloric value of the organic matter constituting various living (producers, herbivores, etc.) and nonliving (humus, leaf litter, etc.) components of the ecosystem.

The third step is to measure the rates of input and output of energy or nutrients across the arbitrary ecosystem boundary and to measure the rates of transfer among components within the system. These measurements describe the total set of processes that result in the state conditions previously determined. A schematic diagram of how this information may be identified and stored is given in Fig. 11–17.

To further illustrate the nature and relationship of the information obtained to this point, let us examine a simple, hypothetical ecosystem (Smith, 1970). In this system we shall recognize three components: the environment (E), producers (P), and herbivores (H) (Fig. 11–18). Of the

System state

Components	1	2	3	4	5.................	N
State condition	X_1	X_2	X_3	X_5	X_5	X_N

System processes

Inflows	a_1	a_2	a_3	a_4	a_5	a_N
Matrix of transfer rates	—	y_{21}	y_{31}	y_{41}	y_{51}	y_{N1}
	y_{12}	—	y_{32}	y_{42}	y_{52}	y_{N2}
	y_{13}	y_{23}	—	y_{43}	y_{53}	y_{N3}
	y_{14}	y_{24}	y_{34}	—	y_{54}	y_{N4}
	y_{15}	y_{25}	y_{35}	y_{45}	—	y_{N5}
	$\vdots$	$\vdots$	$\vdots$	$\vdots$	$\vdots$	$\vdots$
	y_{1N}	y_{2N}	y_{3N}	y_{4N}	y_{5N}	—
Outflows	z_1	z_2	z_3	z_4	z_5	z_N

FIGURE 11–17. System for storage of information on state conditions of various ecosystem components, and on inflow, outflow, and transfer rates for each component or component combination.

various possible system processes, one inflow (a_E), two outflows (z_E, z_H), and four transfer rates (y_{EP}, y_{PE}, y_{PH}, y_{HE}) are assumed to have significant values. Figure 11–18 thus illustrates how measured values for state conditions and for these processes can be identified and stored according to the system presented in Fig. 11–17.

Following the measurement of ecosystem processes corresponding to a given set of state conditions, each of these processes must be analyzed as a function of various other components and conditions of the ecosystem. This represents the first step in the mathematical modeling of the system. Evaluation of a given process as a function of other conditions cannot, however, be done by observation of the system at only one point in time. By observing the system at different times, when state conditions and system processes have different values, such an analysis can be obtained by using techniques of correlation or regression analysis. In making these repeated observations experimental modification of conditions of the ecosystem can provide useful information. The biomass of various components may be modified by introduction or removal of individual organisms. Nutrient conditions may be modified by changing the inflow or outflow of materials across the system boundary, using techniques such as irrigation and fertilization. These experimental modifications may range from very local and temporary *micromanipulations* to changes of greater extent and longer duration, or *macromanipulations*. These experimental modifications may allow observation of a wider range of state

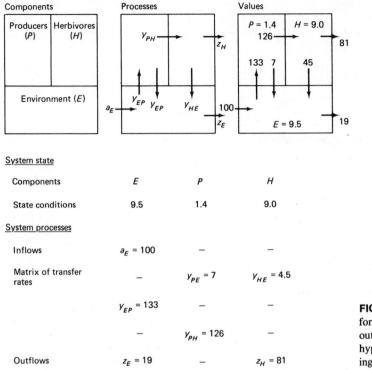

System state			
Components	E	P	H
State conditions	9.5	1.4	9.0

System processes			
Inflows	$a_E = 100$	–	–
Matrix of transfer rates	–	$y_{PE} = 7$	$y_{HE} = 4.5$
	$y_{EP} = 133$	–	–
	–	$y_{PH} = 126$	–
Outflows	$z_E = 19$	–	$z_H = 81$

FIGURE 11–18. Storage of data for state conditions and for inflow, outflow, and transfer rates for a hypothetical ecosystem consisting of three components.

and process values than would otherwise be seen and, thus, give a more complete picture of the functional relationships of specific processes.

To illustrate how such manipulations may be utilized to allow the mathematical evaluation of specific processes as functions of other ecosystem conditions, let us further develop the simple ecosystem model presented earlier (Fig. 11–18). Specifically, let us assume that we are able to manipulate the input (a_E) of this system in order to reduce it to one-fourth the normal value, on one hand, and increase it to four times the normal value, on the other (Fig. 11–19). Examination of the values of various transfer and output processes at the three different levels of input reveals a series of consistent mathematical relationships. For example, the output per unit time from the environmental component (z_E) is consistently twice the amount present in this component at any given time. The output by emigration of herbivores (z_H) is equal to the square of the herbivore standing crop at any given time. Likewise, in this hypothetical example the various transfer processes may be evaluated as simple mathematical functions of the state conditions of the various system components (Fig. 11–19).

A model of steady-state conditions for the three components may be

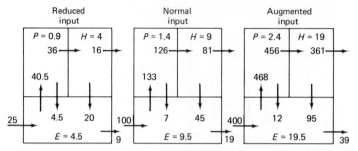

Analysis of system processes as functions of system conditions

y_{EP} (Producer uptake) $\quad = 10EP$

y_{PE} (Producer excretion) $\quad = 5P$

y_{PH} (Herbivore grazing) $\quad = 10PH$

y_{HE} (Herbivore excretion) $\quad = 5H$

z_E $\quad$ (Environmental output) $\quad = 2E$

z_H $\quad$ (Herbivore emigration) $\quad = H^2$

Model for steady state conditions of various components

$$\frac{dE}{dt} = a_E + 5P + 5H - 10EP - 2E = 0$$

$$\frac{dP}{dt} = 10EP - 5P - 10PH = 0$$

$$\frac{dH}{dt} = 10PH - 5H - H^2 = 0$$

FIGURE 11–19. Development of model for steady-state conditions of ecosystem components from data on state conditions, input, output, and transfer rates in systems having normal, reduced, and augmented input levels.

developed from these mathematical relationships (Fig. 11–19). This modeling process is valuable in several ways. First of all, it is capable of predicting the steady-state structure of the system at different levels of input and output. Secondly, it can provide information on the specific relationships within the system which are of greatest importance in determining the behavior of the entire system. In study of an actual system this procedure is equivalent to defining these relationships that must be measured most accurately to allow formulation of an improved model. This technique is termed *sensitivity analysis*. Table 11–7 gives a sensitivity analysis for the hypothetical ecosystem model in Fig. 11–19. Each system process in Fig. 11–19 was represented by an expression with a certain coefficient associated with it. Producer uptake, y_{EP}, for example, was estimated to equal $10EP$. In working with any real ecosystem it is apparrent that the coefficient, in this case equal to 10, would be a system parameter of which we would have only an estimate. Thus, one question that might be posed, once a preliminary model has been developed, concerns the extent to which a certain degree of error in estimation of each of the parameters involved affects the overall model.

TABLE 11-7. SENSITIVITY ANALYSIS FOR THE HYPOTHETICAL ECOSYSTEM MODEL IN FIG. 11-19.*

| System process | Parameter | Component | | | Sum | Relative sensitivity |
		Water	Plant	Herbivore		
z_E	c_1	-0.100	-0.079	-0.106	0.285	0.14
z_H	c_2	-0.426	+0.354	-0.405	1.185	0.59
y_{EP}	c_3	-0.900	+0.068	+0.106	1.074	0.53
y_{PE}	c_4	+0.047	-0.004	-0.006	0.057	0.03
y_{PH}	c_5	+0.853	-1.064	-0.100	2.017	1.00
y_{HE}	c_6	0.000	+0.357	0.000	0.357	0.18

*Modified from Smith, 1970.

Note: The body of the table gives the percentage errors in predicted equilibrium standing crops in the three system components for an overestimation of 1% in each parameter. The sum is taken without respect to sign. The relative sensitivities are obtained from the sums by setting the largest equal to 1.0.

The results in Table 11-7 show the percentage errors predicted for the amounts of material in the three components of the system assuming a 1% overestimate of the parameter associated with each system process. From this analysis it can be seen that errors in the parameter for y_{PE} produce little change in the estimates of E, P, and H. Errors in the parameter for y_{PH} have a strong effect, however, indicating that much care must be taken to obtain as accurate an estimate of this parameter as possible.

Finally, by means of a model of this kind, the effects of factors with which actual experimentation is impossible or unfeasible may be explored (Fig. 11-20). The modeling approach, however, works most effectively when predictions based on a current version of a system model are compared to actual conditions or experimental results, and a revised model developed as a result.

From the complexity of the expressions resulting from the very simple model used in the above illustration it is apparent that study of real ecosystems will require complex analytical procedures, together with computer storage and data processing. Even with the availability of these tools, the theoretical question remains: "Will the measurement of processes involving exchange of energy and a few basic nutrients among a relatively small number of ecosystem components provide a model of actual ecosystem function having useful predictive power?"

There have been few models developed to a point where a fair answer to this question can be given. We will briefly examine one example, however. This model is one developed for nutrient budgets of moist tropical forests in Panama in connection with studies of possible construction of a sea-level canal by nuclear excavation (McGinnis et al., 1969; Golley

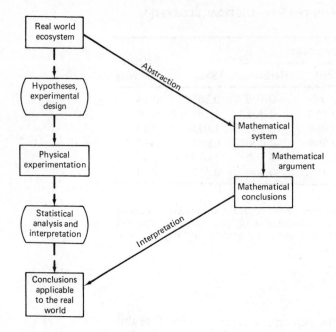

FIGURE 11–20. Two ways of experimenting with ecosystems. One involves the conventional process of formulating hypotheses, designing and conducting experiments, and analysis and interpretation of results. The second involves the abstraction of the system into a model, application of mathematical argument, and interpretation of mathematical conclusions. (From Van Dyne, 1966.)

et al., 1969). Specifically, the objective of these studies was to evaluate the potential significance of various transfer pathways that might carry radioactive fallout materials to man.

The moist forest system was resolved into the series of components (compartments) shown in Fig. 11–21. In this figure the transfer routes that are likely to be followed by mineral nutrients are designated. Measurements were made of the quantities of 13 different mineral elements in each component and estimates obtained of the rates of transfer between compartments. These data are summarized in Table 11–8. In this table the *transfer coefficient* represents the only mathematical description of the transfer process available. This coefficient is simply the daily fractional amount moving out of a compartment by a particular route. These data were, of course, obtained for only a single-system state, rather than for different states (corresponding to the three input levels) in the simple model discussed earlier.

The information that may be obtained from this model is indicated mainly in the column giving turnover times for the various compartments. If we assume that a significant input of radioisotopes occurred, these turnover times suggest the pattern of movement that would occur. The very short turnover times for the animal components of the system suggest that these would quickly become contaminated. Likewise, the long turnover time for the fruit and flower components suggest that contamination of this component would occur only slowly and, thus, would probably not constitute a significant route to man for any of the short-lived radio-isotopes. The very long half-life for the soil compartment suggests that this would constitute the major *sink* for environmental contaminants such as isotopes of the mineral elements.

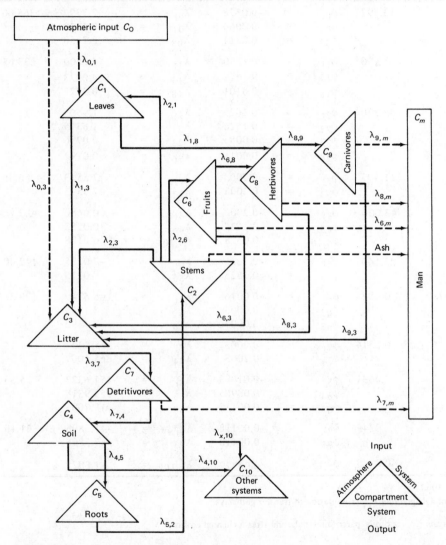

FIGURE 11–21. Compartmentalized model for describing the movement and standing crops of mineral nutrients in a moist tropical forest in Panama. Triangles represent storage compartments, and lines connecting triangles depict pathways. Heavy dashed lines indicate major modes of atmospheric input, solid lines represent normal cycling pathways, and light dashed lines indicate pathways of forest materials utilized by man. (From McGinnis et al., 1969.)

TABLE 11-8. STANDING CROP (CONTENT) AND TRANSFER RATES BETWEEN COMPONENTS FOR MINERAL NUTRIENTS IN MOIST TROPICAL FORESTS OF EASTERN PANAMA.*

	Compartment	Content grams/m^2 C_1			Transfer flux gm/m^2/day			Transfer coefficient		Turnover time-days
0.	Atmospheric	13.797	ϕ_{0j}	=	-0.0378	λ_{0j}	=	−	2.7397†	365.00
			ϕ_{01}	=	0.0064	λ_{01}	=		0.4639	
			ϕ_{03}	=	0.0314	λ_{03}	=		2.2758	
1.	Leaves	37.150	ϕ_{1j}	=	-0.1590	λ_{1j}	=	−	4.2800	233.65
			ϕ_{13}	=	0.1489	λ_{13}	=		4.0073	
			ϕ_{18}	=	0.0101	λ_{18}	=		0.2727	
2.	Stems	83.320	ϕ_{2j}	=	-0.1625	λ_{2j}	=	−	1.9503	512.74
			ϕ_{21}	=	0.1526	λ_{21}	=		1.8315	
			ϕ_{23}	=	0.0083	λ_{23}	=		0.0996	
			ϕ_{26}	=	0.0016	λ_{26}	=		0.0192	
3.	Litter	11.407	ϕ_{3j}	=	-0.2003	λ_{3j}	=	−	17.5593	56.95
			ϕ_{37}	=	0.2003	λ_{37}	=		17.5593	
4.	Soil, 0–30 cm	2,943.000	ϕ_{4j}	=	-0.2003	λ_{4j}	=	−	0.0680	40.2 yrs
			ϕ_{45}	=	0.1625	λ_{45}	=		0.0552	
			$\phi_{4,10}$	=	0.0378	$\lambda_{4,10}$	=		0.0128	
5.	Roots	23.140	ϕ_{5j}	=	-0.1625	λ_{5j}	=		-7.0225	142.40
			ϕ_{55}	=	0.1625	λ_{55}	=		7.0225	
6.	Fruits and flowers	0.287	ϕ_{6j}	=	-0.0016	λ_{6j}	=		-5.5749	179.38
			ϕ_{63}	=	0.0009	λ_{63}	=		3.1359	
			ϕ_{68}	=	0.0007	λ_{68}	=		2.4390	
7.	Detritivores	0.2314	ϕ_{7j}	=	-0.2003	λ_{7j}	=		-865.6007	1.16
			ϕ_{74}	=	0.2003	λ_{74}	=		865.6007	
8.	Herbivores	.0631	ϕ_{8j}	=	-0.01083	λ_{8j}	=		-171.6323	5.83
			ϕ_{83}	=	0.00965	λ_{83}	=		152.9318	
			ϕ_{89}	=	0.00118	λ_{89}	=		18.7005	
9.	Carnivores	.0249	ϕ_{9j}	=	-0.00118	λ_{9j}	=		-47.3896	21.10
			ϕ_{93}	=	0.00118	λ_{93}	=		47.3896	
10.	Other systems	40.3325	$\phi_{x,10}$	=	0.00727‡	$\lambda_{x,10}$	=		1.8025	

*From McGinnis et al., 1969.

Note: See Fig. 11-21 for a diagrammatic representation of transfer pathways.

†Coefficient: $\times 10^{-3}$

‡Contribution to stream from subsoils, parent materials, and stream channel erosion.

REFERENCES

Beeton, A. M. 1965. Eutrophication of the St. Lawrence Great Lakes. *Limnology and Oceanography,* **10**:240–54.

Brewer, R. 1963. Stability in bird populations. *Occas. Papers Adams Ctr. Ecol. Studies,* No. 7. 12 pp.

Chitty, D. 1960. Population processes in the vole and their relevance to general theory. *Can. J. Zool.,* **38**:99–113.

Christian, J. J., and D. E. Davis. 1964. Endocrines, behavior, and population. *Science,* **146**:1550–60.

Cody, M. L. 1966. A general theory of clutch size. *Evolution,* **20**:174–84.

Cole, L. C. 1954. Some features of random cycles. *J. Wildl. Manag.,* **18**:2-24.

Connell, J. H., and E. Orias. 1964. The ecological regulation of species diversity. *Amer. Nat.,* **98**:399–413.

Cox, G. W. 1967. *Laboratory Manual for General Ecology.* Dubuque, Iowa: Wm. C. Brown. 165 pp.

————. In preparation. Species diversity, biotic succession, and the evolution of migration.

Dunbar, M. J. 1968. *Ecological Development in Polar Regions.* Englewood Cliffs, N.J.: Prentice-Hall. 119 pp.

Einsele, W. 1941. Die Umsetzung von zugefuhrtem, anorganischem Phosphat im eutrophen See und ihre Ruckwirkungen auf seinen Gesamthaushalt. *Zschr. Fisch.,* **39**:407–88.

Elton, C. S. 1958. *The Ecology of Invasions by Animals and Plants.* London: Methuen. 181 pp.

Geier, P. W., and L. R. Clark. 1960. An ecological approach to pest control. *Symposium 8th Tech. Meeting Int. Union for Cons. of Nature and Nat. Res.,* Warsaw, 1960. E. J. Brill, Leiden. pp. 10–18.

Golley, F. B., J. T. McGinnis, R. G. Clements, G. I. Child, M. J. Duever. 1969. The structure of tropical forests in Panama and Colombia. *BioScience,* **19**(8): 693–96.

Hairston, N. G., J. D. Allen, R. K. Colwell, D. J. Futuyma, J. Howell, M. D. Lubin, J. Mathias and J. H. Vandermeer. 1968. The relationship between species diversity and stability: An experimental approach with protozoa and bacteria. *Ecology,* **49**:1091–1101.

Hasler, A. D. 1947. Eutrophication of lakes by domestic drainage. *Ecology,* **28**:383–95.

Holling, C. S. 1959. The components of predation as revealed by a study of small mammal predation on the European pine sawfly. *Can. Ent.,* **91**:293–320.

————. 1964. The analysis of complex population processes. *Can. Ent.,* **96**: 335–47.

Hutchinson, G. E. 1948. Circular causal systems in ecology. *Ann. N. Y. Acad. Sci.,* **50**:221–46.

————. 1959. Homage to Santa Rosalia, or why are there so many kinds of animals? *Amer. Nat.,* **93**:145–59.

Kowal, N. E. 1966. Shifting agriculture, fire, and pine forest in the Cordillera Central, Luzon, Philippines, *Ecol. Monog.,* **36**:389–419.

MacArthur, R. 1955. Fluctuations of animal populations, and a measure of community stability. *Ecology,* **36**:533–36.

———. 1960. On the relative abundance of species. *Amer. Nat.,* **94**:25–36.

———, **and E. O. Wilson.** 1967. *Island Biogeography.* Princeton, N.J.: Princeton University Press. 203 pp.

Margalef, R. 1963. On certain unifying principles in ecology. *Amer. Nat.,* **97**:357–74.

———. 1968. *Perspectives in Ecological Theory.* Chicago: University of Chicago Press. 111 pp.

———. 1969. Diversity and stability: A practical proposal and a model of interdependence. *Brookhaven Symposia in Biology,* **22**:25–37.

McGinnis, J. T., F. B. Golley, R. G. Clements, G. I. Child, and M. J. Duever. 1969. Elemental and hydrologic budgets of the Panamanian tropical moist forest. *BioScience,* **19**(8):697–700.

Odum, E. P. 1962. Relationships between structure and function in the ecosystem. *Japanese J. of Ecol.,* **12**:108–18.

———. 1969. The strategy of ecosystem development. *Science,* **164**:262–70.

Olson, J. S. 1958. Rates of succession and soil changes on southern Lake Michigan sand dunes. *Bot. Gaz.,* **119**:125–70.

Pearson, O. P. 1966. The prey of carnivores during one cycle of mouse abundance. *J. of Animal Ecol.,* **35**:217–33.

Pimentel, D. 1961. Species diversity and insect population outbreaks. *Ann. Ent. Soc. Amer.,* **54**:76–86.

Pitelka, F. A. 1964. The nutrient-recovery hypothesis for Arctic microtine cycles. I. Introduction. In D. J. Crisp (ed.), *Grazing in Terrestrial and Marine Environments.* Brit. Ecol. Soc. Symposium No. 4, pp. 55–56.

Salt, G. W. 1957. An analysis of avifaunas in the Teton Mountains and Jackson Hole, Wyoming. *Condor,* **59**:373–93.

Sanders, H. L. 1968. Marine benthic diversity: a comparative study. *Amer. Nat.,* **102**:243–82.

———. 1969. Benthic marine diversity and the stability-time hypothesis. *Brookhaven Symposia in Biology,* **22**:71–81.

Sawyer, C. N. 1966. Basic concepts of eutrophication. *J. Water Poll. Cont. Fed.,* **38**:737–44.

Schultz, A. 1964. The nutrient-recovery hypothesis for Arctic microtine cycles. II. Ecosystem variables in relation to Arctic microtine cycles. In D. J. Crisp (ed.), *Grazing in Terrestrial and Marine Environments.* Brit. Ecol. Soc. Symposium No. 4, pp. 57–68.

———. 1969. A study of an ecosystem: The arctic tundra. In G. Van Dyne (ed.), *The Ecosystem Concept in Natural Resource Management,* New York: Academic Press. pp. 77–93.

Smith, F. E. 1970. Analysis of ecosystems. In D. E. Reichle (ed.), *Analysis of Temperate Forest Ecosystems,* New York: Springer-Verlag. pp. 7–18.

Van Dyne, G. M. 1966. *Ecosystems, Systems Ecology, and Systems Ecologists.* Oak Ridge National Laboratory Report 3957, pp. 1–31.

Watt, K. E. F. 1965. Community stability and the strategy of biological control. *Can. Ent.,* **97**:887–95.

Whittaker, R. H. 1969. Evolution of diversity in plant communities. *Brookhaven Symposia in Biology,* **22**:178–96.

Williams, E. E. 1969. The ecology of colonization as seen in the zoogeography of anoline lizards on small islands. *Quart. Rev. Biol.,* **44**(4):345–89.

Woodwell, G. M. 1967. Radiation and the patterns of nature. *Science,* **156**:461–70.

Wurster, C. F., Jr. 1969. Chlorinated hydrocarbons and the world ecosystem. *Biol. Cons.,* **1**:123–29.

Part VI

CONCLUSION

Chapter 12
THE FUTURE OF ECOLOGY

The environmental crisis has placed ecology in a prominent position. The factors responsible for this are unique in the history of science. Historically, attention has usually become focused on a particular area as a consequence of major breakthroughs in knowledge in that area — recent major examples being in the fields of atomic physics and molecular biology. The current prominence of ecology has quite a different basis. It has arisen from a general concern over environmental deterioration, coupled with a conviction among professional ecologists that they now possess an approach that can lead to an understanding of basic environmental function. This approach — in a sense the rallying point for modern ecology — centers on the concept of the ecosystem (Van Dyne, 1966). Ecologists are only beginning to employ this approach on a scale comparable to that at which human interventions in the environment are occurring. The ecosystem approach, however, is the dominant force in modern ecology, and it is exerting strong influence on other branches of the natural and social sciences. The fruits of the ecosystem approach have yet to be realized in a major way, but its impacts appear certain to influence the development of science and the activities of society in the future.

The need for an understanding of the integrated function of complex ecosystems has become an immediate one. Human technology is resulting in rapidly growing disturbance of terrestrial, freshwater, and marine ecosystems. The multiple-use demands on forest and rangelands are increasing. Agriculture is rapidly becoming a large scale, mechanized activity in which artificial techniques of cultivation, fertilization, and pest control are employed intensively over hundreds of thousands of acres. Fisheries resources are approaching maximum limits of exploitation, or even being over-exploited, worldwide. Industrial development has spawned plans to exploit water, minerals, and other resources and to develop new sources

549

of energy on a scale of magnitude and order greater than now known. Explosive growth of systems of transportation is occurring. The growth and urbanization of human populations is bringing about uncontrolled growth and sprawl of cities and the accompanying loss of farmlands, marshes, and wild areas. Increasingly mechanized recreational activities are defacing deserts, mountains, and other natural areas. The ecological effects of these activities in many cases extend to regional, or even global, aspects of climate, air and water quality, land and water productivity, and environmental esthetics.

In recognition of the magnitude of these problems ecologists have initiated a series of comprehensive research projects aimed at understanding the dynamics of entire ecosystems. Many of these have originated as projects of the International Biological Program (IBP). The IBP, begun in 1966, includes a variety of specific programs of research. One of the most active of these is a Biome Studies Program aimed at understanding the dynamics of representative types of ecosystems. Six major ecosystem types have been designated for intensive study. These include the arctic tundra, the coniferous forest, the temperate deciduous forest, the temperate grassland, the desert, and the tropical forest. Within each biome attention is to be directed both at the terrestrial system itself and the aquatic systems represented within its geographical boundaries. Almost all of these biome projects are now under way. The objectives of these studies include development of a predictive model of ecosystem structure and dynamics. These models will be mathematical in form and are intended to be capable of predicting the response of entire ecosystems to general environmental impacts that have not yet been experienced by the system, as well as to impacts that have been tested experimentally. Models of this capability would serve several purposes. First, they would enable prediction of the probable effects of proposed human interventions in the environment. Second, they would enable evaluation of possible remedies for disruptions that have already occurred. Third, they may suggest patterns of management that may enhance productivity and environmental quality.

Application of this approach has had a far-reaching effect on ecology. It has stimulated the training of a body of ecologists skilled in the techniques of studying complex systems. This training includes familiarization with calculus, statistics, and the use of computer techniques of data analysis and modelling, as well as with basic principles of ecology. The need for individuals having such a background will be a continuing one.

As attempts to model complete ecosystems have begun, several areas of weakness in ecological knowledge have also become apparent. These gaps in knowledge are already stimulating the growth of particular areas of research. For example, one of the difficulties encountered by workers attempting to build preliminary models was ignorance of processes of decomposition in nature. This has stimulated increased research into the quantitative roles of bacteria, fungi, and various invertebrates in decomposition processes. Preliminary modelling attempts have also demonstrated the need for more comprehensive and detailed studies of the physiological responses of organisms to conditions of their environment. The need for physiological ecologists will increase as modelling attempts become more sophisticated. One specific indication of increased activity in this area has been the rapid growth of interest in chemical ecology. We now know that many ecological interactions are governed by the presence of complex chemical substances in the tissues of the organisms involved or by the release of these materials into the environment. The study of such chemical relationships is rapidly becoming an important branch of physiological ecology.

Attempts to understand the integrated function of entire ecosystems have also stimulated interest in evolutionary ecology. It is now known that considerable evolutionary adjustment can occur by species subjected to intense selective pressures, even over relatively short time periods. Furthermore, analysis of co-evolutionary patterns among various groups of organisms is beginning to suggest that considerable mutual adjustment has occurred among species that coexist in ecosystems. Thus, it is likely that an understanding of the processes of evolutionary adjustment will contribute much to an overall understanding of ecosystem dynamics.

The ecosystem approach has also fostered an integration of ecology with a number of other disciplines. In the sciences this is reflected in the virtual disappearance of the distinction between pure and applied ecology. Ecologists are increasingly becoming active in research in forestry, range management, wildlife management, and fisheries biology. This response has occurred with the realization that productivity of particular plant or animal groups is intimately related to overall ecosystem function. In the fields of agriculture and medicine the application of an ecological approach is just beginning. Historically, agricultural science has been pursued in almost complete isolation from ecology. Agricultural research has been largely concerned with evaluating short-term changes in productivity by crop and livestock species. The result has been emergence of long-term effects that threaten to impair not only agricultural systems but also much of the biosphere as well. These include the increasing dependence on intense cultivation and artificial fertilizers, the increasing threat of pest species and the growing use of expensive and dangerous pesticides, and even the increasing restriction of agriculture to marginal lands by the spread of urban areas, industry, and highways. Recognition of these problems has forced the introduction of an ecological approach in agriculture. This can be seen in increased research into topics such as biological control of agricultural pests, design of crop ecosystems to minimize vulnerability to pests, and maintenance of soil fertility with minimum use of highly soluble, inorganic fertilizers. Although application of this approach is only beginning, it seems likely that a field of agricultural ecology will rapidly emerge in an area of major importance.

A similar pattern has existed in medicine. Research into vector-borne and parasitic disease has rarely gone far beyond examination of the species directly involved. Now, however, it is becoming apparent that schistosomiasis, malaria, yellow fever, rabies, and many other diseases are influenced by a complex of biotic and abiotic environmental factors. The ecology of disease promises to be still another subject of major attention in the future.

Basic ecological theory has also become an important ingredient of many of the physical and social sciences. In the physical sciences chemistry and geology have been strongly influenced. Environmental chemistry has appeared as a field concerned with examining the patterns of action, movement, and degradation of chemical substances released into the environment. Geochemistry has intensified its interest in biological processes which influence the structure and composition of the atmosphere, hydrosphere, and crust of the earth.

In the social sciences ecological thought has influenced psychology, geography, economics, and even history. Psychology has begun to examine the relationships of mental health to human population density and the stresses of urban environments. In geography human modes of hunting and agriculture are being examined in the context of the natural ecosystems within which they occur. Economics has begun to consider the implications of human population growth and economic development in the light of ecological constraints. Many historians are directing their attention to the origins of cultural and religious attitudes that have influenced human use of the environment.

The field of landscape architecture has perhaps been the social science most responsive to the spread of ecological thought. Under the leadership of individuals such as McHarg (1969), landscape architects have begun to integrate ecological and sociological concerns with those of economics in the planning of environments for human occupation. This field promises to be the one through which knowledge of the environment gained in both the pure and social sciences must find expression.

The integration of ecology with these fields of physical, applied, and social sciences should lead to a revitalization of the field of human ecology. In theory, human ecology is the discipline that should draw together, from all of the interdisciplinary areas described above, material relevant to the ecology of man. Human ecology has languished, however, primarily as a discipline directed toward examining the influence of environmental conditions of human development and health. Added to this, the developing body of knowledge on the ecology of human populations should foster a true science of human ecology.

The spread of the "new" ecology has been coupled with the appearance of an activist philosophy among many professional ecologists. Many ecologists date this activism from the publication of Rachel Carson's book *Silent Spring* in 1962. This book was the first popularly written, yet scholarly, discussion of the dangers of chemical pesticides to both man and his environment. A committee of the Ecological Society of America concluded that this book stirred "a tide of opinion which will never again allow professional ecologists to remain publicly aloof from public responsibility" (Hardin, 1969).

Activism by ecologists has taken concrete form in public lectures, popular writing, participation in legal actions, local and regional planning organizations, and governmental agencies, as well as many other activities. *The Population Bomb* by Paul Ehrlich (1968) and *Famine—1975!* by William and Paul Paddock (1967) are only two of the many popularly written books directed at what most ecologists consider the most basic of environmental problems—that of the exploding human population. Lecturing and writing on this problem by ecologists have played a role in changing attitudes toward contraception, abortion, and population control by governmental influence. Other activities have been directed toward control of major sources of pollution. The Environmental Defense Fund was organized by concerned ecologists to pursue legal actions aimed at forcing an end to particular types of pollution, such as the use of the insecticide DDT. Activities by ecologists, supported by many concerned citizens, have led to establishment of the Environmental Protection Agency within the U.S. Department of Health, Education, and Welfare. In many parts of the United States ecologists have been instrumental in establishing regional planning organizations, a major activity the result of which has been to encourage retention of open space areas in planning for urban development.

Activity by ecologists has been supported and stimulated by a spreading concern in many segments of society, but especially among younger people. This concern is evidenced in growing memberships of major conservation organizations, one of which, Zero Population Growth, is a direct product of the action of professional ecologists. Many local ecology and conservation groups have been formed. Activities of these groups have ranged from education to the conduct of local programs for recycling materials such as aluminum, glass, and paper. The attention attracted by these groups has become so great that in the minds of many these activities constitute "ecology." The danger here may lie in the possibility that the need for major reforms is obscured by the attention directed at minor reform (Ehrlich and Ehrlich, 1970). Nevertheless, the broad response to the concerns of ecologists gives hope for major reforms.

Much will be required to gain the implementation of sound policies for control of human populations and protection of environmental quality. Governmental actions historically have been guided by short-range considerations; it is unlikely that a sudden reversal of this tendency will occur. Through gradual change, however, we may begin to move in the right direction. Improved environmental education will be required at all levels in order to create an informed public and an informed electorate. Organized activity by groups concerned with environmental quality will be required to counter profit-oriented interests that are insensitive to environmental concerns. Beyond this, mechanisms must gradually be built into all organizations, governmental and private, that are active in the environment. These mechanisms must guarantee adequate consideration of the ecological effects of human actions and insure the implementation of sound practices. If human civilization is to survive, we must eventually achieve a reorganization of human activity equivalent in scale to that caused by the industrial revolution. As Nicholson (1970) has phrased it, we must initiate an "environmental revolution."

REFERENCES

Carson, R. 1962. *Silent Spring*. Boston: Houghton-Mifflin. 368 pp.

Ehrlich, P. R. 1968. *The Population Bomb*. New York: Ballantine Books. 223 pp.

———— , **and A. H. Ehrlich.** 1970. *Population, Resources, Environment: Issues in Human Ecology*. San Francisco: W. H. Freeman. 383 pp.

Hardin, G. 1969. Not peace, but ecology. *Brookhaven Symposia in Biology*, **22**:151–61.

McHarg, I. 1969. *Design with Nature*. Garden City, New York: Natural History Press. 197 pp.

Nicholson, M. 1970. *The Environmental Revolution*. New York: McGraw-Hill. 366 pp.

Paddock, W. and P. Paddock. 1967. *Famine — 1975!* Boston: Little, Brown. 275 pp.

Van Dyne, G. M. 1966. *Ecosystems, Systems Ecology, and Systems Ecologists*. Oak Ridge National Laboratory Report 3957. 31 pp. Reprinted in abridged version in G. W. Cox. 1969. *Readings in Conservation Ecology*, pp. 21–47. New York: Appleton-Century-Crofts.

INDEX*

*Common and scientific names of species listed in tables and in certain figures are not included.